D1698021

Railway Signalling & Interlocking

International Compendium

Editors:
Dipl.-Ing. Gregor Theeg, Technische Universität Dresden
Dr. Sergej Vlasenko, Omsk State Transport University

Authors:
Dr.-Ing. Enrico Anders, Thales Rail Signalling Solutions GmbH,
 Technische Universität Dresden
Prof. Dr.-Ing. Thomas Berndt, Fachhochschule Erfurt
Prof. Dr. Igor Dolgij, Rostov State Transport University
Prof. Dr. Vladimir Ivančenko, Rostov State Transport University
Dr. Andrej Lykov, Petersburg State Transport University
Ing. Peter Márton, PhD., University of Žilina
Dr.-Ing. Ulrich Maschek, Technische Universität Dresden
Dott. Giorgio Mongardi, Ansaldo STS
Dr. Oleg Nasedkin, Petersburg State Transport University
Prof. Dr. Aleksandr Nikitin, Petersburg State Transport University
Prof. Dr.-Ing. Jörn Pachl, FIRSE, Technische Universität Braunschweig
Prof. Dr. Valerij Sapožnikov, Petersburg State Transport University
Prof. Dr. Vladimir Sapožnikov, Petersburg State Transport University
Dr. Andreas Schöbel, Technische Universität Wien
Dipl.-Ing. Eric Schöne, Technische Universität Dresden
Dr. Dmitrij Švalov, Rostov State Transport University
David Stratton MA CEng MIRSE, Alstom Transport
Dipl.-Ing. Gregor Theeg, Technische Universität Dresden
Dipl.-Inform. Heinz Tillmanns, Thales Rail Signalling Solutions GmbH
Prof. Dr.-Ing. Jochen Trinckauf, FIRSE, Technische Universität Dresden
Dr. Sergej Vlasenko, Omsk State Transport University
Dipl.-Ing. Carsten Weber, Technische Universität Dresden
Thomas White, Transit Safety Management

Eurailpress

Bibliographic information published by the Deutsche Nationalbibliothek:
The Deutsche Nationalbibliothek lists this publication in the Deutsche Nationalbibliografie, detailed bibliographic data are available in the Internet at http://d-nb.de

Publishing House:	DVV Media Group GmbH	Eurailpress
	Postbox 10 16 09 · D-20010 Hamburg
	Nordkanalstraße 36 · D-20097 Hamburg
	Telephone: +49 (0) 40 – 237 14 02
	Telefax: +49 (0) 40 – 237 14 236
	E-Mail: info@eurailpress.de
	Internet: www.dvvmedia.com, www.eurailpress.de, www.railwaygazette.com
Publishing Director:	Detlev K. Suchanek
Editorial Office:	Dr. Bettina Guiot, Ulrike Schüring
Advertisements:	Silke Härtel
Distribution and Marketing:	Riccardo di Stefano
Cover Design:	Karl-Heinz Westerholt
Layout and Production:	Axel Pfeiffer
Print:	TZ-Verlag & Print GmbH, Roßdorf
Copyright:	© 2009 by DVV Media Group GmbH	Eurailpress, Hamburg

This publication is protected by copyright. It may not be exploited, in whole or in part, without the approval of the publisher. This applies in particular to any form of reproduction, translation, microfilming and incorporation and processing in electric systems.

1st Edition 2009, ISBN 978-3-7771-0394-5

Printed in Germany

A DVV Media Group publication

Contents

Preface 15

1   Basic Characteristics of Railway Systems and the Requirements for Signalling 17

1.1   Introduction 17

1.2   Specific of Railway Systems 17

1.3   Railway Signalling and Control 18
1.3.1   Definitions 18
1.3.2   The Safety-related Railway Theory 18
1.3.3   Functional Structure 21

2   Safety and Reliability in Signalling Systems 24

2.1   Safety Basics 24
2.1.1   What is RAMS(S)? 24
2.1.2   Safety/Security 24
2.1.3   Availability, Reliability and Maintainability 25
2.1.4   Role of the RAMS Components in the Railway System 26

2.2   Safety Principles in Railway Operation 27
2.2.1   Dealing with Errors, Failures and Disturbances (E/F/D) 27
2.2.2   Analysis of Errors, Failures and Disturbances (E/F/D) by System States ... 29

2.3   Conception of Safety and Reliability of Railway Signalling Equipment 30

2.4   Characteristics of Reliability and Safety 31

2.5   Evaluation of Safety Level of Signalling Equipment 32

2.6   Rating of Safety Parameters 35

2.7   Calculations of the Safety Parameters 36

2.8   Methodology of Safety Case for Railway Signalling Equipment 38

Contents

3	Railway Operation Processes	39
3.1	Historical Background	39
3.2	Classification of Tracks, Stations and Signals	42
3.2.1	Main Tracks and Secondary Tracks	42
3.2.2	The Role of Signals	43
3.2.3	Definitions of Stations and Interlocking Areas	44
3.2.4	Signal Arrangement on Double Track Lines	47
3.3	Movements with Railway Vehicles	48
3.3.1	Train Movements	48
3.3.2	Shunting Movements	49
3.4	Principles of Train Separation	51
3.4.1	Signalled Fixed Block Operation	52
3.4.2	Cab Signal Operation	55
3.4.3	Non Signal-controlled Operation	58
3.5	Dispatching Principles	58
3.5.1	Decentralised Operation	58
3.5.2	Centralised Traffic Control	60
4	Interlocking Principles	61
4.1	Overview	61
4.1.1	Introduction	61
4.1.2	Basic Requirements	61
4.1.3	Basic Principles of Safeguarding a Train's Path	62
4.2	Element Dependences	63
4.2.1	Classification	63
4.2.2	Coupled Elements	63
4.2.3	Unidirectional Locking	64
4.2.4	Simple Bidirectional Locking	65
4.2.5	Conditional Bidirectional Locking	65
4.3	Routes	66
4.3.1	Introduction	66
4.3.2	Extension of Routes and Related Speed Restrictions	67

4.3.3	Basic Route Locking Functions	70
4.3.4	Route Selection by the Signaller	75
4.3.5	Flank Protection	76
4.3.6	Overlaps and Front Protection	79
4.3.7	Route Elements in the Start Zone	83
4.3.8	Life Cycle of Routes	84
4.3.9	Principles of Route Formation in the Track Layout	89
4.3.10	Shunting Routes	91
4.3.11	Automation of Route Operation	93

4.4 Block Dependences ... 94

4.4.1	Introduction	94
4.4.2	Geographical Assignment of Block Sections	96
4.4.3	Classification of Block Systems	97
4.4.4	Process of Block Working in Token Block Systems	99
4.4.5	Process of Block Working in Tokenless Block Systems	100
4.4.6	Locking Functions of Tokenless Block Systems	102
4.4.7	Returning Movements	103

4.5 Special Issues ... 105

4.5.1	Overlaying Block and Route Interlocking Systems	105
4.5.2	Protection of Trains by a Signal at Stop in Rear	106
4.5.3	Several Trains between two Signals	106
4.5.4	Degraded Mode Operation	107

5 Detection ... 113

5.1 Requirements and Methods of Detection ... 113

5.1.1	Introduction	113
5.1.2	Types of Objects	113
5.1.3	Safety Requirements	113
5.1.4	Detection Purposes	114

5.2 Technical Means of Detection ... 116

5.2.1	Classification	116
5.2.2	Spot Wheel Detectors	116
5.2.3	Linear Wheel and Axle Detectors	120
5.2.4	Linear and Area Detectors for Vehicles and External Objects	121
5.2.5	Three-Dimensional Detection	123
5.2.6	Systems with Active Reporting from the Train	124
5.2.7	End of Train (EOT) Detection Systems	127

5.3 Track Circuits 128

- 5.3.1 Basic Structure of Track Circuits 128
- 5.3.2 Geometrical Assembly of Track Circuits 130
- 5.3.3 Treatment of Traction Return Currents 132
- 5.3.4 Additional Functions of Track Circuits 135
- 5.3.5 Immunity against Foreign Currents 135
- 5.3.6 Electrical Parameters and Dimensioning 136
- 5.3.7 Application of the Types of Track Circuits 139

5.4 Axle Counters 143

- 5.4.1 General Structure and Functioning 143
- 5.4.2 The Rail Contact 145
- 5.4.3 Treatment of Counting Errors 146

5.5 Comparison of Track Circuits and Axle Counters 147

- 5.5.1 Advantages and Disadvantages 147
- 5.5.2 Application 147

6 Movable Track Elements 149

6.1 Kinds of Movable Track Elements and their Geometry 149

- 6.1.1 Overview 149
- 6.1.2 Simple Points 149
- 6.1.3 Other Solutions for Connection of Tracks 152
- 6.1.4 Arrangements of Several Movable Track Elements 153
- 6.1.5 Derailing Devices 155

6.2 Safety Requirements at Movable Track Elements 156

6.3 Track Clear Detection at Points and Crossings 156

6.4 Point Machines 157

- 6.4.1 Overview 157
- 6.4.2 Electric Point Machines 159
- 6.4.3 Supervision of Point Position on the Example of SP-6 161

6.5 Point Locking Mechanisms 163

- 6.5.1 External Locking Mechanism: Clamp Lock 163
- 6.5.2 Internal Locking Mechanism 164

6.5.3	Monitoring of Locking Mechanism	165
6.5.4	Mechanical Key Lock	167
6.6	**Circuitry of Point Operation and Control in Relay Technology**	**167**
6.6.1	General Overview	167
6.6.2	Example with Type N Relays: Russian Five-Wire Point Circuitry	168
6.6.3	Example with Type C Relays: GS II DR (Germany)	171
7	**Signals**	**179**
7.1	**Requirements and Basic Classification**	**179**
7.2	**Technical Characteristics of Trackside Signals**	**181**
7.2.1	Structure of Light Signals	181
7.2.2	Optical Parameters	184
7.2.3	Retro-Reflection of Passive Signal Boards	184
7.2.4	Control and Supervision of Signal Lamps	185
7.3	**Principles of Signalling by Light Signals**	**187**
7.3.1	Utilisation of Signal Colours	187
7.3.2	Stop Aspects	188
7.3.3	Signalling of Movement Authorities	188
7.3.4	Signalling of Speed Reductions	192
7.3.5	Combination of Main and Distant Signals	195
7.3.6	Shunting Signals	196
7.4	**Redundancy and Degraded Mode Operation**	**197**
7.5	**Signal System Examples**	**198**
7.5.1	German Mechanical and 'H/V' Light Signals	198
7.5.2	Belgian Mechanical Signals	199
7.5.3	British Light Signals	200
7.5.4	OSŽD Signals	201
7.5.5	Modern Dutch Signal System	202
7.5.6	German System 'Ks'	204
7.5.7	Signal System on Japanese Commuter Lines	205
7.5.8	NORAC Signals	205

8 Train Protection 208

8.1 Requirements, Classification and Conditions for Application 208

- 8.1.1 General Overview 208
- 8.1.2 Cab Signalling Functions 208
- 8.1.3 Supervision Functions 209
- 8.1.4 Intervention Functions 211
- 8.1.5 Role in the Railway Operation Process 212
- 8.1.6 Automation of Train Operation 212

8.2 Technical Solutions for Data Transmission 213

- 8.2.1 Overview over Forms of Transmission 213
- 8.2.2 Spot Transmission 214
- 8.2.3 Linear Transmission 216

8.3 Particular Systems 219

- 8.3.1 Classification of Systems 219
- 8.3.2 Group 1: Systems with Intermittent Transmission and without Braking Supervision ...219
- 8.3.3 Group 2: Systems with Intermittent Transmission at Low Data Volume and with Braking Supervision 223
- 8.3.4 Group 3: Systems with Continuous Transmission of Signal Aspects by Coded Track Circuits 227
- 8.3.5 Group 4: Systems with Intermittent Transmission at High Data Volume and Dynamic Speed Supervision 235
- 8.3.6 Group 5: Systems with Continuous Transmission at High Data Volume and Dynamic Speed Supervision 238

8.4 ETCS 240

- 8.4.1 History + Motivation 240
- 8.4.2 Application Levels and Technical Components 242
- 8.4.3 Functional Concepts 245
- 8.4.4 Operation Modes 249
- 8.4.5 Data Structure 251

9 Interlocking Machines 252

9.1 Classification 252

9.2 Mechanical Interlocking 253

- 9.2.1 Historical Development 253
- 9.2.2 System Safety in Mechanical Interlocking 253

9.2.3	Structure of Mechanical Interlocking Systems	254
9.2.4	Example: British Origin Mechanical Interlocking	255
9.2.5	Example: German Type 'Einheit'	258
9.3	**Relay Interlocking**	**263**
9.3.1	Historical Development	263
9.3.2	System Safety in Relay Interlocking	263
9.3.3	Design of Relay Interlocking Systems	267
9.3.4	Example: SGE 1958 (Britain)	270
9.3.5	Example: SpDrS60 (Germany)	271
9.3.6	Example: UBRI (Russia)	276
9.4	**Electronic Interlocking**	**280**
9.4.1	Historical Development	280
9.4.2	System Safety in Electronic Interlocking	281
9.4.3	Structure of Electronic Interlocking Systems	282
9.4.4	SSI (Britain)	285
9.4.5	SMILE (Japan)	289
9.4.6	Simis and L90 with Derivates (German origin)	291
9.4.7	Ebilock	294
9.4.8	EC-EM (Russia)	298
9.4.9	ACC (Italy)	300
9.4.10	Local-electrical Operated Point Switches (LOPS)	302
9.5	**Hybrid Technologies**	**303**
9.5.1	Hybrid Mechanical and Electrical/Pneumatic/Hydraulic Forms	303
9.5.2	Hybrid Relay and Electronic Forms	305
10	**Line Block Systems**	**306**
10.1	**Classification**	**306**
10.2	**Safety Overlays for Staff Responsible Safety Systems**	**306**
10.3	**Decentralised Block Systems**	**307**
10.3.1	Overview	307
10.3.2	Token Block Systems	308
10.3.3	Systems with Singular Unblocking upon Clearing	309
10.3.4	Systems with Continuous Unblocking	316

Contents

10.4 Centralised Systems for Safety on Open Lines 321
- 10.4.1 Overview 321
- 10.4.2 Centralised Block Systems for Secondary Lines 323
- 10.4.3 Radio Electronic Token Block 323
- 10.4.4 Open Line Controlled from Neighbouring Interlockings 324
- 10.4.5 Train Control Systems for High Speed Lines 325

10.5 Moving Block Systems 326

11 Remote Control and Operation Technology 328

11.1 Remote Control and Monitoring 328
- 11.1.1 Types of Dispatcher Control/Monitoring 328
- 11.1.2 Centralisation of Interlocking Control 330
- 11.1.3 Flexible Allocation of Control Areas 330

11.2 Processes in Operation Control 331
- 11.2.1 Information Input and its Viewing 331
- 11.2.2 Evaluation of Operational Situation 333
- 11.2.3 Command Output 334

11.3 Data Transmission in Remote Control Systems 335
- 11.3.1 Types of Communication between CTC and Stations 335
- 11.3.2 Interface between CTC and Relay Interlockings on the Relay Technology 336
- 11.3.3 Interface between CTC and Relay Interlockings on the Electronic Technology 339

11.4 Operator's Workload 339
- 11.4.1 Influence of Technology 340
- 11.4.2 Influence of Size of Infrastructure 340
- 11.4.3 Influence of Operation Programme 341
- 11.4.4 Influence of Moving Vehicles 341
- 11.4.5 Influence of Disturbances 342
- 11.4.6 Results and Conclusions of Calculating Operator's Workload 343

11.5 Examples for Operation Control Systems 343
- 11.5.1 Centralised Traffic Control in the USA 343
- 11.5.2 Operation Control Centres in Germany 346
- 11.5.3 Operation Control Centres in Russia 349

12	Safety and Control of Marshalling Yards	351
12.1	Principles of Marshalling of Trains	351
12.2	Parts of Marshalling Yards and their Function	351
12.2.1	General Structure and Functioning	351
12.2.2	Layout Variants	352
12.2.3	Automation	354
12.3	Control of Marshalling Yards	355
12.3.1	Introduction	355
12.3.2	Retarders	356
12.3.3	Handling Systems for Freight Wagons	360
12.3.4	Points	361
12.3.5	Sensors	362
12.3.6	Track Clear Detection	363
12.3.7	Yard Management Systems	363
13	Level Crossings	369
13.1	Requirements and Basic Classification	369
13.2	Static Roadside Signs	370
13.3	Passive Level Crossings	371
13.4	Active Level Crossings	375
13.4.1	Overview	375
13.4.2	Dynamic Roadside Safeguarding	376
13.4.3	Opening and Closing of Level Crossings	380
13.4.4	Supervision of Level Crossings	384
13.4.5	Possibilities of Degraded Mode Operation	389
13.4.6	Combination with Road Junctions	390
13.5	Removal of Level Crossings	391
14	Hazard Alert Systems	393
14.1	Hazards in Railway Systems	394
14.1.1	Safety Related Hazards	394
14.1.2	Security Related Hazards	395

14.2	**Solutions for Hazard Detection**	**395**
14.2.1	Ways of Inspection	396
14.2.2	Fault States to Monitor	397
14.2.3	New Approach of Inspection – The Checkpoint Concept	398
14.2.4	Extract of Available Technologies and Products	399
14.3	**Choice of Location**	**405**
14.3.1	Operational Handling	405
14.3.2	Classification of Risky Elements	407
14.3.3	Strategies for Infrastructure Manager	408
	References	409
	Glossary	416
	Explanation of Symbols in Track Layout Schemes	435
	The Authors	437
	Index	443

Assuming that the journey is the reward, how can we make it more efficient?

The Siemens answer: Complete mobility. Integrated solutions for urban and interurban transportation and logistics

Paving the way for people and goods around the world, bringing them to their destinations more safely, profitably and with less environmental impact – that's what Siemens "Complete mobility" is all about: Shaping and efficiently interconnecting the diverse transportation systems for the long term with integrated mobility solutions.
www.siemens.com/mobility

Answers for mobility.

SIEMENS

www.eurailpress.de

High-quality technical know-how for worldwide engineers

RTR – Railway Technical Review

RTR is the Eurailpress international trade magazine for worldwide rail operating managers, engineers, technical staffs and scientifics.

For more than 40 years, RTR has been one of the leading technical mediums for world-wide railway transportation.

RTR's incomparable claim is: Detailed technical articles on all aspects of modern rail technologies – from permanent way and infrastructure to vehicles and rolling stock components, from rail signalling and telecommunication to operating modern railways. The high quality is guaranteed by the renowned publishers and the editor.

You would like to test RTR? Just send us an e-mail: service@eurailpress.de and you'll receive a free copy.

DVV Media Group

Preface

In the era of globalisation, the future success of the railway industry depends significantly on the worldwide sharing of knowledge. That exchange is the only way to find the most effective solutions and to avoid unnecessary parallel developments. Railway signalling is still one of the very few sectors of technology in which national solutions differ substantially.

Up to now there have been no common terms and definitions, and there is a corresponding lack of understanding of the underlying principles. Also, adapting new technologies to different national practices is very costly. In the technical literature, descriptions of railway signalling principles concentrate on the railways of a single country, or group of countries. Even at universities, students learn almost exclusively the signalling technology of their own country.

In the 21st century, this is extremely disappointing. Just to compare different national solutions is not sufficient. What is really needed is a generic description of railway signalling principles that allows the reader to look outside the restrictive national point of view.

This textbook was written with that intention in mind. A team of railway signalling and operations experts from six countries and three continents have written this, the first generic signalling textbook in the world. Instead of beginning with the technical elements, it starts with an analysis of railway operation processes. Due to very different operating philosophies, an understanding of these practices is essential for an understanding of signalling principles. The chapter on operations is followed by a chapter on basic interlocking principles. These two form the basis for the remaining chapters, which describe technical components, signalling and train control systems.

The authors hope that this textbook will become a valuable resource for students at universities and colleges, and for practitioners in the railway environment alike. Comments from readers to be considered in future editions are always welcome. The authors wish to thank the editorial staff of Eurailpress for their support in this venture.

We also thank Mr Aleksej Efremov, editor of the journal *Rail International*, who has supported this book with many advices and pictures from different parts of the world.

The Authors
September 2009

Wheel sensors by Frauscher >

> have the future built-in today.

> **Wheel sensor RSR 123**   > **Wheel sensor RSR 181**

Engineering without compromise

Multiple application options (sometimes in combination):

- > Axle detection, pedal function
- > Axle counting
- > Running direction detection
- > Wheel centre detection
- > Speed detection
- > Wheel diameter detection
- > Vehicle detection
- > Brake detection
- > Protection against vandalism

FRAUSCHER
SENSOR TECHNOLOGY

Frauscher GmbH | Gewerbestraße 1 | 4784 St. Marienkirchen | Austria
Phone +43 (0) 77 11 / 29 20 - 0 | Fax +43 (0) 77 11 / 29 20 - 25 | office@frauscher.com
Please find more info and application data at: www.frauscher.com

1 Basic Characteristics of Railway Systems and the Requirements for Signalling
Jochen Trinckauf

1.1 Introduction

The railway was the first means of mass mechanised movement. Over time the speed of the trains increased, as did the payloads which could be carried, and the length of the trains themselves. Railway development began in the early years of the nineteenth century, and it became the backbone of transportation systems in many countries. Of course, after World War 1 the use of mechanised road transport started to rise, and after World War 2 this became a real competitor.

Nowadays we have railway systems with multiple variations around the world. They can be classified as follows:

1. by rolling stock
   - Trains consisting of one (or more) locomotive(s) and a number of cars for passenger or freight
   - Self contained passenger trains, including multiple units, without a separate locomotive
   Either type may be powered by diesel or electric traction
2. by traffic characteristics
   - long distance
   - regional service
   - urban transport (mass transit)
3. by operation, for example:
   - centralised control
   - fully signalled operation
   - drive on sight
   - driverless automated

There are several other classifications, for instance by public or private ownership, separation of infrastructure and operating companies, main line/secondary line, single track/double track and so on. But all are variations of 'the railway' because they are similar in their principal characteristics and hence their requirements for signalling.

During a long period of gestation, many techniques, technologies and components were developed. The basic characteristics of the railway are described here to help understand the requirements of signalling.

1.2 Specific of Railway Systems

All railway systems have these two identifying features:

1. The path taken by the train is determined by the mechanical guidance system of wheel and rail, and this can be changed only by points. On a single track railway, trains can only pass each other at particular locations, such as loops. Therefore it has to be possible to predetermine the route to be followed and to set the points accordingly. As the vehicle is very closely connected to the guidance system, it can also be termed a linear control system.
2. The steel wheel has a relatively poor braking response on the steel rail, but there is a relatively high running speed. Depending on the braking system, braking distances at as little as 50 to 70 km/h are often longer than the visible and clear route in front of the driver.

Therefore the sight on the route has to be supplemented by other precautions in order to indicate to the driver in good time a clear route or a need to stop. This applies to both an actual stop and a requirement to reduce speed.

These technical problems had to be solved. Consequently, procedures and techniques have been developed to ensure a safe, reliable and effective railway operation.

Starting from the days of simple information transmission, the railway operation and control system has been developed and over generations adapted to the state of the art. These principles generally have been proven and the partial techniques of control systems of the earlier generations are still in use today. The principles are also being put to use in innovative railway systems, where for instance the steel wheel on steel rail is replaced by other guidance systems, or by doing without a driver.

1.3 Railway Signalling and Control

1.3.1 Definitions

The **Railway Signalling and Control System** is needed for the safe control of transport processes in rail traffic.

The objectives and tasks of signalling and control can be defined as follows:

Signalling system
The signalling system ensures the safe control of transport processes. The safety aspect is stressed.

Operation control system
The operation control system ensures optimal control of the sequences of main and auxiliary processes in a traffic system.

Looking at the control loop of the process (see figure 1.1), it has to be ascertained that processes in the sector of the signalling system are triggered by internal events. These events can occur both theoretically and practically at any time. In the sector of the operation control system, railway operation processes are triggered by external influences, such as time-related schedules, response to traffic demands and so on.

Both systems use the means and methods of **information transmission** and **information processing**. Considerations of safety, reliability and availability are important in both systems, even though they serve different purposes:

Signalling systems involve the regulation of traffic and the prevention of accidents, whereas operation control systems have to prevent effective failures.

The technical components of control and signalling systems are similar (industrial computers, electronic controls etc.) but any considerations of safety and availability have to be general considerations that take the entire situation into account.

The subject matter of this book is the railway signalling system.

1.3.2 The Safety-related Railway Theory

The safety related railway theory has been developed to gather the variety of technical solutions in the railway control system. This theory can be demonstrated by means of a control loop, which is also known from the control theory (figure 1.1).

1.3 Railway Signalling and Control

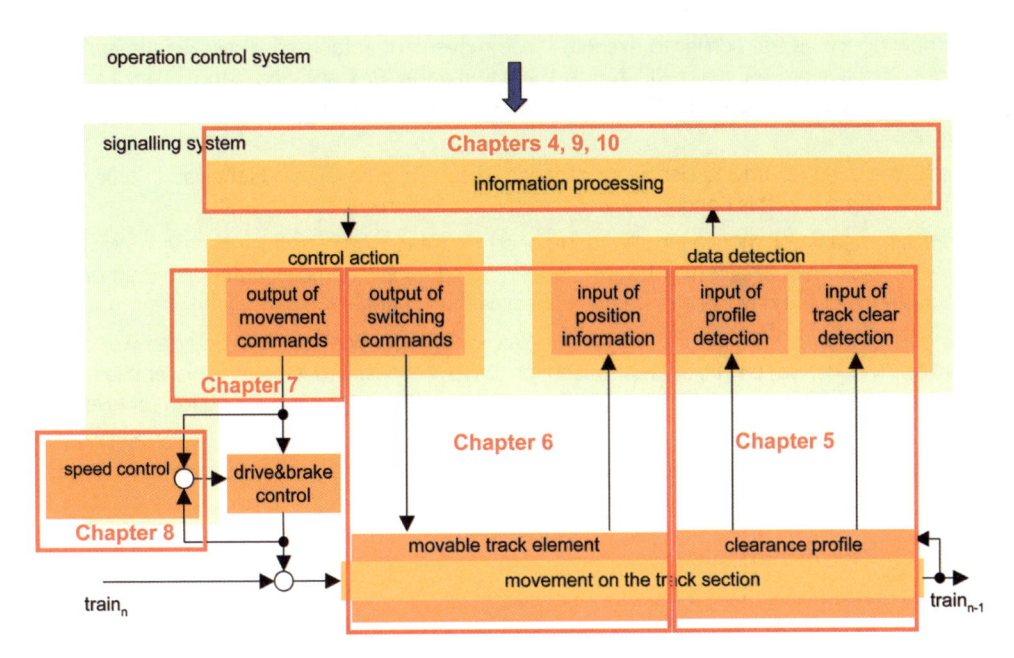

Figure 1.1: Control loop of the railway signalling system

The **train** is defined as all the movements of railway vehicles under local supervision.

The controlled element is 'movement on the track section', which is monitored. A track section is characterised by two significant features, the position of the movable track elements and the state of the clearance profile.

For this theoretical consideration the track sections themselves do not need to be further specified. For instance it could be single track sections within the railway station, entire routes or block sections.

Movable track elements are mainly points. However, there are further movable track elements, which are considered in chapter 6. The control loop is applicable even if the train moves in a track section without movable track elements.

The clearance profile is the space surrounding the rails and of course has to be clear before it can be used by a train. The clearance profile may be or may have been occupied by a train, which has used the same track section immediately before the monitored train, or by an off-track obstruction, for instance road vehicles at a level-crossing. Details are discussed in chapter 5.

A measured value detection takes place in the control loop, in the course of which the following are registered:
- position information of the movable track elements
- present state information of the track clear detection (clear or occupied)
- information on other obstacles (technically or organisationally, if possible)

The measurement values are safely and logically processed. This is the core of such systems, whose logical principles are discussed further in chapter 4 and the technical solutions in chapters 9 and 10.

Finally, control values are issued. They refer to the movable track elements, which have to be positioned as required by the trains. However, they also have to be locked into this position before a corresponding movement authority can be issued to the train (chapter 7).

1 Basic Characteristics of Railway Systems and the Requirements for Signalling

It is now up to the train driver to execute these movement authorities. However, these movement authorities are not just restricted to the commands 'Proceed and 'Stop'. They can also be specified by information on the speed-limit. The stop signal can be interpreted as 'speed limit zero'.

Due to the dynamic driving behaviour of trains, restrictions have to be received in time to be actioned effectively. This concerns lower speed limits in particular.

There are means for monitoring the train speed and acting on the train in case a maximum speed is exceeded. These may be simple train stops, highly complex automatic train protection systems, or the driverless systems, as dealt with in chapter 8.

The control loop is valid for the highly complex systems of control technology as well as for simple solutions, which are merely based on organisational procedures. This applies to all combined forms of organisational and technical processes, at the highest and lowest levels.

Within every control loop the previously described processes are sequences, which depend on each other in a strictly deterministic way. The trigger for the self-controlling system is external (figure 1.2), because the time or occasion for undertaking the control function is determined by operating conditions. Practically speaking, the dispatcher gives the driver a movement authority with the help of the control loop, which is described in the following section 1.3.3.

In summary, the safety-related railway theory results in the following technological basic railway control requirements. They apply generally and are independent from the particular technical solutions. Furthermore they are independent of local peculiarities, which are country-specific and/or related to historical experiences.

1. All track sections in front of the train have to be clear and kept clear, until they have been completely passed by that train.
2. All movable route elements in front of the train have to be held in their correct positions and kept there, until they have been completely passed by that train.

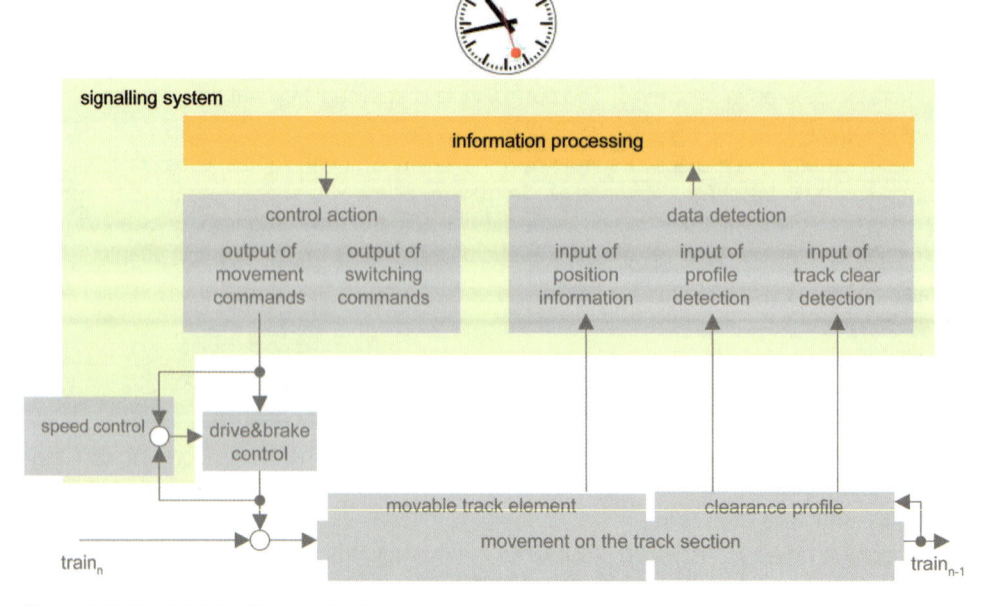

Figure 1.2: Establishing time context

3. Speed changes of the train have to be begun in sufficient time in order to reach the permitted speed at the target speed point.

1.3.3 Functional Structure

Figure 1.3 shows the functional structure of the railway control system, as it has been developed and practically proved over the years. In the course of this development numerous specific solutions have been determined and are discussed later.

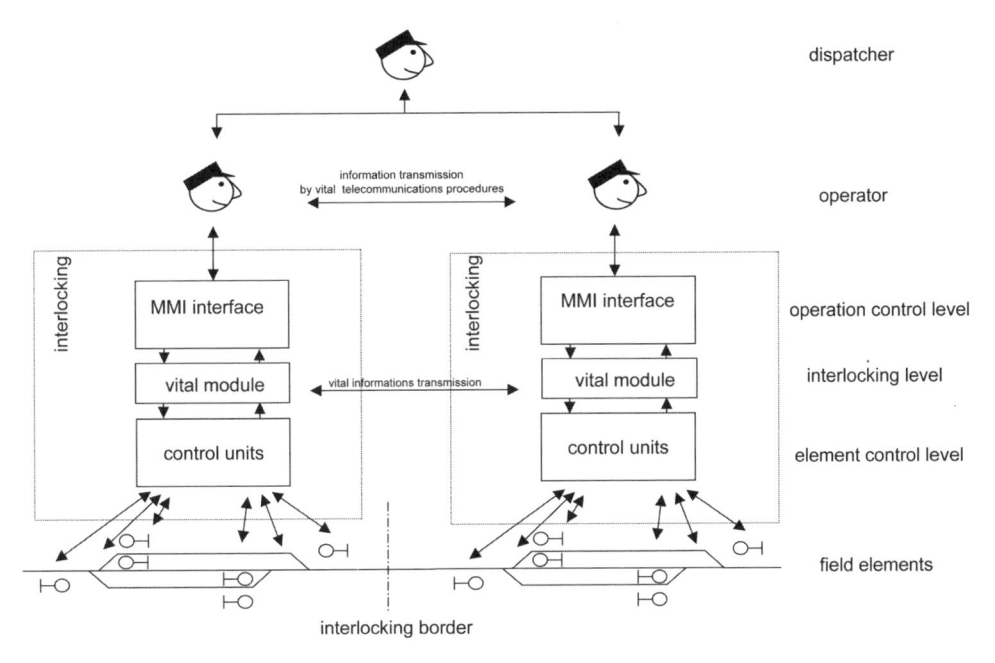

Figure 1.3: Functional structure of the railway control system

Field Elements

Basic field elements are:
- points
- signals
- track clear detection systems
- train activated systems (such as treadles or track circuits)
- automatic train protection systems

These installations are connected to the controlling and monitoring units of the local interlocking station.

Element Control Level

Controlling and monitoring units are arranged on the element control level, defined as that which operates the basic components of the system such as points and signals. From here the components of the field elements are triggered and monitored with regards to their current operating state.

1 Basic Characteristics of Railway Systems and the Requirements for Signalling

Signaller and Operation Control Level (Man-Machine Interface)

The signaller is the person who directly induces the train movement. However, there is a preceding planning stage with regard to the railway operation requirements. The signaller makes the decision on the basis of the corresponding timetable documents, the correct time at which each movement should occur, and with consideration of the current operating situation. He thus has an element of discretion at his disposal. The ease and effectiveness with which he can use this can be very different. It depends on the technical generation of information by the system in general and on the type of the system used in particular.

The standard of the status signals, which informs the signaller about the current state of operation, is also system-dependent.

Interlocking Level

The task of the central vital safety module is to transform the signaller inputs into control commands that are supposed to lead to a proceed signal. It is important that conflicting movements are excluded and route elements for one's own movement (such as points) are locked before the proceed signal is displayed.

Those are the ultimate vital tasks, which are discussed in the immediately following chapters of the book. Basically, it is about the prevention of inadvertent wrong signaller inputs that might lead to dangerous situations.

Interlocking Station

These structural elements are technically and organisationally pooled in an interlocking station. The control distance of an interlocking was in the past limited by technology, such as the

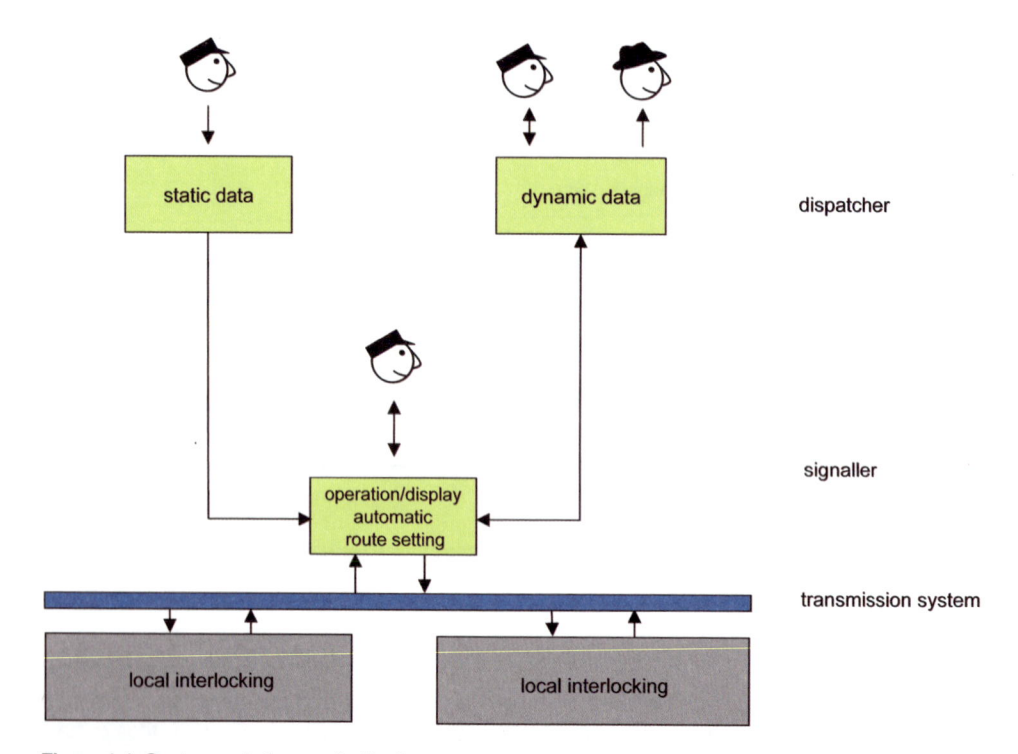

Figure 1.4: System solution centralisation

maximum distance at which a signaller could operate a semaphore signal. For many reasons, control distances have often not been extended beyond the borders of railway stations[1]. In many countries the organisational classification of railway stations and open lines is still held today. However, the technical control distance of modern electronic interfaces are theoretically and practically unlimited and are defined in terms of expedience.

Dispatcher

Basically, the railway operation is communicated to signallers by means of a timetable. However, there are numerous reasons to have a dispatcher level above that of the signallers, which fulfils central and overlapping tasks (figure 1.4). Accordingly, these tasks are supported by railway operation control facilities. There are systems in which interlockings are connected to operation control systems. At the interlocking level, operation then occurs without the local signaller. The operator gets both the signaller and the dispatcher tasks. As a necessary consequence, control processes are strongly automated.

---

[1] Railway stations, as meant here, are railway facilities with at least one set of points, where trains are allowed to start, terminate, cross, pass or reverse by changing tracks (back shunt). Home signals mark the border between open line and stations.

2 Safety and Reliability in Signalling Systems
Valerij Sapožnikov, Vladimir Sapožnikov, Enrico Anders, Jochen Trinckauf

2.1 Safety Basics

2.1.1 What is RAMS(S)?

The rail related norm EN 50126 (CENELEC 1999) carries the term RAMS in its name: Railway Applications – The specification and demonstration of reliability, availability, maintainability and safety (RAMS). This norm defines procedures for railway companies, the rail industry and its suppliers in the European Union, to implement a management system for reliability, availability, maintainability and safety. This management is to be applied during the complete life cycle of a system (figure 2.1) and to permit harmonisation of the technical level of safety. Through this, national restrictions are avoided and competition and interoperability are therefore strengthened.

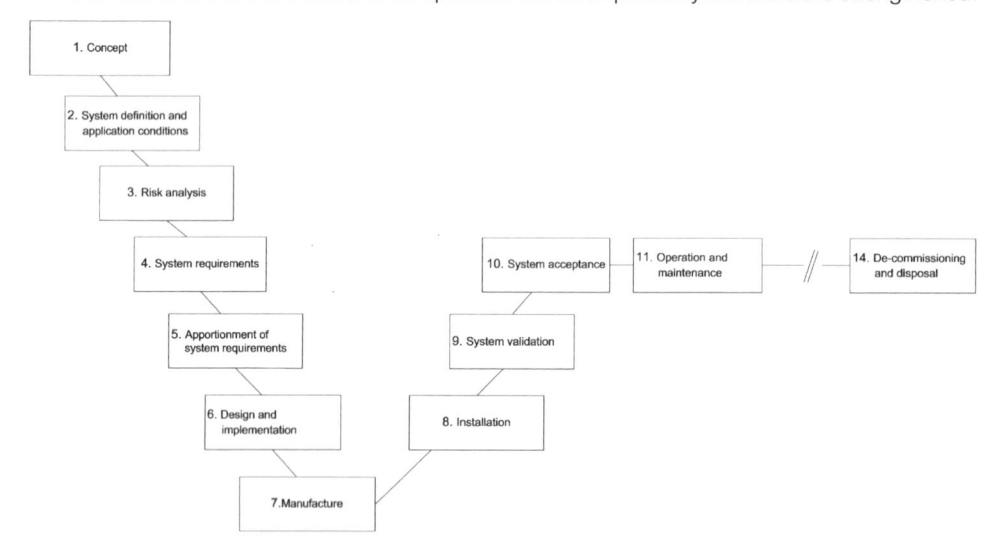

Figure 2.1: Phases in the life cycle of systems in the so-called V-diagram, according to EN 50126

2.1.2 Safety/Security

Frequently, the abbreviations 'RAMSS' or 'RAMS(S)' are used. The two words safety and security need to be defined.

Safety means the functional safety within the system and protection against hazardous consequences caused by technical failure and unintended human mistakes. **Security**, by contrast, protects against hazardous consequences caused by wilful and unreasonable human actions.

A simple example is an emergency exit door. To ensure that it can be released from the inside in an emergency, the door has to be opened by a simple handle with no lock. That is a matter of safety. But to stop intruders, the door has to be locked from the outside. That is a matter of security.

The majority of components in railway signalling, e.g. track clear detection, interlockings, point switching, signals and level crossings, are safety related. But measures to protect buildings containing interlocking equipment from unauthorised intrusion are security related.

2.1.3 Availability, Reliability und Maintainability

This section describes the meaning of 'RAM'. Reliability, availability and maintainability. These interact with each other as shown in figure 2.2.

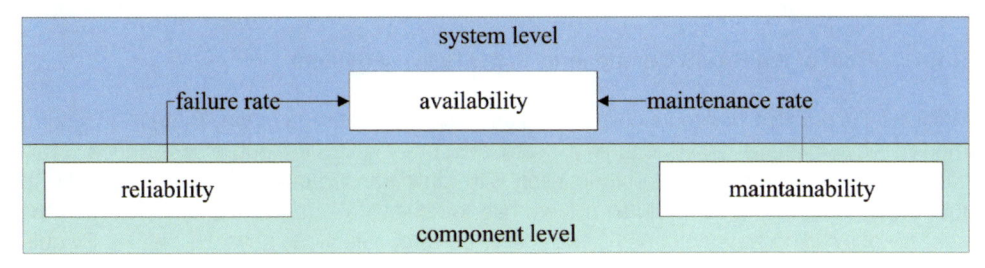

Figure 2.2: Interference in 'RAM'

The term **availability**, according to EN 50126 (CENELEC 1999), is defined as:

The ability of a product to be in a state to perform a required function under given conditions at a given instant of time, or over a given time interval, assuming that the required external sources of help are provided. This means that the system (here called 'product') will fulfil the required tasks (here called 'functions') under the defined framework conditions.

An important function of the railway system is the safe transport of persons and goods. The preconditions to fulfil these functions are the *required external sources of help*. In the case of the railway system, these are reliable functioning technical components such as interlocking between points and signals, the track clear detection and the avoidance of following or crossing trains in one block section, but also reliable performance of railway staff in undertaking their tasks. Therefore, **reliability** is an important factor for availability. Reliability is defined in IEC 61508 (IEC 2001) as follows:

The probability that an item can perform a required function under given conditions for a given time interval (t1, t2).

The phrase *required function* means that a component here referred to as an item works in conformance with its specifications under the precondition that it has done so on entering service and that no maintenance has since been necessary. This results in the requirement of failure-free working of the component during a specified time period.

Therefore, besides reliability, **maintainability** of the used components is an important factor for availability of the system. This term is defined as follows in EN 50126 (CENELEC 1999):

The probability that a given active maintenance action, for an item under given conditions of use, can be carried out within a stated time interval when the maintenance is performed under stated conditions and using stated procedures and resources.

Reliability and maintainability are both probability values which lead to failure rates and maintenance rates respectively, related to a defined time period.

An important requirement of the railway system is high availability. In particular, this is a result of its strong link with safety: The more available a technical system is, the longer it can be operated regularly and the lower is the ratio where it is operated in degraded mode. As systems work on a lower level of safety in degraded mode, high availability reduces the probability of a dangerous error in degraded mode operation.

2 Safety and Reliability in Signalling Systems

From the requirement of high availability, two requirements result regarding reliability and maintainability:
- low failure rates
- high maintenance rates

2.1.4 Role of the RAMS Components in the Railway System

What are the practical consequences of the RAMS components in a railway system? In a common public opinion the railway counts as 'safe' or in other words the public acceptance of railway hazards is extremely low in comparison with other transportation systems like roads. To fulfil these expectations, the railway system has to achieve a high level of safety through the strategy of avoiding accidents. As risk can never be zero, safety can never be perfect. By definition, *safety is the absence of an inadmissible risk*. The railway system has to run in the zone of safety but a certain level of remaining risk cannot be avoided. But it is self-evident that this remaining risk must be low and so the resulting safety as high as necessary.

As shown in figure 2.3, the **risk** is defined as the *product of hazard rate multiplied by damage*. Damage in a railway accident will count as 'high' in most cases so that risk can only be decreased by lowering the hazard rate. This again will be achieved by high availability. As shown in chapter 2.1.3 above both the components of the reliability of system design and maintenance during its operation are able to influence the availability. This is mainly responsible for the safety of a railway system.

D	danger
H	hazard
HR	hazard rate
Da	damage
Ri	risk
R	reliability
A	availability
M	maintainability
S	safety
µ	maintenance rate
λ	failure rate

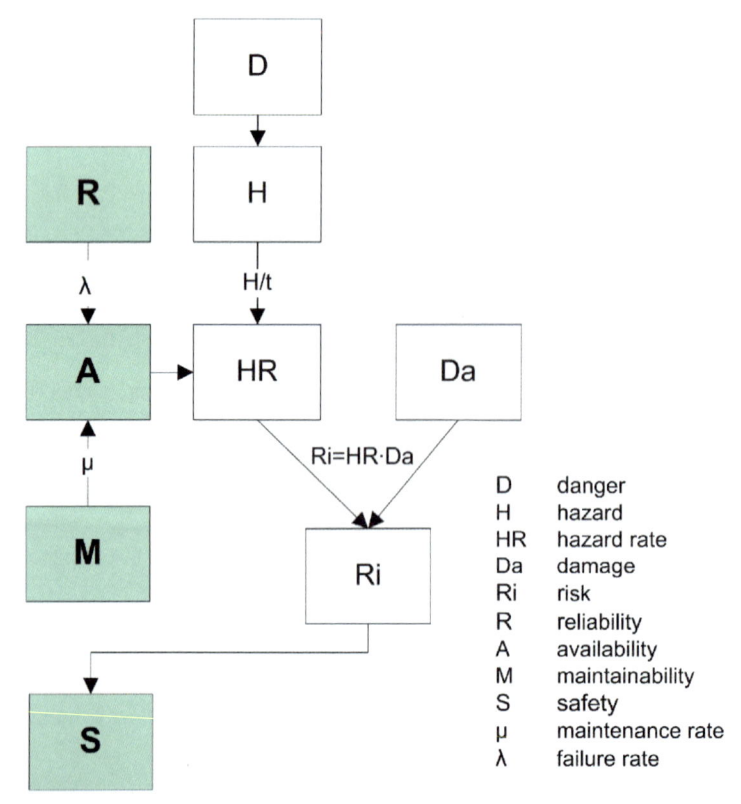

Figure 2.3: Relationship of components of RAMS

2.2 Safety Principles in Railway Operation

2.2.1 Dealing with Errors, Failures and Disturbances (E/F/D)

A basic requirement in safety related systems is that in case of errors, failures and disturbances persons, goods and the environment must not be exposed to (unacceptably high) risks. This is fulfilled as long as the system reacts safely. A profound knowledge on the E/F/D behaviour of the components used in the system is the basis for the definition of the safe state. This knowledge influences the choice of measures which have to be undertaken to reach a sufficient safety level. A systematic error in fabrication can be discovered by careful inspection before putting the system into operation. A spontaneous (random) failure during operation, in contrast, cannot be prevented. However, dangerous consequences of such a failure can be prevented by the design of the system.

To achieve system safety, the following strategies can be applied:
- elimination of E/F/D
- elimination of the consequences of E/F/D
- limitation of the consequences of E/F/D.

2.2.1.1 Elimination of E/F/D

Each system component has defined physical characteristics. Elimination of E/F/D can be assumed if these characteristics are **undetachable**. This means that E/F/D cannot occur due to these characteristics. Components used in the railway system must have high mechanical and/or electrical robustness. This implies that external influences such as vibration and traction return currents must not affect the desired function. This robustness can be achieved by proper materials (e.g. unweldable contact materials), special structure (e.g. mechanical stability) and special production methods (e.g. use of checklists). In a safety case the elimination of E/F/D is proven mostly theoretically or experimentally.

If the probability of E/F/D can be assumed to be sufficiently low, it can be assumed that it is eliminated. This level is specified in safety requirements specifications. If fulfilled, the characteristics are assumed as practically undetachable and therefore E/F/D as practically eliminated.

An example for technical elimination of E/F/D is the impossibility of a relay armature of a type-N-relay to fail to drop down. Non-technical elimination of E/F/D can be provided by forbidding the parking of wagons on certain tracks.

2.2.1.2 Elimination of Consequences of E/F/D

If the elimination of E/F/D cannot be proven, the elimination of dangerous consequences of particular E/F/D has to be proven, which means that the system must go into a safe state (chapter 2.2.2.1). This proof is based on the following basic safety requirements (Fenner/Naumann/Trinckauf 2004):

- non-dangerous singular and multiple E/F/D
- non-propagation of singular E/F/D
- proof of independence of singular E/F/D

Non-dangerous Singular Errors, Failures and Disturbances (E/F/D)

According to EN 50129, in safety systems, assumable singular E/F/D must never lead to a dangerous state, but only to an allowable safe state (chapter 2.2.2.1). In a safety case, the fulfilment of this requirement has to be proven. If this cannot be done, the system architecture has to be changed. In practice, this mostly means the application of redundant structures.

Non-dangerous Multiple Errors, Failures and Disturbances (E/F/D)

Besides non-dangerous singular E/F/D, some safety systems require the same for (dangerous) multiple E/F/D. If a simultaneous malfunction of two or more interfering components can lead to a dangerous state, these components have to be independent from each other. This helps to exclude systematically multiple E/F/D, also called consecutive E/F/D.

Non-propagation of Singular Errors, Failures and Disturbances (E/F/D)

The requirement of non-propagation of (dangerous) singular E/F/D in safety systems is fulfilled by its rapid revealing and treatment. By this, the emergence of multiple E/F/D can be prevented. The time until E/F/D are revealed therefore has a great impact on the level of safety.

Referring to EN 50129 (CENELEC 2003) the reaction time is 'the period of time between failure detection and reaching the safe state.' Some time delay cannot be avoided. However, it must be sufficiently low in accordance with the required safety level. A central requirement in this context is that an E/F/D must be revealed before executing the next safety related function. In railway signalling, the rapid revealing of E/F/D is achieved preferably by checking functions upon activation depending on data flow. Where revealing by circuitry is not possible, inspection at regular intervals is used. In electronic systems, this is achieved by very short supervision cycles, typically far below one second.

Independence of Singular Errors, Failures and Disturbances (E/F/D)

Besides rapid revealing and treatment, the second solution to achieve non-propagation of singular E/F/D is to ensure the independence of components respectively of the E/F/D acting on them. This results in the requirement of redundancy as an appropriate method for compensation of E/F/D. However, redundant structures require special methods to really ensure their independence. Otherwise, a combined malfunction in form of systematic multiple E/F/D could lead to a dangerous system state. Therefore, for example, cables for safety functions are always laid locally separated to eliminate simultaneous E/F/D.

Proving independence can be difficult in practice because one E/F/D can be the common route cause for several E/F/D, although they might have been assumed as independent. For example, simultaneous failure of two redundant control units with different power supply units is considered to be impossible. However, if these power supply units are powered via the same distributor and this distributor becomes unpowered, this improbable case arises.

This example illustrates the high importance of exact definition of the system and of the system borders in safety analysis.

2.2.1.3 Limitation of Consequences of E/F/D

If neither E/F/D nor its consequences can be proved to be excluded, the probability of dangerous consequences in case of occurring E/F/D has to be sufficiently low. The exact value of this acceptable probability has to be defined in safety requirements specifications.

To achieve limitation of consequences of E/F/D, one approach is rapid revealing of E/F/D and the other is limitation of the possible damage. An example in railway signalling for limitation of damage is the limitation of speed in degraded mode operation.

Summing up, the highest goal in railway signalling is the elimination of E/F/D. However, due the technical and economic restrictions, many safety measures in components and systems are based on the strategies of elimination of consequences of E/F/D or, as a minimum, the limitation of consequences. Figure 2.4 illustrates this.

2.2.2 Analysis of Errors, Failures and Disturbances (E/F/D) by System States

2.2.2.1 Safe State

A basic requirement in railway signalling is that E/F/D must lead to a safe state. Therewith, the safe normal state (without E/F/D) is left, but another safe state of degraded mode operation is immediately achieved. Accordingly, the safe state is defined as a state of technical system freedom from unacceptable risk of harm (CENELEC 2003).

Emerging E/F/D must not lead the system to leave the safe state. This feature of safety related systems is called the **fail-safe principle**. Furthermore, this state must be maintained until all E/F/D have been removed. Leaving the safe condition during fault-clearing must only be possible with the participation of specially trained maintenance staff.

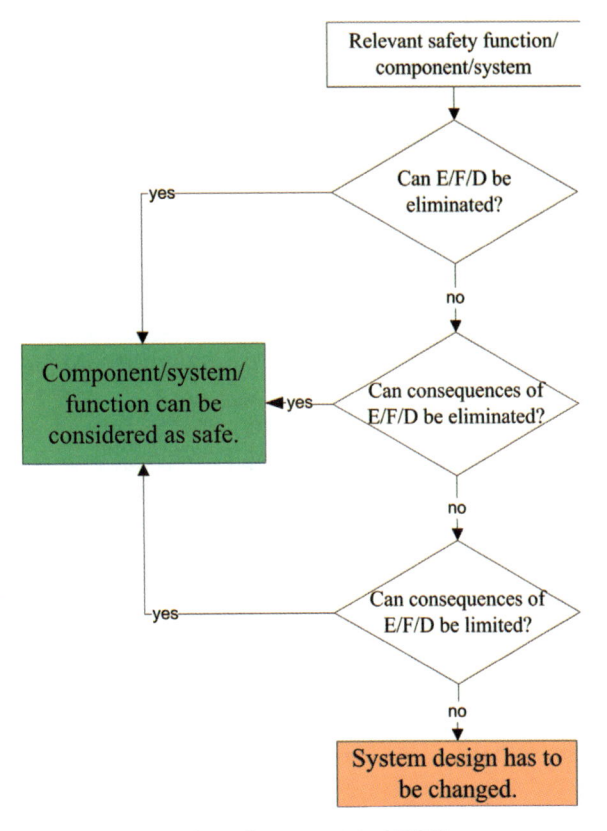

Figure 2.4: Procedures in treatment of E/F/D

In railway operation, the safe state is standstill of traffic. This is achieved by defined components to take an undetachable technical state, e.g. melting of a fuse in a circuit for control of signal lamps.

In the complex railway system, there are several safe states respectively fallback levels for degraded mode operation. Depending on the type of E/F/D, these states are passed until a favourable state has been achieved.

The definition of the safe states is an important part within the planning and development of interlocking components and systems. To define the safe states of all imaginable E/F/D requires a profound knowledge of the processes.

2.2.2.2 Obstructing State

When due to E/F/D a system leaves the safe normal state and cannot immediately return to it, it goes into the **obstructing state**. It remains there until returned to the normal state by staff. Obstructing state is a special kind of the safe state.

In practice, when leaving the obstructing state, often the human has to undertake safety related actions, resulting in higher E/F/D rates. To mitigate this factor, suitable procedures for safe transfer into the safe state have to be planned within the development of the system.

2 Safety and Reliability in Signalling Systems

2.2.2.3 Dangerous State

Systems in which the elimination of E/F/D has been proved cannot lead to dangerous states. However, in practice a dangerous state can only seldom be completely excluded. This means that in safety related systems the application of the fail-safe-principle must reduce the probability of dangerous states to a minimum. If the system goes into a dangerous state nevertheless, it must be transferred into a safe state as soon as possible.

2.3 Conception of Safety and Reliability of Railway Signalling Equipment

Quality of railway transportation is determined especially by speed and safety of conveyance of passengers and goods to the destination. Both of these parameters depend critically on the reliability of signalling systems. Failures of these systems can lead to trains being delayed and, in worst cases, to derailments and collisions.

The reliability of signalling equipment is the capability to provide uninterrupted and safe control of train movement in all specified modes and conditions of operations, maintenance and repair.

The reliability of signalling equipment includes five components (OST 32.17-92):

- failure-free operation,
- safety,
- longevity,
- maintainability,
- storability.

Set of states of the signalling equipment is divided on following subsets (see figure 2.5):

- up states (S_U),
- operable states (S_O),
- disabled protected states (S_P),
- disabled hazardous states (S_H).

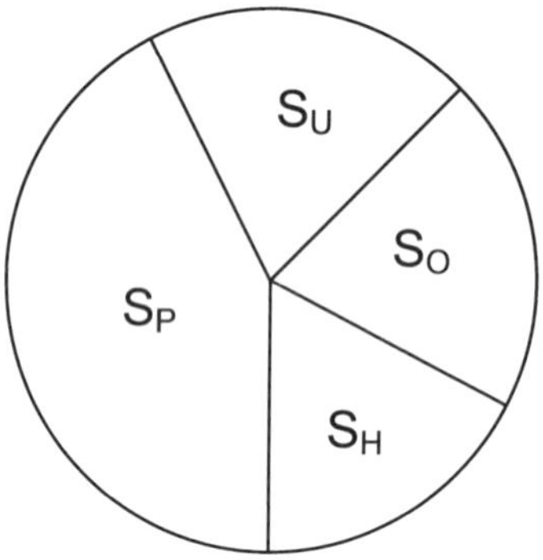

Figure 2.5: Diagram of states for safe system

The **up state** is a state in which the system is completely suited to all functional and technical requirements on application and environment conditions. A system in this state is also operable.

The **operable state** is a state in which all parameters of the system are suited to all technical requirements related to its possibility to fulfil all intended functions.

The **disabled state** is a state in which even one parameter of the system mismatches its requirements.

The **disabled protected state** is a state in which values of all parameters of the system, determining its possibility to fulfil intended safety functions, are suited to all given requirements.

The **disabled hazardous state** is a state in which even though one parameter of the system, determining its possibility to fulfil intended safety functions, is not suited to its requirements.

a) Failure-free operation of signalling equipment is a property of the system to provide continuously its up or operable state during a determined time span or operating time.
b) Safety of signalling equipment is a property of the system to provide continuously its up, operable or protected state during determined time span or operating time.

Protected failure does affect failure-free operation of signalling equipment, but does not affect its safety. Failure-free operation is characterised with sets S_F of up and operable states:

$$S_F = S_U \cup S_O,$$

and safety – with sets S_S of up, operable and protected states.

$$S_S = S_U \cup S_O \cup S_P$$

Comparison of sets S_F and S_S shows that failure-free operation is always greater or equal safety:

$$\text{Safety} \geq \text{FAILURE-FREE OPERATION}$$

In specific case, when any failure in the system is the dangerous failure (i.e. $S_P = \emptyset$), safety is equal to failure-free operation.

The indicated inequality is mainly responsible for considering safety as a separate feature of signalling equipment. Such an approach allows protecting the systems from dangerous failures and increasing safety by the cheapest means, sometimes at the expense of failure-free operation.

2.4 Characteristics of Reliability and Safety

Reliability and safety are represented most adequately by the following quantitative parameters

Quantities $\lambda(t)$ and $\lambda_D(t)$ are determined on the base of statistical data with the following formulas:

$$\lambda(t) = \frac{n(\Delta t)}{N_m \Delta t}; \quad \lambda_D(t) = \frac{n_D(\Delta t)}{N_{sm} \Delta t} \quad (2.1)$$

Indication	Name
P(t)	Failure-free operation probability
Q(t)	Probability of failure
λ(t)	Failure rate
T	Mean operating time to failure
P$_S$(t)	Probability of safety
Q$_D$(t)	Probability of dangerous failure
λ$_D$(t)	Dangerous failure rate
T$_D$	Mean operating time to hazardous failure

Table 2.1: The quantitative parameters of reliability and safety

where:

$n(\Delta t)$, $(n_D(\Delta t))$ – quantity of samples of the system having a failure (hazardous failure) in a given time Δt;

$N_m = \frac{N_i + N_{i+1}}{2}$ – mean quantity of failure free systems in a given time Δt;

N_i, N_{i+1} – quantity of failure free systems in the times $t - \frac{\Delta t}{2}$ and $t + \frac{\Delta t}{2}$;

$N_{sm} = \frac{N_i + N_{i+1}}{2}$ – mean quantity of failure free systems not having any hazardous failures in the time interval Δt (on condition that samples of the system having protected failure are immediately replaced by the new samples);

N_{j+1} – quantity of failure free systems not having any hazardous failures to the time of $t + \frac{\Delta t}{2}$.

As dangerous failures are rare events, the statistical data for determining parameter $\lambda_D(t)$ is a result of long-term investigation which is not to be expected in practical terms. Actually, formula (2.1) uses results of system monitoring over a long period of time of its operations. Therefore, replacement of samples of the system, having protected failure in the time interval Δt, by the new samples (which take place during the system operations), is the condition on calculations over formula (2.1). Such the condition may not be laid down in experiments not related to the operations. Therefore formula (2.1) gives upper evaluation of the $\lambda_D(t)$.

For automatic railway systems the following equations will be correct:

$$\lambda(t) = \lambda = \text{const} \quad \text{and} \quad \lambda_D(t) = \lambda_D = \text{const}.$$

Then, on the base of exponential law of reliability, the parameters $P(t)$, $P_S(t)$, $Q(t)$, $Q_D(t)$, T and T_D can be calculated as follows:

$$P(t) = e^{-\lambda t}, \qquad P_S(t) = e^{-\lambda_D t}, \qquad (2.2)$$
$$Q(t) = 1 - e^{-\lambda t}, \qquad Q_D(t) = 1 - e^{-\lambda_D t} \qquad (2.3)$$
$$T = 1/\lambda, \qquad T_D = 1/\lambda_D \qquad (2.4)$$

The diagram of the function $P(t)$ is shown in figure 2.6. It can be seen that the reliability of the system becomes lower exponentially by time, the rate of decrease is the higher, and that the higher is parameter λ. The same is true for the safety of the system.

If $\lambda t < 0.1$, the following simple approximations can be used:

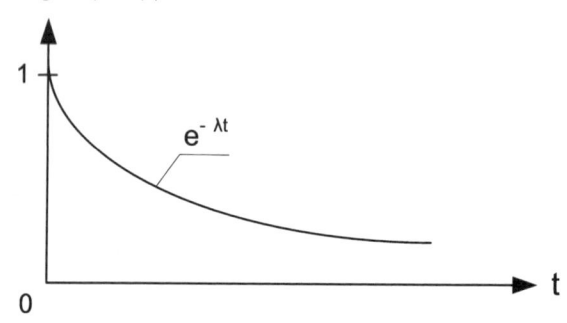

Figure 2.6: Diagram of the function P(t)

$$Q(t) = \lambda t, \qquad Q_D(t) = \lambda_D t \qquad (2.5)$$

2.5 Evaluation of Safety Level of Signalling Equipment

Long experience of the operation of technical systems shows that the random nature of the appearance and actions of hazardous factors does not allow the complete exclusion of their impact on humans and the environment. Therefore, absolute safety cannot be achieved. The desired ideal (absolute safety) and the consummated level of a safety of technical systems and equipment should be clearly separated.

Three concepts can be seen:
- 'reasonably permissible'
- 'risk substitution'
- 'method of normalisation'

At first there exists a conception of a 'reasonably permissible' level of danger (level of risk). The achievement of such a level of risk often requires complex technical solutions and considerable costs.

The value of 10^{-6} (loss the life of one person from 1,000,000 in a year) is broadly acceptable level of danger for humans. Such the value is equal to the probability of death of a person at

2.5 Evaluation of Safety Level of Signalling Equipment

his (her) home from an accident. Such a value should be considered as a criterion for the level of risk stipulated by industrial and transport objectives.

Thus, we can use this criterion and the expression (2.5) for an estimation of λ_D – broadly accepted rate of hazardous events. Taking $Q_D(t) = 10^{-6}$, $t = 1$ year $= 8{,}760$ hours, we can receive λ_D from (2.5):

$$Q_D(t) = \lambda_D\, t \rightarrow 10^{-6} = \lambda_D\, 8760\ h$$

and

$$\lambda_D = 1.1 \cdot 10^{-10}\ (1/\text{hour}). \qquad (2.6)$$

Value (2.6) can be used as the criterion for a set of broadly accepted rates of hazardous events, and safety parameters of various applications are compared with it. Now, consider these parameters for existing signalling equipment.

Statistical data on interlocking operation show that dangerous failures are rare events. The cause of this is strictly keeping within the safety requirements on the stages of system development, manufacture and operation. The technical signalling equipment contribute, on average, only 3% of the statistics on train accidents and emergencies (Sapožnikov et al. 1995).

An evaluation of the real level of safety of systems and their elements can be fulfilled on the basis of comparison of experimental data concerning dangerous failures in large quantity of RACS for long period of time, and expert judgement and safety norms accepted by professionals in area of railway automatics and communications. The tables 2.2 and 2.3 below show statistical data for safety violation collected in the period 1985–1990 years on Russian railway network. The reasons for the safety violation were dangerous failures in signalling equipment.

2 Safety and Reliability in Signalling Systems

The tables contain values of the rate of hazardous events λ_D, mean time to hazardous failure T_D and probability of safety $P_S(t)$.

Equipment	Quantity	Number of dangerous failures				Rate of dangerous failures λ_D, (1/h)	Probability of safe operations during 5 years
		Relay interlocking	Automatic blocks	Automatic level crossing	Total		
Relays	14826350	7	2	-	9	$1,4 \cdot 10^{-11}$	0,9999994
Signals	378853	2	-	2	4	$2,4 \cdot 10^{-10}$	0,9999895
Track circuits	292924	33	1	-	34	$2,7 \cdot 10^{-9}$	0,9998817
Relay racks	95510	4	7	-	11	$2,6 \cdot 10^{-9}$	0,9998861

Table 2.2: The statistical data for safety violation in signalling equipment

System	Indicator	Quantity	Number of dangerous failures	Rate of dangerous failures λ_D, (1/h)	Mean time to hazardous failure T_D, years	Probability of safety $P_S(t)$ during 20 years
Relay interlocking	Railway station	9754	77	$1,8 \cdot 10^{-7}$	634	0,9685
	Points	229571	77	$7,7 \cdot 10^{-9}$	14825	0,9987
Automatic blocks	Signals	44570	18	$9,2 \cdot 10^{-9}$	12408	0,9984
	Line length, (km)	58898	18	$7,0 \cdot 10^{-9}$	16308	0,9988
Automatic level crossing	Level crossing	20279	5	$5,6 \cdot 10^{-9}$	20384	0,9990

Table 2.3: The statistical data for safety violation in signalling systems

Element, equipment	Rate of safety λ_D, 1/hour (max)
Safe element	10^{-12}
Safe relay	10^{-12}
Track circuit	10^{-9}
Signal control circuit	10^{-10}
Circuit of a relay track	10^{-9}

Table 2.4: The norms of safety for signalling equipment

System	Indicator	Rate of safety λ_D, 1/hour (max)
Relay interlocking	Points	10^{-9}
	Computerised control complex	10^{-11}
Automatic blocks	Signals	10^{-9}
Automatic level crossing	Level crossing	10^{-9}

Table 2.5: The norms of safety for signalling systems

When values of λ_D from the tables 2.2 and 2.3 is compared with their known expert values and their calculated values, provided by developers of modern systems, it is possible to determine recommended norms of safety for equipment (table 2.4) and systems (table 2.5) of railway automatics.

2.6 Rating of Safety Parameters

Developers of new signalling equipment need to resolve two main tasks. Firstly, the safety requirements should be defined as the important part of the specifications, i.e. criteria for safety parameters (verification stage). Secondly, after the system has been developed, it is necessary to fulfill the safety case and calculation of the achieved level of safety (validation stage).

In rating of safety parameters of signalling equipment, conventional rating indicators are introduced, as shown in the table 2.6 for a number of primary systems.

System	Conventional rating indicator
Electric and hump interlocking	Set of centralised points
Traffic and station coded interlocking	Object under control
Train control centres	Local interlockings
Channels of telecontrol	1 km of channel
Automatic block	Signal point
Crossing signalling	Level crossing
Automatic in-cab signalling	Signal decoder or in-cab equipment
Semi-automatic block	Track section

Table 2.6: The conventional rating indicator for signalling systems

The second most frequently conception of 'risk substitution' is used for the determination of safety criteria. Thus safety parameters of new systems and equipment should be no worse than the same parameters of replaced systems and equipment.

As dangerous failures are rare events, the statistical data concerning safety of the system in operation are frequently absent. In such cases the statistical data of failure-free operation are used for the rating with the coefficient K_a of failures asymmetry. The coefficient K_a shows the ratio between the rates of dangerous and protected failures:

$$K_a = \lambda_D/\lambda \qquad (2.7)$$

The normative value of coefficient K_a can be evaluated on the base of the ratio between the rates of dangerous and protected failures for safety relay (see table 2.2):

$$K_a = \frac{1.4 \cdot 10^{-11}}{1.3 \cdot 10^{-7}} \approx 10^{-4} \qquad (2.8)$$

Thus, in accordance with (2.8), a safety device should have rate of hazardous failures at least in 10,000 times lower than the rate of protected failures.

The value of K_a can be used as the normative value for determining the accepted criteria of dangerous failures λ_D for complex system of locomotive safety (CSLS). The statistical data of failure-free operation for similar systems are known. Accordingly this data, rate of protected failures, for example, of LZB80 system is no more than $\lambda < 9{,}6 \cdot 10^{-5}$ 1/hour (Sapožnikov et al. 1995). Then the accepted criteria of dangerous failures λ_D for CSLS can be determined as

$$\lambda_D = K_a \cdot \lambda = 10^{-4} \cdot 9.6 \cdot 10^{-5} = 9.6 \cdot 10^{-9} \text{ 1/hour.} \qquad (2.9)$$

2 Safety and Reliability in Signalling Systems

Third 'method of normalization' is used if statistical data for safety and failure-free operation are absent. In particular, this takes place if a new-built device or system has no prototypes between devices or systems in operation. The idea of the method represents the following. The broadly accepted criteria of dangerous failures λ_D for such the systems (devices) is defined from the condition that only one dangerous failure can occur for all set of single-type systems (devices) during all normative period of the system (device) operations.

If, for example, in set of N = 100,000 locomotives, equipped with CSLS, no more than one dangerous failure can occur during a life span of 10 years of locomotive operations. Then:

$$\lambda_D = \frac{1}{N \cdot T_{op}} = 1.15 \cdot 10^{-10} 1/h \qquad (2.10)$$

In the case of (2.10) estimation of λ_D is received more stringent than in the case of (2.9).

2.7 Calculations of the Safety Parameters

The developers of a system fulfil a calculation of safety parameters. The purpose is to provide evidence that the achieved level of safety meet the given criteria. The calculation depends on the structure selected for the system development. At the present time, variants of two-channel and three-channel structures (majority systems) are widely adopted. Let us consider calculation of safety parameters based on the example of a two-channel structure. Figure 2.7 shows a two-channel system (duplex system) with safety comparison. In this system two identical microcomputers, connected in parallel, are working simultaneously. Safety comparison circuit (SCC) compares output signals of the microcomputers and forms a control signal only when both the output signals coincide. Such the systems are named ('2oo2') 'two out of two'-systems.

Table 2.7 shows states of the system depending of the channel states. Using one channel will be named 'one out of one system' ('1oo1'). The circuit SCC is considered as absolutely reliable.

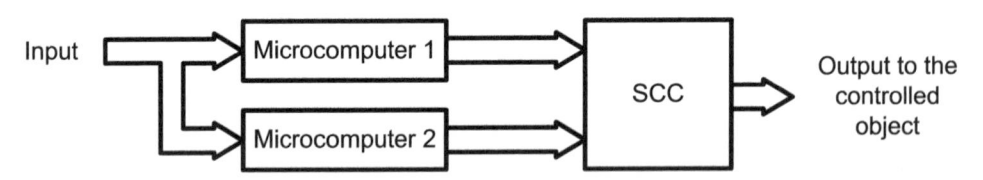

Figure 2.7: Two-channel system with safety comparison

If the failure rate of one of the microcomputers is known, then parameters of failure-free operation for '1oo1' system can be calculated with the following formulae:

n.n	State		State of the '2oo2' system
	Microcomputer1	Microcomputer 2	
1	Operable	Operable	Operable
2	Operable	Disabled	Protected
3	Disabled	Operable	Protected
4	Disabled	Disabled	Dangerous

Table 2.7: The states of the duplex system depending of the channels states

2.7 Calculations of the Safety Parameters

$$P_{1oo1}(t) = e^{-\lambda t}, \quad P_{2oo2}(t) = P_{1oo1}^2(t) = e^{-2\lambda(t)}, \tag{2.11}$$

$$Q_{1oo1}(t) = 1 - e^{-\lambda t}, \quad Q_{2oo2}(t) = 1 - e^{-2\lambda t}, \tag{2.12}$$

$$\lambda_{1oo1}(t) = \lambda, \quad \lambda_{2oo2}(t) = 2\lambda, \tag{2.13}$$

$$T_{1oo1} = \frac{1}{\lambda}, \quad T_{2oo2}(t) = \frac{1}{2\lambda}. \tag{2.14}$$

And safety parameters of '2oo2' system can be calculated, accordingly the table 2.7, with the following formulae:

$$Q_{D2oo2}(t) = Q_{1oo1}^2(t) = (1 - e^{-\lambda t})^2, \tag{2.15}$$

$$P_{S2oo2}(t) = 1 - (1 - e^{-\lambda t})^2 = 2e^{-\lambda t} - e^{-2\lambda t}, \tag{2.16}$$

$$\lambda_{D2oo2}(t) = -\frac{P'_{S2oo2}(t)}{P_{S2oo2}(t)} = \frac{2\lambda(1 - e^{-\lambda t})}{2 - e^{-\lambda t}}, \tag{2.17}$$

$$T_{D2oo2} = \int_0^\infty P_{S2oo2}(t) dt = \frac{3}{2\lambda}. \tag{2.18}$$

Let, for example, $\lambda = 10^{-5}$ 1/hour and $t = 1000$ hour. Then:

$P_{1oo1}(t) = e^{-0.01} = 0.99005; Q_{1oo1}(t) = 0.00995; T = 10^5$ hours $= 11.4$ years,
$P_{2oo2}(t) = e^{-0.02} = 0.9802; Q_{2oo2}(t) = 0.0198; \lambda_{2oo2} = 2 \cdot 10^{-5}; T_{2oo2} = 5 \cdot 10^4$ hour $= 5.7$ year,
$Q_{D2oo2}(t) = (1 - e^{-0.01})^2 = 0.000099; P_{S2oo2}(t) = 0.999901; \lambda_{D2oo2}(t) = 1.97 \cdot 10^{-7}$ 1/hour,
$T_{D2oo2} = 1.5 \cdot 10^5$ hour $= 17.1$ year

Thus, comparison of the systems for $t = 1,000$ hours shows that:
- probability of failure Q(t) for a two-channel system has risen by a multiple of 1.99;
- mean time to failure T has reduced by a factor of 2;
- probability of dangerous failure $Q_D(t)$ has reduced by 100 times;
- rate of dangerous failures $\lambda_D(t)$ has reduced by 50 times;
- mean time to dangerous failure T_D has increased by 1.5 times.

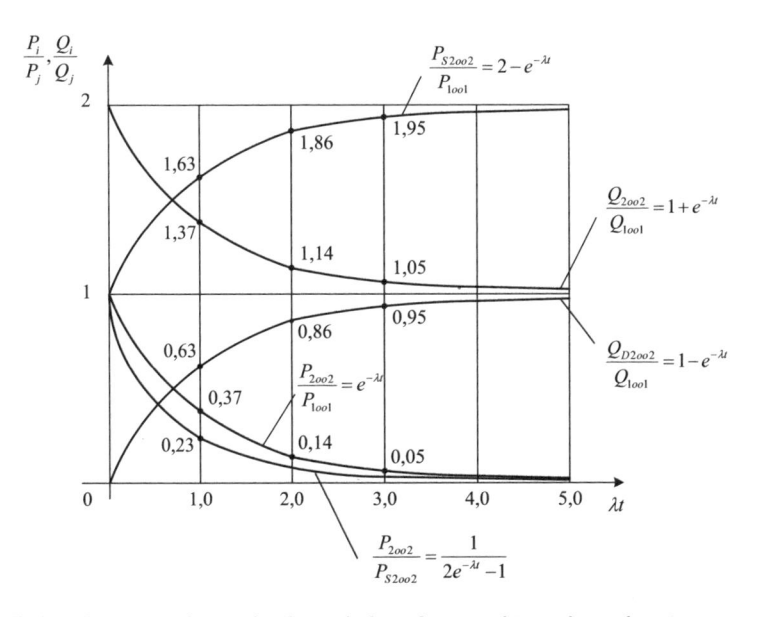

Figure 2.8: Failure-free operation and safety relation of one and two-channel system

2 Safety and Reliability in Signalling Systems

$P_{S2oo2}(t) = 2e^{-\lambda t} - e^{-2\lambda t}$

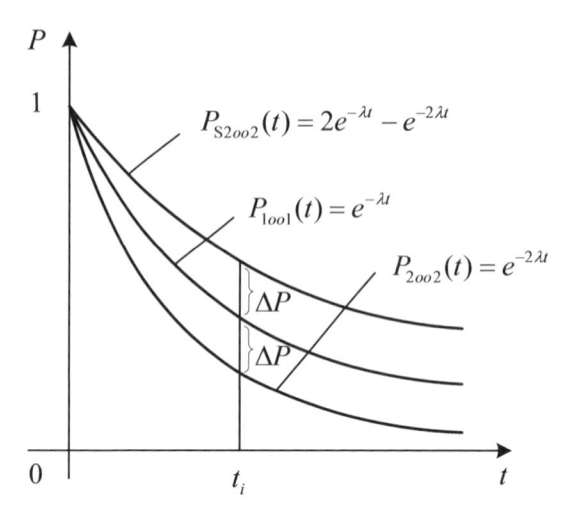

$P_{1oo1}(t) = e^{-\lambda t}$

$P_{2oo2}(t) = e^{-2\lambda t}$

Figure 2.9: Characteristics of reliability of 'two out of two' system

As $\lim_{t\to\infty} P_{S2oo2}(t) / P_{1oo1}(t) = \lim_{t\to\infty}(2 - e^{-\lambda t}) = 2$, then the probability of safety of a two-channel system cannot be higher than the failure-free operation probability of a one channel system than 2 times. For example, when t = 10T then it reduces by the value 1.99955. In order to get higher safety it is necessary to increase the number of channels in a multichannel system (redundancy rate).

Figure 2.8 shows diagrams of ratios of failure-free operation and safety for a one-channel system and for a two-channel system for various values of λt. For a given moment t value of P_{2oo2} is lower than P_{1oo1} in $e^{\lambda t}$ times, and the value of safety P_{S2oo2} is higher than P_{1oo1} in $(2-e^{-\lambda t})$ times.

Figure 2.9 shows relations between parameters of reliability for one-channel and two-channel systems. The following equality takes place:

$$P_{S2oo2}(t) - P_{1oo1}(t) = P_{1oo1}(t) - P_{2oo2}(t) = e^{-\lambda t} - e^{-2\lambda t} = \Delta P.$$

It can be seen that in a two-channel system for any moment t an increase of safety is accompained by an decrease of reliability. Consequently, this shows an important disadvantage of two-channel systems: safety is provided at the cost of reliability.

2.8 Methodology of Safety Case for Railway Signalling Equipment

The main task of a procedure of safety case of signalling equipment is the confirmation that the level of safety achieved in the system meets the specified values of the normative parameters. This procedure has its most important role in the set of certification works on acceptance of systems in accordance with safety requirements.

The following are the main methods to prove safety:
- methods for evaluation and appraisal of the project documentation and calculations on an analytical model;
- simulation methods based on a computer models;
- methods of testing prototypes of a system on the stages of bench development and acceptance;
- methods of testing prototypes of a system in application conditions on the stages of commissioning trials and implementation;
- methods of gathering statistical data concerning malfunctions and failures of single-type systems on the operation and maintenance stages.

The above mentioned methods are listed in the order of their application through the life cycle of a system. Simultaneously, they are listed in an order of increasing of proof of system safety.

3 Railway Operation Processes
Jörn Pachl

3.1 Historical Background

Today, there is hardly any field of technology in which the differences between national rules and procedures are as great as in railway operation and signalling. Of course, the very basic principles of how a railway works are the same everywhere. However, operating procedures differ significantly between different countries or regions worldwide. There are not only differences in some details but in fundamental terms, definitions, and procedures. As a result, a lot of railway education and training concentrates on the national rules. The same is true for knowledge presented in textbooks and other teaching materials. One of the objectives of this textbook is to provide generic knowledge on railway signalling that does not concentrate on the rules of a single country or region.

Despite the differences between the operating procedures of various national railway systems, there are three basic operating philosophies in running a railway that have influenced railway operation worldwide. These are:
- British operating principles
- German operating principles
- North American operating principles

Some countries follow one of these three philosophies in a quite pure form, while other countries use a specific mix of several systems, or added national peculiarities.

The British system is used in Great Britain and Ireland, in all countries of the Commonwealth with the exception of Canada, and in several South American countries. The German system is used in the German-speaking countries, in Luxembourg, in Eastern Europe, in the Balkan countries, and in Turkey. The North American system is used in the USA, in Canada, and in Mexico. The railways of these three countries have achieved a very high level of harmonisation. Today, the whole of North America has a uniform railway system with a high degree of standardisation.

Typical 'mixed systems' are:
- The railways of Western Europe, with the exception of Luxembourg, show influence both from the British and from the German system. French railways also developed specific characteristics that can only be found there. After World War II, the railways of the Netherlands adopted selected principles from North American operation and signalling.
- The railways in Scandinavia mainly follow German operating principles. The interlocking systems developed in these countries show also some influence from British signalling technology.
- The South African railways were originally based on the British principles. However, in the field of interlocking systems, South Africa has moved toward German principles.
- The operating rules of the railways of the Russian Federation and of the countries of the former Soviet Union are mainly influenced by German principles. However, in the field of signalling, these railways adopted a lot of ideas from North America and Britain.
- As part of the Commonwealth, the railways in Australia and New Zealand have followed the British system. However, there is an increasing influence from North America.
- The railways of China originally started with the British system. Later, they adopted a lot of principles from the former Soviet railways but tried to combine them with traditional British principles.

Other mixed systems in any possible configuration can be found in many developing countries.

Regarding operating rules, North American railways differ significantly from the rest of the world. This concerns both the lack of a distinction between tracks within station areas and

3 Railway Operation Processes

sections between stations, and the classification of movements with railway vehicles. Regarding signalling principles, German railways developed specific interlocking and block principles that cannot be found in countries that do not follow the German principles.

These differences, which influence railway operation and signalling significantly even today, have their roots in the nineteenth century. With the construction of the first railway lines, a period of experimenting started in which basic knowledge on the limits of the new system had to be gained. This period, which ended at about 1870, was followed by a three decades known as the 'Founding Years of Railway Signalling'. During that time, with the exception of automatic train protection, all basic principles of railway operation and signalling were developed. These principles are still used today. However, that period was also the start of increasing separation between the North American and the European railways. It was also the start of separate development of the German railway signalling, which more and more departed from the British signalling philosophy.

This development was closely connected to two fundamental inventions in railway signalling that occurred at about same time. Those inventions were the track circuit (*William Robinson* 1871; chapter 5.3) and the interlocked block instrument (*Carl Frischen* 1872; chapter 10.3.3.2). From this time on, the development of German signalling was significantly influenced of the use of interlocked block instruments (in German called Blockfeld). Block instruments of that type were not only used for safe train separation with the positive locking of signals but also for irreversible electric route locking (chapter 4.3.8.1) and other purposes. These instruments must not be confused with the British block instruments (e.g. the instruments invented by *Edward Tyer* and others). They were never used on railways that did not follow the German operating and signalling principles.

From this time on, German signalling was based on the principle that a locking produced at one station cannot be released by that station, but only by electric action from a corresponding station – or by automatic action of the moving train. Although these block instruments can only be found in technology of a past era, that philosophy has influenced German signalling ever since. Even in automatic systems, block signals are not only controlled by 'track clear' detection technology; there is always an overlaid locking procedure. This follows the principle that the signal at the entrance to a block section is kept locked in stop position by the signal at the end of the block section. The locking will only release after the train has passed the signal at the end of the block section. Since that principle does not work on station tracks where trains may start or terminate, there is always a sharp distinction between tracks in station areas and those on the open line (i.e. sections between stations). Block instruments of a similar kind as used for block systems were also used as part of the interlocking systems for electric route locking, and for electric interlocking between adjacent interlocking stations.

As a big difference, the development of signalling technology in North America was mainly driven by the invention of the track circuit. The introduction of automatic block systems had already started in the nineteenth century; interlocked manual block systems with block instruments were never used. In North American automatic block systems, block signals are directly controlled by track circuits without an overlaid locking procedure. This even allowed the wiring of automatic block systems through turnouts, in a way that throwing the points will reset the protecting automatic block signal to stop but without locking the points by a cleared signal (unidirectional locking chapter 4.2.3). These non-interlocked points on automatic block lines require to follow specific rules to ensure safe train operations.

Beside automatic block systems, track circuits were used quite early on interlocking systems. While in Europe visual track clear detection by the local operator was common practice and can still be found today in old installations, track clear detection by track circuits became a standard safety feature in North America from the beginning of the twentieth century. Even in mechanical interlocking systems, the need for visual track clear detection was abolished. Track circuits were not only used for track clear detection when clearing a signal, they were also used for electric route locking. Instead of the German philosophy of using block instruments of a similar kind as developed for the interlocked manual block systems also for electric route locking and release, electric route locking was effected by electric lever locks controlled by track circuits. This led to the principle of approach locking, (chapters 4.3.8.2 and 4.3.8.3) which became a standard feature of all interlocking systems in the English-speaking world.

When the first railways were built, sufficient means of communication for traffic control did not exist. Traffic was controlled by the timetable, using the principle of time spacing (chapter 4.4.3; figure 4.46) (also called time interval working) for train separation. This situation changed with the invention of the electric telegraph. However, from this point, the development of the railways in Europe went into a direction completely different from the railways in North America. In the early 1870s, after the invention of the electric telegraph, train separation based on fixed block sections became the standard form of operation on all European railways.

British railways were the first to introduce block working, but were followed shortly after by the continental railways. The introduction of the fixed block system was combined with the use of lineside signals for the control of train movements. From this time on, lineside signals became an integral part of European railway operation. Since the block systems developed in the nineteenth century were manually controlled by local operators, the same people also effected system control. Later on, for lines with heavy traffic, traffic regulators were established to supervise traffic on a longer stretch of line or within a terminal region. On some railways, they were called dispatchers but with a quite different meaning from the dispatchers on North American railways.

A manually operated block system needs to have all stations staffed by a local operator. With block lengths of between 1 km and 5 km, this requires a lot of local operating staff along the line. In North America, with the exception of some lines in the East, this was simply impossible due to the low population density. Consequently North American railways, even after the invention of the electric telegraph, did not introduce fixed block operation. Signalled fixed block operation became only possible after the invention of automatic block systems based on track circuits.

On the majority of North American railway lines, timetable-controlled operation with train separation by time spacing remained in use. From the beginning, traffic on North American railways line was not controlled by local operators but by a dispatcher who worked in a central office. The electric telegraph allowed the dispatcher to alter the timetable by issuing train orders. The train orders were transmitted by telegraph to staffed train order stations, and delivered by local operators to train crews. This led to the principle of Timetable & Train Order, which became the standard procedure of North American railway operation and remained in use until the 1980s. Then, powerful radio systems became available that allowed the dispatcher to issue movement authorities directly to train crews. A very short description of Timetable & Train Order can be found in (Pachl 2002). In these operations, signals were used only at specific locations that required a higher degree of protection, for instance at intersections or junctions of different railway lines. These signals protected just the intersection or junction, and sometimes they were not even interlocked. However, they did not provide movement authority for the following sections of line. In (MacFarlane 2004), the author makes the very striking statement that the difference from European operating practices was that signals in traditional North American operations were an adjunct to the railway system, rather than an integrated part of it.

The procedure of Timetable & Train Order was even used on most lines with automatic block signals. Although time spacing was replaced by fixed block operation, the traffic was still con-

3 Railway Operation Processes

trolled by train orders. This allowed the railway to have non-interlocked hand-operated points on automatic block lines. A signal-controlled operation in which trains are governed directly by signal indications can only be found on lines with Centralised Traffic Control (CTC). Although CTC is normally associated with remote control of interlocking stations, the basic definition of CTC is that trains are governed by signal indication. This is a frequent cause of misunderstandings between North American and European railway experts, since governing trains by signal indication is the standard form on all European mainlines. This applies even on lines controlled by old mechanical interlocking systems. Formally though, any European line controlled by local mechanical interlocking stations meets the North American definition of CTC.

3.2 Classification of Tracks, Stations and Signals

3.2.1 Main Tracks and Secondary Tracks

For operational purposes, tracks are divided in two main classes, which have different descriptions in the rule books of several railways. However, the basic idea is always the same. First, there are tracks that can be used for regular train movements (for the classification of movements with railway vehicles see chapter 3.4). Here, these tracks are called main tracks. Another term mainly used in the British rules, is running lines (since in the British terminology in operating and signalling rules, a track is often referred to as a line). The tracks of the open line, i.e. the sections between stations and their continuation through stations and interlockings, are always main tracks. Main tracks used for passing and overtaking trains are called loops (figure 3.1).

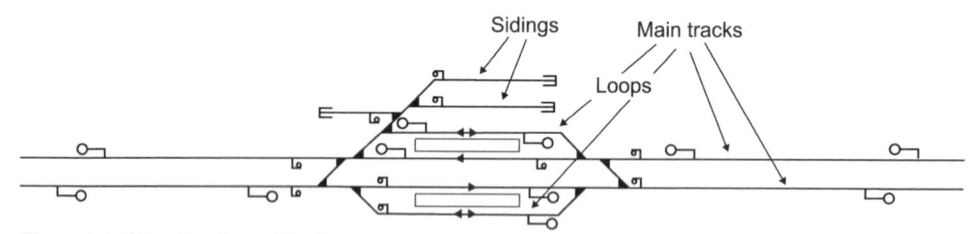

Figure 3.1: Classification of tracks

In a signalled territory, main tracks are equipped with signalling appliances for train movements. Points on main tracks are usually interlocked with signals. Tracks other than main tracks are called secondary tracks or, in British terminology, sidings. These tracks must not be used for regular train movements. An arrangement of secondary tracks for making up trains, storing cars and trains, and similar purposes, is called a yard (figure 3.2).

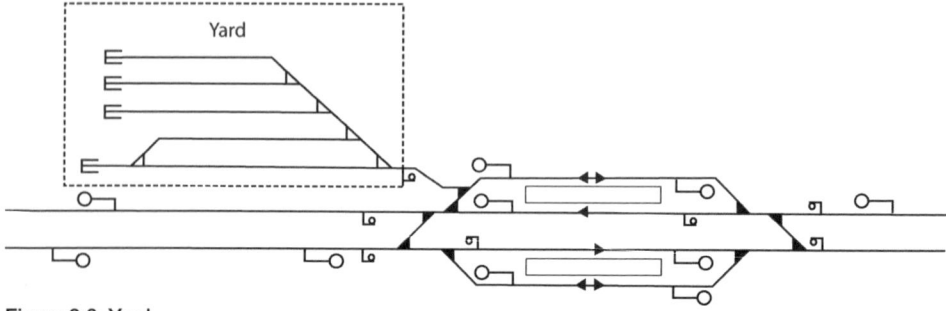

Figure 3.2: Yard

3.2 Classification of Tracks, Stations and Signals

In the North American terminology, loops are called sidings. Depending on the operating procedures, a siding of that kind may be considered as a main track or a secondary track. Even in signalled territory, points in main tracks are not necessarily interlocked with signals (White 2003).

3.2.2 The Role of Signals

Trackside signals indicate if a movement may enter the section of track beyond that signal. On railways where train movements are strongly separated from shunting movements (see chapter 3.4), there are also different kinds of signals for these kinds of movements. Here, main signals indicate if a train may enter a track section. This is typical for almost all railways outside North America. In a fixed block territory with a signal-controlled operation, train movements are authorised by signal indications. Apart from when the approach line has a low maximum speed, a signal that authorises a train movement requires an approach aspect (also called 'warning aspect' or 'caution aspect') at the braking distance. This is because the stopping distance is generally greater than the distance the driver can see ahead. The approach aspect is necessary for safe braking when approaching a stop signal.

In a territory where the distance between signals does not much exceed the braking distance, the approach aspect is usually provided by the signal in rear. In a territory with very long distances between main signals, distant signals are placed at the braking distance in approach to a main signal (chapter 7.3.3). A distant signal warns; it can only provide an approach aspect for the signal ahead. It cannot show a stop aspect.

Shunting signals are used to authorise shunting movements and to protect train routes against shunting movements. Main signals are not cleared for shunting movements. On tracks where shunting movements may pass a main signal, a shunt aspect is also incorporated into the main signal, so that shunting movements may be authorised to pass main signals showing stop. For shunting signals, an approach aspect is not necessary, since drivers are expected to be able to stop short of any vehicle or obstruction (driving on sight).

On North American railways, there is no differentiation between main and shunting signals because of different operating rules for shunting movements. But at some places, a signal may show a special aspect that is used only for shunting movements. Another characteristic of North American railways is the general lack of distant signals. On signalled tracks, the approach indication is always provided by the block signal in rear, regardless of block length. Distant signals are only used in approach to an interlocking on a non-signalled track.

Concerning the control principle, signals may be divided in three classes:
– Controlled signals
– Automatic signals
– Semi-automatic signals

A controlled signal is one that is locally or remotely controlled by an operator. The working of an automatic signal is effected automatically by the trains moving along the line. A semi-automatic signal is a controlled signal that can be switched into an automatic mode. Beside working signals, many railways use inoperable (or fixed) signals in specific layouts to suit their individual operating rules. Such signals always display one indication only. Thus, fixed distant signals may be used in situations where all trains must be prepared to stop at the home signal. Inoperable main signals that always display a stop indication are used on tracks where trains, after having stopped at the signal, may only proceed as a shunting movement.

In a territory with a fixed block system, main tracks are divided into block sections for the purpose of safe train separation. A train must generally not enter a block section until it has been

3 Railway Operation Processes

cleared by the train ahead. The principle is no more than one train, in any one block, at any one time. In a territory with lineside signals, block sections are limited by signals which govern train movements. A signal that controls entry to a block section is called a block signal.

3.2.3 Definitions of Stations and Interlocking Areas

Generally, stations are all places designated in the timetable by name. Regarding the schedule, stations are the points where time applies. On British railways, only places where trains stop for load/unload passengers or freight are referred to as stations. In North American usage, each station is designated by a station sign that designates the specific point at which an instruction using only the name of the station applies.

On European railways, station signs are mainly used at passenger stations for the purpose of passenger information. In extended and complex terminal areas, some railways place station signs at interlockings outside passenger stations to support the driver in local orientation. But this is usually only be done at places without a local interlocking station, so that the station cannot be identified by the interlocking station's ID. Some railways use the term station only for places where trains have regular stops. The term station is not necessarily associated with the term station track which is used by several railways to separate sections of main track where station rules apply from the block sections of the open line. Rules on station tracks are closely related to the interlocking rules of a specific railway.

An interlocking is an arrangement of points and signals interconnected in a way so that each movement follows another in a proper and safe sequence. Signalled routes for trains on main tracks are usually interlocked (chapter 4.3). Signals that govern train movements through an interlocking are called interlocking signals. An interlocking signal can also be a block signal. The points and signals within interlocking limits are controlled either by a local interlocking station or from a remote control centre. Local interlocking stations are called interlocking towers in North America, and signalboxes or signal cabins on railways that follow British principles. The block signals between controlled interlockings are often called intermediate block signals. In Britain, this term is used only in older systems for a block signal that is controlled from the interlocking station in rear.

Concerning interlockings and stations, the railways designated different names and limits in accordance to their individual operating practice. In particular, there is a big difference between North American railways and those elsewhere. In North America, the block system that protects train movements is not interrupted in interlocking areas. There is no station track separated from the open line. Figure 3.3 demonstrates the essential difference at the example of a track arrangement with several loop tracks. In North America, the point zones at both sides of the loop tracks would form separate small interlockings. These are limited by opposing interlocking signals in a way that each interlocking does normally not contain any consecutive interlocking signals. Station names refer to these small interlockings but not to the entire loop track layout. In Europe and on other railways outside North America, the entire loop track layout would be a station designated by name. The tracks between the outer point zones are station tracks. On station tracks, there are consecutive interlocking signals, which form station track sections. Train movements on these sections are protected by the interlocking system but not by a block system. Thus, the entire layout that may even contain more than just two point zones forms one interlocking area.

Figure 3.4 gives a more detailed example of how interlocking limits are established on North American railways. At each track that leads into interlocking limits, there is a signal that may authorise train movements, even at tracks that are not used for regular train movements (this

3.2 Classification of Tracks, Stations and Signals

a) North American Practice

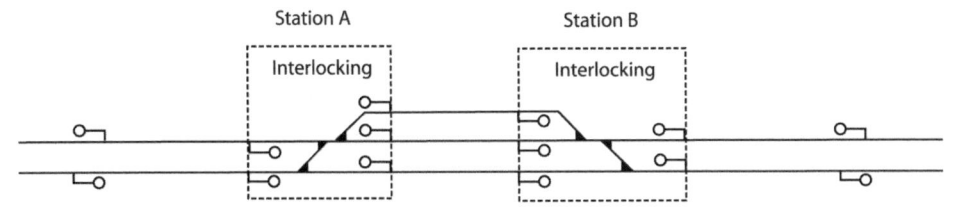

b) European Practice (except modern UK practice)

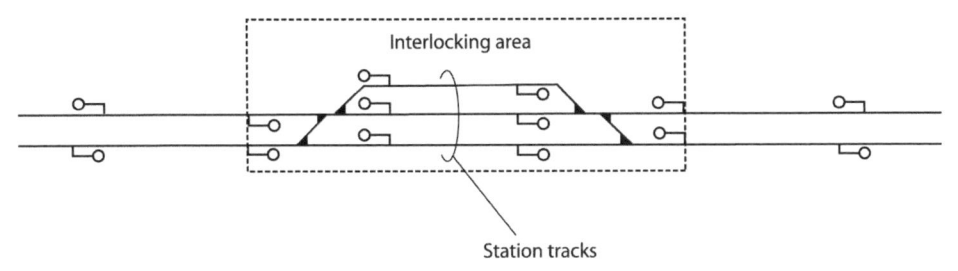

Figure 3.3: Different principles of assigning interlocking and station names to a track layout on North American and European railways

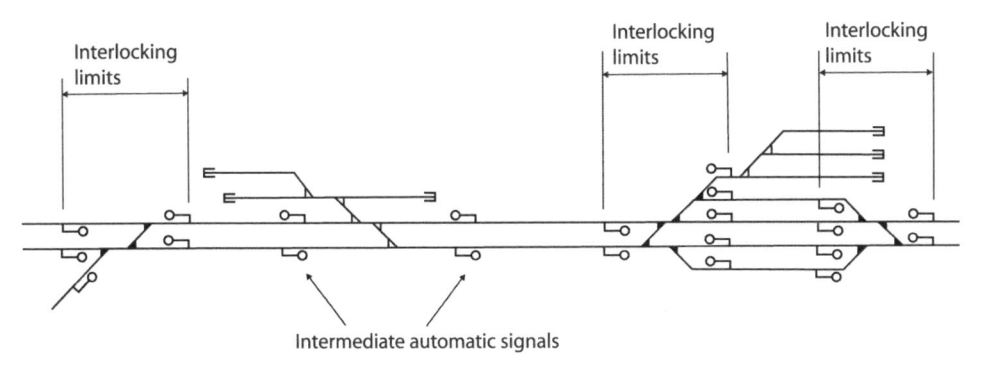

Figure 3.4: Interlocking limits (North American practice)

is an essential difference from European signalling). These signals are called home signals. A speed indication at an interlocking signal applies until the train has passed the first opposing interlocking signal, which is called the exit signal of that interlocking.

On European railways, there is still a difference between the traditional British practice and the continental railways that followed more the German principles. In traditional British interlocking systems, there are designated 'station limits'. Station limits are the tracks between the home signal and the last main signal of the same direction (the section, or starting signal), controlled from the same signalbox (interlocking station). The section signal permits trains to leave the station limits and enter the next block section. There are different station limits for each direction. In most British installations, this signal is placed behind the last points of the interlocking (then also called an advance signal or formerly an advanced starter signal), thus usually requiring additional interlocking signals before the points (figure 3.5).

3 Railway Operation Processes

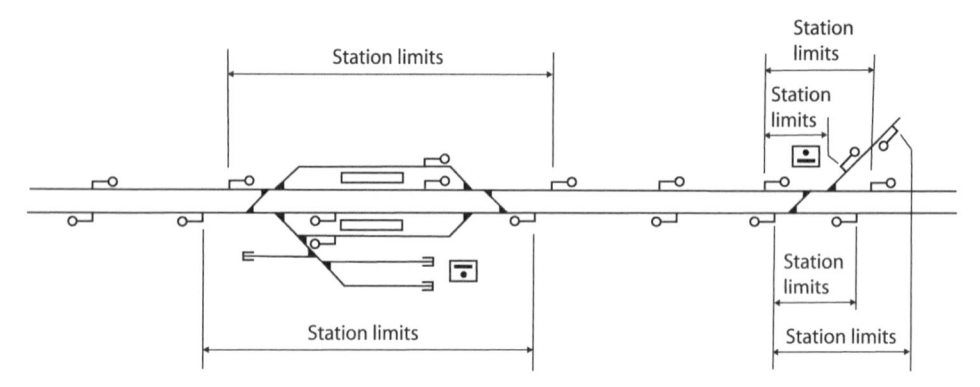

Figure 3.5: Station limits (older British practice)

In an essential difference from the German home signal limits explained below, British station limits are not associated with a track layout but with a single interlocking station. In a track layout that is controlled by more than one interlocking station, each has its own station limits. Station limits are not designated in modern interlockings with a continuous track clear detection. In areas controlled by modern control centres, there is just a difference between track sections on which trains are protected by controlled interlocking signals, and track sections on which trains are controlled by automatic block signals. In those areas, there are no longer any signals referred to as home signals. For more information on modern British signalling practice see (IRSE 1980).

On many other railways worldwide, a track layout with station tracks is limited by the opposing home signals that protect the interlocking area. The main tracks outside the home signal limits are called the open line (figure 3.6). This meets the traditional German practice where such a station area is called a 'Bahnhof' (there is no suitable English translation). In a difference from British station limits, home signal limits of that kind may contain several interlocking stations.

The signals that govern train movements to leave the home signal limits for the open line are called exit signals. Interlocking signals inside home signal limits that are neither home nor exit signals are called intermediate interlocking signals (figure 3.7). Interlockings at junctions and crossovers outside the home signal limits belong to the open line. The signals at these junctions and crossovers are referred to as block signals, but not as home signals. Compared with junctions and crossovers inside home signal limits, an interlocking on the open line is called a junction station or a crossover station.

In areas controlled by control centres, several formerly separated station areas have sometimes been combined into consolidated home signal limits, with a number of consecutive inter-

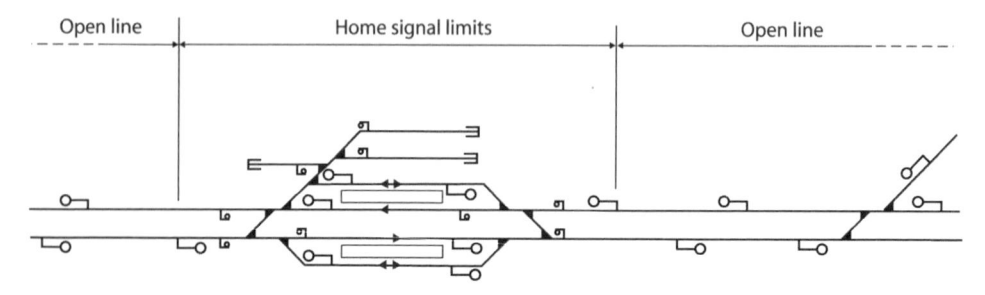

Figure 3.6: Home signal limits of a station area (German practice and other)

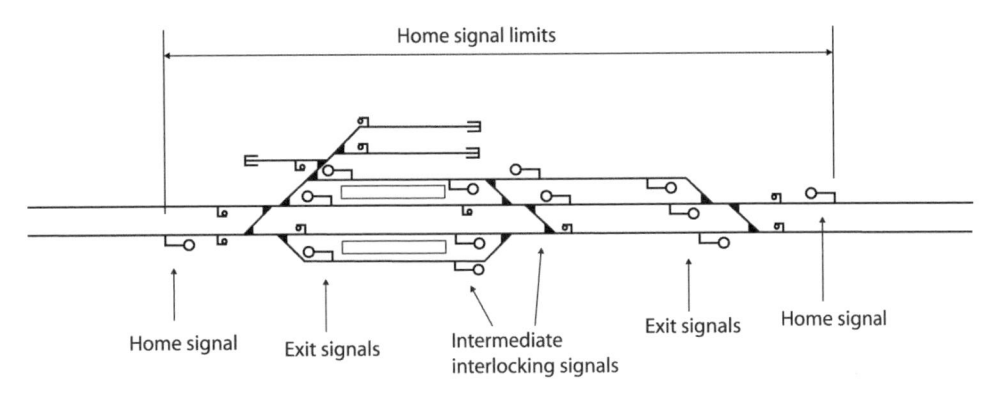

3.2 Classification of Tracks, Stations and Signals

Figure 3.7: Home signal limits with intermediate interlocking signals

mediate interlocking signals. In such an extended station area, a train may even have several scheduled stops without leaving the home signal limits. On some railways, this development has led to a discussion on whether to abolish the traditional distinction between station tracks and tracks of the open line in territories with a high degree of centralisation.

Despite the different limits defined in interlocking areas by the different railways, interlocking main signals can be divided into four main classes (the terms in brackets are suggested names to characterise these signals independently from the operational rules of a specific railway):

- Interlocking signals that lead from a block section into a station track (station home signals)
- Interlocking signals that lead from a station track into a block section (station exit signals)
- Interlocking signals that lead from a block section into another block section (block home signals)
- Interlocking signals that lead from a station track into another station track (intermediate interlocking signals)

Some railways use all of these classes of interlocking signals, while others use a selection or even just one of them.

3.2.4 Signal Arrangement on Double Track Lines

On double track lines, there is usually a specified direction of traffic (in North America also called the current of traffic) for each track. While right-track operation dominates slightly on a worldwide basis, there are significant numbers of countries where left-track operation is the standard form of operation on double track lines. Left-track operation is used in the UK and Ireland, in France, Italy, Belgium, Switzerland, Sweden, Portugal, in most African countries, in Iran, Israel, Japan, China, India, Pakistan, Indonesia and some other countries in that region, Australia and New Zealand, Uruguay, Argentina, Chile, and on broad gauge lines in Brazil. Some countries have mixed systems. Typical examples in Europe are Austria and Spain which use left-track operation on selected parts of their network only.

On lines not equipped with a bi-directional signalling system for two-way working, all regular train movements have to be made with the direction of traffic. On such lines, movements against the direction of traffic (also called 'wrong line moves' or 'reverse movements') have to be authorised by special instructions. On lines that are equipped with a signalling system for two-way working, movements against the current of traffic can be authorised by an interlocking signal. Some railways prefer not to have intermediate block signals for movements against the current of traffic, because on most lines reverse movements are not carried out frequently.

3 Railway Operation Processes

Figure 3.8 shows typical examples of signal arrangements on double track lines. On railways that follow German operating practice, a direction of traffic is only designated on main tracks outside home signal limits.

a) Double track line with one-way working

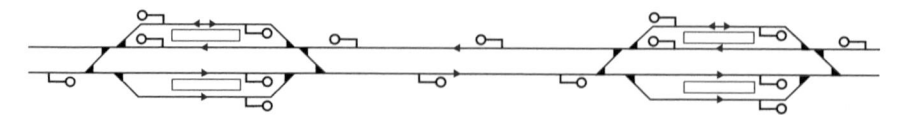

b) Double track line with two-way working without intermediate block signals for movements against the regular direction of traffic

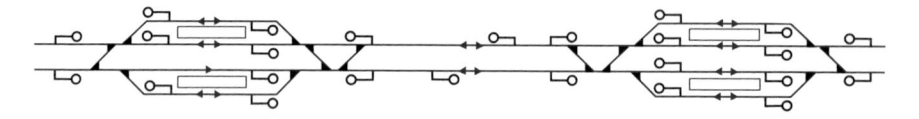

c) Double track line with two-way working with intermediate block signals for movements against the regular direction of traffic

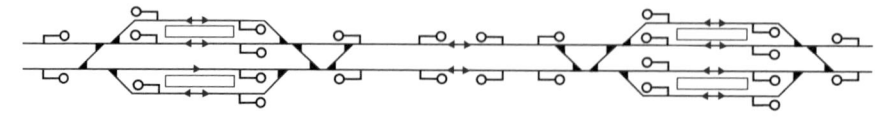

Figure 3.8: Signal arrangements on double track lines

3.3 Movements with Railway Vehicles

Movements of railway vehicles are divided in two classes which are called here train movements and shunting movements. However, the names used for these movements differ from railway to railway. On European railways, for both kinds of movements there are very different rules in use. To a certain degree on North American railways, there are also train and shunting movements, but the rules do not differ as significantly as in Europe. That is why North American railways, as mentioned above, do not use shunting signals.

3.3.1 Train Movements

Train movements are of locomotives or self-propelled vehicles, alone or coupled to one or more vehicles, with authority to occupy a main track to rules specified for train movements. The timetable also contains the maximum speed allowed in the sections along the line. Every train displays rear end markers to enable the lineside staff to check the train completeness. All regular movements on main tracks outside the home signal limits are train movements. In Britain especially, train movements are also called 'running movements'.

The authorisation of a train movement has two elements:
- A valid timetable as the authority to run through the network along a pre-defined route under specified operating conditions,
- A movement authority for every single section of track in the path of the train

3.3 Movements with Railway Vehicles

The movement authority (chapter 7.3.3.1) to enter a section of track is issued by the operator who is in charge of controlling train movements on that section of track. This way, a train is always under the external guidance of an operator. In a signal-controlled territory, the authority for train movements is given by:
- A proceed indication of a main signal
- A proceed indication of a cab signal
- A call-on signal permitting a train to pass a signal displaying a stop aspect under special conditions
- A written or verbal instruction permitting a train to pass a signal displaying a stop aspect under special conditions

In non signal-controlled territory, the authority for train movements is given by:
- Timetable rules
- Verbal or written authorisation

3.3.2 Shunting Movements

Shunting movements are movements for making up trains, moving cars from one track to another, and similar purposes. Shunting movements are accomplished under simplified requirements at a very low speed that allows stopping short of any vehicle or obstruction. Block rules are not in effect. Shunting units may enter occupied tracks. Movements in industrial sidings are also carried out as shunting movements. On some railways, train movements into an occupied track must be carried out in shunting mode. There are also railways that use shunting movements as a fall-back level for regular train movements in degraded mode operations.

The authority of shunting movements is given by:
- A proceed indication of a shunting signal (may be combined with an interlocking signal to authorise a shunting unit to pass an interlocking signal in stop position)
- Verbal permission

In the North American terminology, shunting movements are called 'switching movements' or 'movements of yard engines' and are authorised by yard limit rules (see below).

Railways designate different limits for shunting movements on main tracks according to their individual operating practices.

3.3.2.1 Shunting Limits (European Practice)

On European railways, with the exception of modern British rules, shunting units must not enter main tracks outside the home signal limits. The area behind the home signal that may be used for shunting is usually designated by limit of shunt boards (figure 3.9).

The distance between the home signal and the limit of shunt board equals the overlap of the home signal. Shunts may pass the limit of shunt boards only with written permission from the operator. Before issuing an authority to pass the board, the operator has to make sure that there is no train approaching the home signal.

In British practice, limit of shunt boards are always associated with station limits as used in traditional interlocking systems. On modern lines, instead of boards, limit of shunt signals (inoperable shunting signals permanently displaying a stop aspect) are used at places where no other shunt or main signal exists. This prevents shunting movements from running against the signalled direction onto an automatic block line without bidirectional working. On some other European railways, e.g. Switzerland, shunting movements may leave the shunting limits un-

3 Railway Operation Processes

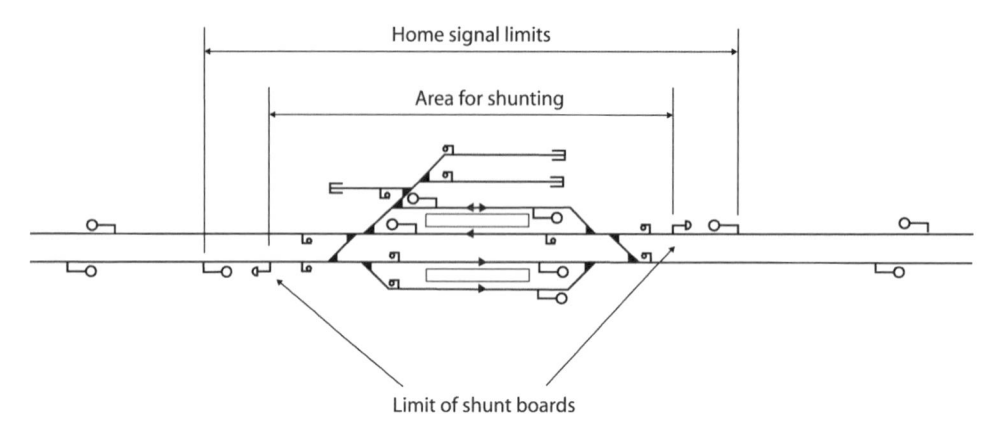

Figure 3.9: Shunting limits (German arrangement as example for typical European practice)

der specified conditions. For this purpose, the rules of those railways define a special class of shunting movements.

3.3.2.2 Yard Limits (North American Practice)

Yard limits are a speciality of railways following North American operating procedures. Here, train movements are usually authorised by a dispatcher who works in a remote office while the shunting is done by local staff. The purpose of yard limits is to simplify the authorisation of shunting movements on main tracks in terminal areas and also to relief the need for flag protection on unsignalled lines while working at small stations in dark territory (see section 3.4.3). In areas designated as yard limits, shunting units may enter main tracks without authorisation by the dispatcher. But at some places, local rules require dispatcher permission to use main tracks for shunting to ensure some control over train movements. Thus, shunting movements can be carried out by using main tracks in the 'yard mode'. On tracks where trains are protected by automatic block signals, yard limits are based on the signalling principle that the automatic block system is wired through hand-thrown points.

At hand-thrown points equipped with an electric lock, the release of the electric lock can be activated by an employee directly at the points after having the points mechanically unlocked. The activation will cause the signals governing the block to display the most restrictive indication. If the block in which the points are located is unoccupied or the points itself are occupied (so that a train leaving the main track may unlock the points), the lock will open immediately on activation. If the block is occupied but the points are not, the lock will release after a timer operation of several minutes – enough for any approaching train to have passed. At hand-thrown points without an electric lock, the signals governing the block go to their most restrictive indication when the point is thrown after having it mechanically unlocked by the employee.

Within yard limits, trains that are not running under a clear aspect of a block signal have to move at a reduced speed that allows them to stop short of any vehicle or obstruction, since there can be shunting units at work on the main track. When a train is approaching, shunting units have to give way by clearing the main track immediately. The beginning and the end of yard limits are marked by special marker boards to be obeyed by the train crews (figure 3.10).

Because on North American railways, train movements are not so sharply separated from shunting movements, they may also be carried out by verbal authorisation within yard limits. On lines operated under the rules of Centralised Traffic Control, yard limits are not used. On

3.4 Principles of Train Separation

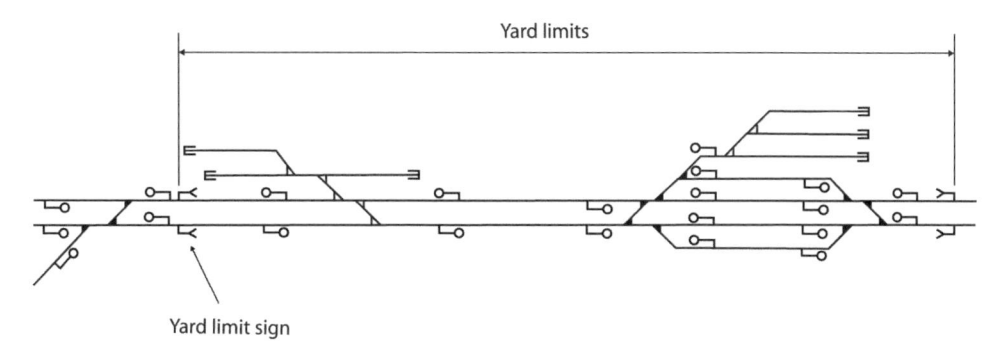

Figure 3.10: Yard limits (North American practice)

main tracks, yard engines are governed by signal aspects in the same way as train movements. However, the dispatcher can establish temporary shunting limits by issuing track and time authority to a train. Within the given limits that are blocked for other trains, the train crew can do shunting work. Remotely controlled points in such areas can be released for local control (dual control points).

3.4 Principles of Train Separation

In a steel wheel on steel rail system, the coefficient of adhesion is on average eight times less than in highway traffic. As a result, the braking force is also eight times less. That leads to braking distances for railway vehicles that may exceed the viewing range of the driver significantly. Thus, train separation by the sight of the driver is only possible when running at a restricted speed. Usually speed limits between 15 and 30 km/h are applied. This is only acceptable for shunting movements and for train movements in non-regular operation. For regular train movements, procedures of train separation are required that work independently from the viewing range of the driver.

The principle used for safe train separation depends on the following criteria:
- How movement authority (chapter 7.3.3.1) is transmitted from track to train
- How the track is released behind a train

If movement authority is only transmitted at discrete points, e.g. at trackside signals, or by written or verbal orders, this will necessarily lead to train separation in fixed block distance. Each movement authority has to cover the entire section up to the next point at which further authority may be received. On lines where trains are governed continuously by a cab signal system, this restriction does not exist. However, continuous transmission of movement authority is not yet a sufficient criterion to abolish fixed block sections.

In addition, the train has to release the track not in fixed intervals but continuously. This requires a permanent train-borne checking of train completeness. Since for traditional railway systems a sufficient solution for that problem has not yet been found, train separation at a fixed block distance is still the standard principle for safe train spacing on most railways worldwide.

The principle of train separation by time spacing is not covered in the following sections since European railways replaced time spacing by fixed block sections in the 1870s. As mentioned in 3.1, on unsignalled lines in North America, time spacing survived much longer as part of the Timetable & Train Order procedure. However, today it is almost completely extinct. For a short description of the basic idea, see (Pachl 2002).

3 Railway Operation Processes

Before explaining the different principles of train separation, another essential feature has to be mentioned. The braking distance of a train does not mainly depend on the weight of the train but on the percentage of the weight that is used to transmit braking force between wheel and rail. Trains with the same braking ratio have generally the same braking distance. For safe train separation, a train must always have a clear track ahead at least as long as the braking distance. Thus, from the viewpoint of capacity, it makes sense to assemble vehicles into trains. All vehicles that form a train do need just one common braking distance for the entire consist (figure 3.11). This will significantly reduce the capacity consumption that is produced by the long braking distances. This is why running whole trains instead of single vehicles is one of the very basic characteristics of a railway system.

a) Single vehicles

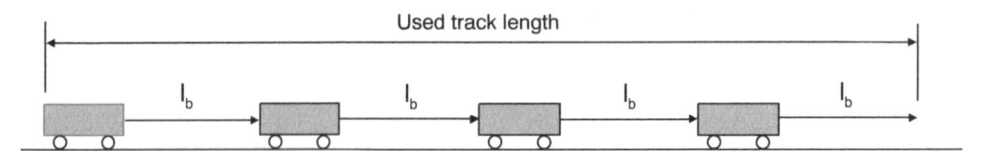

b) Vehicles in a train consist

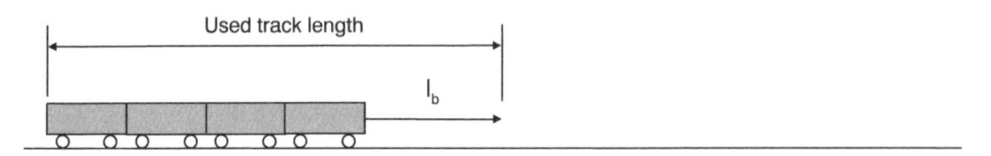

l_b Braking distance including all add-ons for safe train separation

Figure 3.11: Influence of train length on the used capacity

3.4.1 Signalled Fixed Block Operation

On lines where train separation by block distances is used, the track is divided into block sections. A block should be occupied exclusively by only one train. In a signalled fixed block operation, the block sections are limited by signals, which provide movement authority to enter the block section protected by the signals. To clear a signal for a train that is to enter a block section, the following conditions must have been fulfilled:
- The train ahead must have cleared the block section
- The train ahead must have cleared the overlap behind the next signal (only on lines where block overlaps are used)
- The train ahead must be protected from following train movements by a stop signal
- The train must be protected against opposing movements

On railways where block overlaps are not required, the control length of a signal equals the block section. Examples are mainline railways in North America and in Russia. Other railways require a control length of a signal that is longer than the block section (figure 3.12).

The difference is called 'overlap' (chapters 4.3.6 and 4.4.2) because in that area the control length of a signal overlaps with the control length of the next signal. The main purpose of the

3.4 Principles of Train Separation

a) Line without block overlaps

b) Line with block overlaps

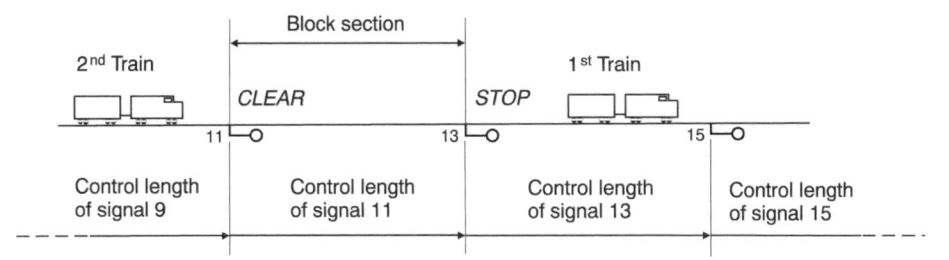

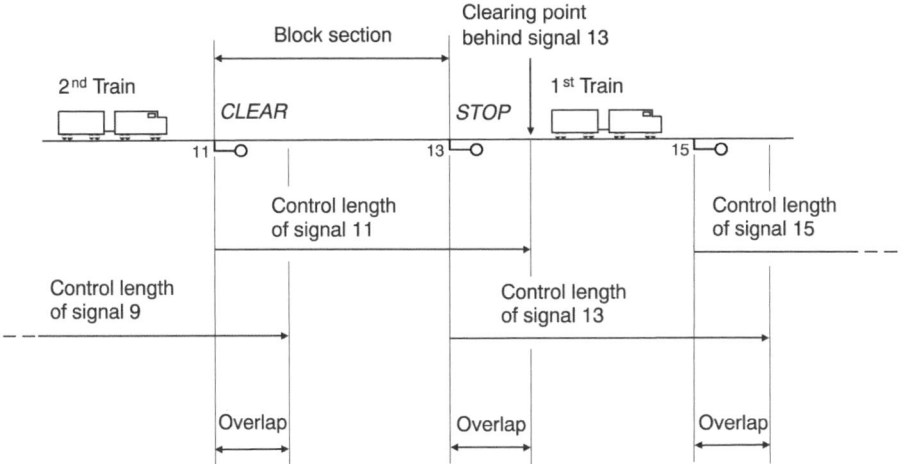

Figure 3.12: Control length of signals in fixed block territory

overlap is to provide additional safety in case the driver fails to brake adequately before a stop signal. A signal may not be cleared until the full control length is clear. Thus, the clearing point behind a signal equals the end of the control length of the signal in rear. Block overlaps are used on all European railways, many railways outside Europe, and also on almost all subways and subway-like electric city railways worldwide.

On a line with a fixed block system, the minimum headway is the time interval between two following trains, and depends on the so-called 'blocking time' (Hansen/Pachl 2008). **The blocking time (from the German term 'Sperrzeit') is the time interval in which a section of track (usually a block section) is allocated exclusively to a train and therefore blocked to other trains.** So, the blocking time lasts from issuing a train its movement authority (e.g. by clearing a signal) to the possibility of issuing a movement authority to another train to enter that same section. **The blocking time of a track element is usually much longer than the time the train occupies the track element.** In a territory with lineside signals, for a train without a scheduled stop, the blocking time of a block section consists of the following time intervals (figure 3.13):

- The time for clearing the signal
- A certain time for the driver to view the clear aspect at the signal in rear that gives the approach aspect to the signal at the entrance of the block section (this can be the block signal in rear or a separate distant signal)

3 Railway Operation Processes

- The approach time between the signal that provides the approach aspect and the signal at the entrance of the block section
- The time between the block signals
- The clearing time to clear the block section and – if required – the overlap with the full length of the train
- The release time to 'unlock' the block system

The approach time does not apply if the train has a scheduled stop at the signal at the entrance of the block section. In such a case, the signal watching time applies at that signal.

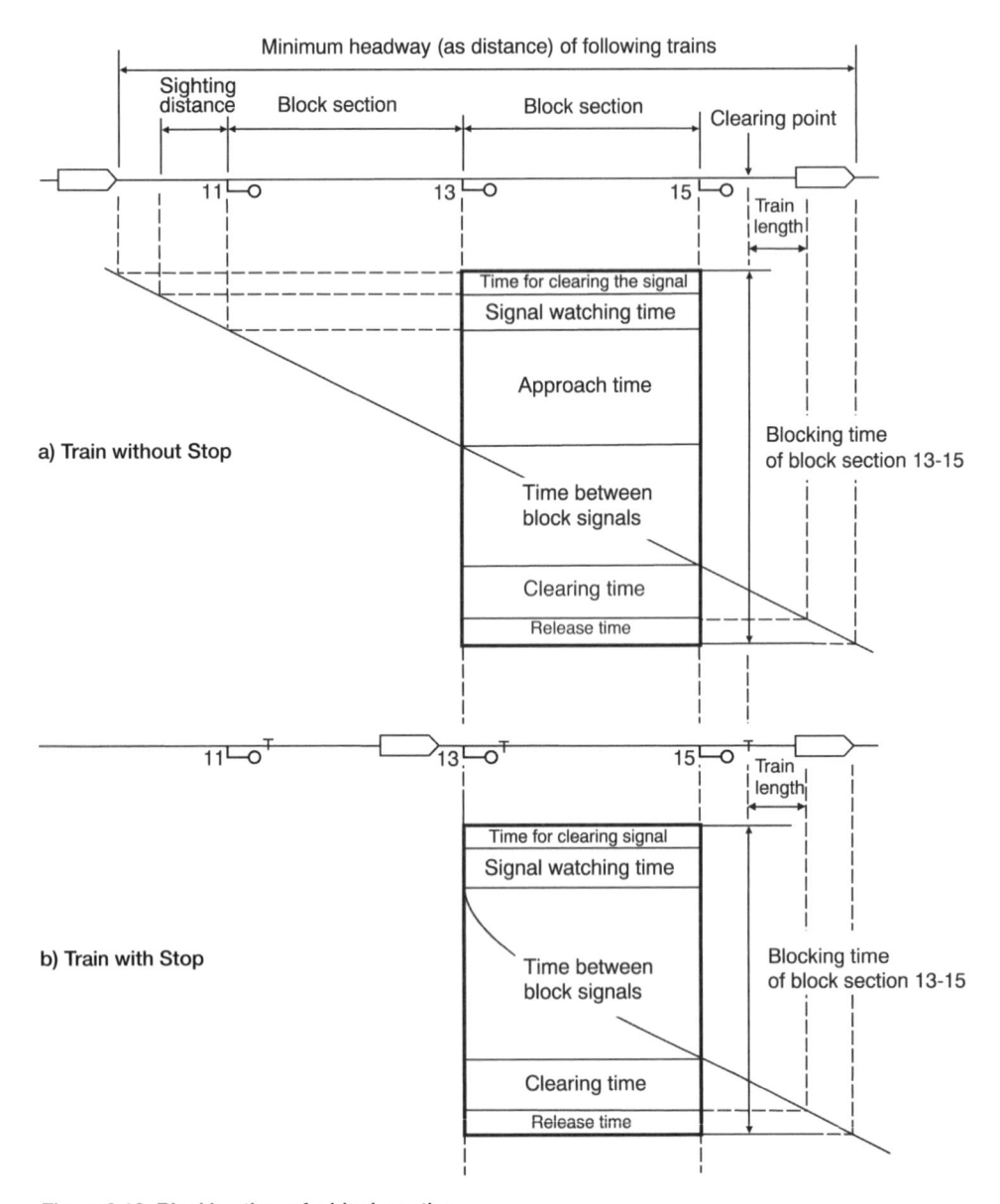

Figure 3.13: Blocking time of a block section

3.4 Principles of Train Separation

In a territory with cab signalling, the approach time is the time the train takes to run through the braking distance that is signalled by the cab signal system. Drawing the blocking times of all block sections a train passes into a time-over-distance diagram leads to the so-called 'blocking time stairway' (figure 3.14). This represents perfectly the operational use of a line by a train. With the blocking time stairway, it is possible to determine the minimum headway of two trains. The blocking times directly establish the signal headway as the minimum time interval between two following trains in each block section.

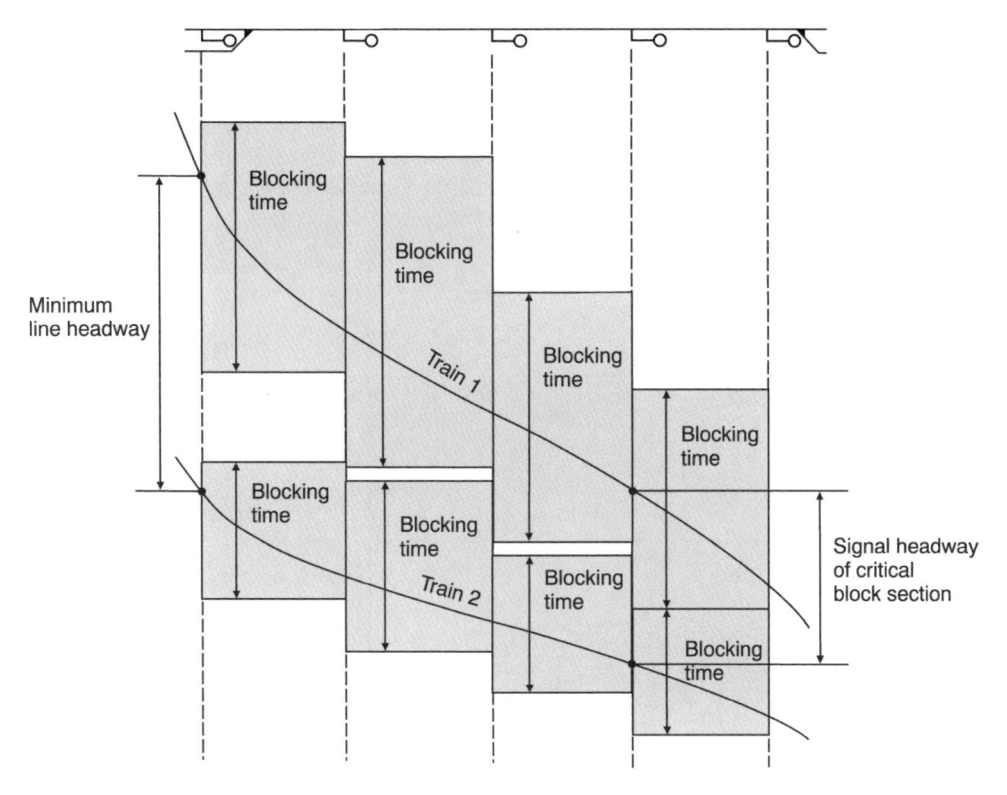

Figure 3.14: Blocking Time 'Stairways'

3.4.2 Cab Signal Operation

On most of the railways that use cab signalling, it is combined with a continuous Automatic Train Protection (ATP) system. However, on some railways, cab signalling is also used as a pure signalling system without brake enforcement (chapters 8.1.1 and 8.1.2). On many railways, cab signal indications are superior to lineside signals, so that trains are directly governed by cab signals. This is typical for all advanced cab signal systems used on high speed lines. This allows the infrastructure operator to remove lineside signals completely. However, some railways keep a reduced number of lineside signals so that trains without operative cab signals can still be governed by lineside signals.

On some railways, there are still older cab signal systems in use that work only as auxiliary systems. On such lines, trains are still governed by lineside signals, but the cab signal indications support the driver in watching the lineside signals.

3 Railway Operation Processes

3.4.2.1 Cab Signalling with Fixed Block Sections

The main reason for having fixed block sections on lines with cab signalling is the need for trackside train completeness checking. The main difference from a fixed block system with lineside signals is the independence from the approach distance of the lineside signal system, is the distance between the signal at the entrance of the block section and the signal in rear that provides the approach indication. The approach time is no longer the running time between these two signals but the running time within the real braking distance based on the supervision curves of the cab signal system (chapter 8.1.3). The other elements of the blocking time do not differ from a system with lineside signals (figure 3.15). Most railways use block marker boards to mark the block limits at places without lineside signals. In some older cab signal systems that do not yet use a speed supervision curve, the approach distance does not differ from the lineside signal system. However, since signal information is continuously displayed at the driver's desk, there is no need to watch the aspect of a lineside signal that provides the approach information. For this reason, the signal watching time may be removed from the blocking time diagram while the approach time will not change.

$s_{B,CS}$ Braking distance of the cab signal system
l_{BS} Block length

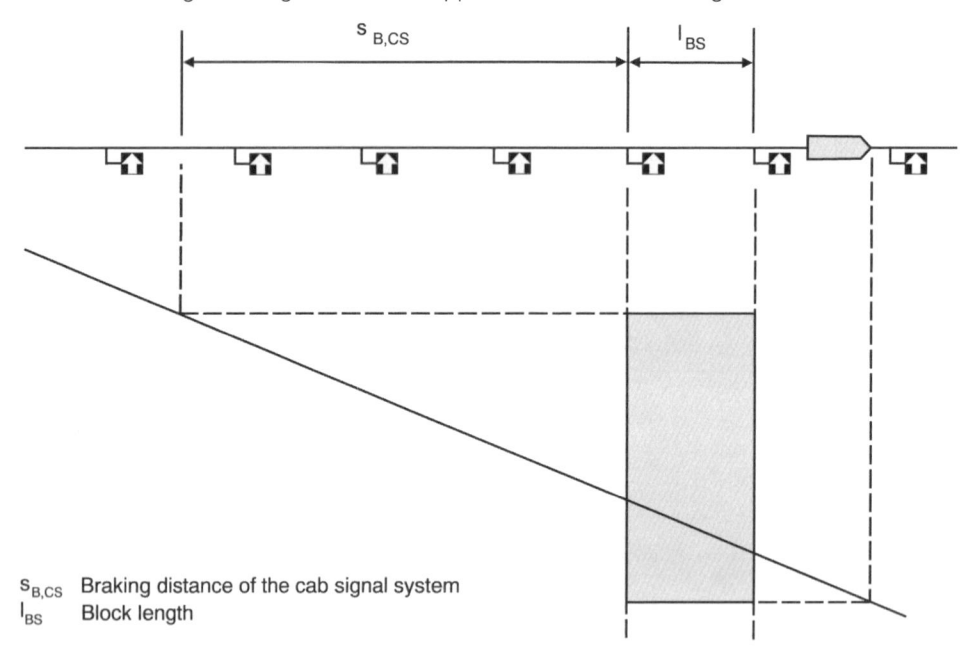

Figure 3.15: Blocking time of a block section with cab signalling

3.4.2.2 Cab Signalling with Moving Block

Although moving block does not yet exist in standard railway operations, it is a frequent point of discussion. The improvement of line capacity possible by the introduction of moving block is still often over-estimated. On a moving block line, the length of the block sections is reduced to zero. That means that the running time between the block signals will be eliminated. But all other components of the blocking time can also be found in moving block. On most lines, the sum of these other components is much greater even than the part of the blocking time that can be eliminated by moving block. That is why, compared with fixed block operation with short block sections, moving block will just lead to a moderate improvement of capacity. The

3.4 Principles of Train Separation

difference from a line with fixed block sections is that only the 'steps' of the blocking time stairway will be eliminated. The blocking time diagram will be transformed into a continuous time channel (figure 3.16). On lines with mixed traffic of trains running at different speeds, the possible improvement is almost negligible.

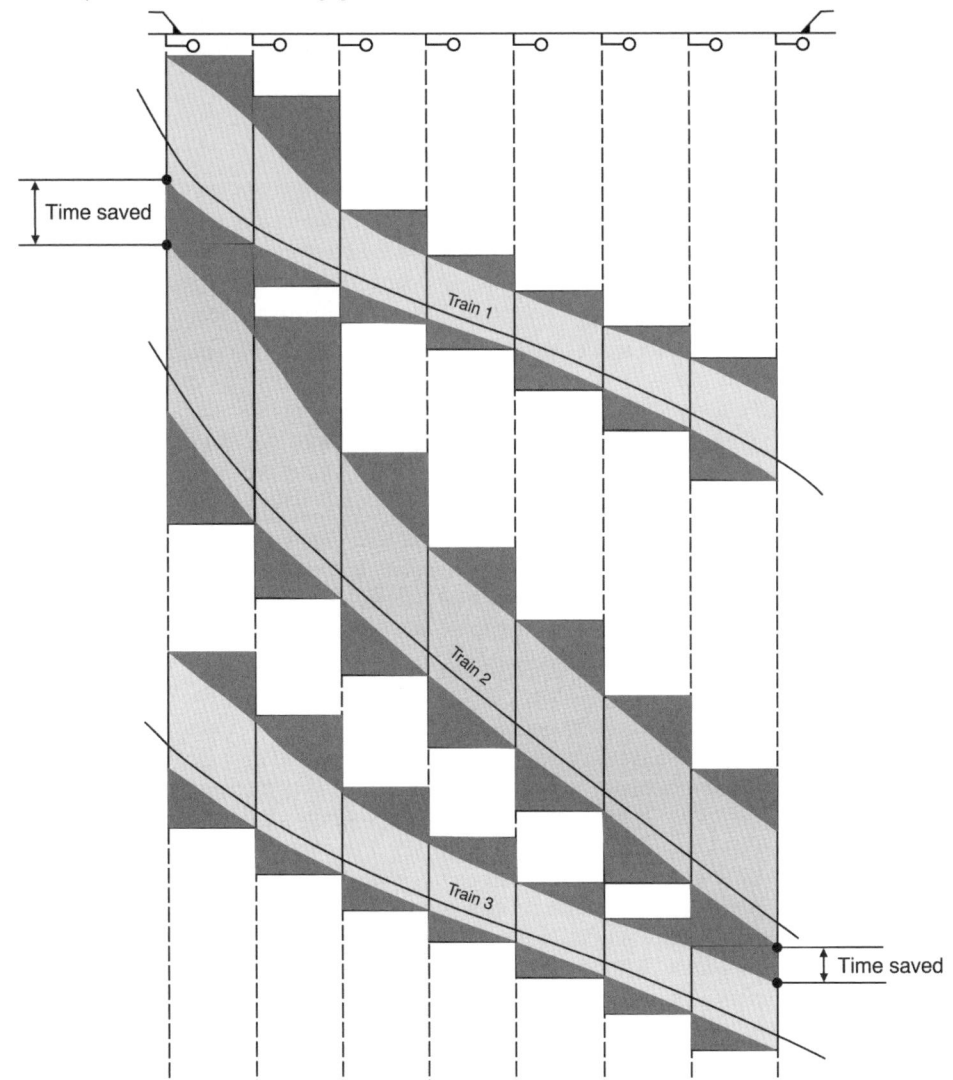

Figure 3.16: Blocking time of moving block compared to fixed block

Moving block as explained above is also called train separation in absolute braking distance. However, there is another moving block principle known as train separation in relative braking distance. Relative braking distance means that the distance between two following trains equals the difference of the braking distances of the trains plus an additional safety distance. Therefore, the braking distances of both trains must be calculated with braking curves as a function of speed. Train separation in relative braking distance leads to a maximum of line capacity.

But there is an essential problem. When running through an interlocking, it is not possible to move points between two trains. When points are to be moved, the second train has to have a full braking distance to the points until the points are locked in the new position. Another problem is that in case of an accident to the first train, the second train has no chance of stopping and is going to collide with the first train. Because of these problems, train separation in relative braking distance terms is a rather academic idea with no realistic chance of being adopted.

3.4.3 Non Signal-controlled Operation

Some railway lines with a very low density of traffic are operated without any signalling system. On many railways, such lines are called 'dark territory'. The movement authorities are issued to the train drivers by the dispatcher in verbal form using radio or telephone. The authority takes effect only after it has been repeated back to and then been verified by the dispatcher. Train crews give information to the dispatcher about the actual location of the train at certain intervals.

In its most simple form, the dispatcher relies completely on a handwritten trainsheet of tabular or graphical design. In the more advanced systems, the dispatcher has a computer-based workstation. Before issuing a movement authority to the train crew, it has to be entered into the control system. The control system would refuse any case of overlapping movement authorities. So, the dispatcher is protected from issuing conflicting authorities ('lap orders').

Some lines operated by that principle may have a simplified signalling system as a safety overlay (chapter 10.2). However, trains are not directly governed by these signals but by the verbal movement authorities. Another solution to improve safety in dark territory is non-signalled token working. To enter a section, the driver must be in possession of a token (either a physical object, or an electronic code). Since only one token exists for each section, two trains cannot legally enter a section at the same time (chapter 4.4.4).

Concerning the logic of how track sections are assigned to trains, there are two basic types known as Track Warrant Control (TWC) and Direct Traffic Control (DTC). Originally, these terms were introduced by North American railways but are today also used by railways on other continents. In TWC, trains may occupy main tracks only on the basis of the possession of a 'track warrant' covering a precisely defined track segment of any length. The limit of movement authority may be any point designated in the track warrant. But often, the limit of the movement authority assigned by the track warrant is the next meeting point. In DTC territory, fixed block sections (marked by lineside block limit boards and often extending from one passing loop to the next) are established. Train crews receive exclusive authority to occupy one or more blocks. The limit of movement authority is always a block limit board at the entrance of a block section the train has no authority to occupy. After leaving the block, the crew releases it to the dispatcher.

3.5 Dispatching Principles

3.5.1 Decentralised Operation

In a decentralised operation, train movements are controlled by local interlocking stations. The operators of neighbouring interlocking stations communicate to each other by means of telecommunication, mostly by simple telephone connections (figure 3.17).

All communications between the local interlocking stations and all train movements are registered by the operators in train records. On North American railways, all lines are controlled by

3.5 Dispatching Principles

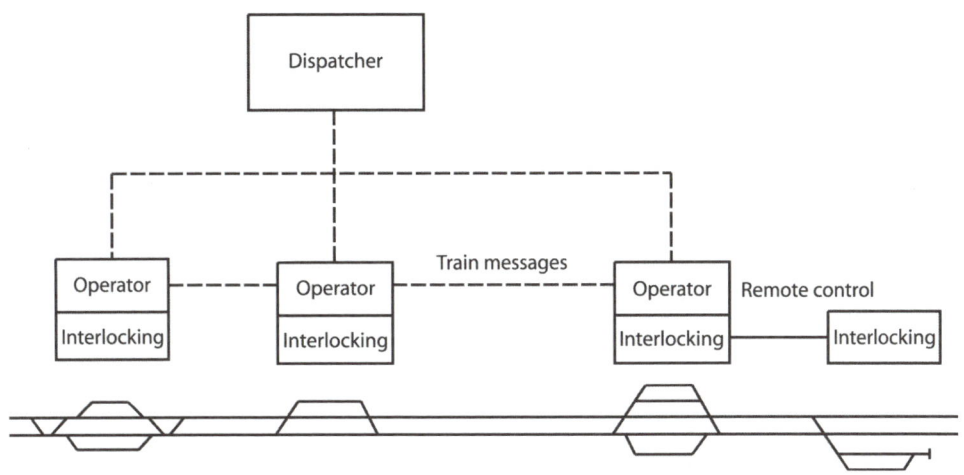

Figure 3.17: Decentralised operation

a dispatcher who works in a centralised office. On European railways, this applies only on lines with a heavier traffic, while lines with low traffic are operated without a dispatcher.

Because of the different operating procedures of North American and European railways, the role of the dispatcher is also very different. Compared with European railways, in North American practice the dispatcher has a much higher authority. The dispatcher is the person who is-

We put safety on the Track

Our comprehensive services cover your needs from planning through to maintenance. We support you through the whole product life cycle.

Our accreditations comprise:
- Inspection Body (EN ISO/IEC 17020), scope: Signalling, Train Control and Management Systems, Rail Automation, RAMS, Automation, Rolling Stock, Fire Protection.
- Rail Automation Test Laboratory (EN ISO/IEC 17025)
- Rolling Stock Test Center (EN ISO/IEC 17025)

Our promise:
The expert team of TÜV SÜD Rail guarantees a wide spectrum of services in the areas of certification, testing, assessment and engineering support.

TÜV SÜD Rail GmbH • Ridlerstr. 7 • 80339 Munich • Germany • Phone: +49 89 5190-1473   TÜV®

Choose certainty.
Add value.

www.tuev-sued.de/rail

3 Railway Operation Processes

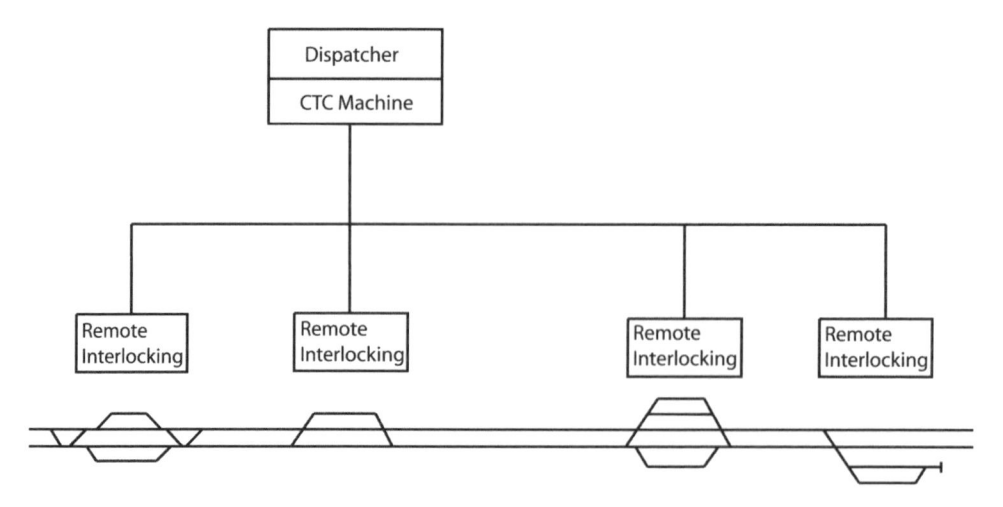

Figure 3.18: Centralised Traffic Control

sues movement authorities to trains. The local operators are in some way only the 'lengthened arms' of the dispatcher to set up routes, clear signals, and transmit orders in compliance with the dispatcher's instructions. On European railways, the movement authorities are issued by the local operators. Because the local operator is the 'authority person', the dispatcher is only responsible for watching the traffic and for solving scheduling conflicts to avoid delays and congestion. Thus, the dispatcher supports the local operators in an efficient operation.

Since the dispatcher-controlled railway is a North American invention, the American term 'dispatcher' was directly adopted by many railways in non-English speaking countries. It is even used in the Russian language. However, this term is not used in the UK. In British terms, such a position is called a 'controller' or an 'operator', while the operator who authorises train movements by clearing signals is called a 'signaller'. For more information on British rail traffic control see (IRSE 1991).

3.5.2 Centralised Traffic Control

In Centralised Traffic Control (CTC), all points and signals inside the controlled area are directly controlled by the dispatcher (figure 3.18). On British railways, that position is called a 'signaller' like in decentralised operation. All train movements are governed by signal indications. The local interlockings are remote-controlled without local staff. In CTC territory, all main tracks must be equipped with track clear detection. CTC technology has a long tradition on railways that operate long lines in territories with a very low population density and long distances between stations. Typical examples are lines in North America and in Russia.

With the introduction of CTC, some of the essential differences between railways which follow North American and European operating procedures partly disappeared. In some ways, CTC brought the two worlds together. To the Europeans, CTC brought centralised control, to the Americans it brought signal-controlled operation.

4 Interlocking Principles

Gregor Theeg, Ulrich Maschek, Oleg Nasedkin, with support from David Stratton, Jörn Pachl, Giorgio Mongardi, Heinz Tillmanns, Jochen Trinckauf, Thomas White

4.1 Overview

4.1.1 Introduction

Interlocking fulfils the function of 'information processing', as shown in figure 1.1 (the control loop). Interlocking is the central function to ensure that trains move safely in technical terms. To achieve this, the interlocking obtains information about track occupation (by rail vehicles and other objects; chapter 5) and the position of movable track elements (chapter 6). It then evaluates this information and permits movements via the signals (chapter 7). Amongst others, two basic principles are enforced technically by interlocking functions:

- A signal can only permit a train movement if all movable track elements are in proper position and locked **(dependence between points and signals)**, and the elements must remain locked as long as they are being used by the train.
- With train spacing by **fixed block** (chapter 3.4.1), a train can only be permitted to enter a section which is clear of other rolling stock; and no other train may be permitted to enter that section.

Amongst the different interlocking systems (mechanical, relay, electronic), the same logical principles of interlocking are applied to a large extent, though there are large variations of detail between the countries and the technical systems.

This chapter 4 deals with the logical principles of interlocking, whereas chapter 9 describes the technical solutions for route interlocking and chapter 10 those for line block systems. The primary focus is on systems where safety is provided technically, although occasionally non-technical systems also have to be considered. Within this, the focus is further systems using the 'fixed block' as a principle of train separation (chapter 3.4.1); those with moving block or time intervals are not considered.

The interlocking principles are connected with the operating philosophies and, based upon their historical development, can be devided roughly into those influenced by British, German and North American principles respectively (chapter 3.1). For discussion on the geographical distribution and mixed forms, see chapter 3.1. Concerning interlocking, there are larger similarities between the British and North American principles, than there are with the German. The following explanations will occasionally refer to these three basic groups.

4.1.2 Basic Requirements

In contrast to other modes of transport, railways have two decisive characteristics that determine their safety requirements (figure 4.1):

- The friction is low and therefore the braking distances are long. This decreases the ability of the driver to prevent collisions with other rail vehicles, moving units of crossing ways (e.g. at level crossings with roads) or any external obstacles.
- Rail vehicles are guided by the rails, and maintenance of this guidance is essential for safety. Thus derailment has to be prevented, and this includes the non-continuous guideway locations at (for example) points. Also the driver has no means of evading obstacles.

4 Interlocking Principles

From this, it may be concluded that the following protective functions for railways have to be incorporated (figure 4.1):
- Protection against following movements (described in this chapter 4)
- Protection against opposing movements (described in this chapter 4)
- Protection against flank movements (described in this chapter 4)
- Safety at movable track elements (described in chapter 4 and 6)
- Definition of permitted speed (described in chapters 7, 8)
- Speed control and supervision (described mainly in chapter 8)
- Protection at level crossings (described in chapter 13)
- Protection against external objects (described mainly in chapter 5)

Characteristics of the Railway Systems:							
low friction resulting in long braking distance				guidance			
Dangers:							
collision with ...				derailment at ...			
... moving units of crossing ways	... external environment	... rail vehicles		... non-continuous guideway locations		... continuous guideway	
Protective Function:							
protection at level crossings	protection against obstacles	following protection	opposing protection	flank protection	safety at movable track elements	speed targeting	speed control & supervision
Application Examples:							
clearance profile detection systems		point locking and supervision					
level crossings		block systems					
route interlocking							
							train protection

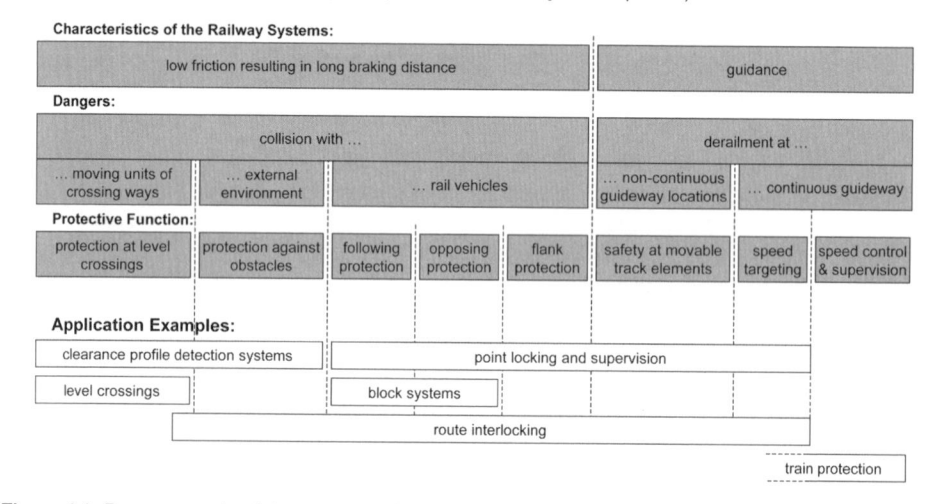

Figure 4.1: Dangers and safety measures in railway operation (issues relevant to signalling only)

4.1.3 Basic Principles of Safeguarding a Train's Path

There are two basic principles regarding the methods of safeguarding the way, which can be distinguished and defined as follows:
- **Route.** The whole path including the positions of movable track elements and track clear detection is only checked upon request, normally when setting the route before clearing the signal (and it is then supervised until the train enters the route). The principle 'route' can provide almost all the protective functions of figure 4.1. Typically, these are following train, opposing train and flank protection and safety at movable track elements, but it also contributes to speed targeting at movable track elements. It can even incorporate level crossings and obstacle detection.
- **Block Information.** After a train has cleared the block section, the message confirming this is generated and transmitted to the entry point of the section. There, this information is stored, to permit the later entry of another train. The principle 'block information' can only provide following and opposite protection. Therefore it is basically applicable to open line sections.

Historically, this difference emerged from track clear detection being by the observation of the signaller: Within the control area of a signal box (an interlocking), the signallers who were responsible for setting the routes proved the track clear directly by sight before clearing the signal. In contrast, between two neighbouring signal boxes with a longer portion of line in between which was not visible from either of them, this principle could not be applied and other solutions had to be found. Here the clear status of the line section was concluded from information about trains entering and completely leaving the section.

However, the difference between the principles 'route' and 'block information' lost much of its operational importance with the introduction of continuous technical track clear detection. With this, all block system functions can also be provided by routes. Thus in recent years the principle of 'block information' has been superseded in some countries, with the principle of 'route' being used also on open lines.

In most countries where train and shunting movements are determined separately (chapter 3.3), shunting movements are restricted to particular areas (e.g. station areas). The principle 'block information' is basically applied to train movements. The principle 'route', in contrast, can be applied to both train and shunting movements.

4.2 Element Dependences

4.2.1 Classification

Dependences between individual elements are the simplest form of interlocking. Elements which can be interlocked are:
- Movable track elements such as points, derailers and others (chapter 6.1)
- Signals
- Other elements such as level crossings

Element dependences can be distinguished by different criteria, one of which is the **number of elements** to be locked:
- Interlocking between two elements
- Interlocking among three or more elements, also called 'conditional locking'

According to the logical arrangements, element dependences can be devided into the following elementary locking functions:
- **Coupling** of two or more elements (chapter 4.2.2)
- **Unidirectional locking** of two or more elements (chapter 4.2.3)
- **Bidirectional locking**:
  - between two elements ('simple locking'; chapter 4.2.4)
  - with three (or more) elements ('conditional locking'; chapter 4.2.5)

The principles are explained in the following section. By AND- and OR-combinations of several of these elementary locking functions, complex route interlocking systems can be built up. This is used particularly in British and North American influenced interlocking principles.

4.2.2 Coupled Elements

Coupled elements are operated by the same operational element and can only be switched together in regular operation. The most typical case is that of two movable track elements giving flank protection to each other, such as the two sets of points of a crossover (figure 4.2) or a set of points and a derailer protecting that set of points.

This kind of locking is typical in British and North American influenced interlocking logic, but not for the German. Often both coupled sets of points have the same identification number, possibly distinguished by the suffixes A and B. A disadvantage is that if the detection of one element fails, both cannot be switched by normal operation command. Another disadvantage in mechanical interlocking is that the signaller must use twice as much effort to switch them.

4 Interlocking Principles

Figure 4.2: Switching of two coupled elements

4.2.3 Unidirectional Locking

In unidirectional locking with two interlocked elements, one independent and one dependent element exist. The independent element can be moved freely (unless locked by other functions). The dependent element can only be set to a certain position if the independent element is also in a certain position, and leaves this immediately when the independent element leaves its position. With more than two interlocked elements, the position of the dependent element depends on combinations of positions of the independent elements. In the following, the case of two interlocked elements is described further, as it is the most frequent case.

The most typical case is the dependence between a main signal and its associated distant signal (figure 4.3). As the ability to bring the main signal to Stop at any time must be preserved for safety reasons, the main signal can be switched without considering the position of the distant signal. The aspect of the distant signal, however, depends on the aspect of the main signal. Only if the main signal shows a proceed aspect, the distant signal can show Clear. However, it can also be kept at Caution by other locking functions. Reasons might be points in the wrong position between the main and the distant signal, or another main signal located at the same place as the distant signal being in Stop position. But if the main signal is returned to Stop, the distant signal follows immediately.

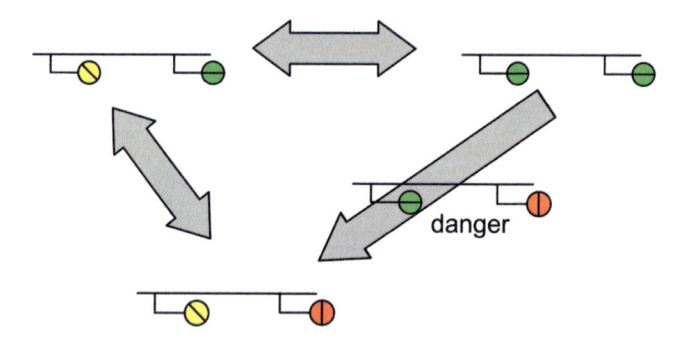

Figure 4.3: Unidirectional dependence between main and distant signal

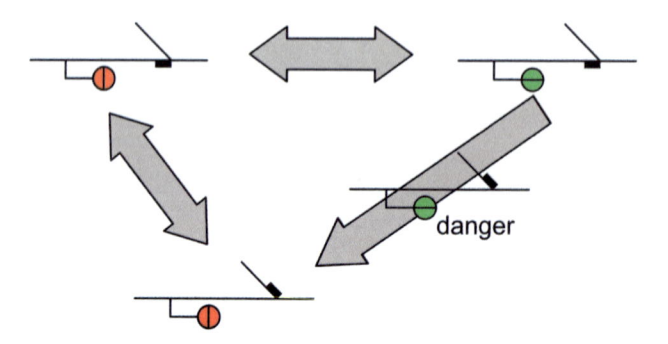

Figure 4.4: Unidirectional locking between points and signal

4.2 Element Dependences

In North American interlocking, the principle of unidirectional locking is also applied to the dependences between points and signals in some situations with manually operated points (figure 4.4, chapter 3.3.2.2). The signal can only show a proceed aspect if all the points in the path are in their correct positions. But once the signal shows a proceed aspect, the points are not locked by the signal. If the points leave their end position (meaning: the manual key lock is unlocked), the signal is immediately returned to Stop. To ensure a certain level of safety if a train is already approaching, this arrangement is related to the obligation to wait a defined time between unlocking the points by the key and actually switching them.

4.2.4 Simple Bidirectional Locking

In bidirectional locking, two or more elements are interlocked that way so that one combination of positions is impossible and each element is locked if the others are in these respective position(s).

A typical example here is the most frequently used form of dependence between points and signals (figure 4.5): The signal is locked in the Stop position if the points are diverging in the example of figure 4.5, and the points are locked in the straight position if the signal shows a proceed aspect. This means that a certain combination of positions (signal at Proceed and points diverging) is impossible. The elements have to be switched in a defined sequence. First, the points have to be straight, then the signal can be cleared. Thus this kind of locking is also named 'sequential locking'.

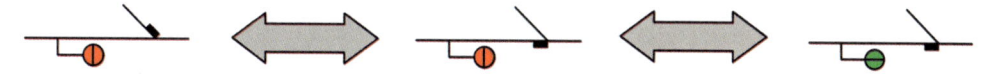

Figure 4.5: Sequential dependence between a set of points and a signal

The same principle is also applied to points and derailers (figure 4.6) or to two sets of points offering flank protection to each other, particularly in German influenced interlocking logic. The purpose is the same as for coupled switching (chapter 4.2.2) to ensure flank protection, and systems differ in which of the two elements has to be switched first and which second.

In mechanical systems, sequential locking can be solved by manual key locks, with a key unlocking either the one or the other element from the locked position, or by tappets (chapter 9.2.4.4).

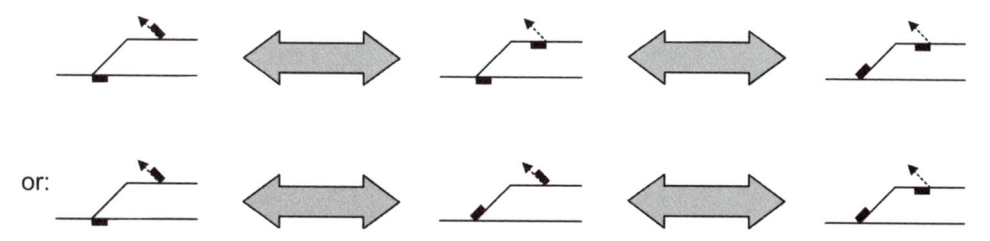

Figure 4.6: Sequential dependence between a set of points and a derailer

4.2.5 Conditional Bidirectional Locking

Conditional bidirectional locking is similar to the simple, but for three or (rarely) more interlocked elements.

4 Interlocking Principles

In figure 4.7, signal 1 can only be cleared if there is a safe path ahead. One condition (not considered further here) is that points 2 must be locked in either end position, which is in fact an OR-combination of two simple locks, one for each end position. However, if points 2 are straight, there is another safety condition: points 3 must also be locked straight. The commonly used term in Western European countries 'conditional locking' (Retiveau 1987, Such 1956) mean in the example of figure 4.7: 'If points 3 are in the diverging position, then signal 1 at Proceed locks points 2 in the diverging position (and points 2 straight lock signal 1 at Stop)'.

This is equivalent to 'Signal 1 at Proceed, points 2 straight and points 3 diverging cannot occur at the same time'.

Combinations of several simple and conditional locks form the basis for cascade route locking (chapter 4.3.9) in British and North American influenced interlocking logic.

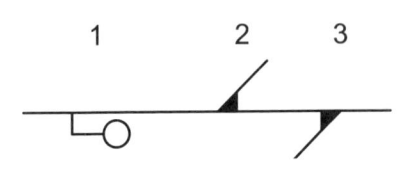

Figure 4.7: Conditional bidirectional locking

4.3 Routes

4.3.1 Introduction

4.3.1.1 Safety Functions Provided by Routes

Routes are a form of safeguarding a complete path of a train over the complete length over which the movement is permitted. An important basis of a route is dependences between elements, e.g. between points and signals (chapter 4.2), but also checking that the tracks are clear before permitting the movement is an important precondition.

Routes can fulfil or contribute to the following protective functions (figure 4.1):
- Prevention of derailment on non-continuous guideway locations: Movable track elements have to be in their proper positions
- Prevention of derailment on the continuous guideway by choice of the proper speed according to the line geometry and radii at divergences
- Protection of following, opposing and flank movements
- Optional protection at level crossings if they are included in the route (chapter 13.4.4.1)
- Optional protection against obstacles, if these detection systems are included in the route dependences (chapter 5.1.2)

These functions can also include a certain protection against human errors, such as overruns of Stop signals (chapter 4.3.6).

4.3.1.2 Definitions of Terms

When all the movable track elements are in proper positions so that a train can take a certain direction, this way is called **'path'**.

Before a train movement is allowed to traverse the path, certain preconditions have to be proved. The most important of these preconditions are:
- The movable track elements of the path have to be in the correct positions and locked against being switched while the train is passing over them.
- The tracks have to be clear and conflicting movements at the same time prevented. The objects to be detected can differ: Often the detection is only for rail vehicles, not for external obstacles (which are actually not allowed to be in the tracks), leading to a certain loss of safety.
- The route needs a distinct target, e.g. signal at Stop (chapter 4.5.2).

4.3 Routes

The path can be safeguarded either technically or non-technically under the responsibility of staff. A technically safeguarded path with a defined entrance and exit point is called a **'route'**.

In countries where train and shunting movements are distinguished as different classes of movements (chapter 3.5), routes can be distinguished into train routes and shunting routes. This refers mainly to countries which follow German or British operational principles. Often (particularly in older systems following German principles) separate shunting routes are not provided, but shunting is done completely under the responsibility of staff ('free shunting'). Where separate shunting routes are provided, their interlocking functions are often less complex than these for train routes. For these reasons, in the following the interlocking principles will be described primarily regarding train routes. Shunting routes are considered in chapter 4.3.10.

4.3.1.3 Safeguarding of the Path by Human Operators

If paths are safeguarded fully or partly under the responsibility of staff, the responsible persons can be one of the following, or both with shared responsibility:

– The ground staff who have to prove all conditions before authorising a movement, must not switch elements under or in front of a moving train. They must also ensure protection against other movements.
– The train driver whose movement authority is only valid with the restriction that he has to check the tracks. The driver has to go slowly on sight, looking for other rail vehicles and external obstacles, and be able to stop at any obstacle. In some cases, the driver even has to check the position and locked status of each movable track element, which is displayed to him by special point position indicators.

Until now, safeguarding the path with staff taking responsibility for safety is often applied in situations with low speed or reduced safety requirements. Examples are:

– For shunting. This applies also in countries where no distinction between train and shunting movements is made (North American operational principles), for movements fulfilling the same purposes. According to different definitions of the terms 'train movement' (also called 'running movement') and 'shunting movement', this can also include all movements into occupied tracks, even if they are defined as train movements.
– In many countries on secondary lines and urban railways with reduced requirements due to low speed and/or low traffic volume (non signal-controlled operation).
– To maintain traffic in the case of equipment failure (degraded mode operation, chapter 4.5.4).

4.3.2 Extension of Routes and Related Speed Restrictions

4.3.2.1 Topological Extension

Each route has a defined start and a defined target position, usually called route entrance/exit position. The entrance is at the signal permitting the movement, or a distinct point in proximity to that signal. In case the route is signalled without trackside signals, e.g. by cab signalling, it is a comparable point as 'virtual signal'. The route exit can be one of the following:

– A main signal or another signal which can indicate a stop aspect
– An end of a track (usually equipped with a fixed signal)
– The end of the interlocking area. This can be the exit to the open line (whose exact location is defined differently among the railways) or the entry into an unsupervised shunting area

4 Interlocking Principles

4.3.2.2 Topological Parts of a Route

Parts of the route which have to be safeguarded are (figure 4.8):
- The **running path**, which means the tracks which will regularly be occupied by the train during its traverse of the route. It includes sections both with and without movable track elements.
- An **overlap** which can extend over a short section to protect against a minor overrun of the train caused by misjudgement of braking (chapter 4.3.6), or which can even accommodate the whole braking distance from maximum speed to Stop.
- The **flank areas** which protect the train against unauthorised movements from the flank (chapter 4.3.5).
- The **end** of the protected section, to protect against opposing movements (chapter 4.3.6).
- The **start section** which is situated in rear of the route entrance signal, but either occupied by the starting train or situated between the head of the starting train and the route entrance signal (chapter 4.3.7).

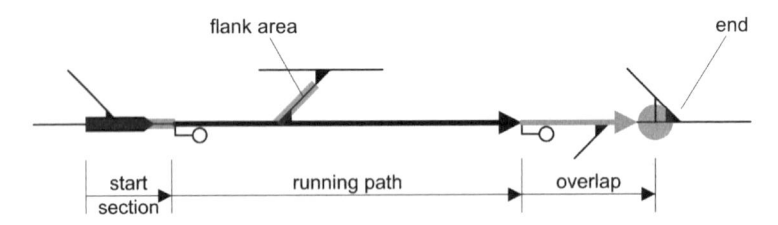

Figure 4.8: Logical parts of a route

The practical consideration of these parts for interlocking differs between the railways. While locking the running path is essential, there are large differences regarding the other parts:
- Some railways do not provide overlaps, arguing that good train protection systems prevent overruns safely.
- Other railways do not provide particular flank protection, arguing that all movements traverse routes, and therefore assume that flank movements cannot occur.
- Front protection is considered to different extents.
- Locking of the start section is of importance only if there are movable track elements within this section.

4.3.2.3 Local Validity of Speed Restrictions

When using route signalling with trackside signals (chapter 7.3.4.2) or modern cab signalling systems (chapters 8.3.5, 8.3.6), the speed limit can be individually adapted to each track element. This is applied only in some of those systems. However, when using speed signalling with trackside signals (chapter 7.3.4.2), which is the most frequent case, usually one speed limit is assigned to the whole route. Particularly in stations with two or more consecutive main signals, the regulations about the local validity of the speed limits differ much between the railways. There are two basic solutions, which for simplicity of explanation are described for stations with entry and exit routes only and without intermediate signals (figure 4.9):

1. The speed indication of a station home signal is valid through the whole station until the rear end of the train has cleared the last movable track element of the station (figure 4.9.1). Therefore, at the station exit signal often no speed indication is given, but the indication of the home signal in rear is still valid although only green is displayed on that signal. The advantage of this solution, which is applied in France, Belgium, Russia and other countries,

4.3 Routes

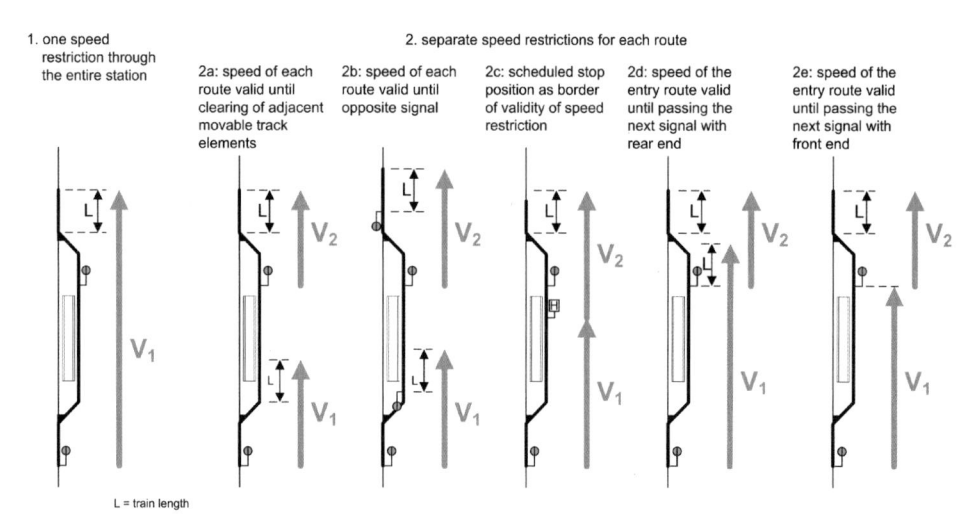

Figure 4.9: Local validity of speed restrictions in stations

1. one speed restriction through the entire station
2. separate speed restrictions for each route
   - 2a: speed of each route valid until clearing of adjacent movable track elements
   - 2b: speed of each route valid until opposite signal
   - 2c: scheduled stop position as border of validity of speed restriction
   - 2d: speed of the entry route valid until passing the next signal with rear end
   - 2e: speed of the entry route valid until passing the next signal with front end

L = train length

is less effort for speed signalling and simpler operational rules, particularly in special cases like points situated in the start section of an exit route (chapter 4.3.7). However, the task of informing a driver who starts a train in the station track or has a longer stop there has to be considered. Other difficulties occur if different speeds could be achieved for the entry and exit routes (figure 4.10). For these cases, either the lower of both speeds is valid through the whole station, or exceptionally solution 2 has to be used.

2. Different speed restrictions are indicated for the entry and the exit route (figure 4.9.2 a-e). This solution is applied by the majority of railways. The advantages are more flexibility in track layout planning (no need to plan symmetrical stations), fewer difficulties in the case of non-symmetrical train movements (figure 4.10) and greater safety (not forgetting the previous speed indication). However, this is the more complicated case, as it is necessary to define the exact 'border' between these two speed limits. Some solutions are described in the following:
   - In the case 2a, each speed is valid from the route entrance signal until the rear end of the train has passed all movable track elements belonging to the route. In the section between, different definitions can be applied (e.g. the higher of both speeds). An example is in the Czech Republic.
   - In the case 2b, each station throat is defined as a separate interlocking area, limited by the opposite home signals. In the section between (defined as open line section), a separate line speed is fixed. This is the solution in North America.
   - In the case 2c, the border between the two speed restrictions is the regular stop position of a train. This is only applicable for trains that have a regular stop. An example

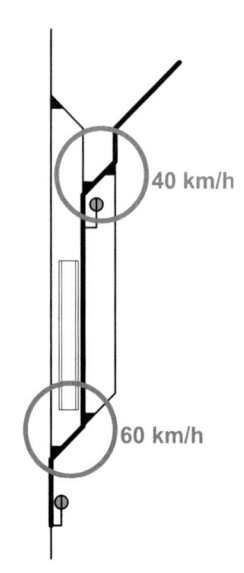

Figure 4.10: Example for non-symmetrical entry and exit routes

69

4 Interlocking Principles

is Germany. For trains without stop, the rule of case 2d applies. This solution is easy to understand for the train driver, but can cause additional difficulties as the scheduled stop position receives safety importance.
- In the case 2d, the speed of the home signal is valid until the rear end of the train has passed the next signal. However, this causes problems for the drivers, who have to memorise the speed displayed at the signal in rear after a scheduled stop, or if the train starts in the station track. This is the reason why in Germany, for trains with stop, the rule of case 2c applies.
- In the case 2e, the speed of the station entry or intermediate signal is valid until the front end has passed the next signal, with the same disadvantage as 2d.

In chapter 4.3.7, the solutions 2a – 2e of figure 4.9 are further analysed in a context where the differences are particularly important.

4.3.3 Basic Route Locking Functions

4.3.3.1 Overview

To fulfil the safety requirements described in chapter 4.3.1 and other operational requirements, the following functions of routes are to be considered:
- Locking of movable track elements in the route (chapter 4.3.3.2). Besides points, movable track elements include all other elements which disrupt the running rail or protrude into the clearance profile, such as crossings, derailers, movable bridges, cranes etc. (chapter 6.1.1)
- Interlocking of conflicting routes (chapter 4.3.3.3).
- Track clear detection before permitting a movement. This refers to occupation by rail vehicles (chapter 4.3.3.4), in some cases also external objects.
- The route has to be proved to have a distinct target to ensure the local limitation of the movement authority. Therefore, most railways prove the route exit signal to be illuminated within the route locking functions (chapter 4.5.2).
- In some cases, level crossings with roads are locked in the route in a similar way to movable track elements (chapter 13.4.4.1).
- In some cases other functions are included in route locking. Examples are preventing an electric locomotive from entering a non-electrified track, or preventing a too long train from entering a station track (chapter 4.3.3.5).

Over time, interlocking systems were continuously improved, and several new features were developed independently in different countries. Later developments then had to be adapted to the existing systems of each particular country. Besides, different priorities were set in the conflict between safety and availability of the systems. This resulted in many variations, though the basic principles are much the same.

4.3.3.2 Locking of Movable Track Elements

For protection at movable track elements, the following two functions of interlocking between signals and moving trains on the one hand and movable track elements on the other side have to be fulfilled:
- Whenever a signal is cleared, the elements belonging to the route have to be locked and proven in the correct position as long as the signal remains in a clear position (dependence between points and signals). The elements included in the locking can differ, especially in flank protection and overlap elements (chapters 4.3.5, 4.3.6).
- After a signal has been cleared and a train enters the section, each element has to be kept

4.3 Routes

locked until the train has cleared this section, even if the signal has returned to Stop position in the meantime (chapter 4.3.8).

4.3.3.3 Interlocking of Conflicting Routes

Conflicting routes are different routes which use a common piece of infrastructure. The following two methods of interlocking two conflicting routes against each other can be used (figure 4.11):

- **Simple route interlocking (by track elements):** Routes which require at least one movable track element to be in a different position are interlocked against each other by the position of this element and do not need to be interlocked separately. Such simple route interlockings are accepted by most railways today, otherwise special interlocking has to be provided in all cases.
- **Special route interlocking:** Conflicting routes which do not differ in the position of any movable track element have to be interlocked against each other by special means. This method requires additional effort in the interlocking logic.

The most important cases of particular interlocking are (figure 4.11):

- The **opposing** case: Unless moving on sight, opposing routes into the same track must always be excluded. For shunting movements and train movements on sight, the requirements differ among the countries and with the length of the track section. In some cases these movements are excluded, whereas others aim to achieve faster operation when assembling trains. One solution to fulfil this interlocking function is the individual locking of routes, another is the use of separate locking of up and down direction in each movable track element.
- The **consecutive** case: In some technical and operational situations, consecutive routes need to be interlocked against each other to prevent through movements using a certain track (chapter 4.5.2). The requirements also differ for a train route and a consecutive shunting route. In some situations and by some railways, exclusion is required to prevent a moving unit which entered as a train to continue without stop as a shunting movement. A

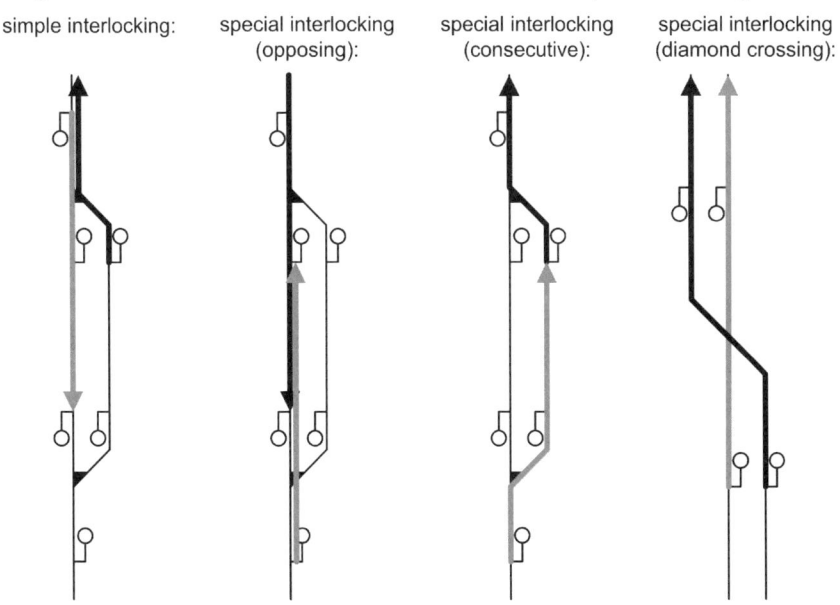

Figure 4.11: Simple and Special Route Interlocking

4 Interlocking Principles

typical example is a reception route for freight trains and, following from the same track, a route for pushing wagons over the hump. In other cases, and for different reasons, a certain sequence of setting the two routes is required.
- The **crossing/overlapping of profile** case: Even if the diamond crossing has no movable parts, movements on both tracks at the same time have to be prevented. The same refers to any situation where clearance profiles of neighbouring track overlap. This situation can be handled either by special interlocking between individual routes or by defining the crossing as a movable track element with two virtual end positions.

4.3.3.4 Track Clear Detection and Movement on Sight

To prove the tracks clear is a necessity before permitting a movement unless movement on sight is commanded, in which case the train driver fulfils this function.

In most modern interlocking systems, track occupancy is supervised continuously by track circuits (chapter 5.3) or axle counters (chapter 5.4). This is included in the route locking conditions, at least on main tracks. In many older (mechanical and electro-mechanical) interlocking systems, some of which are still in operation, no technical detection was used. Instead, checking track occupancy before permitting a train movement was the responsibility of the signaller, which is an important safety shortcoming of these systems, but can be appropriate as a cheap solution for reduced requirements.

Where the driver is responsible for track clear detection, two possibilities of **movement on sight** can be distinguished (figure 4.12):
- movement to the full sighting distance, if no opposing movement is possible,
- movement to half the sighting distance, if opposing movements have to be considered.

The cases can be distinguished by different signal aspects or (in most cases) by local or general regulations.

In some cases, the train is proven by track occupation to be stationary or moving at a very low speed before on sight operation is permitted (IRSE 1999).

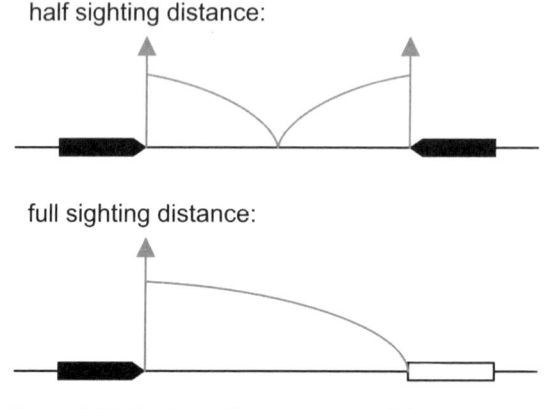

Figure 4.12: Versions of movement on sight

4.3.3.5 Other Locking Functions

Besides the above mentioned basic functions, several other locking functions are provided by some railways. Some of them are safety related, others serve only to prevent operational hindrances. They are discussed below.

Most railways use some form of **prevention of premature release** of routes due to errors of track clear detection, which would result in dangerous consequences. This is particularly important where track circuits are used, as the probability of false occupation is relatively high. Different strategies are applied (figure 4.13):
1. The route can only be released after the sequential occupation and clearing of at least two neighbouring sections of track, in an appropriate order.

2. A combination of a rail contact (which detects the occupation safely) with a short track circuit (which detects the clearing safely) is used (chapter 5.2.3.2).
3. Time delays are applied between the passage of the train and the release of the related route section. This only prevents dangerous consequences of detection error of short duration, but not for longer enduring errors.

At borders between electrified and non-electrified areas or between differently electrified areas, the **entry of an electric motive power unit into an inappropriate area** is not desirable. One solution is additional route indicators at the wayside signal which indicate the direction of the route to the driver. If the signaller has set a wrong route by error, the driver can recognise this error in time. Another measure is a special input command required from the signaller to confirm that he is aware of the non-electrification of the route target, and has checked that no unfitted train is approaching. Complete (non-safe) technical solutions can be found in connection with a train describer system (chapter 4.3.11) which knows the category of an approaching train.

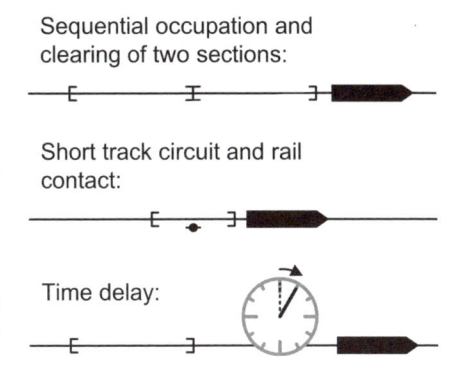

Figure 4.13: Methods to prevent premature release of a route

Preventing deadlock situations can be a reason for additional non-safety locking functions (figure 4.14). A typical case is in installations where the principle 'route' is applied to open lines to ensure that only trains travelling in the same direction can be between two neighbouring train sequence stations.

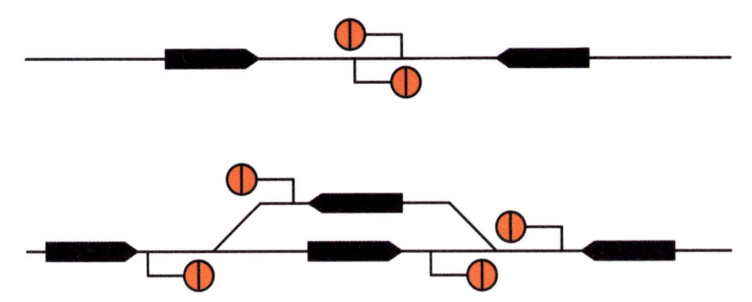

Figure 4.14: Examples of deadlock situations

Another useful feature is the prevention of **entry into a too short track in stations**, which would also obstruct traffic. Train describer systems or special signal aspects can help to prevent that. A third solution is length measurement of trains by track clear detection: Let us assume that in figure 4.15 section E is already occupied by another train, D is clear and D has the same length as B. Then an additional locking function can be provided which permits signal 1 to clear only if section B is occupied and section A clear. As the train nevertheless has to enter the station slowly (or even on sight) due to partial occupation of the target track, the loss of time by forcing the train to stop at signal 1 is only slight. In Britain, this kind of control is called 'Lime Street control' from the name of the station in Liverpool where it was first applied (IRSE 1999).

4 Interlocking Principles

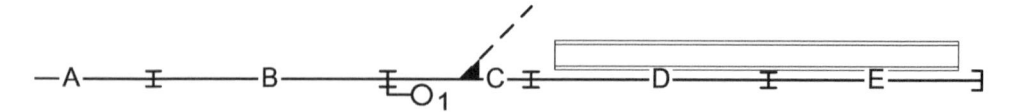

Figure 4.15: Length measurement by track clear detection

In some cases, authorisation by special persons such as customs officers is required for certain routes. This authorisation is given by a special operation element, e.g. by depressing a button which is only accessible to these persons. Examples can be the authorisation of border police forces to cross the border or the authorisation of maintenance staff to leave the workshop.

4.3.3.6 Duration of Locking

The route must be locked before permitting a train on a route and remain locked either until cleared by the train or until the trains has come to a stop. Although the route is usually set up and locked as a whole, the release (unlocking) of the route takes place either as a whole after the release conditions are fulfilled for the whole route or in sections when the release conditions are fulfilled for the respective route section. With sectional route release, elements which are no more required can be used for another route, increasing the capacity of the infrastructure. The release conditions are usually defined in the following ways (figure 4.16):

– For elements in the running path. They have to remain locked until cleared by the train. For the prevention of premature release, see chapter 4.3.3.5. A special situation in stations refers to elements situated in the target track of the route, which remain occupied by the train standing in the station. They can be released following the same principles as for the overlap.
– For elements in the overlap (if they are included). They have to remain locked and checked clear until the train is stationary in the target section of the route, or until it has travelled further onto a continuation route beyond the route exit signal.
– For flank protection. These have to be locked for as long as the corresponding running path element or overlap element, to which they offer flank protection.

Obtaining the confirmation that a train is stationary, which is required for overlap release, offers some difficulties: There are various solutions, the most important of which are:

– The stop is observed visually by a person (mainly applied in old systems).
– The stop is assumed after a defined time has elapsed since the train has completely reached the target section (mainly applied in modern systems).

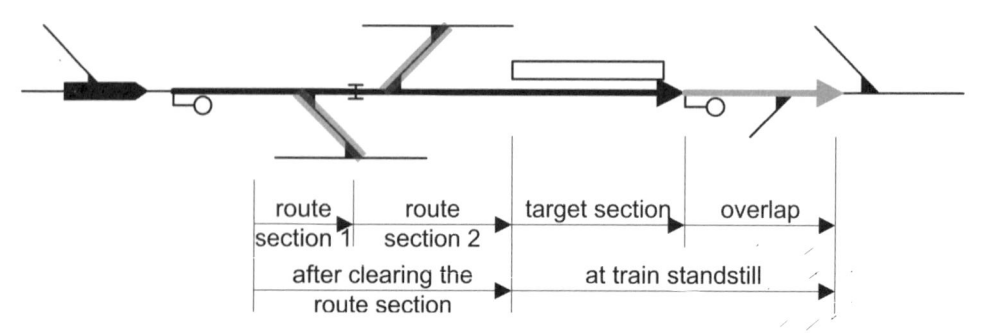

Figure 4.16: Duration of locking of elements

4.3 Routes

- The stop information is transmitted from the train by an advanced train control system. Only few modern systems (e.g. LZB, ETCS Level 2) can enable this (chapters 8.3.6, 8.4), but even here this possibility is not used in all applications.

4.3.4 Route Selection by the Signaller

In the operation of the interlocking, for setting up a route by the signaller, there are two principles:
- Individual point operation. Each movable track element is set individually and if all movable track elements are in the correct positions and locked, the route locking functions can be applied. This method is used in mechanical interlocking technology but also in some, especially older, relay interlocking systems. In other relay and electronic interlocking systems it is applied in degraded mode operation.
- Entrance-Exit (NX) operation. The signaller selects the route by operating only the entrance and exit elements. The movable track elements are then set automatically by automatic point setting (chapter 4.3.11).

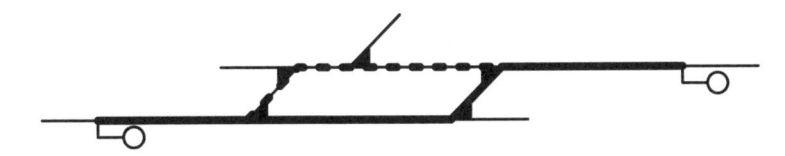

Figure 4.17: Alternative routes between the same entrance and exit

An interesting issue in entrance-exit operation is in case there is more than one route available between an entrance and an exit (figure 4.17). Usually, one of the routes is considered as the priority route, which will be set if possible and if no other selection has been made. To select an alternative route, one of the following criteria can be employed (figure 4.18):

- If the priority route is not available, the alternative route is automatically chosen. Where more than two alternative routes are available, an order of preference has to be defined.
- Selection of the alternative route by the signaller. Depending on the interlocking system, this can be achieved by pushing an intermediate button, thus selecting a sequence of partial routes, or by individually setting or locking one or more decisive sets of points.

The selection of alternative overlaps (chapter 4.3.6.5) for the same route can be done in a similar way, with the additional possibility of selection of direction by a continuing route.

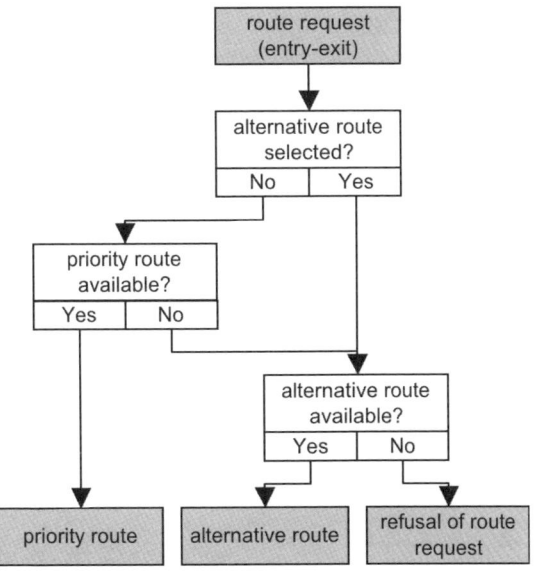

Figure 4.18: Route selection in case of alternative routes (with two possible routes only)

4 Interlocking Principles

4.3.5 Flank Protection

4.3.5.1 Methods of Flank Protection

Dangerous flank movements can originate from following types of activities:
- Trains and shunting movements on conflicting routes. This is excluded by interlocking logic (chapter 4.3.3.3)
- Shunting without route
- Unsecured parked vehicles which then roll away

Most railways provide flank protection at least for train routes. The area between the running path and the protective element is often proved clear. Flank protection can be given by the following protective measures (figure 4.19):

1. Movable track elements such as points, derailers and catch points (chapter 6.1.5) which direct offending movements away from the route or derail them. This can be solved by linking the points of a crossover (chapters 4.2.2–4.3.5) or by including the protective elements into the route locking functions.

	protection against flank violation by:			strength of protection	operational restraints	installation costs
	trains/ shunting on routes	shunting without route	rolling away of vehicles			
points: 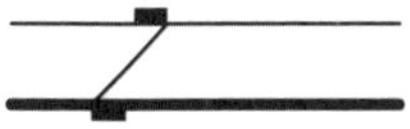	by normal route interlocking	yes	yes	strong	no	very high
derailer, catch points: 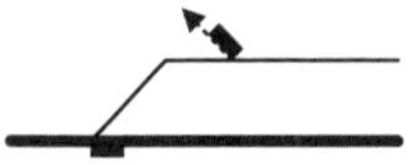		yes	yes	strong (but danger by derailing of vehicles)	no	high
signal: 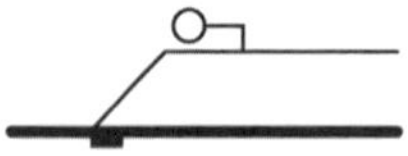		yes	no	medium	no	high
shunting forbidden 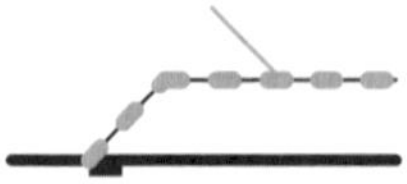		yes	yes	relatively weak	yes	low
parking forbidden 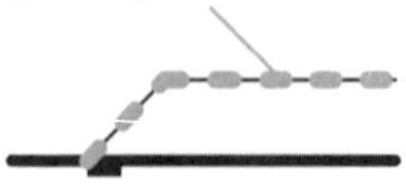		no	yes	relatively weak	yes	low

Figure 4.19: Methods of flank protection

2. Signals locked in the Stop position. To enable full protection, the red signal lamp has to be proven alight. With reduced safety, some railways also accept a dark (unlit) signal as flank protection (examples: Poland, Russia), as train driver is obligated to consider this extinct signal as Stop.
3. Regulations which prevent flank movements. Such regulations can forbid shunting or parking vehicles on certain tracks or even forbid free shunting (without shunting route) in general (example for the latter: Netherlands).

Of these methods, (1) is the strongest, whereas the others are weaker because they depend on people obeying the rules. Not all kinds of measures can protect against all kinds of endangering movements (figure 4.19). Therefore, often a combination of two methods is used, e.g. a signal and a prohibition on leaving vehicles on the track concerned.

The requirements for application and the chosen form of flank protection differ. An especially controversial issue is whether to provide flank protection for the overlap or not, as this is only relevant if two independent errors occur at the same time. These are the overrun of the train and an unauthorised movement of other vehicles into the overlap, which is very unlikely.

4.3.5.2 Track Clear Detection of Flank Areas

The railways differ in whether to prove the flank areas, which are the tracks between route points and corresponding protective elements (figure 4.8), technically clear or not. Under certain circumstances, the requirement of proving the flank area clear can reduce the usable length of a station track significantly (figure 4.20). On the other hand, renouncing this requirement decreases safety and/or requires additional operational rules to ensure safety against vehicles rolling away unintended. Such rules can forbid shunting or the parking of vehicles, or define special requirements concerning the parking brakes and the supervision of parked vehicles.

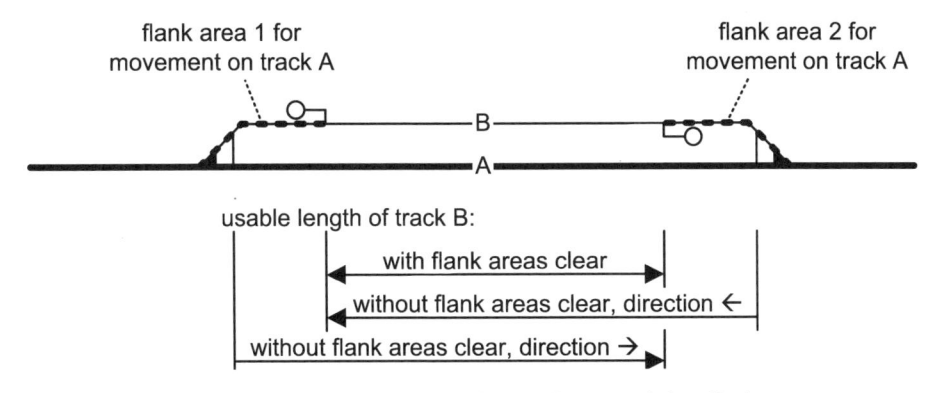

Figure 4.20: Usable track length with and without the requirement of clear flank areas

4.3.5.3 Transferred and Branched Flank Protection

If the adjacent track element to that requiring flank protection is not able to give it, the request is transferred to the next element, further away. Thus the flank area which has to be proven clear becomes longer. Two typical cases are explained in the following:
1. **Transferred (secondary) flank protection:** In the case of figure 4.21, if the element which should normally give flank protection cannot do so, it transfers the request to the next ele-

4 Interlocking Principles

ment (in this case the signal). Application examples are dual protective requests (chapter 4.3.5.4) and defective elements. If in this case also the signal is not operating properly, track section B may give flank protection, which means it has to be clear. This is implemented in some situations. An additional possibility is to transfer flank protection even in case an already active route the flank protection element becomes disturbed.

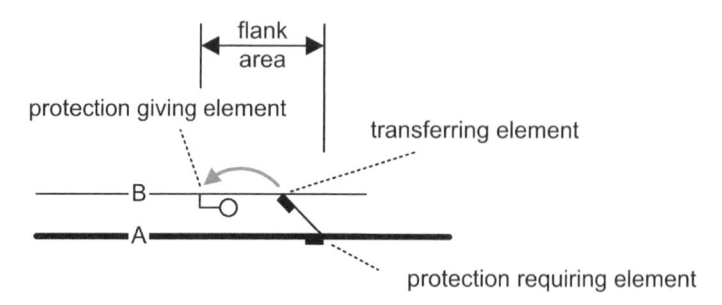

Figure 4.21: Transferred flank protection

2. **Branched Flank Protection:** In the case of figure 4.22, points 2 cannot give flank protection for points 1 due to the track layout. Therefore, points 2 have to transfer the flank protection request toward both trailing ends. In the one branch, points 3 can give protection against a movement from track C, whereas in the other branch signal 4 can give protection against movements from track B. The flank area extends, which means that points 2 also have to be detected clear.

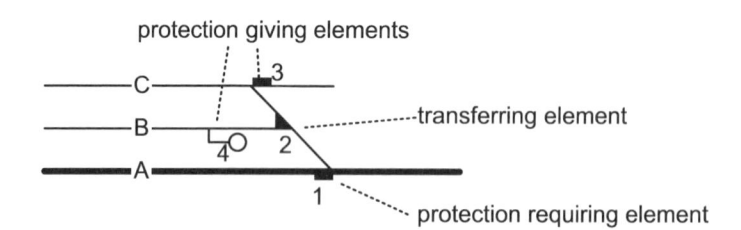

Figure 4.22: Branched flank protection

4.3.5.4 Dual Protection Requests

A particular case of transferred flank protection is conflicting requests by two routes (or by the same route). Examples are shown in figure 4.23. Cases of conflicting requests by two non-consecutive routes (figure 4.23a) can be solved as follows, when using the topological principle (chapter 4.3.9):

1. The first requested route gets flank protection by the dual protective points.
2. The second requested route gets remote flank protection by another element further away from the route to be protected.
3. When the first requested route (which means the partial route for which the dual protective points give flank protection) has been released, the dual protective points can switch and give flank protection for the other route. This feature is optional and its implementation differs between the railways and systems.

Some interlocking systems offer the possibility of defining priorities between the two conflicting flank protection requests. The prioritised route always gets flank protection from the dual

4.3 Routes

a) b)   remote protective element

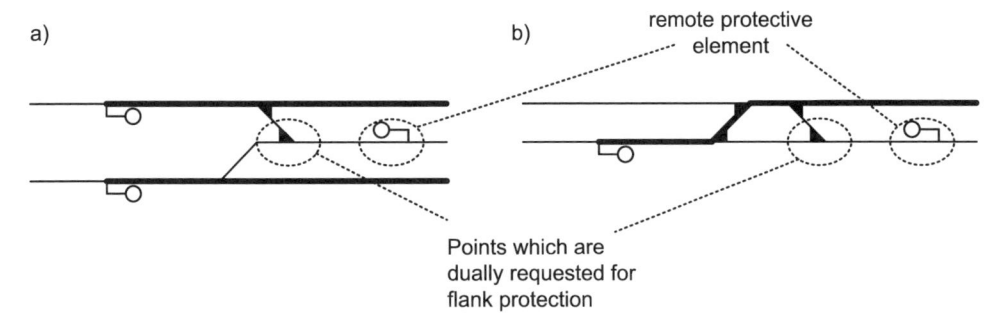

Points which are dually requested for flank protection

Figure 4.23: Conflicting flank protection requests by two routes (left) and by the same route (right)

protective elements, whereas the other route gets flank protection by the dual protective element only if no request from the priority route exists. Otherwise, it gets remote protection from another element, further away. For this reason, the dual protective element is not locked in the non-priority route. Thus it can be switched if a flank protection request is issued by the priority route.

In the case of conflicting requests by the same route (figure 4.23b), two solutions (including variants and mixed forms) are possible:

– A priority is defined and the non-priority flank protection request gets remote protection. The priority can be defined by different criteria, such as the length of the flank areas.
– The situation is solved dynamically. Initially, the dual protective element protects the running path element which is the first in the direction the train is moving. After the train has cleared this part of the route, the dual protective element switches to give flank protection for the other running path element until the train has cleared this also. As the time is usually short, this solution is seldom applied.

In special cases, flank protection requests to a dual protective element can generally be transferred to another element further behind (e. g. a signal) to avoid extremely frequent switching of the element. This is particularly suitable if both flank protection requests occur very frequently and usually alternately, e. g. for points in parking and reversing tracks between the main tracks on metropolitan railways, with fixed interval timetables. This can be the case in figure 4.23a. In metropolitan railways where no non-tractive vehicles are parked, but only fixed coupled units with at least one tractive vehicle, the probability of rolling away is low and therefore the loss of safety by giving flank protection by a signal instead of points is relatively slight.

4.3.6 Overlaps and Front Protection

4.3.6.1 Purpose

Overlaps are applied by most railways and give an additional protection against a minor error by the driver in target braking. In some cases, they even accommodate the whole braking distance of the train in case the driver doesn't brake at all and the train protection system can't effectively prevent this. Besides, some railways include ideas of front protection against opposing movements into their overlap locking functions. Railways also differ in whether or not to provide flank protection for their overlaps (chapter 4.3.5.1).

4 Interlocking Principles

4.3.6.2 Overlap Length

The required length of the overlap is determined by the following:
- Kind of danger point: Distinction can be made according to the type of element to be protected (e.g. facing points, fouling point of trailing points, standing rail vehicles etc.). Another possibility is to distinguish between the kind of movement (train/shunting) beyond the danger point to be protected.
- Speed of the train. The faster the train approaches the braking target point, the higher the probability of an overrun and the longer the expected length of the overrun is. Therefore, many railways determine the length of the overlap by the approach speed.
- The kind of applied train protection system (chapter 8). A main purpose of train protection systems is to prevent overruns or, in the case of an overrun, to limit its length. Some systems are more effective than others in this respect. Therefore, several railways distinguish the length of the overlap by presence or absence of train protection or by particular systems.

Based on these ideas, the railways have defined overlap lengths as being between nil and 400 metres for protection against minor overrun on conventional railways, or even up to a few kilometres in some older high speed signalling systems. Some railways use constant length overlaps, but others vary the overlap length by the above criteria. Some railways (e.g. the Dutch) have abandoned overlaps in areas equipped with advanced train protection.

4.3.6.3 Locking of Movable Track Elements in the Overlap

Railways differ as to whether points in the overlap have to be locked with the route or not. In some countries, all points in the overlap have to be locked. But in other railways, either facing or trailing points in the overlap remain unlocked: Reasons for that are the following:
- The issue of **front protection**. If an unauthorised movement in the opposite direction enters the overlap, danger can be prevented by setting any trailing points in the overlap away from the route (figure 4.24). Therefore, trailing points in the overlap can remain unlocked to give the signaller who notices this danger the possibility of averting it. To increase this protection, trailing points in the overlap or at the end of the overlap can even be locked in the 'wrong' position (as done e.g. in Britain), given the damage this will cause to a non-trailable point machine in the case of an overrun.

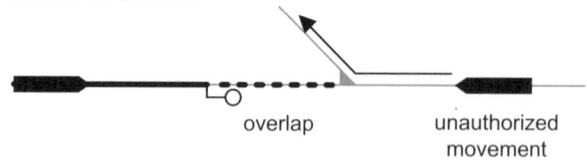

Figure 4.24: Front protection against unauthorised opposing movements

- Possibility to **change the overlaps** of an active route: If overlaps can be changed while a route is active, this offers more flexible operation and enables route setting by the signaller in a progressive sequence without circuitous overlap selection commands. These procedures, so-called swinging overlaps, are described in chapter 4.3.6.6.
- **Shared overlaps:** As the probability of the overrunning of two trains occuring at the same time is very low, it can be acceptable (and is accepted by some railways) for two different

Figure 4.25: Shared overlaps

routes to use the same portion of track for overlap (figure 4.25). There the trailing points where both overlaps meet cannot be locked for both routes.

4.3.6.4 Radii in the Overlap

It can be assumed that in the case of an overrun, overlap elements are traversed at very low speed. Therefore radii in the overlap usually have no influence on the speed of the route.

4.3.6.5 Selective Overlaps

Selective overlaps (figure 4.26) increase the flexibility of railway operation. Depending on the operational situation, the signaller can choose between two or more overlaps. Selective overlaps can be distinguished as follows:
- Overlaps in different directions (e. g. cases a and b in figure 4.26) differ in the position of at least one set of facing points. The motivation can be the choice of the continuation route or conflicts with other routes.
- Overlaps of different length in the same direction (e. g. cases a and c in figure 4.26) can permit different speeds on the route. The shorter overlap can be selected when necessary due to route conflicts, otherwise the longer overlap is selected to enable higher speed of the route.

Figure 4.26: Different overlaps for the same route

4.3.6.6 Alteration of the Overlap of an Active Route

In many cases, the possibility of changing the overlap while the route is effective can be helpful for flexibility in railway operation:
- When the reason for a conflict in the overlap of an effective route has ceased, but the train has not yet entered the route, the overlap can be **switched to a longer one** to enable higher speed.
- On the other hand, a **reduction of the overlap** length of a route can enable another movement to be made earlier.
- The **change of the direction** (swinging) of the overlap of an effective route can help to enable a continuing route for the same train, or a route for another train which is in conflict with the old overlap.

On the other hand, changing the overlap of an effective route can cause a reduction in safety in the case of an overrun in two situations:
- In the case of a direction change of the overlap, if the points in the overlap do not reach their new position in time. This causes additional requirements for the point machines to reach the new end position or return to the old one safely.
- In the case of shortening the overlap, should the driver be unable to reduce speed sufficiently. This prevents reducing the length of the overlap if there is no possibility of informing the driver of the reduced safe speed in time. With trackside signals, reducing the length of the overlap is therefore equivalent to releasing the route and setting a new route with a shorter overlap. It follows that shortening the overlap of an active route is of no practical importance.

4 Interlocking Principles

The following solutions regarding overlap alterations can therefore be found in practice (figure 4.27):
- **Static overlaps.** Once the route has been set, the overlap cannot be changed.
- **Extendable overlaps.** The overlap of an effective route can be lengthened if the reason for limitation (usually a conflicting route) has ceased. But shortening of the overlap or changing its direction is not possible. This solution is used in Germany in modern interlocking systems with selective overlaps. An advantage which increases line capacity is that the route can be set when the running path and the shortest overlap have been cleared by the previous movement, but speed can be upgraded later as soon as the longer overlap becomes available. In an extreme case, which is applied on some metropolitan railways if only following movements need to be protected, the route is first set without overlap and full speed is signalled, but the overlap is locked few seconds later after the train beyond has cleared it (figure 4.28). This solution takes into account that in the case of following movements, the probability that the previous train comes to standstill in the overlap (far in rear of the signal beyond) is very low. Only a combination of two improbable situations (the previous train stands in the overlap and the following train overruns) can cause danger.

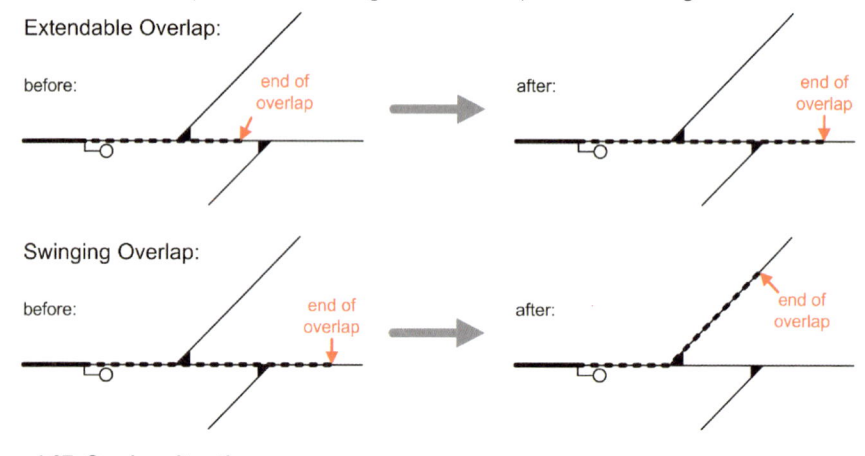

Figure 4.27: Overlap alterations

- **Swinging overlaps** (applied in Britain). Facing points in the overlap are not locked and the direction of the overlap can be changed, provided that the new overlap is clear. Another precondition for swinging the overlap is that the train is not yet too close to the route exit signal to make sure that the points reach the new end position in time. If the preconditions are fulfilled, first the new overlap is reserved, then the decisive points are moved to the new position and finally the old overlap released. The point machines as well as the processing units have to be reliable enough to reach the new end position safely, or return to the old position before an overrunning train can reach the points. An advantage of swinging overlaps is that the signaller can set consecutive routes for the same train in the 'correct' sequence. Without swinging overlaps, special action is needed to select the overlap before clearing the respective signal.

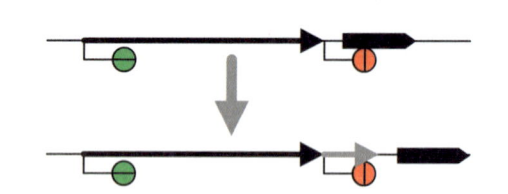

Figure 4.28: Extendable overlap in some metropolitan railways

4.3 Routes

4.3.7 Route Elements in the Start Zone

In some cases, movable track elements which are situated in rear of the route entrance signal in the start zone have to be included into the route functions (figure 4.29). In particular, this occurs with station exit routes in situations where the train starts the route from a scheduled stop, with the previous route of this train already released (chapter 4.3.3.6). Many railways try to avoid such situations in track layout planning by not placing movable track elements in tracks where trains will stop regularly. But particularly in areas with a restricted availability of space, these situations cannot always be avoided.

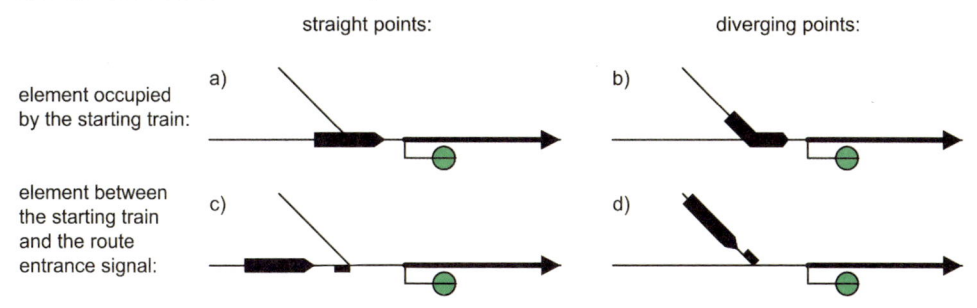

Figure 4.29: Examples of route elements in the start zone

In cases where the movable track element is already occupied by the starting train (figure 29 a and b), locking functions are already fulfilled by the track occupation: However, depending on the interlocking system, additional locking of this element in the route can be applied. In cases where the element can be situated between the train front end position and the route entrance signal (figure 4.29, c and d), this element must always be included in the route locking functions. Additional special requirements in interlocking logic can occur to determine the exact position of the starting train in this context, and in the case of converging tracks to determine the track from which the train starts. The solutions are particular for the interlocking systems and are not be discussed in detail here.

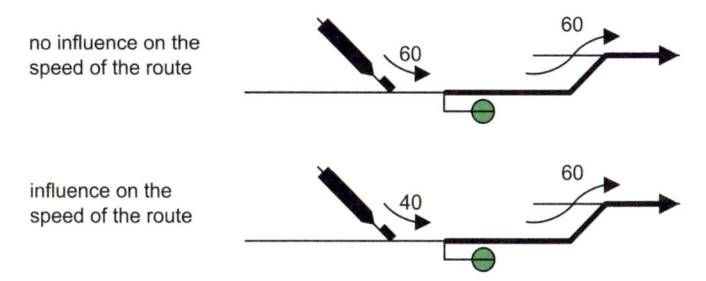

Figure 4.30: Points of the start zone influencing the route speed

A controversial problem of speed signalling occurs if the train uses an element in the start zone in the diverging track (figure 4.29, b and d) and the speed permitted by this element is lower than the speed the route would permit without this element (figure 4.30). The solutions depend on the regulations about the local validity of the speed indication of the signal in rear and which speed has been signalled there (chapter 4.3.2.3). The case that the speed signalled at the station home signal is valid through the whole station (figure 4.9, case 1) is the simpler case in this context. In case of separate speed restrictions for each route (figure 4.9, cases 2a-e), particular attention has to be paid to the element of the start zone concerning speed regulation:

4 Interlocking Principles

- In the case 2a, the driver is obliged to obey the speed restriction of the station entry route until the rear end of the train has passed the last movable track element between the station home and exit signals. This includes the element discussed. However, there is a danger of the driver forgetting this speed restriction when starting the train after a stop, if a higher speed restriction is displayed at the station exit signal. The same problem occurs with starting trains.
- In case 2b, the problem does not occur, as diverging points in the route only appear in interlocking areas. However, the case can occur that a train is in two interlocking areas at the same time (in one with the front end and in the other with the rear end) and then has to obey the lower of both speed restrictions.
- In the cases 2c and 2e, the element has to be considered for speed selection of the exit route. In unfavourable cases, the speed of the whole exit route is determined by an element which is occupied only by the last wagon of a starting train. Considering different train lengths, additional requirements regarding the detection and evaluation of the exact position and length of the starting train can be necessary.
- In the case 2d, the element only has to be considered in the exit route in cases of a train starting in the station track, because here there is no speed of the entry route which would be additionally valid until the exit signal is passed completely. However, the danger of the driver forgetting can be a problem anyway.

4.3.8   Life Cycle of Routes

The content of this section is the process of setting up a route before a train can move, and of releasing that route behind the train.

4.3.8.1   Steps in the Life Cycle of a Route

Generally, the following steps in the life cycle of a route can be distinguished (with some variations in the sequence):

- **Route calling.** The request for a certain route is identified and the route is searched in the system. This is an especially interesting process in systems working with topological principles (chapter 4.3.9).
- **Route checking** (optional). The availability of the route is checked. The purpose is to avoid the unnecessary switching of movable track elements and the obstruction of other routes in case the route is not available.
- **Commanding of movable track elements** into the required position and checking their end position.
- **Route locking.** All elements belonging to the route are locked in their proper positions and the route itself is interlocked to prevent it being released improperly and to protect other routes.
- Checking of the **clear** status of the **tracks**.
- **Signal selection and control.** Determined by route factors such as the radii of diverging points, overlap length and occupation by other vehicles, the signal aspect is selected and the main, distant and (if present) repeater signals are cleared. Before clearing the signal and as long as the signal shows a proceed aspect, several safety conditions are checked.
- **Route supervision.** During the time that the route entrance signal is cleared and/or a train is traversing the route, the route is supervised permanently or on a cyclic basis. If endangering occurrences such as flank movements, points leaving their positions, overruns, failure of the route exit signal or other problems are detected, the signal must be immediately put

back to Stop. This often leads to the operation being continued in degraded mode (chapter 4.5.4). Also, the signaller retains the possibility of placing the signal to Stop. There are different views concerning whether after the loss of supervision due to a technical failure of short duration, the route can be restored or not.

- **Signal returning to Stop.** When the front end of the train has passed the signal, the signal can be returned to Stop. This must not normally occur within the sight of the driver, even if in some locomotives the driver's cab is not at its front end. In some cases and interlocking systems, releasing the signal not too late can have a safety importance, to prevent following movements. One case is described in chapter 4.5.2, another is splitting of a train into two in the start section in rear of the signal (figure 4.31).
- **Route release.** After the train has traversed the running path, the route is released. Release can either be done in one step by the train clearing the route or coming to a halt on the target track, or sequentially for each route section which bas been cleared (**sectional route release**, chapter 4.3.3.6).

In old interlocking systems, each of these steps had to be performed manually by the signaller. In electronic and most relay interlocking systems, steps are automatically initiated one after the other following the entrance-exit settings of the signaller (chapter 4.3.4) and partly controlled by the moving train.

Route locking divides into two functions and both are applied in parallel in modern interlocking systems (figure 4.32):

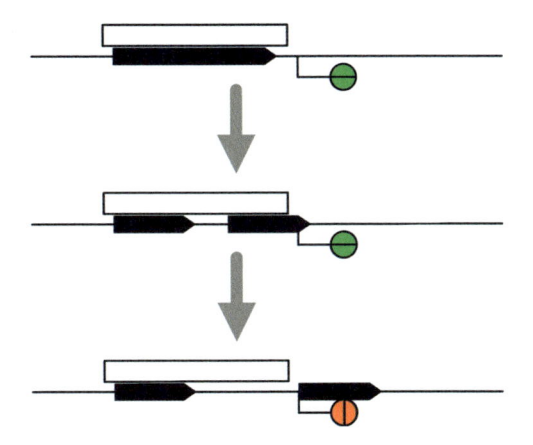

Figure 4.31: Following protection after splitting of trains

1. **Reversible route locking** guarantees the dependence between points and signals. This function ensures that the signal can only be cleared if all movable track elements are in their proper positions, and the movable track elements remain in this position as long as the signal shows a proceed aspect. It is called reversible here as it can be manually released by the signaller without special safety precautions when the signal is returned at Stop. Historically, reversible locking is the older form, but alone it is not sufficient for safety: It can be released after returning the signal to Stop even if the train is still traversing the route. Reversible locking can be

reversible locking function

reversible and irreversible locking functions

reversible and irreversible locking functions

only irreversible locking function

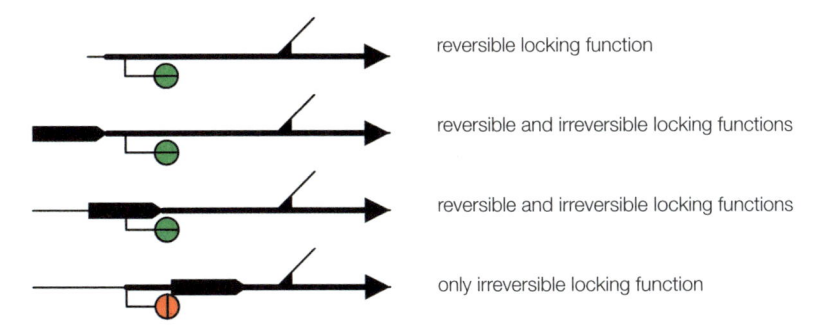

Figure 4.32: Reversible and irreversible route locking

4 Interlocking Principles

effected either by direct element dependences between movable track elements and signals (chapter 4.2; British and North American interlocking principles) or by a particular route locking function (German interlocking principles). The reasons for this difference are historical and the safety effect of both solutions is the same.

2. **Irreversible route locking or 'holding the route'** is an additional function which maintains the route locked even after signal release, until the route or route section has been cleared by the moving train. It is called irreversible here as it can only be released either by the moving train, or manually with special safety precautions such as time delay and/or registration.

4.3.8.2 Sequence of the Steps

The order in which the steps described in chapter 4.3.8.1 are performed differs between the railways. The most important differences are:

- The moment of proving **track occupation**. Some railways and interlocking systems prove the complete route clear already before setting the points by automatic point setting. Others check track clear detection only for those points which need to be switched, but prove the complete route only immediately before clearing the signal.
- The moment of searching and locking **flank protection** and commanding protective elements into the required position. In some systems this is done together with the running path elements; in other systems it takes place later in the process, when the running path elements are already locked.
- The moment of closing **level crossings**, if they are integrated into the route (chapter 13.4.4.1). A simple solution, which was applied mainly in old interlocking systems, is to close them together with movable track elements. However, short closing times for road traffic are required today. Therefore, most modern interlocking systems close level crossings immediately before clearing the signal or even after clearing the signal when the train has reached a particular approach point for the level crossing. The latter case requires the safe closing of the level crossing.
- The order of **irreversible route locking and signal opening**. This issue, which is more complex, is now discussed.

Irreversible route locking can be achieved in two different ways (figure 4.33). Both solutions are applied with different variations in several countries:

- **Irreversible locking before signal opening** (German influenced interlocking principles). Safety is ensured by the signal only being able to be cleared if the irreversible route locking function has already been effected. A disadvantage from the point of operational flexibility is that if the signal is returned to Stop without a train having passed it, it cannot be cleared again and the route must be released by special, safety critical commands (chapter 4.5.4.3). However, this can be avoided by delaying irreversible route locking and signal clearing until the train approaches, which is performed in some relay and electronic interlocking systems (chapter 4.3.11).
- **Irreversible route locking after signal opening** (British and North American influenced interlocking principles). The signal can be cleared when all the moveable track elements are in their proper positions and the tracks are clear, but without irreversibly locking the route. The cleared signal locks the route elements, but the signal can be repeatedly cleared and restored to stop, thus locking and unlocking the route. Only when the train approaches the route will it be irreversibly locked and remain locked until the train clears the route or the respective route section. To ensure safety, the irreversible route locking function must safely occur early enough that the driver can prepare for the stop. This increases the requirements for the detection system and the locking function.

4.3 Routes

irreversible locking after signal opening

- default position (no route locked)
- route calling & checking
- commanding of elements
- signal opening with reversible locking
- irreversible (approach) locking
- **train enters the route**
- signal release
- **train clears the (partial) route**
- route release

irreversible locking before signal opening

- default position (no route locked)
- route calling & checking
- commanding of elements
- reversible route locking
- irreversible route locking
- signal opening
- **train enters the route**
- signal release
- **train clears the (partial) route**
- route release

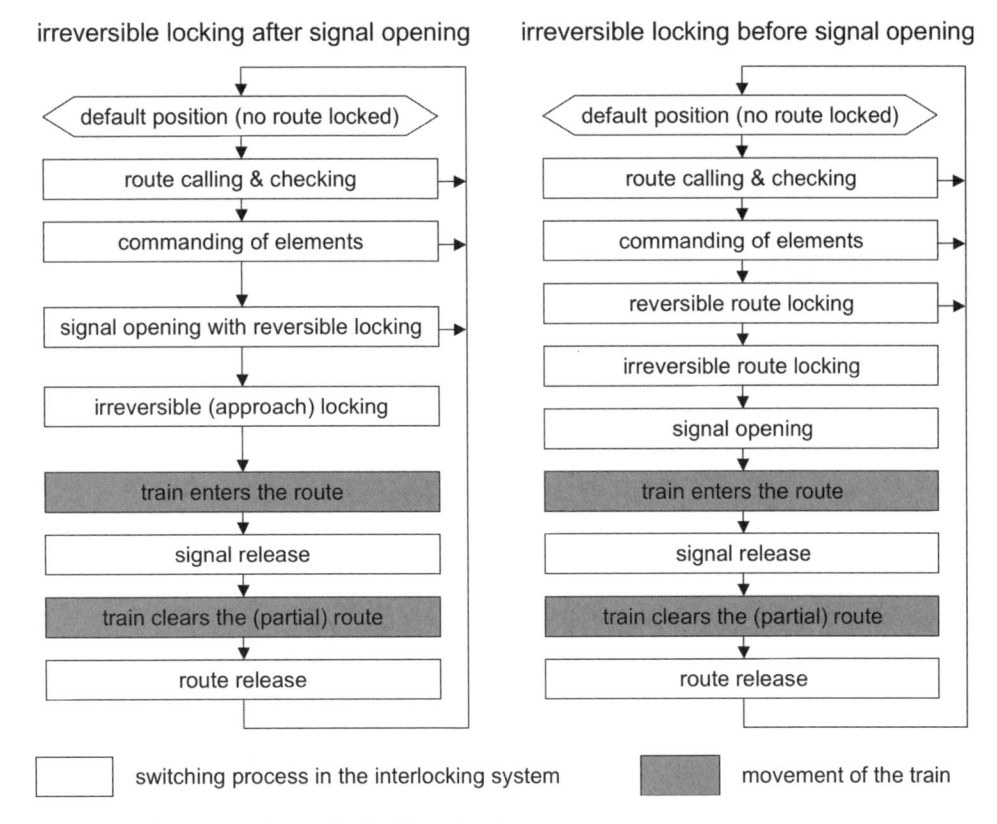

☐ switching process in the interlocking system   ■ movement of the train

Figure 4.33: Sequence of steps in the life cycle of a route

However, in both solutions the signal can be returned to danger by manual operation or automatically at any time to prevent hazards. The difference is in the possibility of clearing it again with normal operations.

4.3.8.3 Example: British Signalling

In the following, the life cycle of a route in British signalling with irreversible locking after signal opening is described briefly using the example shown in figure 4.34 in British relay interlocking. The route to be referred to starts at signal 5 and reaches up to signal 7.

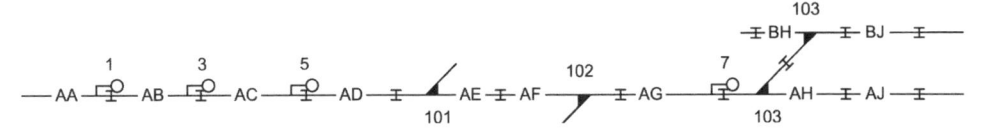

Figure 4.34: Example for British approach locking and cascade release

After bringing all movable track elements including flank protection and overlap to their positions, the signal is cleared, which locks the movable track elements. Hereby, each partial route (which means each track circuit in this example) is route locked. Up to this stage, as no train is approaching, the route can be released manually without time delay.

4 Interlocking Principles

The second step is **approach locking**, which is an irreversible lock. No later than when the train has reached the sighting point of the first warning signal, which would be the occupation of track circuit AB in three-aspect signalling (chapter 7.3.3.2) and AA in four-aspect signalling (chapter 7.3.3.4) in the example, the route is approach locked. As manual release of an approached route is potentially dangerous, it is bound to a time delay of typically around two minutes for train routes. If the train enters the route within this time, the lock will be maintained, otherwise is will be released. To ensure safety, this requires special operational rules for the driver in case of very slow running of the train. In this case, the driver cannot rely on still having a locked route in front of him, even if he has seen a green distant signal.

When a train enters a locked route (which means it has passed the entrance signal), so-called **back locking** as the second part of irreversible locking becomes effective. The route locking is held by the track occupation until the respective section has been cleared. The logic of train operated route release (TORR) is cascade-shaped, which means that the route locking of each track section is held by the occupation of this section or by the section in rear being locked. Following that, each track section releases individually behind the train in the appropriate sequence and in case of a single detection error no dangerous situation can occur. In the example, route release is performed as follows:

– Section AD can release when the train has entered the route and track circuit AD is clear.
– Section AE can release when AD has been released and track circuit AE is clear.
– Section AF can release when AE has been released and track circuit AF is clear.

In relay technology, the route locking functions are solved by type N relays (chapter 9.3.2.1). Circuits are designed that way so that the respective relays are dropped down in the locked status and picked up in the unlocked status, ensuring a fail-safe working.

Another feature to be mentioned is the utilisation of swinging overlaps which are described in chapter 4.3.6.6.

More information on the British example can be found in (Nock 1982, Hawkes 1969).

4.3.8.4 Example: German Signalling

In contrast to the British logic, in the German logic the reversible route locking is not performed by direct element dependences between movable track elements and signals, but via a particular route locking function which locks all elements belonging to the route in their proper positions. In mechanical interlocking, this was performed by a particular route drawbar (chapter 9.2.5.4). After reversible route locking, irreversible route locking has to be applied before clearing the signal. The irreversible route locking is an electrical lock, even in mechanical interlocking, because the detection of the train could not be performed mechanically.

A disadvantage for the flexibility of operation in comparison with the solution with irreversible route locking after signal clearing can be that once a route has been locked, the signaller cannot change it if the operational situation changes, although the train is still far in rear. To improve the situation, some interlocking systems are provided with a form of approach locking as an automation function (chapter 4.3.11).

In mechanical interlocking, sectional route release was not normally provided, which means that the route could only be released as a whole. This causes a reduction of node capacity. In newer systems with the tabular principle (chapter 4.3.9), partial routes are defined separately. In systems with topological principles (chapter 4.3.9), naturally each section for track clear detection is a partial route, defined by the relation to its predecessor and successor, and releases individually behind the train. Also here, route release is performed by the evaluation of occupation and clearing of at least two track sections to ensure safety also in case of detection error.

4.3.9 Principles of Route Formation in the Track Layout

In this section, the principles that define which elements belong to a particular route are discussed. There are three basic principles of route formation: the tabular, the cascade and the topological, the latter often also called geographical principle. All three principles are applicable for systems with irreversible route locking before and after signal clearing (chapter 4.3.8), although not all combinations are of practical importance.

The **cascade principle** was the oldest principle in Britain and North American interlocking logic in mechanical systems. Today, it can only be found in old installations with lever frame machines. The basic principle is that each movable track element is locked by decisive moveable track elements or the signal in rear. Thus each element can usually be released individually behind the train. Due to its low importance today, it is not described here in detail. Cascade locking is described exactly in (Pachl 2002).

The **tabular principle** is the traditional form originating from mechanical interlocking in German-influenced interlocking logic. One historical reason for its development was the necessity for free shunting (without shunting signals and with freely switchable points) in Germany, which was not possible with cascade locking. The tabular principle is still widely applied in mechanical, relay and electronic interlocking worldwide. All possible routes are predefined in a matrix, indicating exactly which elements belong to the respective route in which position. Figure 4.35 and table 4.1 show an example. Routes which cannot be active at the same time are also predefined, distinguished between simple and special route exclusions (chapter 4.3.3.3).

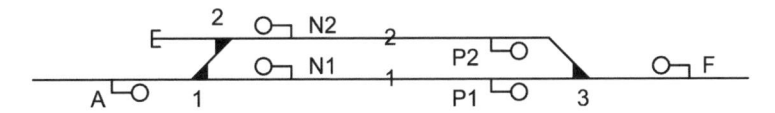

Figure 4.35: Layout plan for route locking matrix of table 4.1

Routes	Conflicting Routes								Points			Flank Prot. Signals		
	A/1	A/2	N1	N2	P1	P2	F/1	F/2	1	2	3	N1	P1	P2
A/1		|	||	|			||		+	+				
A/2	|		|	||		||		||	-	-		S		
N1	||	|			|				+	+				
N2	|	||	|					||	-	-		S		
P1						|	||	|			+			S
P2		||			|		|	||			-		S	
F/1	||			||	|	|		|			+			S
F/2		||		||	|	||	|				-		S	

A/1	Route beginning at signal A into track 1
N1	Route beginning at signal N1 (if there is only one route from this signal)
+	Point in normal (here: straight) position
-	Point in reverse (here: diverging) position
S	Signal at Stop
|	Simple exclusion of routes by movable track element in different positions
||	Special exclusion of routes

Table 4.1: Route locking matrix (example)

4 Interlocking Principles

The **topological (geographical)** principle, firstly appearing in Germany in the 1950s, was developed for relay interlocking and, until today, it is used internationally for relay and electronic interlocking on a large scale. The elements of the track layout are defined with their neighbourhood relations to each other (figure 4.36). When a particular route is to be set, the running path, overlap and flank protection are searched by these topological relations. Therefore, all routes which are possible according to the track layout can be selected automatically, unless they are deliberately suppressed.

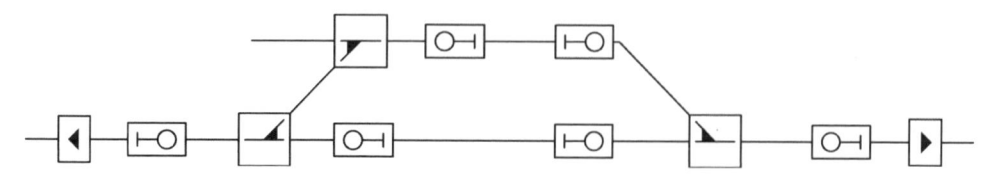

Figure 4.36: Simplified relay set connection map for topological interlocking

As the characteristics, advantages and disadvantages of the topological and the tabular principle become most obvious in relay interlocking, they will be compared first with relay technology (table 4.2).

In topological relay interlocking systems, the relays are positioned in relay sets, each representing a certain element in the track layout. The relay sets are standardised and can be manufactured and tested automatically in the factory. This reduces the necessity of wiring on the construction site to a minimum of connecting the relay sets by standardised cables. This facilitates wiring and testing processes as well as the adaptation of the interlocking to alterations in the track layout. This advantage over the tabular principle grows, the more elements that are included in the interlocking area. On the other hand, a disadvantage of the topological principle is the large amount of effort to be put into the relay sets. The wiring within the sets is generally more complicated, and due to standardised relay sets, a large number of relays which are not required for the particular case have to be physically present.

Figure 4.37 illustrates the effort which is necessary for each principle in relay technology, depending on the size of the station/junction to be interlocked. As can be seen, the tabular prin-

	tabular principle	topological principle
efforts for relay system development	relatively low	high
relay equipment efforts once per interlocking area	low	high (central functions)
relay equipment efforts for each additional switching unit	high	relatively low
efforts for later changes in relay interlocking	relatively high	relatively low
efforts for later changes in electronic interlocking	high	high
adaptability to seldom special cases	mostly easy	difficult
efficient application for …	small stations (relay) low number of applications of the system (relay, electronic)	large stations (relay) high number of applications of the system (relay, electronic)

Table 4.2: Comparison of tabular and topological principle

4.3 Routes

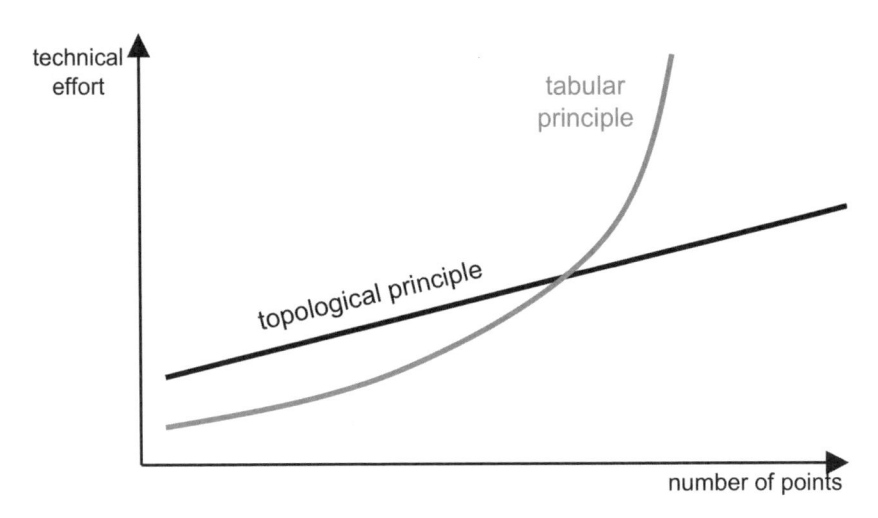

Figure 4.37: Technical effort for tabular and topological principle in relay interlocking

ciple is advantageous for smaller schemes, whereas the topological principle gains with larger interlocking areas.

In electronic interlocking, some of the advantages and disadvantages of both principles lose much of their importance (table 4.2). Therefore, both principles are suitable for small and large interlocking areas. A difference remains concerning the number of installations of a certain type. The development of a new interlocking type is more complex with the topological principle, but once developed, projecting a new installation is simpler. Altogether, both principles are applied in electronic interlocking, differentiated by manufacturers and types of interlocking.

4.3.10 Shunting Routes

4.3.10.1 Interlocking Functions

Most railways, other than those following North American operational principles, distinguish between at least two classes of movements. These are train movements and shunting movements (chapter 3.3). The exact definitions of what constitutes a train or a shunting movement differ, but generally shunting movements are performed at a lower speed (mostly on sight) and without a timetable. Due to the more limited requirements, shunting is often performed on shunting routes with a reduced interlocking function or even by verbal permission without interlocked routes. Nevertheless, the trend in modern technology is to provide even shunting routes with advanced interlocking functions. Higher requirements are often set for areas where shunting with occupied passenger coaches or dangerous goods occurs frequently. Where shunting movements traverse routes, the main differences to train routes can be:

- Omitting or reduction of **track clear detection**. As the purposes of shunting imply the entry into occupied tracks for the joining up of rolling stock units, the permission for a shunting movement cannot generally require the route to be clear. Therefore, either no detection is provided at all, or all track sections except the target section are checked for being clear.
- Omitting or reduction of the **overlap**. As shunting movements move slowly, the probability and the expected length of an overrun is that much lower. Thus overlaps are not usually provided.

4 Interlocking Principles

- Omitting of flank protection. As vehicles without passengers are usually moved at low speed, the expected severity of accidents is lower and the ability of the driver to prevent the danger is that much greater. Therefore, flank protection is often not used for shunting routes.
- Permission for **opposing shunting movements**. In some cases, the possibility of moving vehicles onto the same track at the same time can be needed for efficient operation. An example is a train where coaches are to be attached to the rear end and a locomotive to the head at the same time. However, if opposing movements are permitted, they have to be able to stop in half the range of vision. A solution to avoid this is to waive the requirement for opposing locking only after the target track has been occupied for a specific time. After that, the occupying vehicles can be assumed to be stationary. This is applied in Britain, for example (Nock 1982).

Signals for shunting can show at least one stop and one proceed aspect. They can stand alone or be part of train signals (chapter 7.3.6). As an option, different proceed aspects can be displayed, depending on the safety conditions of the particular shunting route, e. g.:
- Movable track elements in the shunting route locked or unlocked
- Tracks clear or occupied
- Aspect of the next shunting signal

4.3.10.2 Shunting Areas under Dual (Central and Local) Control

To reduce the work load of the signaller, many railways permit shunting operations in defined areas to be carried out under temporary local responsibility. In these cases, repeated shunting movements are possible without any action by the signaller.

In European practice, this feature is assigned to certain areas with precisely defined boundaries. The points for entering and leaving the locally operated area are locked and shunting signals are permanently cleared, omitting the locking of opposing signals (figure 4.38).

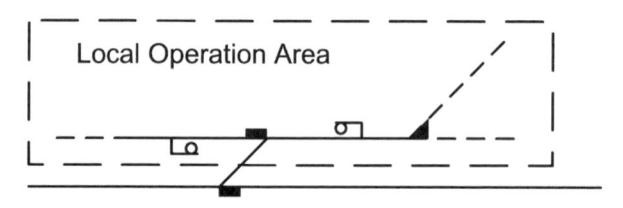

Figure 4.38: Shunting areas under temporary local responsibility

North American interlockings are generally equipped with dual control points. In contrast to European practice, where the manual operation of normally electrically controlled points is usually only applied in the case of technical failure, in North America it fulfils the same purpose as European local operation areas. The temporary permission to operate points by hand in the case of repeated shunting movements can be given from the dispatcher to the local signaller verbally or by written instruction.

4.3.10.3 Reversing Shunting Routes

A special problem of shunting routes is those which require reversal (figure 4.39). Suppose a shunting unit moves consecutively on two shunting routes in opposite directions. Often it is not helpful for the shunting unit to move until the end of the first route is reached and return there, but it can return after passing the signal which permits the second route. As a result, a section of the first route often remains unused (track circuits D and E in figure 4.39), so the first shunting route does not release in the normal way.

4.3 Routes

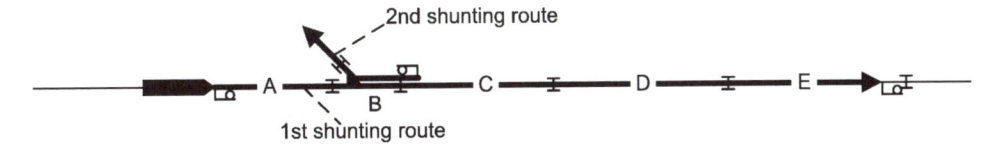

Figure 4.39: Reversing Shunting Movement

To avoid the necessity of an emergency release of the remainder, specific interlocking functions can be applied to release the remainder of the first route depending on the status of the second shunting route. These functions have to be projected for the particular case on tracks where such situations are expected to occur frequently. One solution is that the remainder of the first route is released together with the first section of the second route. By this, it can be ensured that the shunting unit has actually reversed and does not continue on the first route. Another solution is a time delayed release of the remainder.

4.3.11 Automation of Route Operation

To reduce the workload of the signaller, to relieve him from mundane tasks, to reduce the personal efforts of ground staff and to quicken railway operation, several features have been developed to automate the calling and setting of routes. They are applied differently by the railways.

Automatic point setting. This is a standard feature in most relay and electronic interlocking technologies: The route is selected normally by entrance-exit operation (chapter 4.3.4), and all movable track elements are automatically switched to the required positions. However, safety when working in degraded mode operation (chapter 4.5.4) often requires the switching off of automatic point setting for the whole interlocking area, for defined parts of it, or for an individual element.

Approach locking and signal clearing. This non-safety automation function is to be clearly distinguished from the approach locking after signal clearing (chapter 4.3.8.2). It is an additional non-safety automation function in systems with irreversible route locking before signal clearing (chapter 4.3.8.2). The route is reversibly locked immediately after receiving the command from the signaller, but irreversible route locking **and** signal clearing are postponed until short before the train driver can see the first warning signal and is then done automatically (figure 4.40). This enables the signaller to cancel a route which is not approached without safety critical commands if the operational situation so requires.

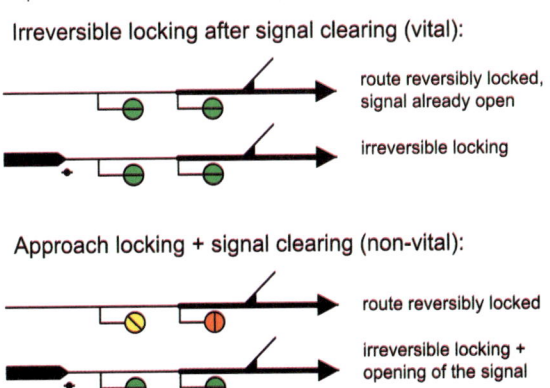

Figure 4.40: (Irreversible) approach locking as a safety function after signal clearing and as a non-safety automation function before signal clearing

4 Interlocking Principles

Long routes. The signaller is able to set two or more consecutive routes for the same train by one entrance-exit-operation over the whole path (figure 4.41).

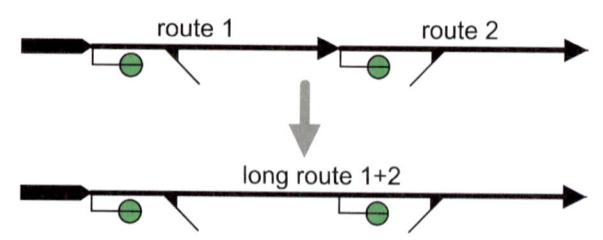

Figure 4.41: Long route

Route queuing. If a route which a signaller attempts to set up is not available, it is preselected and will be set later when it has become available. However, the queuing of more than one route can lead to the wrong route being set at the wrong time and therefore cause operational problems.

Fleeting and Automatic Route Calling: With **fleeting**, after a route has been completely traversed by the train, it remains locked and can be used by a following train. **Automatic route calling** is similar to fleeting in its effect. However, here the route is released behind the train and will be automatically called again, on the same path, when the next train approaches. Fleeting and automatic route calling are especially useful in situations where the vast majority of trains traverse the station on the same path, such as often occurs in night hours. In the case of fleeting, the signals along the route can be used like permissive block signals (chapter 7.3.2) under certain circumstances, whereas in automatic route calling the signals remain as absolute stop signals.

Automatic route setting/Automatic train routeing. Trains are routed through different tracks based on stored information such as timetable data and on forwarding train numbers with the moving train by a train describer system. Irregular situations usually have to be solved by staff.

Route commanding by train. The train itself, either by a command input from the driver or automatically, sends a route calling command to the interlocking. Thereupon, the other steps in the life cycle of the route (chapter 4.3.8.1) are performed by the interlocking system.

4.4 Block Dependences

4.4.1 Introduction

4.4.1.1 Safety Functions of Block Systems

Block dependences, in contrast to routes, can provide for only two safety functions (chapter 4.1.2):
– Protection of following movements
– Protection of opposing movements

Therefore, the principle 'block information' is applicable to open line sections only. The principle 'route', in contrast, can be applied in interlocking areas and on open lines. When applying it to open lines, only a part of the possible safety functions is used.

4.4.1.2 Definitions

A **block section** is a section of an open line track in which, unless moving on sight, only one train is permitted at any one time. The block section is limited by **block signals** as entrance and exit signals of the block section. Also, signals at stations and junctions can serve as block signals for the adjacent open line sections.

The block signal together with the related evaluation units for block messages is called a **block station**. Block stations can be staffed or unstaffed. Each block station, in cooperation with the neighbouring block stations, regulates following movements.

To avoid deadlock situations, only these block stations are actively involved in opposing protection where the order of trains can be changed, such as stations, loops, junctions and crossovers. In the following, they will be called **'train sequence station'**. Protection of opposing movements is only regulated between neighbouring train sequence stations, whereas intermediate block stations only forward the related messages, are only informed about it or are even not involved at all in these processes. The section between two train sequence stations is called **'train sequence section'** (figure 4.42).

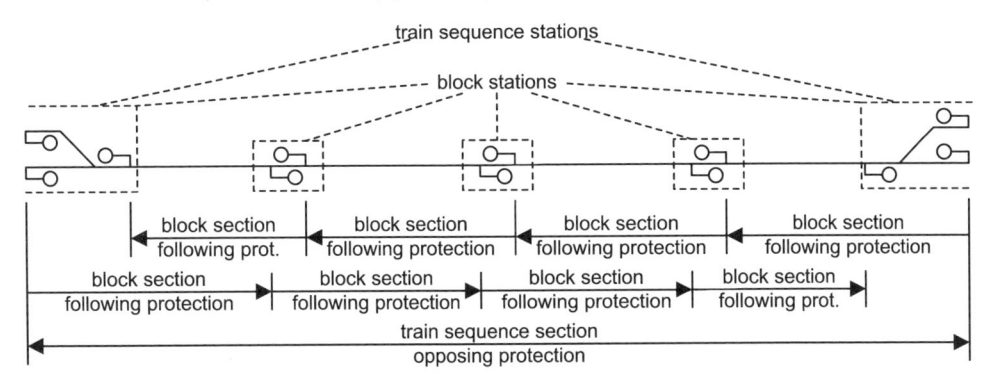

Figure 4.42: Train sequence stations and intermediate block stations

4.4.1.3 Systems of Non Signal-Controlled Operation

As non-technical ancestors of block systems, the following – each with different variants – are the most important forms (see also chapter 3.5.3):

- **Telephone block** (applied mainly in countries following German operational principles): It can be considered as the non-technical ancestor of tokenless block systems (chapters 4.4.3, 4.4.5). The line is divided into fixed block sections limited by trackside signals. Each section is blocked and unblocked by telephonic correspondence between the block stations and the written notations of this correspondence. The signaller at the entrance of the block section must not open the signal for a following train before receiving the clearing message telephonically from the block station ahead.
- **Direct Traffic Control** (DTC; applied in North America): It has certain similarities with token block (chapters 4.4.3, 4.4.4). The line is divided into defined block sections marked by block marker boards. A central dispatcher issues a permission to the train to enter a section. After clearing the section, the train has to give the permission back. Likewise in telephone block, the correspondence is noted in the protocol book.
- **Track Warrant Control** (TWC). This originated from North America, but today is applied worldwide in different versions. In Europe, derived versions may be found particularly on secondary lines, for cost effectiveness. TWC has similarities with direct traffic control, but any identifiable location can serve as an end of a movement authority. This enables the system to handle the length of the block sections flexibly, enabling shorter headways, though it includes an increased workload for the persons involved.

4 Interlocking Principles

Such systems are still applied on lines with low requirements and in degraded mode operation in case of failure of the technical block system.

4.4.2 Geographical Assignment of Block Sections

The supervised length of a block signal begins at that signal. The end of the supervised length differs:
- Some railways require no block overlaps. Here the supervised length of the signal equals the block section. Examples are mainline railways in North America and in Russia.
- Other railways require an additional portion of track in advance of the section exit signal to be clear before the section entrance signal can be opened (figure 4.43). The difference is called 'overlap' because in that area the control length of a signal overlaps with the control length of the next signal. Block overlaps are used on most European railways, many railways outside Europe and also on almost all subways and subway-like electric city railways worldwide.

The clearing point beyond a signal equals the end of the control length of the signal in rear.

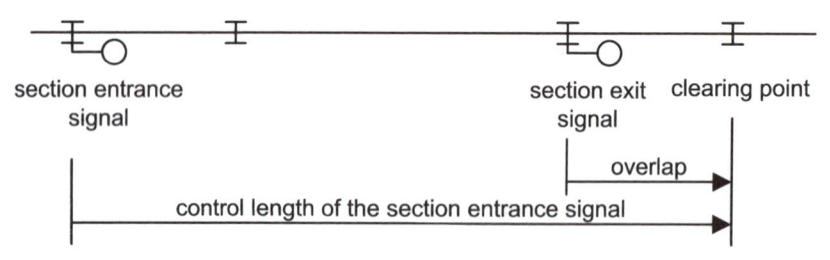

Figure 4.43: Control length of block signals

The main purpose of block overlap, likewise route overlaps (chapter 4.3.6) is to provide additional safety in the case of driver's failure in target braking. Those railways who don't use block overlaps, but do use overlaps for routes in interlocked areas, can argue that in case of an overrun of a block signal, the probability and the severity of a possible accident would be much lower: The accident would be a collision of the overrunning train which moves at low speed with a standing train. Even this case is relatively improbable, unless the length of the standing train equals the length of the block section in advance (figure 4.44).

Figure 4.44: Normal standing position of a train at a block signal

Some railways, for economic reasons, do not detect that the overlap is clear as a separate exercise. Instead, they extend the borders of track clear detection by the length of the overlap beyond the signal (figure 4.45). Then the distance immediately beyond the signal is not super-

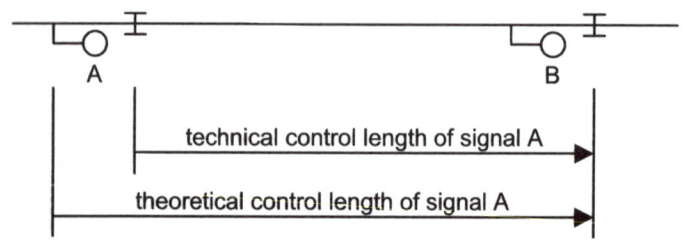

Figure 4.45: Displaced detection borders in block systems

vised technically by this signal. However, the probability that it is occupied when the rest of the block section is clear is very low. And even if it was occupied, no following train can approach it without having entered an occupied block section beforehand. But in the case of splitting trains, this can become safety critical.

4.4.3 Classification of Block Systems

Figure 4.46 shows a basic functional classification of systems for train separation on the open line. As described in chapter 3.4, the principles 'on sight', 'time interval' and 'space interval' can be distinguished. The latter can further be distinguished into movement in fixed block, absolute braking and relative braking distance. The principle of movement in time intervals is almost extinct, absolute braking distance is now used only in a few instances and relative braking distance has no practical importance at all for the railway safety function.

The principle of fixed block distance is the regular principle for train movements and the basis for the block systems described here, whereas movement on sight has to be considered in block systems in the form of permissive driving and in case of technical failure.

The logical principle 'block information' has to be clearly distinguished from the principle 'route' (chapter 4.1.3). Systems using the former can be further classified (figure 4.46):

– **Token block systems.** In token block systems, the presence of the train in a defined block section is permitted by the presence of a particular object, called a 'token' on the train.

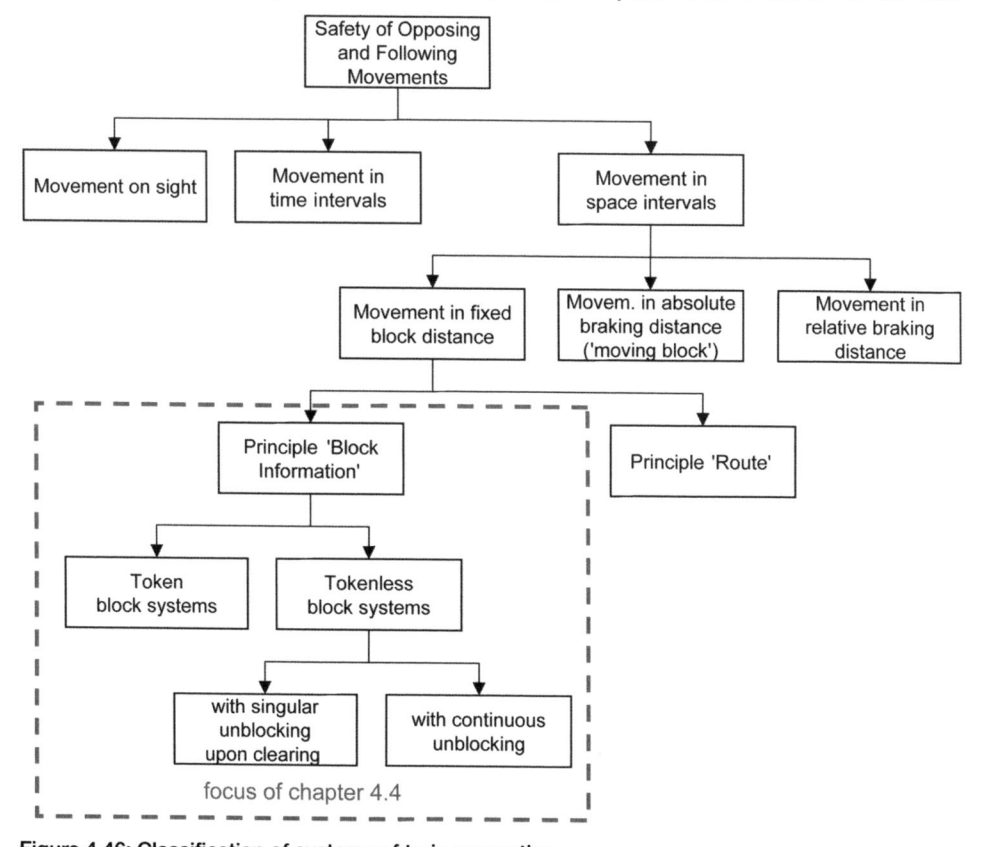

Figure 4.46: Classification of systems of train separation

4 Interlocking Principles

This token can be a physical object, a person (so-called 'pilotman') or virtual information simulating the physical object. It is handed over to the train driver before entering the block section and has to be returned to another train or ground staff after clearing. Each token is assigned to one block section and only one token for the same block section can be in circulation at any one time.

- **Tokenless block systems.** Blocking and unblocking processes are regulated by information exchange between different trackside entities about entering and leaving trains, without a token being employed. When the train has left the block section, the complete clearing of the block section is detected by the receiving block station and the section is unblocked.

Tokenless systems predominate in Europe today. There are two types, as shown in figure 4.46.

- **Block systems with singular unblocking** upon clearing. The unblocking information is transmitted from the exit to the entrance of the block section only once after the train has cleared the (if other conditions are fulfilled). Later unauthorised entry does not normally lead to an occupation, although some systems provide a separate alarm status for this case. Systems of this type may or may not have continuous track clear detection.
- **Block systems with continuous unblocking.** The unblocked status is supervised continuously. Entry of a moving unit into a clear section leads to an occupation of the block. This necessarily requires continuous technical track clear detection and implies a slight increase of safety, considering the (improbable) case of unauthorised entry.

This distinction will play an important role in chapter 10.

Absolute Block:

Permissive Block:

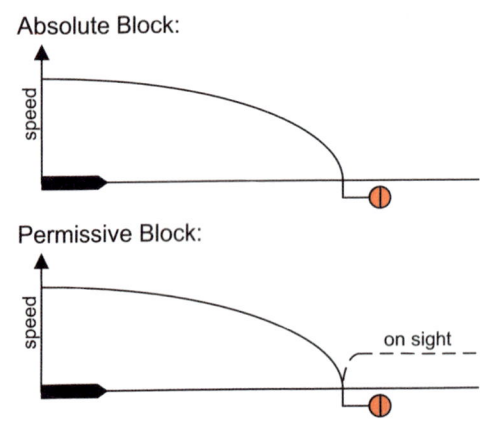

Figure 4.47: Absolute and permissive block

Tokenless automatic block systems can be further divided into **absolute block** systems where (other than during technical disturbance) train operation is always maintained in fixed block intervals, and **permissive block** systems where the most restrictive signal aspect is 'Stop and Proceed' (chapter 7.3.2). This permits any train to enter an occupied section on sight (figure 4.47). Permissive block systems predominate in many countries following North American and British interlocking principles, whereas absolute block systems predominates in countries following German principles (except metropolitan and suburban railways, which mainly use permissive block).

Another criterion to classify block systems according to the management of opposing protection is into those with **neutral direction** and those with **placed direction**. In systems with neutral direction, in normal position (if the line is clear) both train sequence stations have equal rights to send a train, whereas in systems with placed direction, only one train sequence station can send trains at any time. The functioning of these two principles is described in chapter 4.4.5.2 and 4.4.5.3 and illustrated in figure 4.49. Block systems with neutral direction are traditionally more used in Western Europe (British influenced interlocking logic), but partly also in Eastern Europe. Block systems with placed direction are traditionally more used in Central and Eastern Europe (German influenced interlocking logic). But there is no sharp borderline, as in many countries different systems belonging to one or the other categories exist to a different extent and there are also sub-variant of both. In the USA, traditionally opposing protection is excluded from the block system and done centrally by the dispatcher (chapter 4.5.1) in a logic similar to neutral direction.

4.4.4 Process of Block Working in Token Block Systems

In the simplest form, the **one-train-staff system** (table 4.3) only one token exists for each block section, thereby exclusivity of the authority to enter is ensured. A major disadvantage of the one-train-staff system is that trains can only move if the directions of trains alternate. This limits flexibility in operation.

To overcome this disadvantage, in the **train staff and ticket systems**, consecutive movements in the same direction are permitted by the additional rule that a driver is shown the token but may be issued with a ticket (written permission). This allows him to enter the section. In the version with pilotman, the pilotman issues the tickets. Another solution is the use of an overlaying tokenless block system which can only handle following movements.

Both the one-train-staff and the train-staff-and-ticket system follow the principle of placed direction, as the token is present at one or other end of the block section unless it is on a train. A disadvantage of both is that the token can only be moved to the other end of the section by a train and this is the only means of changing the block direction. This reduces flexibility in operation, particularly if the token is present at one end of the block section or on the way there, and unexpectedly a train requests entry from the other end.

To overcome this disadvantage, the **electric token block** (chapter 10.3.2.1) was developed. A stock of tokens is present on each station, but they are electrically interlocked in a way that only one token can be out of the token instruments at a time. Thus in normal operation, each station has equal rights to take out a token (neutral direction). A disadvantage remains that in case of uneven movements all tokens gather at one end, and have to be redistributed by maintenance staff.

A further development is the **radio electronic token block (RETB)** (chapter 10.4.3), where no physical tokens are handed over, but the respective permissions (virtual tokens) are transmitted by radio between the train and a central token processor and displayed in the driver's cab. RETB also works with neutral direction.

Table 4.3 compares the different forms of token block.

	one-train-staff-system	train-staff-and-ticket-system	Electric Token Block	Radio Electronic Token Block (RETB)
nature of the movement authority	physical token (unique per section)	physical token or written permission	physical token (several interlocked)	electronic information
possibility of following trains of same direction	No	Yes	Yes	Yes
possibility of unequal number of trains for both directions	No	Yes	with additional maintenance efforts	Yes
flexibility regarding train sequence	No	No	Yes	Yes

Table 4.3: Comparison of different forms of token block

4 Interlocking Principles

4.4.5 Process of Block Working in Tokenless Block Systems

4.4.5.1 Following Protection

First the simple case is described, where the block system is only responsible for following protection. This is applied on double lines with unidirectional signalling. The sequence of actions is as follows:

- When the train enters the section, the **line is blocked**, which means that it is reserved for that train. The blocking message can either be transmitted from the entrance of the block section to the exit and used there as a precondition for unblocking. Or it can be stored in the entrance only (applied in some systems with continuous unblocking, especially in North America). In both cases, the blocking prevents another train from entering the section by locking the signal in the entrance. The safety problem to prevent is that if the system fails to block the line, it will cause danger by permitting a second train to enter.
- After the train has cleared the block section, the block system is **unblocked** by the receiving end. Several preconditions have to be checked to permit unblocking (chapter 4.4.6). Then the block system returns to normal position. In systems with singular unblocking, the unblocking message is transmitted once upon arrival of the train and checking of further criteria. In systems with continuous unblocking, it is transmitted continuously as long as the block section is clear and other criterions fulfilled (chapters 4.4.3 and 10.1).

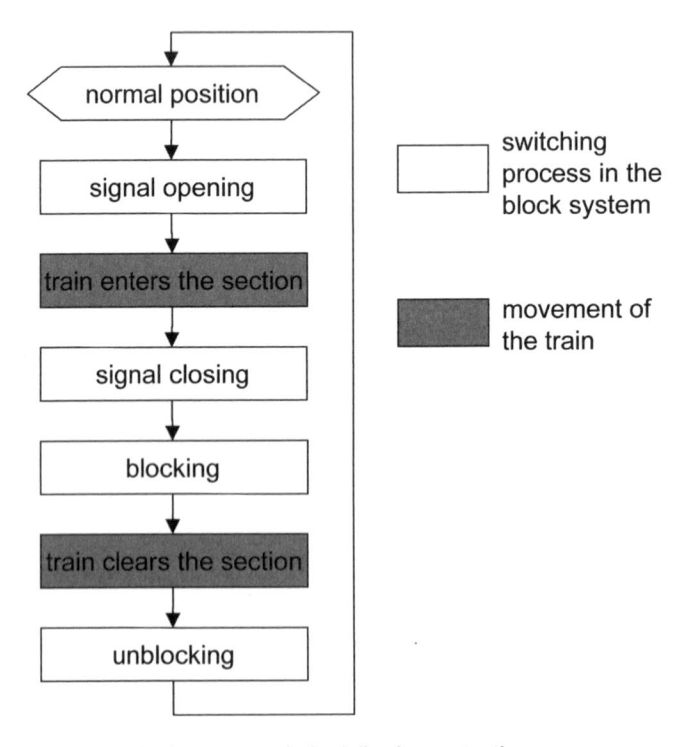

Figure 4.48: Sequence in a block system only for following protection

4.4 Block Dependences

4.4.5.2 Systems with Neutral Direction

In the following, the sequence of actions in tokenless block systems for traffic in both directions is described. Systems can be distinguished into those with neutral and with placed direction. In systems with neutral direction, the sequence of actions is basically the following (figure 4.49):

- In normal position, no train can enter the train sequence section.
- **Acceptance** of a train: The main function of acceptance is that only trains running in the same direction can be within one train sequence section at the same time. Each train has to be offered by the sending train sequence station and, if certain preconditions are fulfilled, accepted by the receiving one.
- After acceptance, the section between two train sequence stations is ready to be entered by one train. In some systems, the acceptance can be cancelled as long as no train has entered.
- While one or several trains are traversing the train sequence section, the blocking and unblocking actions are performed between the block stations as described in chapter 4.4.5.1. New acceptance actions in the block system can be necessary or not for each individual train, depending on the system.

4.4.5.3 Systems with Placed Direction

In block systems with placed direction, the sequence of actions is principally the following (figure 4.49):

- In normal position, one train sequence station holds the **direction permission** and is therefore able to send trains; the other is not.
- In some systems, the train sequence station which wants to send a train onto the train sequence section has to offer the train telephonically to the receiving train sequence station and wait for acceptance. In regular operation, this is not safety-critical, but serves for operational purposes.
- Only if the train sequence section is clear can the **direction** be **reversed** in normal operation. Either the receiving train sequence station gives it or the sending one takes it, depending on the arrangements of the system.

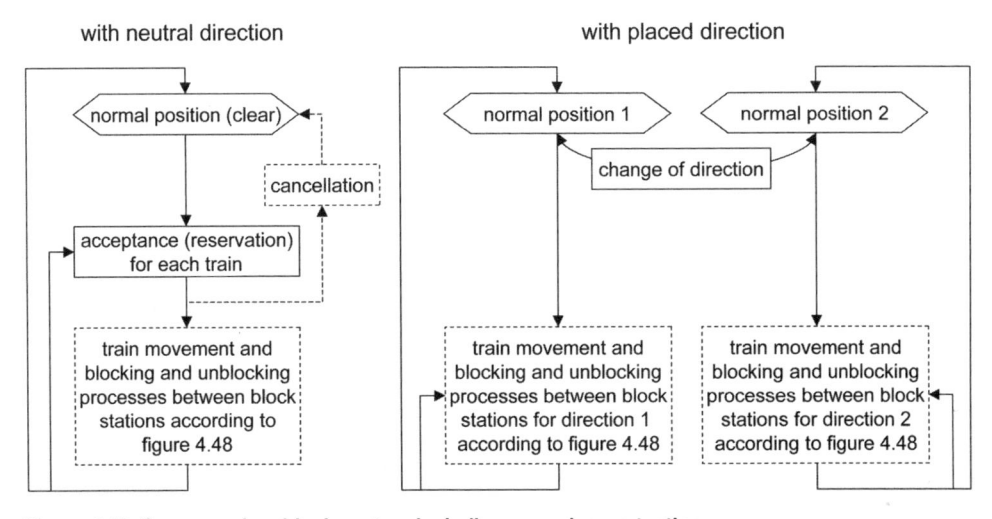

Figure 4.49: Sequence in a block system including opposing protection

4 Interlocking Principles

- Following movements are regulated between neighbouring block stations as described in chapter 4.4.5.1. Several trains can traverse the train sequence section without special technical acceptance of each individual train.

In several cases, automatic changing of the direction can be provided in systems with placed direction. A typical example on double lines is to give the direction permission for contraflow traffic for one train only and then shift it back automatically to the other direction as soon as the train sequence section is clear.

4.4.6 Locking Functions of Tokenless Block Systems

To achieve the block conditions described in chapter 3.4.1, several locking functions are used in the block systems. While token block systems imply only simple interlocking functions and rely more on the correct behaviour of people involved, tokenless systems imply more complex functions, described here.

The technical interlocking functions between the block instruments and the signals differ between the systems. In the earliest block systems in Britain with neutral direction, there was no technical interlocking and the signaller was fully responsible for the correct order of actions. In modern systems, safety is guaranteed by full technical interlocking. In the group of systems with continuous unblocking, some are controlled directly by the track clear detection, whereas others use overlaying interlocking functions (chapter 10.3.4).

Among those interlocking functions related to the sending of a train onto a section, the following are the most important:

- Locking of the signals of a train sequence station leading onto the train sequence section in stop position if the direction permission is not present at the respective station (systems with placed direction) or the train has not been accepted (systems with neutral direction).
- Locking of the signals of a block station in closed position if the block section ahead is already occupied.

The other group of locking functions in the block system prevents the unblocking of the line in certain circumstances. The most important of these functions are:

- Before unblocking the line, it must have been occupied. This is checked in most block systems, but not in all systems with continuous unblocking.
- Before unblocking the line, the respective section including the overlap at the receiving station (figure 4.43) must have been cleared by the train. To obtain this information, there are different possibilities:
  - With continuous track clear detection, this information can be obtained directly from the detection system.
  - With track clear detection in the areas of the receiving block station, but not continuously on the open line, the consecutive occupation and clearing of two or more sections can be evaluated for this purpose (figure 4.50). Any wagons which become detached from the train will not be detected, so staff are responsible for observing the end of train marker, or it has to be detected by additional technical measures (chapter 5.2.7).
  - Without any technical track clear detection, a point at or beyond the end of the overlap has to be proved to have been occupied and cleared in this sequence. A frequently used technical solution in countries influenced by German interlocking principles is a combination of a short track circuit and a wheel detector (figure 4.50). Also here, the end of train has to be observed by staff.

4.4 Block Dependences

- Another solution with less effort is the British berth track circuit. Technical detection was not applied continuously, but only on several metres in rear of the signal (figure 4.50). This gives relatively good protection against the most frequent case of inappropriate unblocking when a train is standing at the home signal. However, the berth track circuit gives no protection if the train is still moving inside the block section. Berth track circuits are also used in station tracks and prevent the signaller from clearing a signal into a track on which the berth track circuit of the next signal is occupied.
- Before unblocking the line, it has to be ensured that the block signal of the receiving station has reverted to stop after the train has passed to always protect the train by a signal at stop in rear. Different strategies are applied to achieve this (chapter 4.5.2).

By consecutive track circuits:

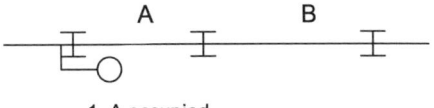

1. A occupied
2. B occupied
3. A cleared (safe)
4. B cleared (safe)

By combination of short track circuit and rail contact:

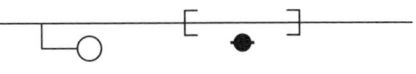

1. Track circuit occupied
2. Rail contact passed (safe)
3. Track circuit cleared (safe)

With berth track circuit:

Figure 4.50: Line clearing conditions

In systems with continuous unblocking (chapters 4.4.3 and 10.1), the unblocking conditions are supervised continuously. This means that if an unblocking condition, usually the clear status of the block section, becomes violated, the entrance block signal is immediately restored to stop. In systems with singular unblocking, this can optionally be achieved by an additional alert status in case of unauthorised entry into the block section. However, in most of these cases the reason for a block signal at switching to stop is a technical defect, e.g. of signal lamps or track circuits. Railways differ in whether in such situations the block signal will be cleared again, automatically, if the unblocking conditions are fulfilled again, or if it can be cleared only by special command, implying frequent working in degraded mode operation (chapter 4.5.4).

4.4.7 Returning Movements

In some situations, there is a need for a movement not to pass through the block section completely, but to change direction inside the block section and return to whence it came (figure 4.51). Reasons, among others, can be:
- An industrial track branching from the open line has to be serviced
- A train with a technical defect has to move back or be hauled back by another traction vehicle
- An engineering train enters the line and leaves it in the same direction

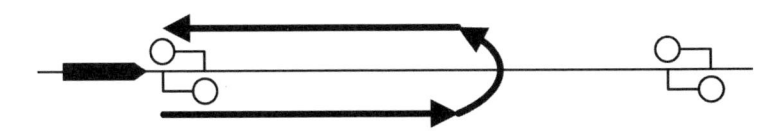

Figure 4.51: Returning movement in a block section

4 Interlocking Principles

To handle these cases, there are different possibilities, of which these are some:
- The block section is closed for traffic and the movement is permitted by methods of degraded mode operation (chapter 4.5.4). However, for regular movements, this is not a realistic solution.
- An extra train sequence station is defined within the section. This is only possible if trains reverse regularly at exactly the same place and requires additional efforts.
- Auxiliary reversing of the block direction is provided. This enables normal operation of the blocking and unblocking functions. The location is variable. However, this option has to be provided additionally in the block logic and is possible only in some block systems.
- A special kind of permission is provided for the train to enter the block section and to reverse. This can be issued to the train by an additional signal aspect or by a token. As long as the train has the permission, the line is blocked for other trains.

Concerning the particular case of industrial tracks to be operated via the open line, locking of movable track elements is not a genuine function of block systems (chapter 4.1.3), but has to be built in additionally. Different solutions are possible. The most important are (figure 4.52):

1. The points are locked by a key lock. In normal operation, the key is locked in one of the neighbouring train sequence stations and can only be unlocked by closing the line section to normal operation. The driver who requires access to the industrial track obtains the key from the signaller and uses it to switch the points. After returning to the train sequence station, the key is locked again and the block becomes operable. The disadvantage is that as long as the industrial track is being served, the open line track cannot be used by any other train.
2. In addition to the above solution, the driver has the possibility to lock the key in a key lock connected with the block system while serving the industrial track. This becomes possible after returning the points and the flank protection device to normal position and enables other trains to use the line.
3. For trains moving along the line, the industrial connection is safeguarded like a normal junction with interlocked signals. Moving units serving the industrial connection can open the points by a key under certain restriction. Obtaining this key locks the signals in stop position. While working on the industrial tracks, the key can be locked into the key lock.

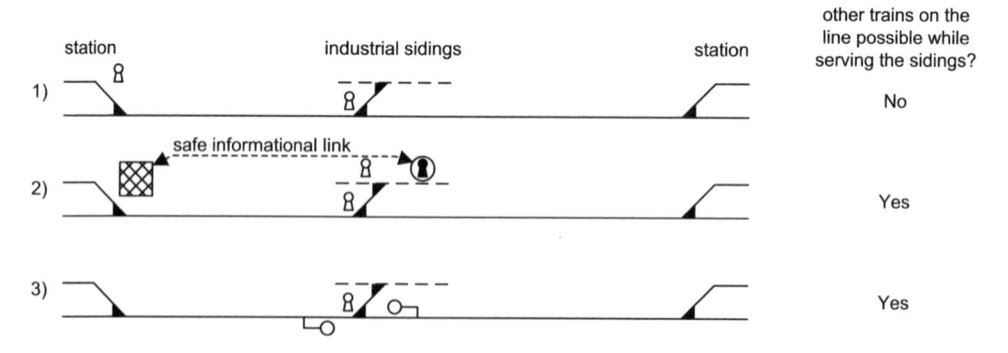

Figure 4.52: Industrial sidings served from an open line section

4.5 Special Issues

In the following, some particular problems which refer to both principles, 'route' and 'block information', are discussed.

4.5.1 Overlaying Block and Route Interlocking Systems

For the treatment of the block system in station areas with more than one consecutive signal, there are two basic possibilities (figure 4.53):

- The station disrupts the block system. This means that for the block system, the station is considered like one block station, whereas the station itself is safeguarded by route interlocking only. Opposing protection is integrated into the block system. This principle is connected with a sharp distinction between stations ('Bahnhof', chapter 3.2.3) and open line sections in the operational rules and is therefore applied mainly in countries influenced by German operational and interlocking principles (e. g. Central and Eastern Europe).
- The block system of the open line runs through the station and provides track clear detection functions, whereas the route interlocking system locks the movable track elements. This principle dominates in North America and some European countries. The block system is either only applied to the straight tracks in the station, or the block section has more than one entrance or exit signal. The movement authority in station areas is the sum of at least two different permissions, one from the route interlocking and one from the block system. Whereas in modern systems, all permissions are usually combined into the same signals, in older systems each permission is often separately given to the driver (by signals or verbally) and the driver has to obey the most restrictive. French mechanical signals, for example, provide the 'Carré' signal with absolute stop for route interlocking and the 'Semaphore' signal with permissive stop for the block (chapter 7.3.2). Until now, this has been represented by two different stop aspects. The driver is allowed to move permissively in a station if the route is locked, but tracks are occupied.

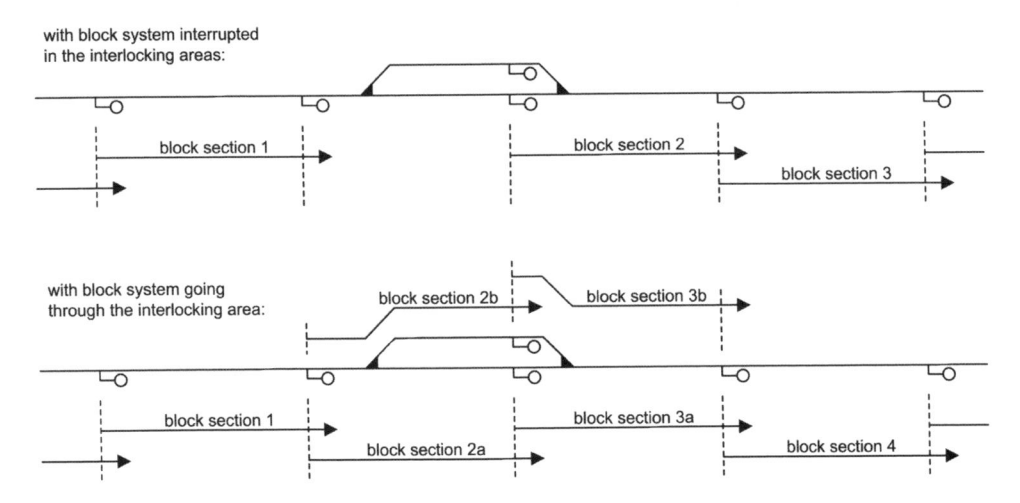

Figure 4.53: Line block in interlocking areas

4 Interlocking Principles

4.5.2 Protection of Trains by a Signal at Stop in Rear

A   B
(remained open erroneously)

Figure 4.54: Dangerous situation caused by signal remaining open erroneously

A basic principle of the protection of following movements is that a train always has to be protected by a stop signal in rear. In this sense, danger can occur when a signal remains in the clear position behind the train erroneously, while another train follows (figure 4.54). To prevent this situation, the following strategies were developed and are applied additionally to the normal route and block conditions:

1. Signal B turns to stop safely before the rear end of the train has cleared the end of the supervised section of signal A. This solution requires additional technical efforts, e.g. the use of type N relays (chapter 9.3.2.1) in relay technology or closing of the signal by redundant information. It is widely applied in route locking systems with clearing the signal before irreversible route locking (chapter 4.3.8.2) and in block systems with continuous unblocking (chapter 4.4.3).
2. If safe placing of the signal at stop is not provided, logical solutions in the sequence of clearing and the placing at stop of consecutive signals have to be defined. They are particularly important for railways influenced by German interlocking principles. Variants are:
   a. Two consecutive main signals must never show a proceed aspect at the same time. An important disadvantage of this strategy is that in case of short block sections, the driver is always shown caution aspects, which makes free running traffic impossible. However, this strategy is applied in station tracks where no through-routes are required.
   b. Signal A can only be cleared when signal B is in Stop position. However, this does not prevent signal B from clearing while signal A shows a proceed aspect (unidirectional dependence, see chapter 4.2.3). This solution requires a strict sequence of operation input commands and is therefore difficult in practical terms, but it is used in several applications.
   c. Signal A can only be cleared if since the last clearing of signal A, signal B has some time shown a stop aspect. This is a widely used strategy for modern route interlocking and block systems following German principles. It is operationally flexible, but requires more effort than solutions a and b.

4.5.3 Several Trains between two Signals

In the following situations (and others), two trains can be situated in the same block section or route at the same time:
- Permissive block systems (chapter 7.3.2), which means that the driver is allowed to pass signals at stop, on sight, at his own responsibility
- A train has been authorised to pass a signal at stop in the case of technical defect
- A train has been authorised to enter an occupied section on sight on a regular basis
- A train has been regularly split into two trains in a station track

Dangerous cases can occur where the driver of a train which moves in an occupied section sees the 'Proceed' signal in front of him, but not the train between him and the signal. An example where this can occur frequently is desert areas if the signal can be seen over several

4.5 Special Issues

kilometres and the first train consists of empty flat wagons. To prevent this danger, in permissive driving the driver is obliged to drive on sight until the front end of the train reaches the next signal.

But even when moving on sight through the whole block section, the following dangerous situation can occur (figure 4.55): The second train approaches signal B, which has remained at proceed after the first train has passed it. The driver sees the proceed signal and accelerates, thinking the block section ahead is clear. This can occur in the following situations:

- The point where the signal regularly goes to stop is planned far enough beyond the signal to accommodate a whole train between the (still clear) signal and this point, or if the same is effected by time delays in the signal release. This can be prevented by the obligation to proceed on sight even a defined distance beyond the proceed signal (e. g. 400 m in Germany).

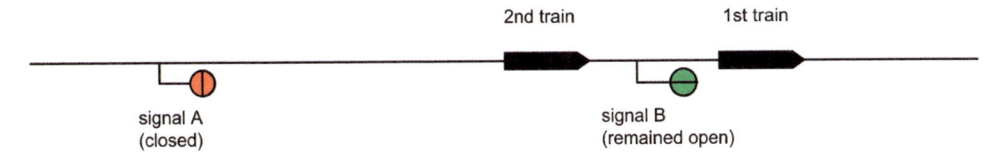

Figure 4.55: Dangerous situation caused by two trains in the same block section in permissive block systems

- The signal has remained at proceed due to technical failure. This situation can only occur in systems without safe putting to stop of the signal (see chapter 4.5.2). This can be prevented by the obligation to proceed on sight always over two block sections (figure 4.56).

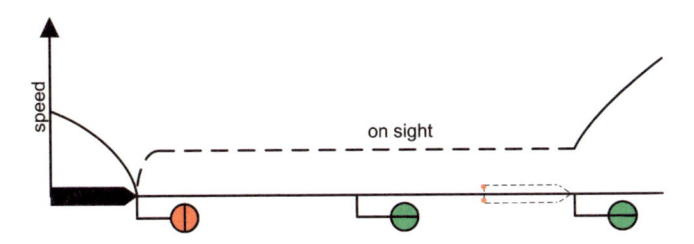

Figure 4.56: Obligation to drive on sight over two block sections in permissive block

4.5.4 Degraded Mode Operation

4.5.4.1 Purpose

In this section, methods are described to maintain railway operation in case the normal interlocking functions are not applicable. Reasons can be the following:

- **Technical failure** in the signalling equipment which causes a fail-safe reaction: The failure can be in obtaining information about track occupation and positions of movable track elements, in the evaluation unit or in giving out commands to the train (signal failure). Operation has to be maintained by bypassing the normal interlocking functions.
- Special **operational situations** where the standard functions of the interlocking system are not applicable: Economic considerations mean that a technical system cannot reasonably be designed for any possible situation. Therefore, situations which occur only infrequently often have to be handled manually by overriding the technical system. The most frequent

4 Interlocking Principles

case of such situation is an exceptional movement on a path where no route is provided or, in a block system, where no technical block is provided for this direction.
- A train has to **reverse**, unplanned. Examples can be in case of technical failure of the train or in emergency situations such as fires.
- **Operational reasons** which require the cancellation of an already irreversibly locked route without a train having traversed it. As this causes potentially dangerous situations, it needs to be avoided. The restrictions under which emergency route release for these purposes is permitted are defined differently between the railways. Railways using the registration method (see chapter 4.5.4.2) for emergency commands tend to issue stricter preconditions than railways using the time delay method.

For routes, the most important actions in degraded mode operation can be summarised into the two groups of emergency release of routes or parts of routes (chapter 4.5.4.3) and authorizing movements by bypassing interlocking functions (chapter 4.5.4.4). Degraded mode operation for block systems is described in chapter 4.5.4.5.

4.5.4.2 Safety in Degraded Mode Operation

In most cases of degraded mode operation, a person gets safety responsibility which is on the technical system in normal operation. Due to the higher human error rate in comparison with technical systems, the level of safety is generally lower. Several modern systems try to limit this loss of safety by leaving as much of safety responsibility as possible at the technical system also in degraded mode operation. For example, the signaller gets safety responsibility only for that route element which is affected, but not the whole route. Another strategy is to involve several persons in safety related actions, e.g. by the necessity to authorise a special command by a second person.

Safety responsibility in degraded mode operation can be given to one of two groups of people, or shared between them:
- **Train staff** (e.g. drivers)
- **Ground staff** (e.g. signallers)

Giving safety responsibility to the driver enables a high grade of centralisation in operation, whereas giving safety responsibility to the signaller causes difficulties in this respect. As some preconditions can only be proved locally, additional communication between a centrally located signaller and local ground staff or train drivers is necessary.

The railways differ in their preferences for giving this additional safety responsibility to the one or the other group of staff. In countries following German interlocking principles, it is preferably given to ground staff, whereas railways influenced by British or North American principles tend more to the train staff.

The issue of safety responsibility of ground staff is connected with the safety of display and command functions of the man-machine-interface (MMI): The more safety the MMI can guarantee, the greater the opportunities for staff to take safety responsibility.

4.5.4.3 Auxiliary Route Release

If auxiliary release of a route becomes necessary for the reasons described in chapter 4.5.4.1, the first step is always to return the signal to Stop. It must be possible to do this immediately at any time, to prevent any danger which the signaller has perhaps recognised. After releasing a signal of an irreversibly locked route without a train having passed it, normal route release by the traversing train is not possible and the route has to be released manually.

4.5 Special Issues

This is a potentially dangerous action, as a train which has already entered the route or is positioned shortly before the route entrance signal without the opportunity to stop at the signal might be endangered by traversing unsafe tracks. Two different solutions to increase safety have been developed.

- **Time delayed route release** is used basically in those countries following British and North American principles (chapter 3.1) including Russia, France and most countries in Southern Europe and Scandinavia. A time delay determined by factors such as route length, permitted speed and sighting distance of the signals is applied between the operation action and actual route release to give the train enough time to leave the route or come to a halt. A special form is the inhibition of any action in the respective interlocking area during the time period. The time value can vary by country, interlocking type, type of route, local situations and the status of the route (train already entered or not). Typical values for train routes are around 2-3 min, for shunting routes shorter. If the train occupies the route during this time, the route continues to be locked. The time delay method implies the passing of a part of the safety responsibility to the driver: If the train comes to a halt or moves unusually slowly, the driver must not rely on his still having a safe route, as it might have been released in the meantime.
- **Recording** of the safety critical action is used in most of those countries following German operational principles, mainly in Central and Eastern Europe. The idea of this method, where the signaller gets full safety responsibility, is to encourage the signaller to think further about the action by forcing him to justify it, and thus to reduce the probability of human error. Check lists are often provided to the signaller as to which circumstances have to be proved before undertaking the safety critical action.

The following differences between the methods can be pointed out:
- In the time delay method, other trains are held up until the time delay has elapsed, which reduces capacity and affects punctuality. In the recording method, these effects are much less, determined only by the additional time which is needed to prove the conditions and carry out the safety critical actions.
- The approach of the time delay method is to eliminate harmful consequences of human error (chapter 2.2.1.2). In contrast, the idea of the registration method is to increase discipline and therefore to reduce the probability of human errors (chapter 2.2.1.1).

In some interlocking systems, especially modern electronic systems, both the time delay and the recording methods are applied in parallel or as alternatives. The auxiliary release can be made either by the signaller or by the train driver after authorisation by the signaller.

After auxiliary release, depending on the interlocking system, the equipment can go directly to normal condition, which means that other routes can be set normally. Alternatively, there are certain restrictions which require the next train(s) to proceed on sight.

4.5.4.4 Bypassing Defective Elements when Setting a Route

A second important situation which requires actions in degraded mode operation in route interlocking areas is that preconditions for clearing the signal (such as elements in the correct position and tracks clear) are fulfilled, but this cannot be proven technically due to a technical defect. As the normal signal cannot be cleared due to missing preconditions, other methods have to be used to authorise the movement. These might be a written instruction or an auxiliary signal (chapter 7.4).

Additional safety responsibility can be given to the signaller or to the driver, or shared between both of them (chapter 4.5.4.2):
- Giving safety responsibility to the driver means that the affected sections have to be passed

4 Interlocking Principles

on sight to prove the track clear. Under certain circumstances, even checking of position of movable track elements can be delegated to the driver.
- Giving safety responsibility to the signaller means that the signaller, by authorising the movement, guarantees a safe path. Beforehand, the signaller has to prove the status of the track sections and to lock movable track elements by manual locks. In most railways, these actions have to be recorded. The method requires the signaller to be present on site and is therefore only possible with decentralised operation.
- Mixed solutions mean that the signaller obtains information about the safety preconditions from train drivers or other staff who are on site and then makes the decision to authorise the movement. It is used mainly when due to the centralisation of operational control the signaller is not on the site of his control.

To enable these different solutions, different types of auxiliary signals can be defined. The exact regulations differ among the railways, but the two basic groups are the one which requires movement on sight and the other which does not. As the possibilities of displaying exact speed restrictions by the auxiliary signal are limited, for the point zone the most restrictive of all possible speed restrictions has to be obeyed by the signaller. The equipment of stations with auxiliary signals and with the one or the other type of it varies between the railways and between locations.

In older technology and in several railways until now, staff take full safety responsibility for the whole route when using the auxiliary signal. However, the tendency in modern interlocking systems is to provide as much safety as possible by using reduced interlocking functions even when using the auxiliary signal to minimise the probability of creating danger by human error. It can be achieved by particular interlocking functions which prove all conditions which can be proved technically, giving safety responsibility to the signaller only for the defective element (figure 4.57). The fact to be checked by the signaller can be, for example, the position of one movable track element or the occupation of one track section.

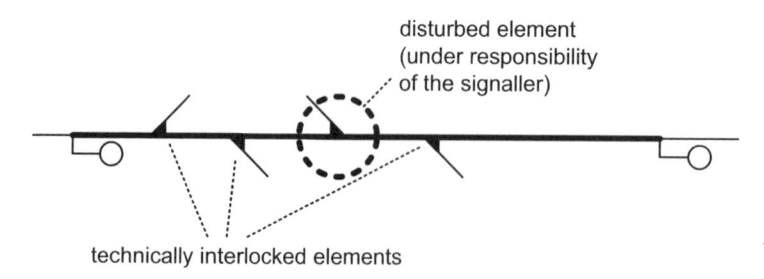

Figure 4.57: Partial interlocking in case of failure of single elements

4.5.4.5 Degraded Mode Operation in Block Systems

Reasons preventing the block system from working normally can be:
- Failure of the block system or problems originating from the adjacent interlocking areas
- Special operational situations, such as a train not traversing the block section completely

Different methods are applied to maintain railway operation on the affected line:
- **Permissive driving** (responsibility of train staff). The driver is permitted to pass any signal at stop which only protects following movements, not only in cases of failure, but in his own responsibility and on sight. Safety is ensured by the fact that the only possible dangerous situation is a train in the same direction in the block section, which can be solved by driv-

ing on sight. Permissive driving is often applied in automatic block systems with continuous unblocking (chapter 10.3.4). It is a favourable solution where no local ground staff is present due to high centralisation of operation control (modern systems) or where communication links have low reliability (older systems). Often, instead of permissive driving, the permission to pass a signal at Stop is bound to more or less restrictive preconditions to avoid entering a track which really is occupied (chapter 7.3.2).
- **Auxiliary authorisation by ground staff.** Safety functions of the block system can be bypassed in case of failure and replaced by manual actions. The exact procedures differ among the railways and block systems. Examples are auxiliary unblocking of a block section from the exit block station if the section has not unblocked in the normal way, and auxiliary change of direction in a falsely occupied train sequence section.
- In the case of complete technical failure (and often combined with the above methods), non-technical measures can be introduced. This can be the telephone block (chapter 4.4.1.3) with permission to the driver given by written instructions or auxiliary signals, or token block, e.g. with a paper with the name of the section written on it as a token.

Advantages and disadvantages are listed in table 4.4 and analysed more detailed in (Pachl 2000–I). The railways differ in their preferences towards the one or the other class of methods.

	permissive driving	auxiliary authorisation by ground staff
special requirements in system design	no special requirements	increased requirements for communication link and presence of ground staff
line capacity	high	reduced
train speed	sharply decreased	slightly decreased
possibilities of accidents in case of human error	too high speed (driver's error)	inappropriate authorization (signaller's error)
severity of accidents in case of human error	lower (low speed)	higher (normal speed)

Table 4.4: Comparison of degraded mode operation in block systems

Besides the above solutions, to avoid degraded mode operation, some systems offer the possibility of excluding a single block station from block working and connecting two adjacent block sections to work as one (figure 4.58). The signal between the two is switched out. Loss in line capacity is a result.

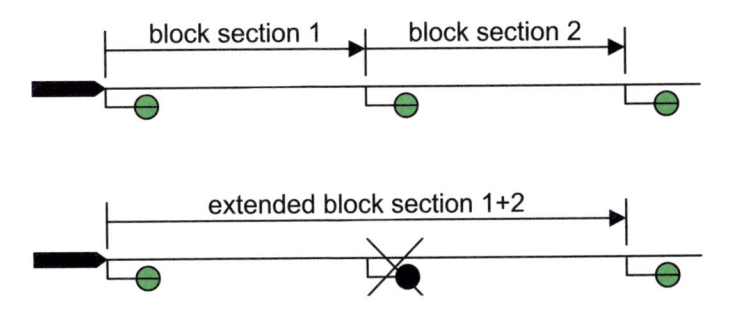

Figure 4.58: Extended block sections in case of signal failure

4.5.4.6 Technical Failure of Short Duration

Failures considered here are technical problems of short duration which lead to fail-safe reactions in the interlocking or block system (signal to Stop), but remove themselves without action by a person after few seconds. Typical examples are loss of track clear detection (particularly in track circuits) or of supervision of point positions for short duration. There are different possibilities to deal with these problems:

- The problem leads to a failure state of the interlocking/block system which has to be solved by degraded mode operation. A disadvantage is that, especially for frequently occurring failures, this can lead to frequent use of degraded mode operation with lower levels of safety and reduced speed or capacity. But it is an appropriate solution in highly available systems.
- The reaction to the problem comes only after a defined time. The disadvantage is that in case of a really safety critical occurrence (e. g. actual occupation by a flank movement), the danger increases. Therefore, this is no appropriate solution in most situations.
- If the problem doesn't endure longer than a defined time, the failure status can be 'healed', which means that the signal returns to 'Proceed' after the technical problem has ceased, and the previous status of the route or block system is restored.

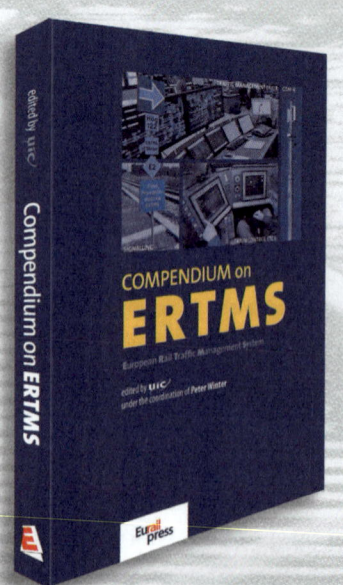

5 Detection
Gregor Theeg, Sergej Vlasenko

5.1 Requirements and Methods of Detection

5.1.1 Introduction

The general purpose of detection is to gain information about the positions of movable objects on or about the railway network. Detection systems include the reception of that information, its transmission and its evaluation. The contents of chapter 5 are as follows:
- The following parts of chapter 5.1 will contain some theoretical considerations on detection.
- Chapter 5.2 describes the most important technical principles for detection of objects.
- In chapters 5.3 to 5.5, the two most important solutions for track clear detection, which are track circuits and axle counters, are described and compared.

5.1.2 Types of Objects

The following groups of objects can be distinguished:
- **Railway vehicles** include all those which are assigned to the railway system, irrespective of type. Rail guidance facilitates their detection.
- **Moving units on crossings** are people or objects not associated with the railway but are temporarily occupying jointly used areas. This mainly refers to road users (persons, vehicles, animals) on level crossings.
- **Obstacles** are all other objects including persons and road vehicles on railway land which is not assigned for common use.

Moving units on crossings and obstacles can also be summarised as **external objects**. The main focus of detection in railway systems concentrates on railway vehicles, but in many situations external objects also become relevant.

When considering the detection of rail vehicles, the following particular information can be relevant (figure 5.1):
1. A **moving unit** (train or shunting movement) **has reached a certain point** with its front end.
2. A **moving unit has passed a certain point** with its current rear end. This does not necessarily imply that the moving unit is complete, so no information on 'lost' wagons is given.
3. A particular track section is **clear** of vehicles.

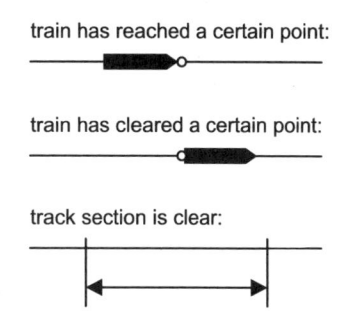

Figure 5.1: Information content of detection of rail vehicles

5.1.3 Safety Requirements

Important factors in determining the design and the technical efforts of a detection system are the safety requirements. Two groups can be distinguished:
- **Safety related** purposes are those whose failure can cause accidents, either alone or in combination with other technical failures or human mistakes. The requirements for component reliability are very high to ensure safe working in all possible conditions such as dif-

ferent speeds of trains, different weather conditions etc. The safety related purposes can be further distinguished in terms of whether the erroneous non-detection or the erroneous detection at the wrong time can cause danger.
- **Not safety related** purposes are those where errors can result in disadvantages such as delays, wrong information to passengers and economic losses, but cannot cost human lives or damage equipment.

5.1.4 Detection Purposes

There is a large variety of possible detection purposes, the most important of them being classified in the following by the criteria of chapters 5.1.2 and 5.1.3.

5.1.4.1 Moving Unit Reaching a Certain Point

If a train reaching a certain point with its front end is to be detected from the trackside, spot and short linear detectors are suitable (chapters 5.2.2 and 5.2.3). The most important purposes are:
- Closing of a level crossing to road vehicles when a train approaches. If no feedback information about the proper closure is given before permitting the train movement over the level crossing (chapter 13.4.4.1), this detection is safety related with the non-detection causing danger. Otherwise it is not.
- Restoring a signal to danger after the train has passed it. This is safety related with the non-detection causing danger if the proper functioning of the signal is not checked before clearing the signal in rear (chapter 4.5.2) or in particular situations such as after splitting a train into two. Otherwise it is not safety related.
- Irreversible route locking when a train approaches. If the signal has already been cleared before irreversible locking (e.g. British logic), this detection is safety related with the non-detection causing danger, otherwise (e.g. German logic) it is not, see chapter 4.3.8.
- Approach control of signals (e.g. British junction signalling) (chapter 7.5.3). In most cases, this detection is safety related with the erroneous detection at a wrong time causing danger (too early upgrading of signal aspect). A frequent North American item is approach lighting: For lamp conservation, the signal becomes illuminated only when a train is approaching.
- Control of dynamic passenger information, a not safety related automation function.
- Support of operational decisions, e.g. whether to wait for a delayed train for passenger interchange or not (not safety related).
- Announcement of an approaching train to the signaller for preparation of route setting (not safety related).

5.1.4.2 Moving Unit Passing a Certain Point with its Rear End

The information that a particular moving unit has passed a certain point with its current rear end contains two messages and their respective status to be monitored:
1. The moving unit has reached that point.
2. The moving unit has cleared that same point afterwards.

In the following, examples for this detection purpose are listed, all of which are safety related with erroneous detection at the wrong time causing danger:
- Opening a level crossing when the train has cleared it.
- Releasing a route after the train has cleared it.
- Unblocking the block after the train has cleared the section.

5.1 Requirements and Methods of Detection

For track based detection with this purpose, one suitable solution is a combination of a spot detector (chapter 5.2.2) (which detects the train reaching that position safely) with a short linear detector (chapter 5.2.3) (which must be longer than the longest permitted distance between axles and be able to detect the clearing safely). Another solution is an inductive detector for the wagon body (chapter 5.2.4.3), if it can detect both safely.

5.1.4.3 Track Clear Detection

The main purposes of track clear detection, often also called track occupancy detection, are to prove the following:
- The clearance profile of a track has to be confirmed before permitting a train movement.
- Moveable track elements such as pointwork have to be clear of rail vehicles before being switched.

Track clear detection is always safety related. The safety philosophies of railways differ as regard to whether or not external objects have to be included in track clear detection. As their detection (chapters 5.2.4 and 5.2.5) is technically much more difficult than the detection of rail vehicles, the presence of external objects is often excluded by just forbidding it, which implies a certain loss of safety. However, in some situations, detection of external objects is required. Definitions vary, but generally the most important situations are:
- The probability of the presence of external objects is high. Examples are platform tracks with high passenger volume on the platform or mountainous areas with a high risk of avalanches.
- The severity of possible accidents is high. Therefore, high speed lines are more often equipped with obstacle detection than conventional lines.
- The chances that in case of a hazard an accident can be prevented by the intervention of a person are low. The unlikelihood of a driver being able to stop at an obstacle on a high speed line is the second reason for their preferred use there. In automatically driven systems, detection of external objects is mandatory. For level crossings with roads, some railways require the detection of road users in the case of full barrier closure, but not in other cases (chapter 13.4.4.4).

If continuously applied, track clear detection devices for rail vehicles can also provide for the detection of trains reaching and passing a certain point (chapters 5.1.4.1 and 5.1.4.2). Therefore, particular detectors for these purposes often become obsolete if continuous track clear detection is applied.

Objects can be detected either directly or indirectly. Track circuits (chapter 5.3) are the most used technology for direct detection of rail vehicles. In indirect detection, the presence of an object is concluded from its entering and leaving the area concerned. Axle counters (chapter 5.4) are the most used technology here.

As an alternative to track clear detection, **exclusion of conflicts** can be mentioned in this context. Examples are:
- One-train-operation (exclusion of conflicts with other rail vehicles).
- Construction of the line without level crossings (exclusion of conflicts with moving units of crossing ways).
- Fencing-off the railway line including overbridges and other structures (exclusion of conflicts with obstacles), though this is never likely to be wholly successful.

5 Detection

5.2 Technical Means of Detection

5.2.1 Classification

Detection systems can be classified according to different criteria. For detectors for rail vehicles, one important criterion is the specific object of detection. This object can be (figure 5.2):
- **Axles** or **wheels** of the rail vehicle.
- The **body** of the rail vehicle.
- Other passive parts of the rail vehicle. An example is the pantograph of a tram.
- Particular **active communication devices** on the train. Systems using these can be further distinguished into systems which know an ID number or other individual identification of the train and systems where the trains move anonymously in the network.

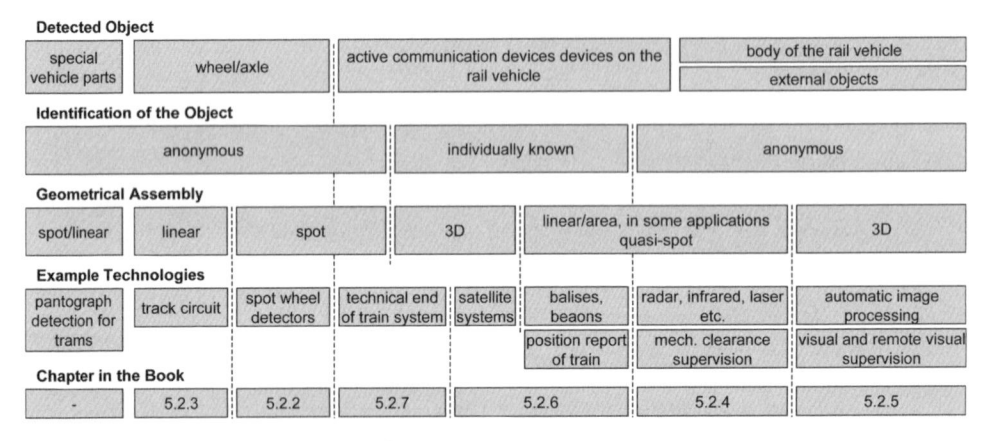

Detected Object							
special vehicle parts	wheel/axle		active communication devices devices on the rail vehicle		body of the rail vehicle		
					external objects		
Identification of the Object							
anonymous			individually known		anonymous		
Geometrical Assembly							
spot/linear	linear	spot	3D	linear/area, in some applications quasi-spot			3D
Example Technologies							
pantograph detection for trams	track circuit	spot wheel detectors	technical end of train system	satellite systems	balises, beacons	radar, infrared, laser etc.	automatic image processing
					position report of train	mech. clearance supervision	visual and remote visual supervision
Chapter in the Book							
-	5.2.3	5.2.2	5.2.7		5.2.6	5.2.4	5.2.5

Figure 5.2: Classification of Detection Systems

The technical components for detection can be classified into:
- Track-based detectors where installations in or near the track are the active element. This category includes most of the detection systems.
- Train-based detectors where installations on the train are the active element. This category includes few systems, such the positioning of trains by satellites (chapter 5.2.6) and (partly) the End of Train detection systems (chapter 5.2.7).

Detection systems can be further subdivided into
- spot detection
- linear detection (in some applications working as quasi-spot)
- area detection
- three-dimensional detection

5.2.2 Spot Wheel Detectors

Spot rail contacts are a suitable solution when the passage of a rail vehicle at a certain location is to be detected, but also for axle counting as indirect track clear detection. They use the guidance function in rail transport and are only capable of detecting rail vehicles, but not of external objects.

5.2 Technical Means of Detection

5.2.2.1 Mechanical Detectors

Mechanical detectors are the historically oldest form. In English speaking countries, such detectors are also known as treadles. Usually an arm is mounted to the rail on the inner or outer side (figures 5.3 and 5.4). When a wheel passes the detector, the arm is depressed and therefore the wheel detected. The types of treadles and the railways in the world vary in whether they consider these devices to be safe or non-safe. As each wheel depresses the arm, this form of mechanical detectors is capable of axle counting (chapter 5.4), if working safely. If the arm is mounted in a way that permits movements in two dimensions, the detector can distinguish the direction of travel. Otherwise, second or third arms have to be installed if direction selectivity is required.

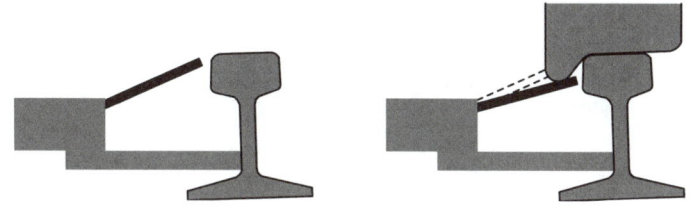

Figure 5.3: Principle of mechanical detector

Figure 5.4: Double mechanical detector with detection of direction

Another seldom used principle is the arm being touched by the body of the vehicle.

Mechanical detectors were historically widely used, but due to high maintenance costs were already being replaced by the end of the nineteenth century in many countries. Due to their simplicity, the main field of application today is temporary installations such as warning equipment for track workers. However, in some countries they are still used for permanent installations to a large extent. Modern systems with an oil dashpot were developed in recent years.

5.2.2.2 Hydraulic and Pneumatic Detectors

Hydraulic and pneumatic detectors (Fenner/Naumann/Trinckauf 2003) are installed below the rails and detect the slight bending of the rail caused by the presence of a large mass (figure 5.5). This force is amplified by a liquid or gas, which finally switches an electrical contact in the evaluation installation.

5 Detection

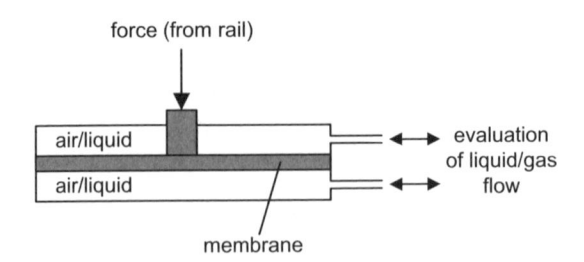

Figure 5.5: Principle of hydraulic/pneumatic detector

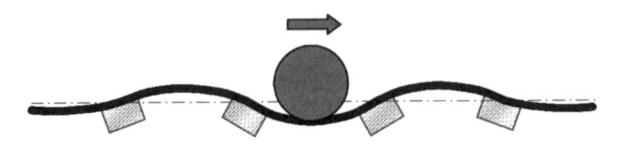

Figure 5.6: Bending line of the rail if passed by a wheel

Hydraulic and pneumatic detectors have some significant disadvantages:
- Under unfavourable circumstances, extremely light vehicles are not detected.
- The detectors have to be adjusted frequently, which increases the maintenance costs and decreases the reliability.
- Depending on which fluids are used, including the technology to transform the liquid/gas flow into an electrical signal, problems of environmental pollution can occur (e.g. mercury).
- As the rails are bent in the form of a wave under each axle, these detectors often detect secondary amplitudes as well (figure 5.6).

The last mentioned makes these detectors incapable of axle counting. However, an advantage of hydraulic and pneumatic detectors for detecting whether a train has reached a certain point is the resistance against operation by unauthorised persons. This is much higher than for any other type of detector. Nevertheless, hydraulic and pneumatic detectors are seldom used today.

5.2.2.3 Magnetic Detectors

Detectors using the magnetic principle (Fenner/Naumann/Trinckauf 2003) contain a magnetic circuit formed by a permanent magnet with a gap at the rail (figure 5.7). This magnetic circuit is adjusted that way so that if no wheel is present there is no magnetic flux at the electric contact, which is therefore open. When a wheel enters the gap, the magnetic field is deformed and therefore switches the electric contact by magnetic force.

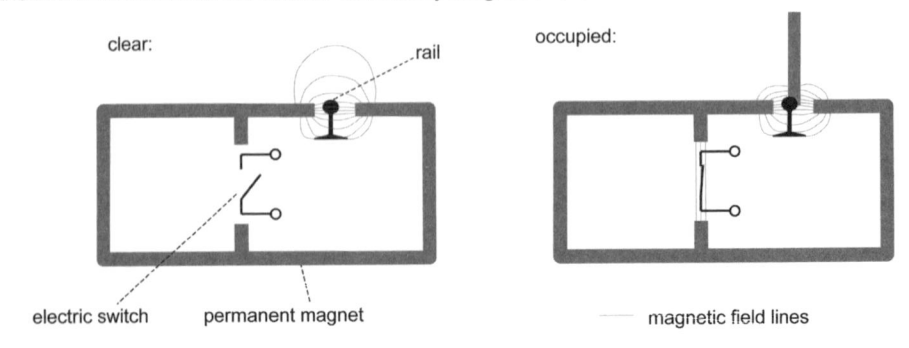

Figure 5.7: Principle of magnetic detector

5.2 Technical Means of Detection

5.2.2.4 Inductive Detectors

Most detectors for rail vehicles in modern installations are based on electromagnetic induction. Different products are on the market, using variations of the basic functional principle. Inductive detectors use an electromagnetic field around the rail. Due to adjustable sensitivity, inductive detectors are flexible and may be applied to different situations. As the electromagnetic field is permanently applied in most products, these detectors provide a continuous self-check.

Many inductive detectors are suitable for axle counting (chapters 5.1.4.3, 5.4) as well as for detection of the train reaching a certain position (chapter 5.1.4.1). In the following, out of the large number of inductive detectors, two examples will be described.

The detectors AzL of *Thales* (figure 5.8) (Fenner/Naumann/Trinckauf 2003) consist of a sender situated at the outer side of the rail and a receiver at the inner side. The sender continuously emits a magnetic field. The presence of a wheel with its iron mass changes the lines of magnetic force. The orthogonal line to the receiver coil and the tangent to the lines of magnetic force cut in an angle which changes between positive and negative with the presence and absence of a wheel. Therefore, the polarity of the voltage induced in the receiver coil changes. The point where that voltage is zero is at a distance between the wheel and the sender-receiver-pair of about 20 cm.

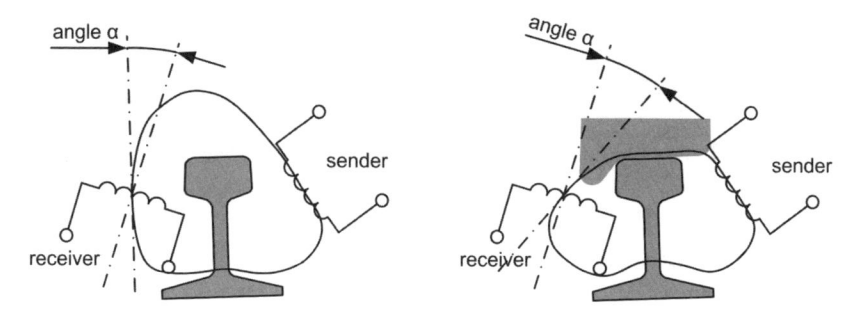

Figure 5.8: Principle of inductive detector (example AzL)

The detectors ZP 43 of *Siemens* (Siemens 2001) consist of pairs of corresponding resonant circuits on both sides of the rail (figure 5.9). The inductivities of both resonant circuits are coupled magnetically. With a wheel inside the sensing range of a detector, the magnetic coupling between sender and receiver and therefore the amplitude of induced voltage on the receiver side increases.

The main field of application of the mentioned detectors is axle counters (chapter 5.4).

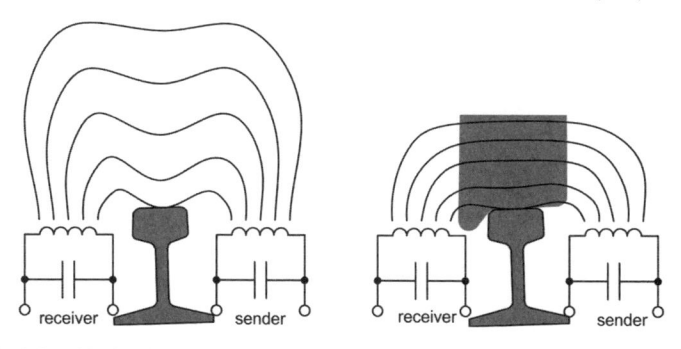

Figure 5.9: Principle of inductive detector (example ZP 43)

5 Detection

5.2.3 Linear Wheel and Axle Detectors

5.2.3.1 Mechanical Detection Bars

In mechanical signalling, mechanical detection bars (figure 5.10) are used by different railways. It can be considered as a treadle as in chapter 5.2.2.1, with a linear extension along the rail. The bar is normally in the low position. It must rise in order to release the interlocked function, which is prevented if a wheel is present.

In a typical application, it extends for several metres in rear of a signal in mechanical signal boxes and is used to detect the presence of a train standing at the signal. As the signaller should be able to see the train, the device usually acts as a reminder. Another typical application is the locking of points in the current position in case of occupation by a vehicle.

Figure 5.10: Mechanical detection bar for prevention of switching points under a train

5.2.3.2 Galvanic Detector (Track Circuit)

Today, the most used linear detector is the track circuit. Caused by an electric shunt between the rails formed by wheels and axle, the presence of vehicles in an isolated track section can be detected (figure 5.11). The isolation can be solved either by cutting the rail and fitting an isolating material, or by electrical means without physical disruption of the rails, the so-called **jointless track circuit** (chapter 5.3.7.4).

The main field of application is track clear detection which is described in detail in chapter 5.3.

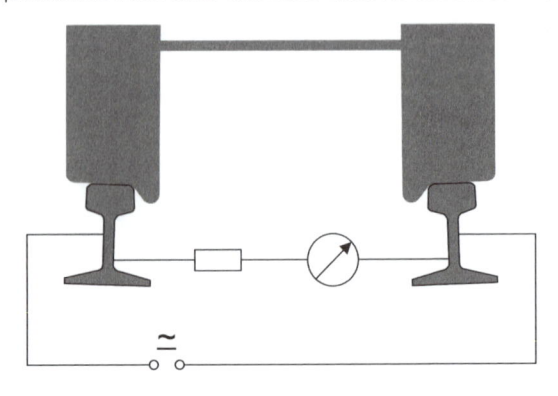

Figure 5.11: Principle of electric (galvanic) detector

5.2 Technical Means of Detection

Another field of application is the use of a combination of a spot rail contact and a short track circuit (which must be longer than the longest permitted distance between two axles) to gain the information that a moving unit has cleared a certain point at the track after reaching it (chapter 5.1.4.2). A route cannot be released or a block section cleared before this information has been obtained. However, this does not guarantee the completeness of the train. Such combinations are applied in Europe, but with continuous track clear detection they become obsolete.

Another historical application of track circuits was the first axle counters in Switzerland without capability to distinguish directions. These systems used track circuits which were shorter than the shortest distance between two axles to detect each individual axle. (Oehler 1981)

5.2.4 Linear and Area Detectors for Vehicles and External Objects

Area detectors detect objects which are present in or move through an area. In contrast to the wheel and axle detectors described above, area detectors are capable of detecting not only rail vehicles, but also external objects.

5.2.4.1 Active Electromagnetic Wave Systems

Systems of this group actively send electromagnetic waves in form of microwave radar, laser, infrared or others and use the reflection or absorption by objects to detect objects with a certain minimum size. Some typical examples are described here.

In one example, a net of several static (often parallel) rays is applied (figure 5.12). Each ray has a transmitter on one and a detector on the other side. Objects are detected by one or more rays being disrupted. The sensitivity of the system can be regulated by the density of barriers and the number of neighbouring barriers which must be disrupted to cause an occupation status.

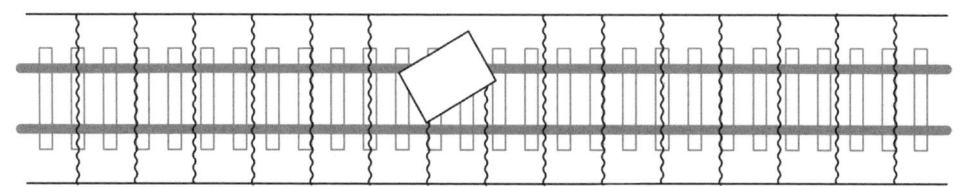

Figure 5.12: Net of electromagnetic barriers to detect obstacles

In another example, an area is supervised by a ray circulating along this area (figure 5.13). Reference mirrors calibrate the position information. An object in the area partly reflects the radiation back to the receiver. The resulting image is compared with an image of the clear status and the object is thereby detected.

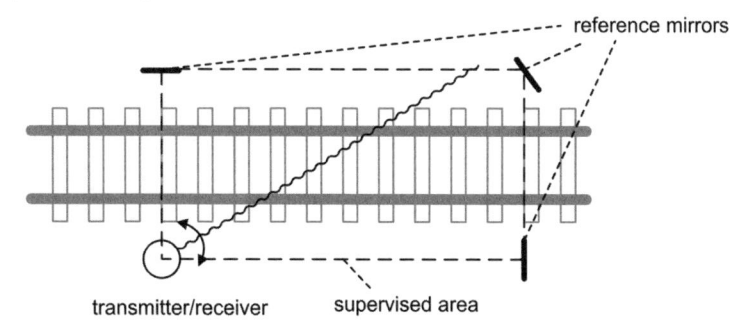

Figure 5.13: Radar scanner

5 Detection

These technologies are mainly applied to detect occupancy of level crossings by road users (chapter 13.4.4.4) and in other cases with increased probability of objects other than rail vehicles occurring, e.g. at platform tracks, particularly in automatic metropolitan railways.

In other applications, systems of this type are also suitable for indirect detection of occupancy by trains. An example is the disruption of a ray with sender and receiver on opposite sides of the track by each wagon of a train. The evaluation principle is similar to that for axle counters, but wagons are counted instead of axles. This principle is applicable mainly to metropolitan railways (Barwell 1983).

By using the Doppler effect, the speed of trains and, calculated from speed and occupation time of a certain position, the train length can be measured (Fenner/Naumann/Trinckauf 2003).

5.2.4.2 Mechanical Technologies

Mechanical supervision of the limiting areas of the clearance profile is applied by some railways. A net of wires or a single horizontal wire stretches outside the limits of the clearance profile and carries a low voltage current (figure 5.14). If an object of not too small extension breaks through this area with a certain minimum force, the wires break, disrupting the current. This disruption of current is evaluated by the interlocking or block system which can hold signals at red in this case. As repair works are necessary after such events, this technology is only useful to detect occurrences which seldom occur. Examples are:

– Detection of avalanches and earthslides in mountainous areas.
– Detection of road vehicles fallen from a bridge above a railway line, as applied on French high speed lines.
– In situations where a railway is situated in proximity of an airport, to protect against an aircraft overshooting the runway and obstructing or destroying the railway line.

Figure 5.14: Example of an installation to detect avalanches mechanically (France)

Another mechanical technology is contact mats placed on and beside the track to detect the presence of persons, vehicles or other objects by their weight. This technology is applied on some automatic metropolitan railways, e.g. in Vancouver.

5.2.4.3 Magnetic Inductive Loops

Such detectors consist of a resonant circuit with the inductivity situated in the track (figure 5.15). When a rail vehicle passes over the loop, the inductivity L changes due to the iron mass of the vehicle. According to the formula of the resonant circuit $(2\pi \cdot f)^2 = \dfrac{1}{L \cdot C}$ (with f being the

5.2 Technical Means of Detection

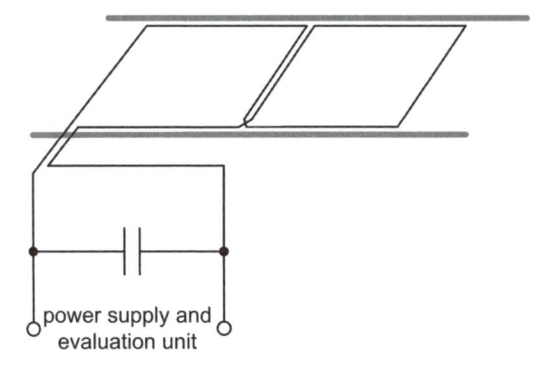

Figure 5.15: Inductive loop for vehicle detection

frequency, L the inductance of the coil and C the capacity of the capacitor), this changes the frequency of the resonant circuit. This shift of frequency is evaluated to detect the vehicle.

To compensate inductive effects of traction return currents in the rails, symmetrical double loops are normally used (figure 5.15).

By this technology, directions cannot be distinguished (unless using two double loops) and axles cannot be counted. It is applied for initiating the opening and closing of level crossings in some systems. It can also be used to detect road vehicles on level crossings, with the disadvantage that due to lack of iron mass, pedestrians and animals are not detected and cyclists rarely. In road traffic management, such loops are widely used.

5.2.5 Three-Dimensional Detection

5.2.5.1 Visual Observation

The simplest and historically oldest form of detection is visual observation of the respective track by staff. The ability of human to also evaluate unexpected observations is the main advantage over all technical systems. Disadvantages are the relatively high probability of human error and the high costs of staffing. Therefore, the usage of visual observation is decreasing, especially in highly developed countries.

Another version is remote visual observation via camera and monitor (figure 5.16). The number of people required for observation can be much reduced by this method. It is used especially in situations where not only rail vehicles have to be detected. Examples are the conflicting areas of level crossings with roads, the tracks in platform areas, but also passenger areas for security purposes.

Figure 5.16: Remote visual supervision of a level crossing

The results of visual and remote visual observation can be used for all kinds of detection purposes described in chapter 5.1.4 as well as for hazard alert purposes (chapter 14). Concerning track clear detection, both the direct and the indirect principle are used:
- Direct track clear detection means that the track section concerned must be completely observed from one or more observation points (which can be signal boxes).
- Indirect track clear detection means that the end-of-train markers are observed and the clear status of the track section is concluded from the facts that the last train was complete on leaving the section and no other train had entered.

5.2.5.2 Automatic Image Processing

An alternative to remote visual observation is automatic image processing. It is a development of recent years, as it requires very high capacities for data processing which can only be achieved by modern electronic computer systems. This is especially applicable when objects other than rail vehicles need to be detected. Cameras can be placed at the trackside or on the train.

The most common evaluation method is differential image processing, evaluating the difference of the current image with a fixed reference image or with a image taken short time before, giving an alarm or automatically stopping trains in case of critical results.

A disadvantage of the image processing method is that in certain circumstances optical effects such as headlights of road vehicles or shadows of persons can be mistaken for an obstacle on the track. One solution is the usage of stereo cameras: By observing the same area using two cameras from different viewpoints, 'flat' optical effects can be distinguished from real objects. As an example, this was developed for level crossings in Japan (Ohta 2005).

5.2.6 Systems with Active Reporting from the Train

5.2.6.1 Fields of Application

Several methods are applied and under development where the train reports its position actively and is thereby in most cases individually known by the trackside equipment, e.g. by a train number. There are many systems where this information is used for purposes of rolling stock disposition, statistics etc. But the number of systems is increasing where such information is also evaluated for safety purposes such as track occupancy, particularly for secondary lines in different parts of the world. These systems often appear in combination with train-based end of train detection systems (chapter 5.2.7). In low-tech applications, the position report is sent telephonically by the train driver. Modern systems can be distinguished into two categories:
- Systems where the train carries a sender or transponder and is thereby identified by the trackside when passing fixed locations (chapter 5.2.6.2).
- Systems where the train measures its position autonomously and reports this to the trackside (chapter 5.2.6.3). The position can be measured relatively, e.g. by odometry or Doppler radar, or absolutely by satellite positioning. Relative position measurement has to be calibrated at certain intervals, e.g. by balises or radio beacons.

5.2.6.2 Systems with Train Identification at Fixed Locations

In the following, systems are described where the train or its parts carry particular senders, for example microwave senders, which send an ID number by passing interrogation points. These systems are widely used for positioning of certain locomotives, wagons or cargos.

As an example without safety application, the Russian system SAIPS has interrogation points on the entry to and exit from stations. The approaching train is detected by a special track

5.2 Technical Means of Detection

circuit or axle counter, whereupon the interrogator sends a powerful signal. The train transponder uses this power to send the ID-number of the vehicle back (figure 5.17). The data are sent from interrogator to the dispatcher centre through Ethernet (Belov/Geršenzon/Kotlecov 2003). Similar systems are in use in the USA ('Automatic Equipment Identification', chapter 12.3.5) and other countries.

In recent years, this detection principle has also been used for interlocking purposes. An example is the Japanese system COMBAT (computer and microwave balise-aided train control system). Interrogators, consisting of a sender/receiver unit and a responder, are located at exit signals (figure 5.18). In the absence of a train, the sender/receiver sends the signal to the responder and receives the answer to check the system to be effective. The interrogators controller checks the serviceability through two independent channels and sends the data to the control centre. The train has two transponders for direction selectivity. The second is situated at the rear end of the train and serves also for end of train detection. Thanks to the train ID number, the route will be released after the same train has cleared the route completely. Train data in the control centre facilitates other dispatcher operations, too (Nishibori/Sasaki/Hiraguri 2002).

Figure 5.17: Interrogator of SAIPS

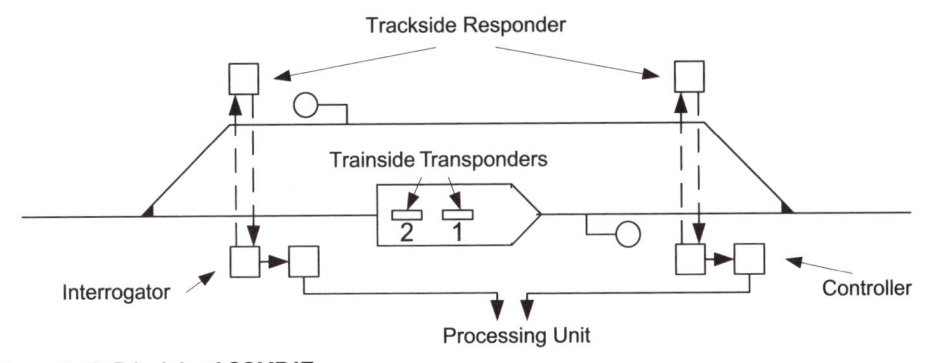

Figure 5.18: Principle of COMBAT

5.2.6.3 Systems with Train Measuring its own Position

On the first railways, a rider on horseback galloped in front of the train, checked the track and warned ground staff about its approach. This way of notification was quickly forgotten. But newer communication systems enabled the train driver (or later automatic systems on the train) to send the position of the train in certain points of the way himself.

The method of intermittent notices by the driver is suitable for low traffic volumes and speeds (due to the high probability of human error) and trains where unrevealed detachment of wagons cannot appear. It is also suitable in combination with train-based end of train detection systems (chapter 5.2.7). But this method may also be used on lines with modern track-based signalling equipment if other systems fail. Telephones on the trackside or mobile communication can be used.

Thanks to the development of data communication systems, today the train can automatically transmit its position in short intervals to a stationary block centre by digital radio. The train can determine its position:

- relatively, e.g. by odometer or Doppler radar,
- absolutely by trackside fixed points (e.g. balises, beacons, crossings of cable loops, track circuit borders or characteristics of the track itself) or
- absolutely by satellite positioning.

Odometry measures the position through the number of wheel rotations and systems with Doppler radar via speed and time. In all **relative systems**, measurement errors (e.g. in odometry by the train slipping and sliding) sum up, therefore the position has to be corrected at absolute control points.

Absolute methods with trackside fixed points are used in several advanced train protection systems for braking curve calculation, e.g. in LZB (crossings of cable loop as fixed point; chapter 8.3.6), TVM 430, Japanese Digital ATC (track circuit borders as fixed points; chapter 8.3.6) and ETCS (balises as fixed points; chapter 8.4.3.1). In several cases, besides the pure train protection function by dynamic speed calculation, the position information is sent to the trackside signalling equipment for different purposes, such as the evaluation of standstill of the train for overlap release. In ETCS Level 3, it is intended for track clear detection, too.

Characteristics of the track itself can be measured by eddy current sensors on the vehicle in connection with a track atlas. The eddy current sensor detects all metal parts along the track, e.g. clips which attach the rails to the sleepers, or at pointwork. It can only be used in limited networks, for which the vehicle has a track atlas. In a European test application in Tatra railways in Slovakia, this method is combined with satellite positioning and serves for train protection purposes to supervise the obeying of Movement Authorities.

In systems based on **satellite positioning**, the receiver on the train calculates the train's position without needing fixed control points. The signals from satellites with their codes are synchronously transferred to all earth receivers. Through signal codes and delay times, the receiver determines the distance from each satellite and, with signals from a minimum of three different satellites received, its absolute position on the Earth. Such systems are tested on different railways, and show a high detection error (up to 30 m), which makes them unsuitable to distinguish on which of several parallel tracks a train is standing.

The principle of Differential GPS (D-GPS) uses special trackside reference points to reduce this error. To enable safe track selective detection, an error below 2 m is necessary. Russian Railways are going to use the satellite positioning data by the locomotive (e.g. for calculation of the braking curve) and by the interlocking as an additional safety level (figure 5.19) (OAO RZD 2007). In USA the principle of satellite positioning for interlocking purposes is used in the system Positive Train Control (PTC).

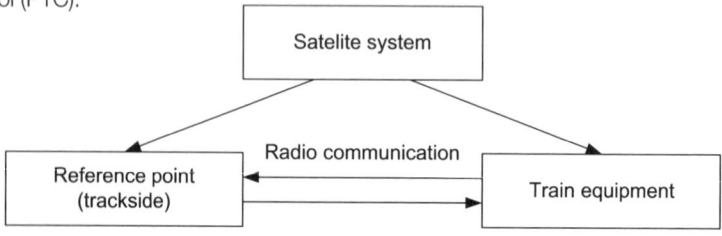

Figure 5.19: Correction of train position

There are several problems limiting the use of satellite positioning for train spacing as a replacement of trackside track clear detection by track circuits or axle counters. One is that all trains in the network have to self-check their completeness (figure 5.20). However, this can be easily solved on secondary lines where only short passenger trains are used. Another problem is that in case of disruption of communication with only one train, the whole network would have to be considered as occupied. This makes such systems (likewise these systems which employ a track atlas) particularly suitable for separated networks with a low number of trains moving inside the network.

5.2.7 End of Train (EOT) Detection Systems

The historically oldest form of end of train detection is the use of rear end marker as a vehicle signal to be observed by staff (chapter 5.2.5.1). Every train must carry this signal, usually one, two or three red lamps or a mechanical signal including the colour red, at the end of the last wagon (figure 5.21).

Figure 5.20: GPS-Receiver at the wagon (photo: *DB AG/Claus Weber*)

Figure 5.21: Different forms of End of Train markers

Later, technical solutions for the detection of the rear end have been introduced on some railways. They can be categorised as train-based or track-based solutions.

Train-based rear-end detection systems are widely used in countries following North American operational principles, whereas track-based systems are to a decreasing extent used in countries following German operational principles (chapter 3.1).

Train-based systems consist of a sending unit at the rear end (figure 5.22) and a receiving unit in the driver's cab, both working together as a matched pair. One solution is to measure the distance between both units by radio waves. Another solution is to measure the pressure of the air brake at the rear end and to transmit the data to the cab unit. There are different possibilities for evaluating this information. Some of them are:

- Alarm in the cab to warn the driver in case of critical values together with special rules about driver's actions in this case.
- Automatic initiation of an emergency stop of the train.
- Transmission of the result to the interlocking system or the operator's workplace and safety measures there.
- Refusing of unblocking in an electronic token block system (chapters 4.4.4, 10.4.3).

Track-based systems consist of a receiving unit mounted in the track and a unit installed to the coupling at the end of the train. To prevent dangerous mistakes by shunting staff (a trainside End Of Train unit forgotten in the middle of the train), it is designed so that either this unit or another wagon, but never both, can be attached at the same time. In the applications, the contact is either mechanical or electrically inductive. Inductive train side devices can either be provided with autonomous power supply or use the transponder principle to use energy sent from the trackside. When the train side unit passes over the trackside, this information is transmitted to the interlocking system, permitting release of the respective track section for further traffic.

Figure 5.22: Rear end device of a train-based EOT detection system in USA (photo: *Reiner Decher*)

The use of continuous track clear detection can be considered to make end of train detection redundant. However, on most railways trains still carry optical rear end markers for following reasons:

- in case they enter an area without technical track clear detection;
- for the benefit of staff other than the signallers (e.g. track workers can see that the train is moving away from them, not towards them);
- for degraded mode operation in case of technical failure.

5.3 Track Circuits

5.3.1 Basic Structure of Track Circuits

The basic functional principle of the track circuit, also called 'rail circuit', is described in chapter 5.2.3.2. The track circuit was invented in second half of the nineteenth century by the American civil engineer *William Robinson*. The first track circuit had a source of current and a detection device at the same end (figure 5.23). If the section is clear, the circuit is open and the relay dropped down, therefore it is called 'normally open' track circuit. Upon entry of a train the relay gets a current. Advantages of this form of track circuit are simplicity and low costs. Another safety advantage is that, as relays are usually faster in picking up than in dropping down, the occupation as the more safety critical case will be detected faster. However, the decisive disadvantage is the non-fail-safe behaviour: In the most frequent error cases (for example breakage of the circuit or low voltage), a track circuit with a train can falsely show track clear status. Due to this disadvantage, this kind of track circuits is applicable for purposes with low safety requirements, e.g. at marshalling yards in some countries.

In the normally closed (fail safe) track circuit (figure 5.24), the track relay is connected with the power feed only via the rails. If no train is present, the relay is picked up, if a train is on the sec-

5.3 Track Circuits

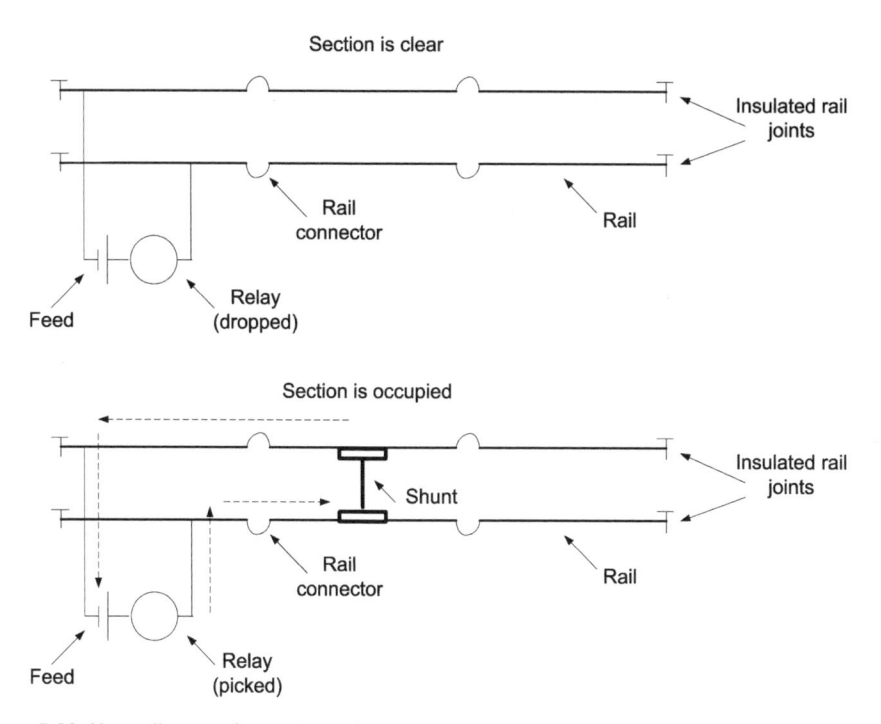

Figure 5.23: Normally open (non-fail-safe) track circuit

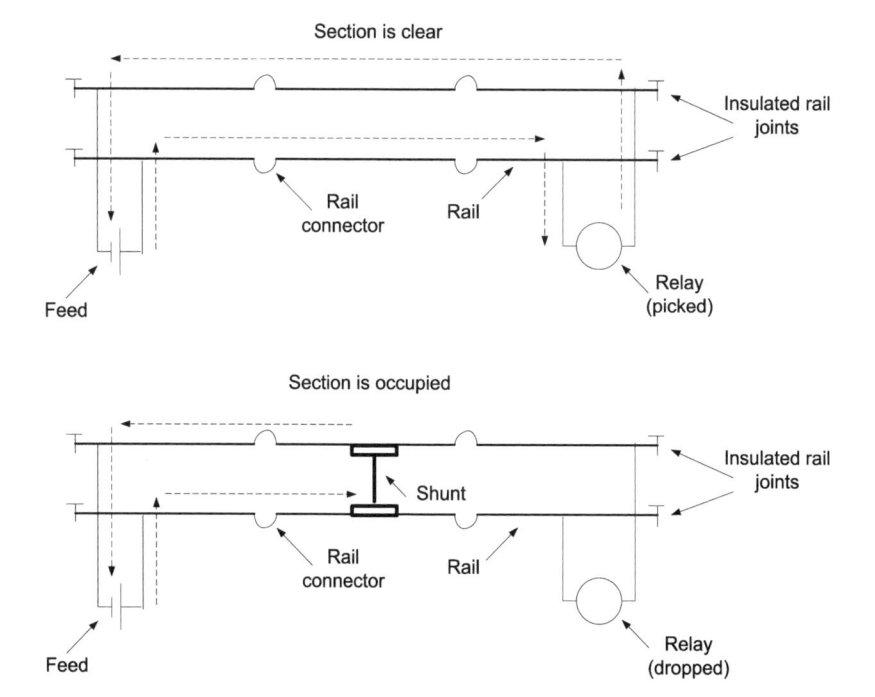

Figure 5.24: Normally closed (fail-safe) track circuit

5 Detection

tion, the track relay is short-circuited and the relay drops down. This track circuit, however, has one important disadvantage: If the train has low axle loading or the rails are rusty or covered with sand or wet leaves or similar, the detection of the train can be lost. This is also known as 'poor shunting' ('failure to shunt' in North America): The resistance between the rails if a train is on the section (so-called shunt) is too high, therefore the relay gets too much feed and does not drop down. To prevent this, there are different solutions. In some countries the sensitivity of track circuits is regularly checked with a normative shunt (for example 0,06 Ω in Russia; Bryleev/Šišljakov/Kravcov 1966), in other countries there is a defined minimum time interval in which the track has to be traversed by a rail vehicle (for example 24 hours in Germany; Naumann/Pachl 2002), otherwise the track circuit has to be taken out of the interlocking dependence and considered as permanently occupied.

The advantages of the normally closed track circuit dominate: The most frequent error cases lead to fail-safe reactions. Therefore, this type of track circuit became the basis for further development and use. Occasionally some special forms are used (figure 5.25), with worse safety behaviour than the standard normally closed arrangement.

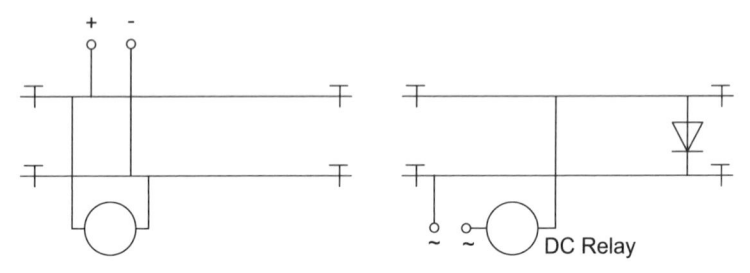

Figure 5.25: Special arrangements of track circuits (examples)

Regarding the form of current, track circuits can be distinguished into the following groups (table 5.1):
- direct current (DC) track circuits
- alternating current (AC) track circuits
- alternating current (AC) track circuits with high (audio) frequency, usually jointless (chapter 5.3.7.4)

A special type is **track circuits with high voltage impulses** in which undesirable isolation between the rail and the wheel is broken by short pulses of a high voltage (for example in France and Britain; Bailey et al. 1995). According to the given classification, these track circuit types will be described in detail in chapter 5.3.7.

Track circuits can have either passive or active receivers. In the latter case the track circuits carry only a weak signal which will be amplified by the receiver. The advantage is low power consumption (up to some tens of Watts) in comparison with other types (up to some hundreds of Watts), but the disadvantage is higher sensitivity to electromagnetic influences.

5.3.2 Geometrical Assembly of Track Circuits

On open lines there are track circuits with one or two receivers. The reason for the application of the latter, so-called centre-fed track circuits (figure 5.26), is that the distance between the receiver and the transmitter is limited to about 1,000 to 1,500 m because of rail resistance and external electromagnetic influence. By using track circuits with one transmitter in the middle and two receivers on both ends, the general length of a track circuit can be doubled.

5.3 Track Circuits

In stations in some countries (e.g. former USSR), branched track circuits with different receivers are used, too (figure 5.27). If one of the relays has no current, the track circuit is considered as occupied. Track circuits with many receivers are difficult to regulate. Therefore one branched track circuit, as a rule, has not more than 3-4 relays. If a branch is short, the relay on this end can be absent. But if a rail connector (long jumper bond) is broken and a train is on this part of the section, the track circuit is not safe. Therefore, for safety redundancy the short branch without relay needs doubled long jumper bonds (figure 5.28).

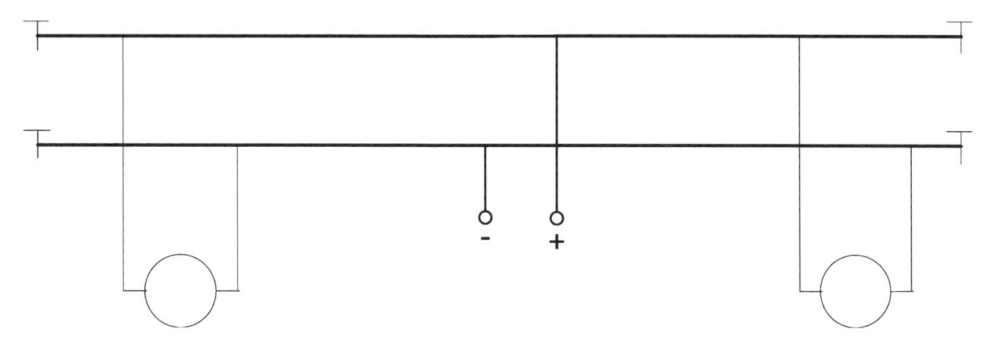

Figure 5.26: Centre-fed track circuit

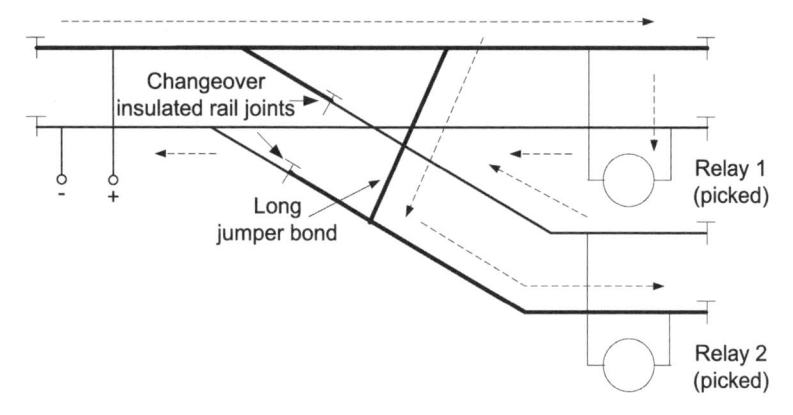

Figure 5.27: Track circuit with long branch

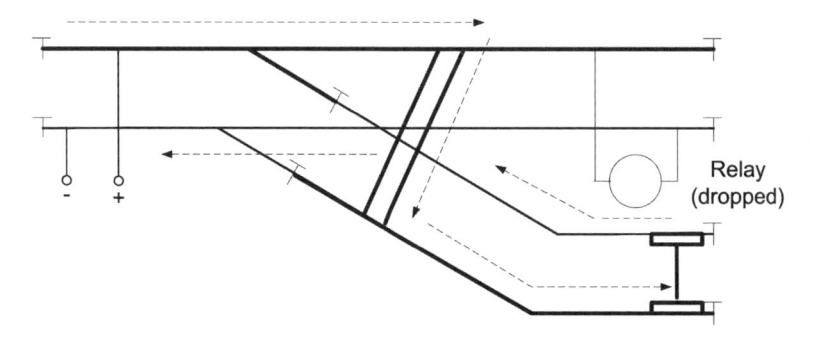

Figure 5.28: Track circuit with short branch

5 Detection

In other countries (e.g. Britain, Germany) the station track circuit usually has only one receiver, and the current proceeds to the relay through all branches (figure 5.29) (Nock 1982). A disadvantage of these track circuit is that the train is not detected in the case of breaking of the other rail (thin line) or of its connectors. However, not all railways require the detection of broken rails anyway.

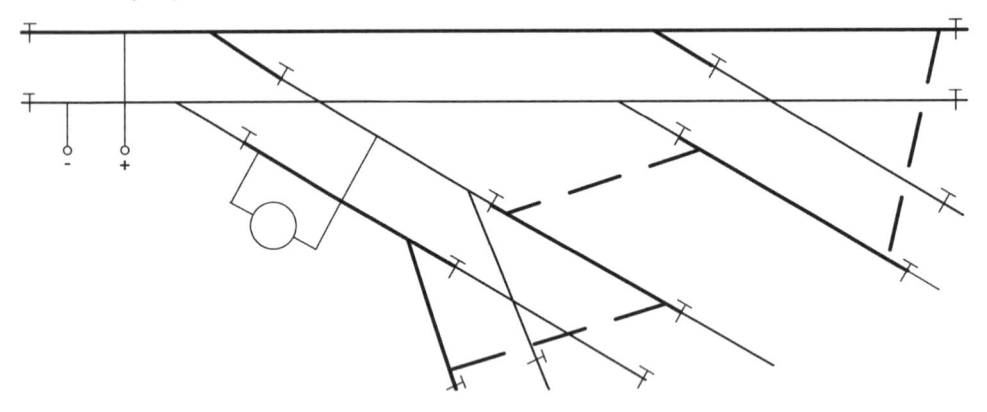

Figure 5.29: Typical series bonding of station track circuits with one relay

In certain locations the probability of derailment is high (e.g. on catch and trap points). To prevent an erroneous and dangerous clear detection due to complete derailment, track circuit interrupters can be installed (figure 5.30). In the act of derailment or when a rail vehicle enters a section which cannot be used for a legal movement, the leading wheel breaks the interrupter which switches the track circuit to permanently occupied by disconnecting the track circuit cable.

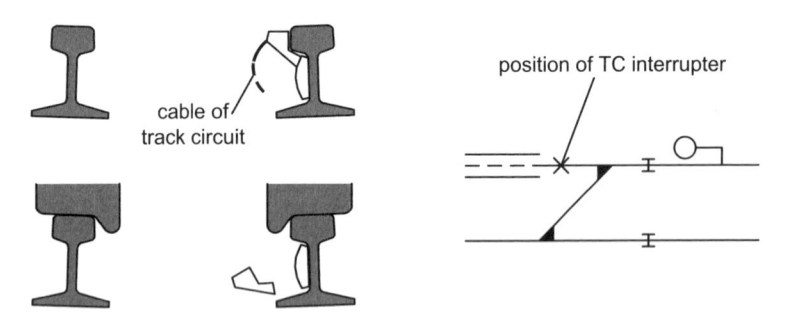

Figure 5.30: Track circuit interrupter at trap points (principle)

5.3.3 Treatment of Traction Return Currents

In electric traction areas a particular problem is the treatment of traction return currents. Besides, some railways transport return currents from interior energy consumers in passenger vehicles to the locomotive via the rails. Two basic solutions are applied (table 5.1):

1. Only one rail carries the return currents. For this purpose, either one rail is constructed without insulated rail joints and the other rail (the signal rail) with (figure 5.31, a), or both rails have isolated rail joints and are alternately used for return currents and connected by diagonal connectors (figure 5.31, b). The advantage of the latter is higher safety in case of overmilling of the insulated rail joint. These solutions are used mainly for DC track circuits in AC traction areas.

5.3 Track Circuits

	direct current track circuits	track circuits with high voltage impulses	alternating current track circuits			audio frequency jointless track circuits	
return of traction supply	through single rail (only for AC traction system)	through single and through double rail with impedance bond	through single rail and through double rail with impedance bond			through double rail	
track circuit supply	continuous	impulse	only impulse	impulse	continuous	modulated	continuous
					onephase relay	polyphase relay		
safety protection against disturbance by means of	–	decoder scheme	decoder scheme	decoder scheme	frequency	frequency and phase	modulation and frequency	frequency
provide own information for cab signalling	–	pulse code	–	pulse code	frequency code	–	frequency code	
control of insulated rail joints by means of	change of poles	decoder scheme	decoder scheme	decoder scheme	change of phase	change of phase	–	
receiver	passive (relay)	passive (relay)	passive (relay)			active (amplifier with relay or with electronic detection device)	
safety measures if the train has low axle load	–	high-voltage impulses	–			high-frequency current	

Table 5.1: Comparison of different types of track circuits

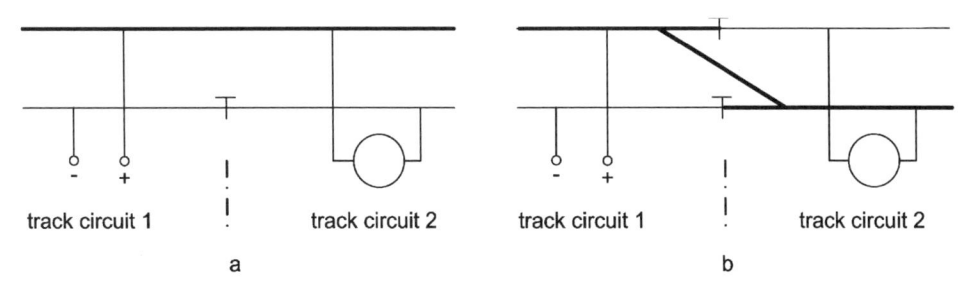

Figure 5.31: Return of traction supply through single rail

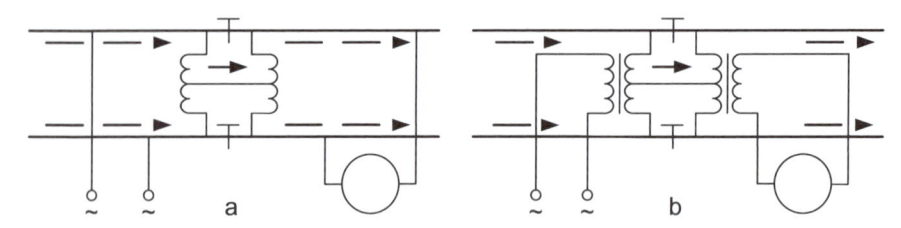

Figure 5.32: Return of traction supply through double rail

2. Both rails carry the return currents in equal parts. In DC traction areas an impedance bond is used in the form as drawn in figure 5.32, a. This impedance bond connects the rails at each end of every track circuit. The bonds offer a high resistance to the AC track circuits and therefore effectively isolate one rail from the other. The DC traction current, however, can flow through the bonds to and from their centre points without hindrance (Nock 1982).

In AC traction areas of North America, a capacitor is connected to the impedance bond. Resonance circuit LC of the impedance bond with the capacitor has minimal resistance for the traction current and maximal resistance for the track circuit which has a higher frequency (IRSE 2008). For all (AC or DC) traction supplies there is a form of impedance bonds with transformer inside which use their second side for the feeding and receipt of the track circuit signal (figure 5.32, b). This solution is widely used in Europe and Asia.

The traction return currents come into one coil of the impedance bond from opposite sides, and the second (or signal) transformer side would get no influence due to the equality of current in both rails (figure 5.33). But in practice there is no exact equality of currents in opposite rails; their asymmetry is usually up to 10-12 percent (Dmitriev/Serganov 1988). Therefore, to prevent influencing the track relay by traction return currents, the track circuits cannot have the same frequency as traction supply or its harmonics. Track circuits 25 Hz (former Soviet Union), 75 Hz (East Europe), 83 1/3 Hz or 125 Hz (Central and West Europe) and others are used for AC traction supply 50 Hz. The track circuits 42 or 125 Hz are usually used with AC traction supply 16 2/3 Hz. All AC track circuits can be applied for the DC traction supply; the track circuits 50 Hz are mainly used here.

The impedance bond must have low (less then one Ohm) resistance and pass big cur-

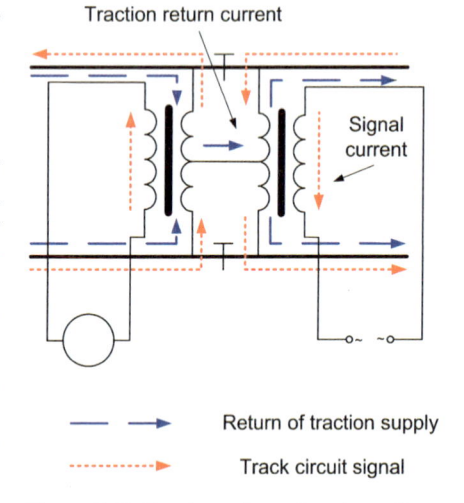

Figure 5.33: Two impedance bonds on the border of track circuits

Figure 5.34: Two impedance bonds and insulated rail joint in Czech Republic

rents (e.g. more than 1000 A in a DC traction area) (Dmitriev/Serganov 1988). The impedance bond can be in the ground between the rails or stand away (figure 5.34). The two transformers belonging to the same track circuit border can be mounted in one casing.

Neighbouring track circuits with different frequencies need no insulated joints and traction return currents can use both rails without impedance bond. Exceptions are cases where an exact physical border between two track circuits is necessary (e.g. for short station sections). Impedance bonds can also be installed to reduce asymmetry of traction return current or to provide current if any rail connector is broken.

5.3.4 Additional Functions of Track Circuits

Track circuits can be used not only for train detection, but also for one or several additional functions:
- to transmit information for cab signalling and train protection,
- to transmit signal information from the exit of the block section to its entrance,
- to detect broken rails.

The transmission of block information by coded track circuits helps to reduce costs for cabling. The main types of track circuits used also for cab signalling and transmission of block information are pulse coded and frequency coded. Almost all impulse track circuits give an impulse signal depending on active routes and track occupation and are used together for detection and cab signalling. Some types of AC track circuits with a continuous feed can change frequency and simultaneously be used for detection and cab signalling. Other types of non-impulse AC track circuits use only one frequency for detection. If a train is on the section, these track circuits give a second signal (mainly impulse coded) used only for cab signalling and train protection (chapter 8).

5.3.5 Immunity against Foreign Currents

The control of insulated rail joints between track circuits with the same frequency is necessary; otherwise a feed of one circuit can come into the other one and cause dangerous detection errors. For this purpose various methods are used which are described in table 5.1. Besides, each receiver is necessary to be protected against dangerous and disturbing influences from foreign currents. The sensitiveness to foreign influences is higher in track circuit with active than with passive receivers (chapter 5.3.1). The AC impulse track circuits and track circuits with polyphase relays have good immunity.

Impulse track circuits have a decoder scheme which can compare impulses in neighbouring track circuits. If the same impulses are in both circuits at the same time, the scheme detects the breakage of the insulated joints. If a continuous signal comes into the track circuit, the scheme detects the disturbance and switches off the relay. The frequency filter in the receiver provides additional protection against influences.

The polyphase (or vane) relay also guarantees high immunity to effects of false feeds and leakage from the traction supply and neighbouring track circuits by working in a frequency and phase discriminating manner.

It is in effect an AC motor limited to a movement of one quarter of a revolution. It has an armature (the 'vane') which rotates in special low-friction bearings about a horizontal axis with its centre of gravity below the axis (figure 5.35). Gravity returns it to the 'dropped down' position in case of current failure.

5 Detection

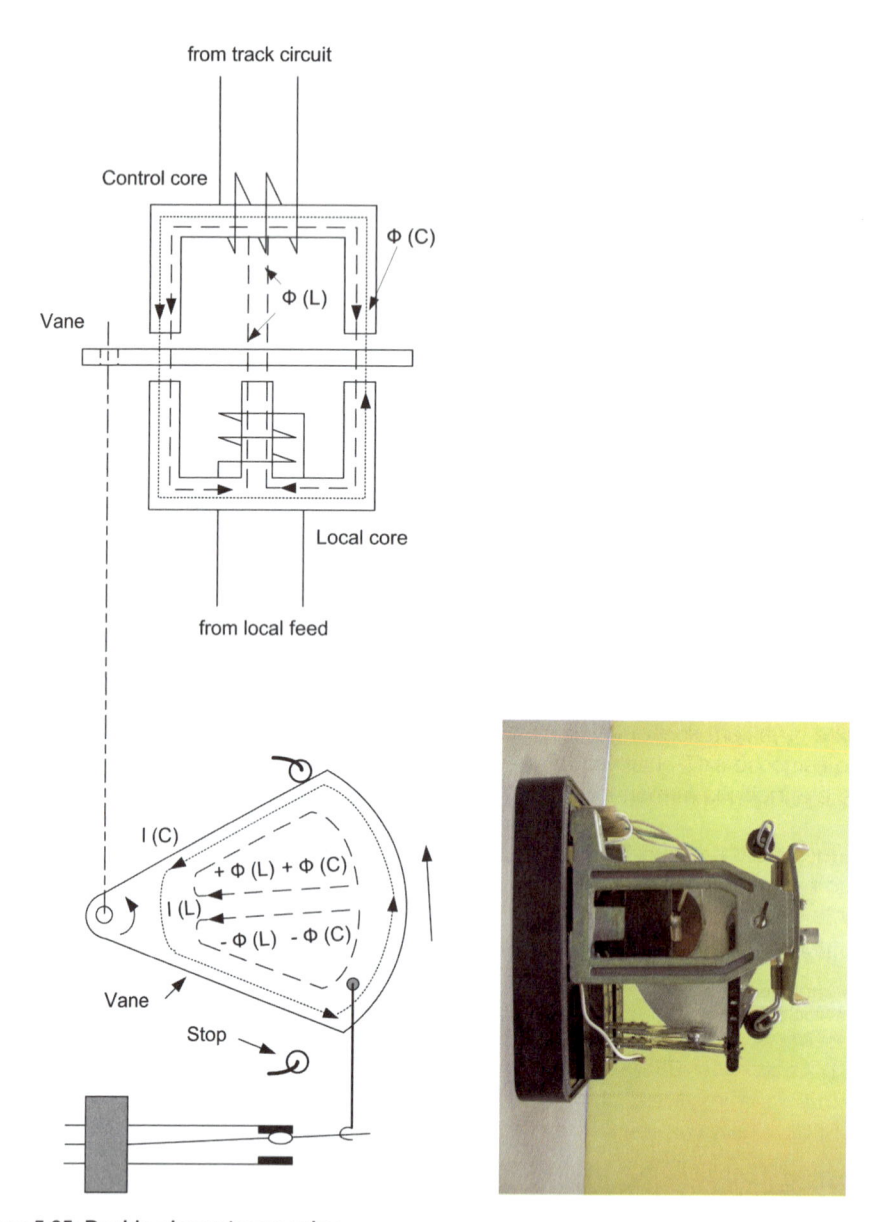

Figure 5.35: Double-element vane relay

It has two coils. The local coil is energised by a local supply. The other is the control coil. The vane relay will only operate if the energisation of the control coil corresponds to that of the local coil in both frequency and phase.

5.3.6 Electrical Parameters and Dimensioning

Rails are a good conductor for electrical current; therefore the transfer of signal electricity through rail needs relatively small voltage. But signal wires between the interlocking and the rail

5.3 Track Circuits

are much thinner. Therefore the transfer of signal electricity through wires needs a higher voltage, and to increase the effectiveness a transformer near the rails is used. A resistance at the transmitter end is also necessary for two reasons:
- protection of the track circuit equipment against a short circuit when a train is on the transmitter end,
- voltage adjustment to climatic conditions.

As equipment like masts or casings for switching units in the field has a protection earthing through the rail; in the case of an overload (lightning or short circuit of the traction supply), a short high-voltage pulse acts in the track circuit. To protect the transmitter and the receiver of the track circuit against the overload, protection elements are necessary. They filter out the high-voltage pulse by short-circuiting through a voltage-depended resistor (varistor) or transfer it to the earth by arrester. This makes the track circuit a complex device which has three parts (figure 5.36).

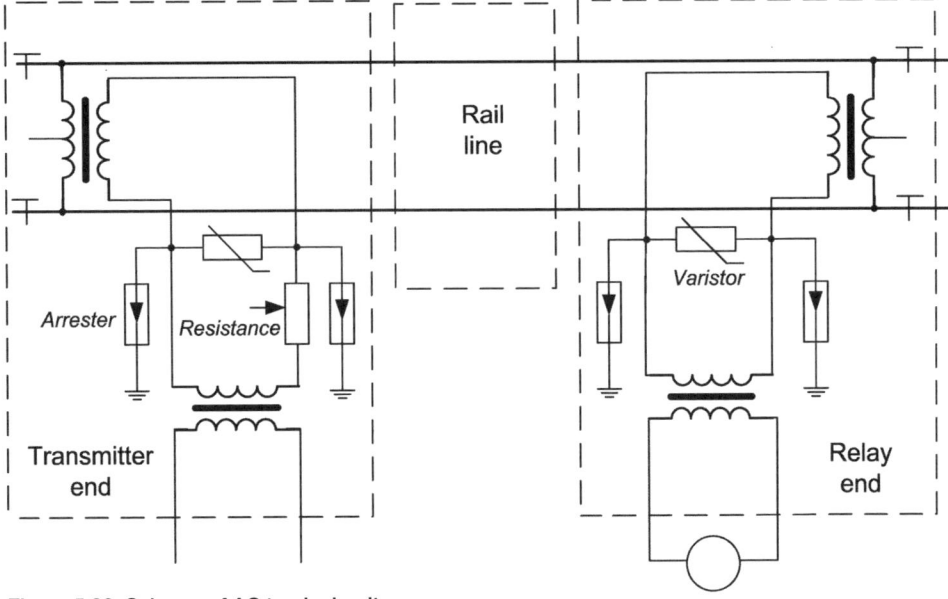

Figure 5.36: Scheme of AC track circuit

The part 'rail line' has only rails, rail connectors and ballast, but is the most compound part of track circuit (figure 5.37). The DC resistance of the rail is very low, around 0,035 Ω/km, although this is increased to approximately 0,25 Ω/km by the relatively higher resistance of galvanised iron

Rail impedance
$R+j\omega L$

Ballast resistance
$1/(G+j\omega C)$

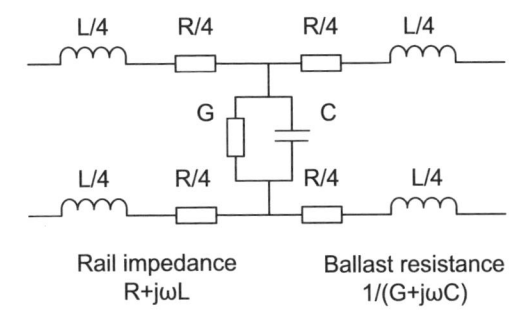

Figure 5.37: Physical sense of the rail line

5 Detection

bonds in jointed track. The inductance of rail can raise the overall impedance per rail to approximately 0,3 Ω/km (50 Hz), 2,5 Ω/km (400 Hz) or 10 Ω/km (2 kHz) (Railtrack 1994).

The ballast resistance (with G the conductance in DC operation) is the resistance between the two rails of a track and comprises of leakage between the rail fixing, sleepers and earth. The value of this resistance is dependent upon the condition of insulations, cleanliness of the ballast and the prevailing weather conditions. Changes of the ballast resistance (from infinity down to less than 1 Ω · km) and also allowable voltage fluctuations must not prevent the correct functioning of the track circuit.

There are therefore five basic operational situations to check the correct functioning of a track circuit at the worst combinations of varying factors (table 5.2):

- The scenario 'clear' means that the track circuit will show clear by worst deviations of these parameters.
- The scenario 'occupied' means that the circuit will detect an occupation by all deviations which are worst for this scenario.
- The scenario 'short circuit' requires that the transmitter end will also function if the resistance between rails is 0 Ω.
- If the track circuit is used to detect broken rails, other worst conditions have to be considered in the scenario 'broken rail'.
- If the track circuit is used for train protection (scenario 'cab signalling'), the current in the rails shall be enough for the locomotive receiver.

The shunt of each axle is normalised depending on country or railway or track circuit's type. The normative shunt (usually taken from 0,06 to 1,5 Ω) needs to determine the worst permissible conditions, when the track circuit is able as yet to function correctly. A dangerous situation occurs when the train shunt is higher than normative shunt, because of the following imperfections: rust films, leaf residue, coal dust or sand between the rails and the axle, other environmental impacts or characteristics of the vehicle (light vehicle, disk brakes).

Scenario	Means	The worst conditions				Required result
		ballast resistance	voltage	rail impedance	other	
clear	section is free	min	min	max	–	receiver is energised
occupied	section is occupied	max	max	min	one train axle is on the feed or receiver end	receiver is de-energised
broken rail	rail is broken	at a special rate	max	min	–	receiver is de-energised
short circuit	train is on the feed end	–	max	–	axle resistance is 0 Ω	feed end is in function
cab signalling	current for cab signalling	min	min	max	one train axle is on the receiver end	current through axle is enough

Table 5.2: Calculation of operating scenarios of track circuits

5.3.7 Application of the Types of Track Circuits

5.3.7.1 Direct Current Track Circuits

The first track circuits were of the direct current (DC) type because at that time, the only available reliable power supply was provided by batteries. High voltage interconnection lines were not then in existence.

DC track circuits have been used in the USA since 1871 and in Europe since 1895. In the USA, from the beginning track circuits were used to detect whole block sections continuously. If a block section was longer than the possible length of a track circuit, several track circuits were used with repeater relays in between. In Europe, initially the track circuits had a length of only few tens of metres and detected the occupation of a short section in rear of the signal (berth track circuit; chapter 4.4.6). DC track circuits for continuous train detection were used later, e. g. in Germany since 1913 (Naumann/Pachl 2002). Today, many countries in Europe (Sweden, Norway, Switzerland, Denmark and Britain), Asia and America apply DC track circuits with continuous supply. These are being replaced by coded track circuits in some countries.

5.3.7.2 Alternating Current Track Circuits with Low Frequency

Due to improved power networks and the possibility to transform AC, the AC track circuits have gained superiority over the DC track circuits and are used now in almost all countries. The track circuit's frequency must not be the same as traction frequency or its harmonics. On open lines, the AC coded track circuits usually use one carrier frequency and some code combinations for information transfer to cab signalling (example LS in Czech: signal frequency 75 Hz, from one to four impulses in a cycle depending on permitted speed). Track circuits with a continuous supply transfer this information by change of the circuit's frequencies (example: underground lines in Russia) (chapter 8.3.4). The station AC continuous track circuits (with polyphase relay, too) mostly have only one frequency; the cab signalling is transferred separately. The AC track circuits with carrier frequency above 200–300 Hz are named audio frequency jointless track circuits (chapter 5.3.7.4).

5.3.7.3 Track Circuits with High Voltage Impulses

Track circuits with high voltage impulses HVI are similar to DC impulse track circuits. The differences are as follows (example: type Jeumont) (Nock 1982, IRSE 1999):
- very short impulse length (3 milliseconds)
- high voltage peak value (100 V)
- the short impulse signal can be sent through a transformer similar to AC signal
- the impulse recurrence frequency is constant (three impulse in one second) and cannot give information for the cab signalling
- broken rails can be detected by deformation or weakening of the pulse.

This type of track circuit has been developed for bad shunting where the rails have been little used or where vehicles have been standing for a considerable time. Basically, these track circuits apply high voltage pulses to the rails to break down the rust film. The receiver detects the specific asymmetric waveform of the signal and operates the fail-safe relay (figure 5.38). The track circuits are insulated from each other by insulated rail joints, their maximum length is 2000 m for all types of electrified lines and 3500 m for non-electrified lines (Bailey et al. 1995). The track circuits with high voltage impulses are available for short lengths up to 150 m, as an overlay track circuits, mostly used on level crossing applications. These track circuits are centre fed.

5 Detection

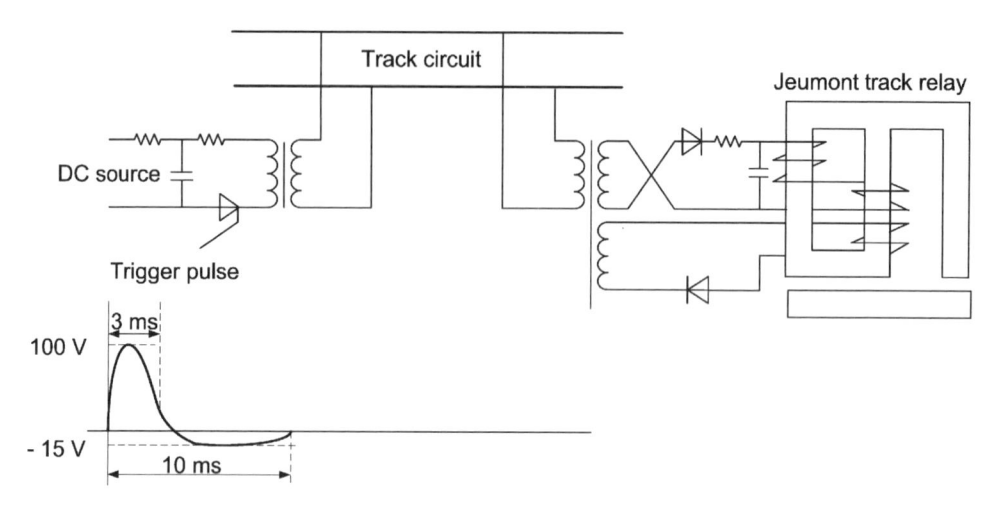

Figure 5.38: Jeumont high voltage impulse track circuit

There are other types of HVI track circuits, e.g. Lucas type. The equipment operates from a 4-volt DC supply and may be used with trickle-charged battery supply. The output is 20–40 volt pulses; the Lucas type is not suitable for electrified areas or for long track circuits.

HVI track circuits are used mainly in Britain and some other countries of the British Commonwealth and in France.

5.3.7.4 Audio Frequency Jointless Track Circuits

Insulated rail joints and impedance bonds cause difficulties for railway operation. Modern developments in electronics give the possibility of using track circuits without conflicts with traction.

When neighbouring track circuits use different frequencies and do not disturb each other, insulated rail joints are not necessary for the separation of track circuits from each other. For this purpose the band of audio frequencies can be used. At increase of frequency ω (chapter 5.3.6),

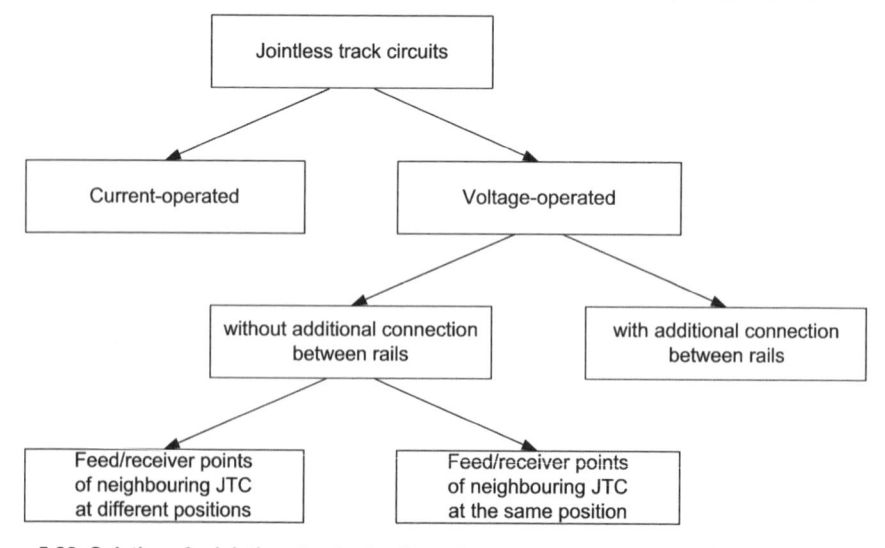

Figure 5.39: Solutions for jointless track circuit receivers

5.3 Track Circuits

the ballast resistance will decrease, but rail impedance will increase. This means that on high frequency the track circuit signal is weakened more with the length. This has two advantages: The physical border between track circuits is more precise and variation of weather conditions (e. g. reduction of ballast resistance 1/G because of rain) is less relevant. A disadvantage is a stronger limitation of length of the track circuit: up to 1 km between feed and receiver for frequencies of some hundred Hz and up to 100 m for frequencies higher than 20 kHz. Therefore almost all high frequency track circuits are centre-fed: one generator energises two lengths and the track circuit is double length. The basic complexity is to establish more exact borders between jointless track circuits (JTC). For this purpose there are various solutions (figure 5.39).

Current-operated receivers detect the presence of a current flowing through the rails. This current will induce a voltage in the coil (figure 5.40). An example is JTC '*Alstom*', working on frequency band from few hundred Hz up to 10 kHz and with two receiver coils. If the section is clear, current flows through the JTC and the ballast of the neighbouring track section. If a train is between the feed and receiver ends, there is no current of the respective frequency in the coils and the track circuit detects occupation. Reed JTC, as another example with frequency band 363–384 Hz and 8 frequency values, has a resonant shunt adjusted to the frequency of this track circuit. If the section is clear, current flows through the rails and is read on the wire of the resonant shunt (Nock 1982). Advantages of the latter type are big length of JTC and precise borders.

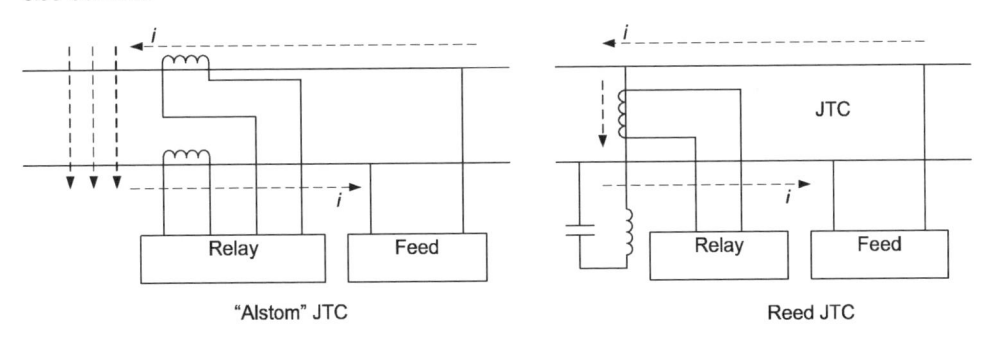

Figure 5.40: Current Operated '*Alstom*' and Reed JTC

Voltage-operated track circuits can have a physical border or be without it thanks to high dumping of audio frequency in rails. The track circuit TI21 (*Bombardier*) is used **without additional connection between rails** (figure 5.41). JTC TI21 works with eight different frequencies in the band 1550-2600 Hz. The length of the dumping zone of each JTC depends on several factors (ballast conditions, shunt value etc.) and is about 10-15 metres. Therefore **the receiver points of neighbouring track circuits are distanced** by 20 metres. Each receiver transformer unit presents low impedance to the frequency of the neighbouring JTC and therefore limits its effective length (Bombardier 2003).

JTC TRC **with receiver points of neighbouring track circuit at the same position and without additional connection between rails** (figure 5.41) is another example of a JTR with voltage operated receiver. It has been design by '*Elteza*' and is used for new installations in countries of the former Soviet Union. JTC TRC has no exact borders, and the overlap is some tens of metres. However the transfer of two receiver signals through a common wire pair is an advantage. But because of non-exact borders between sections, in stations this type of track circuit can be used only with insulated rail joints. There are five frequencies (420, 480, 580, 720, 780 Hz); each frequency is modulated either 8 or 12 Hz (Dmitriev/Minin 1992).

5 Detection

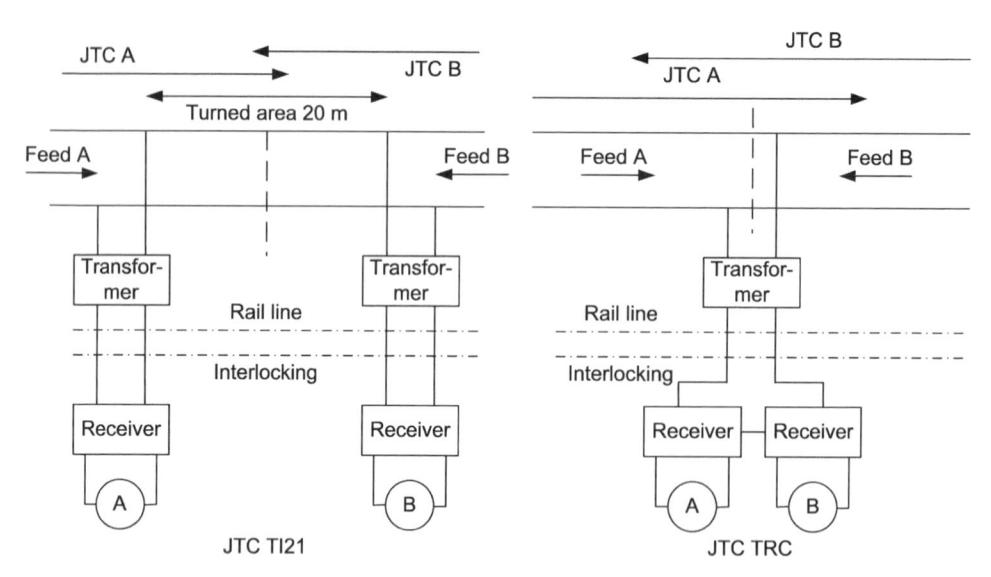

Figure 5.41: Voltage operated JTC without additional connection between rails TI21 and TRC

In JTC with **additional connection between rails**, the precise border between circuits is set by resonant shunt or jumper. In the French JTC Aster 'Type U', the transformers for feed and relay end are located in the overlapping area of two adjacent track circuits (figure 5.42). Borders of JTC B are the two outermost resonant shunts, acting for frequency B as short-circuit. JTC A and C have their own resonant shunts. The employed frequencies are in the band from 1,7 to 2,6 kHz. JTC UM71 *(Union Switch & Signal)* is similar and uses four frequencies of the same band. The frequency of track circuit has dampening on the track air-core inductor (figure 5.42) and short-circuit in the transformer unit of the next JTC (Nock 1982, Retiveau 1987, Ansaldo 2002).

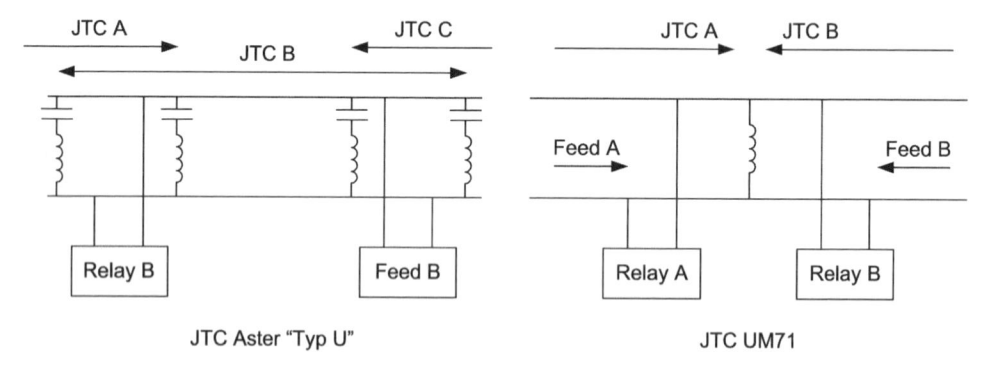

Figure 5.42: Principle of JTC Aster 'Type U' and UM71

JTC FTG S *(Siemens)* uses 12 frequencies of two bands: from 4.75 to 6.25 kHz for the long and from 9.5 to 16.5 kHz for the short track circuits. The bond between the rails is S-shaped (figure 5.44). This jumper bond has inductivity and ensures the short overlapping zone thanks to the resonant circuit LC (figure 5.43). The length of the overlapping zone depends on the frequency and varies between 7 and 19 m. JTC Aster 'Type 1 Watt' has a similar principle and uses six frequencies of the band 1,6–2,8 kHz (Nock 1982).

5.4 Axle Counters

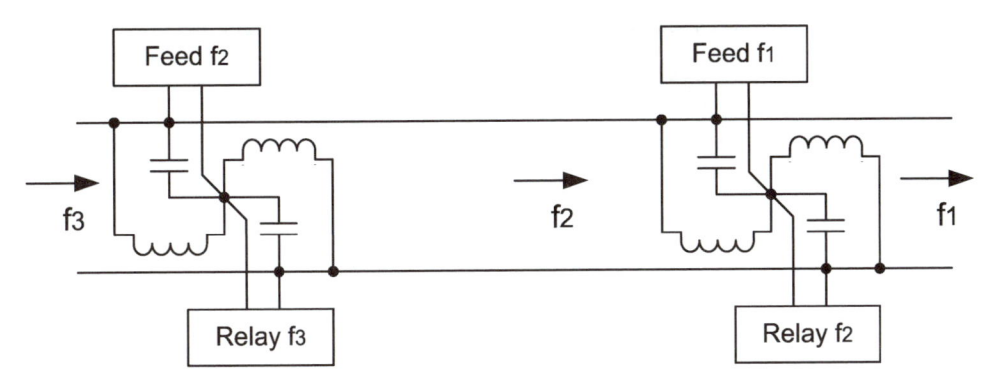

Figure 5.43: Principle of JTC FTG S

Any types of audio frequency track circuits which do not have short circuit devices between rails can be used in combination with other track curcuits (usually DC or AC) as overlay track circuits for different purposes, e.g. activation of level crossings.

A special type of audio frequency track circuit is the one with feed and receiver at the same end. The value of the current can detect not only the presence of a train, but also the distance of the train from the source point. This type of audio frequency track circuits is used in new electronic devices for level crossings (Modern Railways 2004), tracks of marshalling yards (Šeluchin 2005) and for moving block (Watanabe/Takashige 1989).

Figure 5.44: S-shaped connector in FTG S (photo: *Siemens*)

5.4 Axle Counters

5.4.1 General Structure and Functioning

In contrast to the track circuit, the principle of axle counting systems is the indirect detection: If at the beginning of a time period a track section has been clear, and during the period the same number of axles entered and left, then the section must be clear also after this time period. If this condition is not fulfilled, the track section is considered as occupied.

Figure 5.45 shows the basic structure of a modern axle counting system for one track section in the simplest form with one track section to be detected only:

- A **rail contact** (figure 5.45) detects the passage of axles. For detection of direction, it is doubled.
- A **digitaliser** or trackside connection box (figure 5.46) transforms the analogous signal from the rail contact into digital information. This digitaliser is normally situated close to the rail contact to avoid adulteration of the analogous signal. The digitaliser usually consists of different components for amplifying, filtering and transforming functions. An example

5 Detection

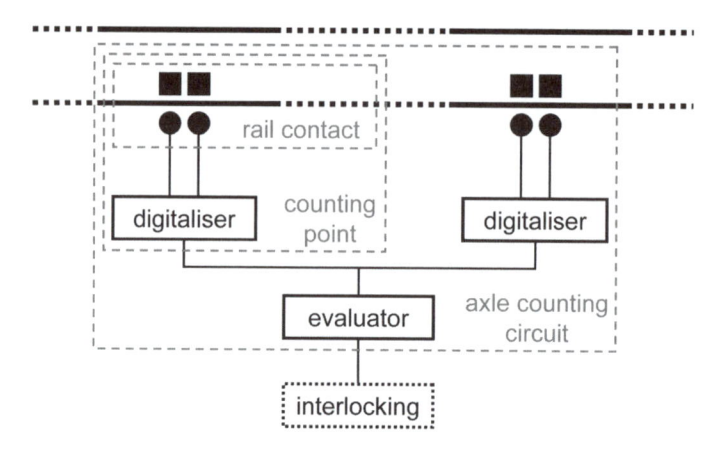

Figure 5.45: Elements of a simple axle counting system

is displayed in figure 5.47. In the evaluation unit for ZP43 (chapter 5.2.2.4), the amplitude changes and their sequence are evaluated (figure 5.47): After passing two amplifiers and a noise filter, the received signal is transformed into a rectangular voltage whose frequency depends on the voltage of the original signal. Finally, a band pass filter filters out the frequency assigned to the activated condition, whereas the signal of the non-activated condition passes the filter.

– An electronic **evaluator** calculates information such as 'track clear' or 'track occupied' from counting results. In modern systems, this is a safe, redundant microcomputer. Depending

Figure 5.46: Rail contact and digitaliser for axle counter produced by *Thales*

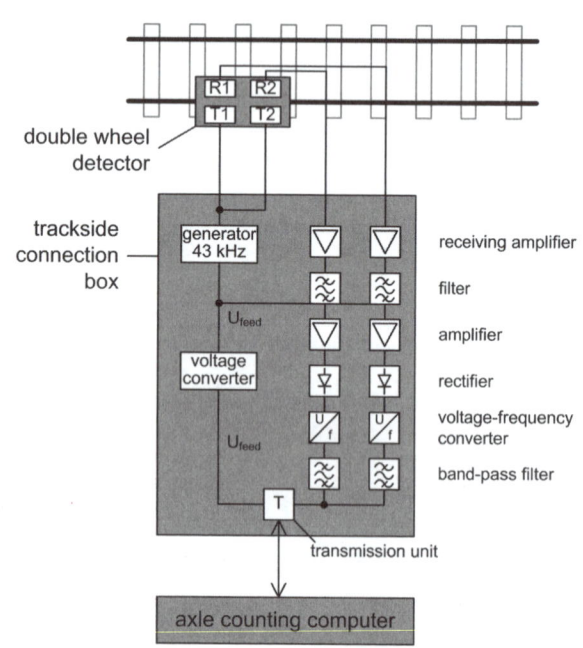

Figure 5.47: Evaluation of axle counter signals in ZP43 (*Siemens*)

5.4 Axle Counters

on the system, parts of the electronic evaluation can be situated in the wayside unit together with the digitaliser, too.

The electronic components and data transmission between the components, with neighbouring axle counting systems, with the interlocking system and with manual data input devices all have to work safely. This is often solved by safety redundancy. If in doubt, the section is considered as occupied instead of clear. The requirements of fail-safety include in particular (Fenner/Naumann/Trinckauf 2003):

- Each entering wheel detected must be evaluated as a line occupation.
- Even if only one out of the two detector heads (chapter 5.4.2) detect an axle (and therefore direction cannot be distinguished), this must be evaluated as an occupation.
- The section must only be evaluated as clear after axles having left in the same number as having entered before.
- No error is allowed to lead to a false clear evaluation.
- The axle counter must be safe also in case of swinging (oscillating) axles.
- The axle counter must be safe against foreign influences such as from traction return currents or magnetic fields from traction vehicles and magnetic brakes.

In the simplest case, an axle counter is formed by two counting points with the intermediate section of track to be proven clear. However, in most modern systems, one evaluation unit is connected with a larger number of rail contacts in geographical proximity and serves for detecting several track sections (figure 5.48). The number of rail contacts wired to one evaluator is different in each type, usually it is up to few tens of counting points.

To ensure continuous track clear detection at the borders between the areas of responsibility of two evaluators, the following solutions are applied as alternatives:

- double connection of one counting point to two evaluators by use of additional modules in the digitaliser; or
- information exchange between two or more evaluators, often using additional modules in the evaluator.

5.4.2 The Rail Contact

In the early days of axle counting, mechanical and electrical galvanic contacts were used. Today, modern axle counters follow inductive principles with the wheel influencing an electromagnetic field (chapter 5.2.2.4). The detailed functioning varies between different manufacturers.

To make a detector useful for axle counting, special requirements apply. In particular, these additional requirements are:

- The rail contact must be able to detect each axle individually.
- The rail contact must be able to distinguish direction.

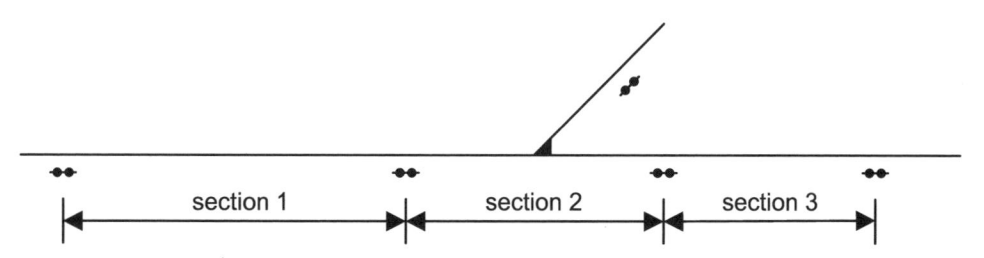

Figure 5.48: Axle counter for multiple sections of track clear detection

5 Detection

A detector which is suitable for axle counting can also be used to obtain information on a train having reached a certain position (chapter 5.1.4.1). Modern detectors are able to perform a permanent self-check, which is necessary to ensure safe detection of the occupation of a track section.

The detection of direction is practiced by doubling the rail contact, with overlapping activation locations of the two sensors (detector heads) (figure 5.49). This overlapping is necessary to detect the direction clearly especially in case of swinging (oscillating) wheels: The sequence of actions of a wheel passing the detector from left to right (figure 5.49) is the following:

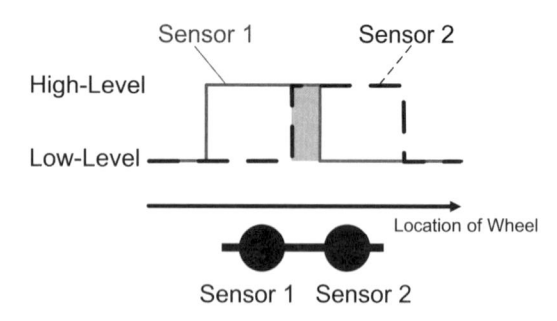

Figure 5.49: Detection of direction by overlapping sensors

1. Sensor 1 activated
2. Sensor 2 activated
3. Sensor 1 deactivated
4. Sensor 2 deactivated

5.4.3 Treatment of Counting Errors

Reliability of axle counters is high in modern technology and counting errors occur seldom. The failure rate of modern axle counting points is not more than 1 failure per 10^8 (Naumann/Pachl 2002). Nevertheless, counting errors have to be considered. Typical error cases can be:
- An axle which passes the axle counter is not detected.
- An axle is counted twice.
- An axle which did not pass the axle counter is counted.
- An axle is counted for the wrong direction.

In first consequence, the fail-safe-principle requires the evaluation of the doubtful sections as occupied. Nevertheless, two simultaneous errors can result in a dangerous situation, e.g. the axle counter counting out two axles in exceed of the real number, and the train loses the last wagon with two axles in the section. Due to high reliability of axle counters and the low probability of two errors which match together to a dangerous situation occurring on the same time, these unrevealed cases are very improbable and need not to be assumed.

In second consequence, operation has to be maintained after such failure. This usually requires human safety critical actions, which reduces the level of safety of the system.

The most important method of failure correction is manual reset of parts of the axle counting system, usually an individual track section. As this action affects safety to a large extend, erroneous resetting of an occupied track has to be prevented. Therefore, the resetting action is bound to several restrictions and obligations which are regulated differently in the operational and technical regulations of the railway companies. Examples of such restrictions and obligations are:
- Only authorised persons are permitted to reset axle counters: These persons can be the signaller or the maintainer, or even both acting together.
- Resetting a section is technically possible only after axles leaving, not after entering the respective section.
- Before resetting, the person concerned is obliged to prove that the section is clear.

- That action is registered and has to be justified in writing by the person executing it.
- After resetting the axle counter, no normal signal operation through the respective section is possible. The first train after resetting has to proceed on sight.

For example, the axle counting system Az S 350 U distinguishes between two resetting functions, applicable depending on the requirements set by the railway infrastructure company (Siemens 2001):

- Immediate resetting function: The section is considered as clear immediately after the reset command by the signaller. However, this input command is only possible if the last counted axle related to the respective section left (and not entered). Otherwise, additional registered safety critical commands are necessary.
- Preparing resetting function: The number of axles is set to zero, but the section is still considered as occupied until the passage of the next (one or more) moving units. For this passage, different requirements can be defined in branched sections. Either two different counting points or all counting points limiting the respective section must be passed, or exactly these two which were used for the last movement before the failure occurred.

As an alternative to manual resetting, automatic correction procedures have been developed (Fenner/Naumann/Trinckauf 2003). These are based on comparison of several consecutive counting points. According to the knowledge of the authors, such procedures did not come into practical use due to reliability of axle counters which have already achieved a very high level (see above).

If a certain counting point cannot be relied on to count correctly, it has to be considered as defective. In some systems, a defective counting point can be taken out of the system and both adjacent sections logically connected to one (Fenner/Naumann/Trinckauf 2003). This usually results in losses of capacity.

5.5 Comparison of Track Circuits and Axle Counters

5.5.1 Advantages and Disadvantages

Both technical systems have advantages and disadvantages (table 5.3). The preference of railways for the one or the other solution for track clear detection varies. If railways wish to make use of the additional capabilities of track circuits such as detection of broken rails, transmission of block, cab signalling and train protection information, the usage of track circuits is favourable. In other cases, axle counters are preferred in recent time due to higher reliability and due to lower requirements regarding the railway superstructure and the treatment of return currents.

5.5.2 Application

The historically older form of technical track clear detection is the track circuit, which until few decades ago was used in the vast majority of cases for technical track clear detection.

The Swiss railways with a large percentage of steel sleepers used in their network were the first to use axle counters already in the first half of twentieth century. In the technical principles of the detectors, mechanical and electric-galvanic solutions dominated (Oehler 1981).

Since the 1950s, usage of axle counters increased, beginning in Central Europe, but also in many other parts of the world. As an example, *German Railways (DB)* use axle counters for all newly built infrastructure, replacement and major refurbishment since around 1995, but in their existing network the share of track circuits is still high. On the other hand, a large number

5 Detection

	track circuit	axle counter
non-detection of railway vehicles	completely derailed vehicles	vehicles newly put on the track
detection of obstacles	only in few cases	no
detection of broken rails	partly	no
vehicle requirements	electrically conducting wheels and axles	ferromagnetic wheels
track requirements	electrical isolation	no special requirements
treatment of traction return currents	special measures necessary	not necessary
excessive voltage problems (e.g. lightning)	present, due to earthing of many devices to the rails	slight
sensitivity to climatic influences	relatively high	low
length of track sections	electrically limited	unlimited
frequency of dangerous failure	extremely low (if rust on rails is prevented)	extremely low
frequency of hindering failure	relatively high	low
possibility of staff preventing danger	stop trains by short-circuiting the rails	no comparable possibility
usability for other purposes	detection of train reaching (in combination also clearing) a certain point; transmission of block information; train protection + cab signalling	detection of train reaching or clearing a certain point

Table 5.3: Comparison of Track Circuits and Axle Counters

of railways, such as in Western Europe, Russia, the USA and Japan, still prefer track circuits. These railways, in the majority, use track circuits also for other purposes than track clear detection, namely for transmission of block information and in their train protection and cab signalling systems.

Nevertheless, several of those railways preferring track circuits do use axle counters in special situations like on steel bridges or on low ballast resistance. For long block sections (e.g. in Russia), axle counters can be considered as favourable due to the unlimited extension of an axle counting section.

6 Movable Track Elements
Andrej Lykov, Gregor Theeg

6.1 Kinds of Movable Track Elements and their Geometry

6.1.1 Overview

Movable track elements are those with movable parts, as follows:
- Elements which interrupt the running rail. Examples of such elements are points in different forms, turntables, traversers, catch points and movable bridges.
- Objects which reach into the clearance profile. Examples are derailers, movable buffer stops, gates and water cranes.

Other elements such as diamond crossings and gauntlet tracks do not always contain movable parts, but if they do (e.g. crossing with movable frogs), the requirements and methods of safeguarding are similar as for points. Therefore, crossings are also a topic of this chapter.

The purpose of movable track elements can be, amongst others:
- transfer of rolling stock units from one track to another
- crossing of tracks
- protection against unauthorised movements
- protection of areas against weather conditions and unauthorised access (gates)

In the following, the most important kinds of movable track elements are described in detail.

6.1.2 Simple Points

6.1.2.1 Structure

The purpose of points is to split a single track into two or more different tracks, or to provide connections between crossing tracks. The basic structure and functioning of points is similar in almost all railways. Figure 6.1 shows the structure of a simple **set of points** comprised of the blades area, the connecting rails and the frog area (Konarev 1994, Uzdin et al. 2002). The transfer of the trains from one track to another is carried out by the blades (2) that are moved by the point machine (1). The blades can be one of the following types:
- Pivoted type: The rail is rigid and has a hinged joint. Historically it is the older form, with the disadvantage of low passenger comfort and of permitting only low speeds. Today this type is little used and mainly for low requirement applications, e.g. in industrial and mining railways.
- Elastic type: The rail bends. The disadvantage is the high force needed for switching. However, this is the predominating type today.

The connection of tracks is provided by the stock rails (3) and connecting rails (4, 5). The guard (or check) rails (6) ensure the guidance of wheel flanges in the area of discontinuous rail between the frog core (8) and the wing rails (7) and therefore prevent accidents by derailment.

The **fouling point** is the intersection of the outer limits of the clearance profiles of both tracks, which means that location up to which a vehicle can stand on the one track without endangering a movement on the other track of the points. In most railways, it is marked by a fouling point indicator (FPI) (9) which can have different shapes (figure 6.2).

6 Movable Track Elements

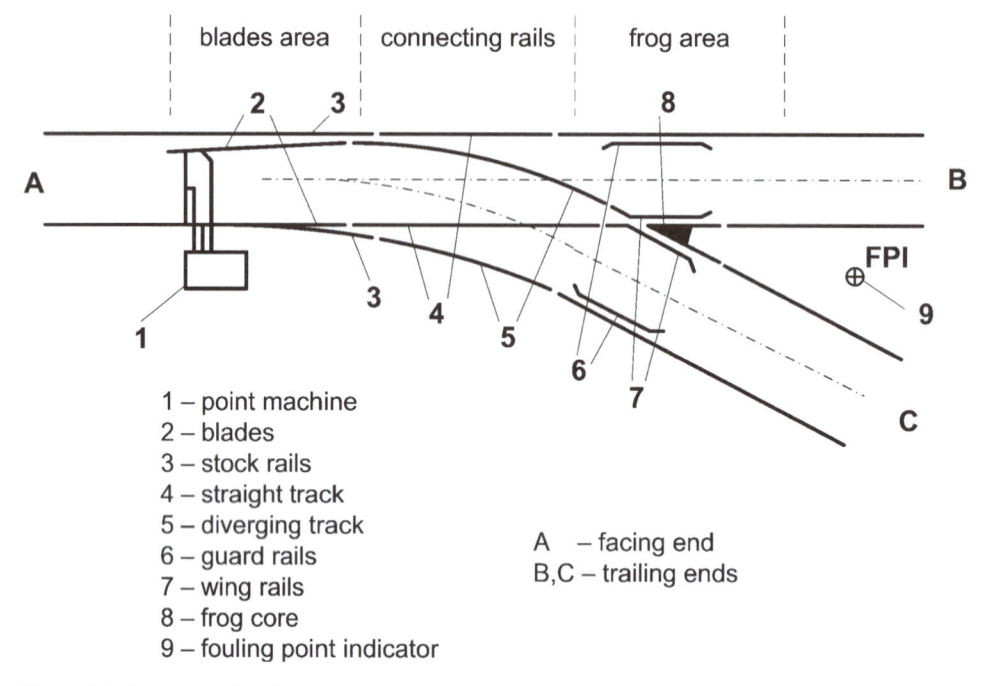

Figure 6.1: Structure of a simple set of points

1 – point machine
2 – blades
3 – stock rails
4 – straight track
5 – diverging track
6 – guard rails
7 – wing rails
8 – frog core
9 – fouling point indicator

A – facing end
B,C – trailing ends

Figure 6.2: Different forms of fouling point indicators (left: Slovakia, middle: Germany, right: Italy)

6.1.2.2 Movements at Points

Points can be positioned in three ways: two of them are called end positions, the other one the intermediate position. Traditionally, on most railways one of the two end positions is referred to as normal (plus) position, the other one as reverse (minus) position, usually according to the position in which the points are used more frequently and to the position of the point lever in mechanical interlocking. It is thus incorrect to identify a point's plus position always with the straight position. Today in several countries the nomenclature 'normal/reverse' or 'plus/minus' is no longer used. Instead, a 'left' and a 'right' position is spoken of, seen from the facing end.

In the position shown in figure 6.1, the points enable traffic in the A ↔ B direction. The lower blade is fitted to the stock rail, whereas the upper one is retracted from it to a suitable distance that ensures the unimpeded passage of a wheel flange. To enable the movement in the A ↔ C direction, the blades are shifted in such a way that the upper blade is fitted to, whereas the lower one is held away from the stock rail.

A movement in the direction in which the tracks diverge at the points is called a **facing move** (A → B/C). A movement in the direction in which they converge is called a **trailing move** (B/C → A). A trailing move from the direction opposite to the one in which the points are set will cause the blade to be moved by the wheel flanges of the train, i.e. the points will be trailed. With some types of point drive or locking mechanisms the blades cannot be trailed, and a vehicle attempting such a movement unauthorised would be derailed and/or point damage would occur. Other point locking mechanisms are trailable to avoid derailment and damage.

Besides the unintended trailing, in some cases points and interlocking systems are intentionally designed for the switching of the points by moving rolling stock. The main fields of application are tramways, secondary lines and shunting areas. So-called **spring points** have a defined normal position and a mechanism which automatically returns them to their normal position if not occupied by rolling stock. They are particularly useful in applications such as a station with two tracks on a single line, which is only used for the crossing of trains.

6.1.2.3 Geometrical Parameters

Depending on the local situation, the traffic density and the speed of the trains that is acceptable in the diverging track, different points with different parameters are applied.

When a moving unit moves in the diverging track of a set of points, the sequence of geometrical elements to be passed is in most cases: straight line → arc with constant radius → straight line (figure 6.3). The speed in the diverging track is limited by the radius of the arc, the so-called turnout radius, for safety, maintenance and comfort reasons. Besides, in some situations, speed limits can also be caused by the kind of locking of the points. If the turnout radius is the reason for the speed limit, in most cases the limiting effect is the jerk (change of centrifugal force) and thereby the change of curvature at the ends of the arc. As the radius is related to the length of the points and the turnout angle, regulations in some countries consider the turnout angle (α in figure 6.3) as the limiting factor: The smaller that angle is, the higher the acceptable movement speed usually becomes. Other parameters like the tangent length are connected with the parameters mentioned above.

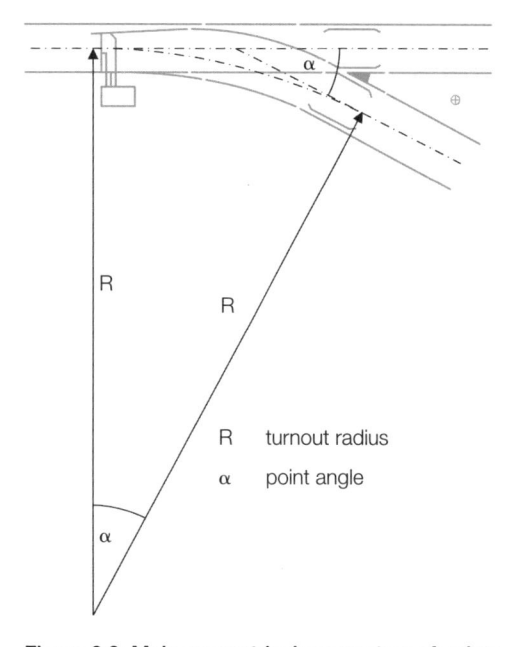

R turnout radius
α point angle

Figure 6.3: Main geometrical parameters of points

Most railways have defined standardised types of points which are used in their networks. For example, on the Russian and North American railways, the inclination of tracks in the frog area is considered as the defining parameter. In Russia, points with frog inclinations 1:9, 1:11 and in high speed areas – 1:18 and 1:22 are defined. In North America each railroad has its own standard.

In Germany, as another example, the turnout radius and the inclination at the trailing end of the points are considered as the defining parameters. To increase flexibility in planning the track

6 Movable Track Elements

layout of a station, usually two or more different turnout radii are defined with the same inclination and different length of the arc. For example, points 190-1:9 (with straight frog) permits speed up to 40 km/h, whereas points 300-1:9 (with curved frog) permits speed up to 50 km/h. Altogether, inclinations 1:7.5, 1:9, 1:14 and 1:18.5 and radii 190, 300, 500, 760 and 1200 m are defined for turnout speed between 40 km/h and 100 km/h, not considering special high speed points.

6.1.2.4 Special Forms of Simple Points

For **high speed points**, the simple geometrical layout straight line → arc with constant radius → straight line is not sufficient to provide high speed in combination with an acceptable length of the point construction. Instead, segmental, clothoid or parabolic arches with changing radius are used for speeds up to 200 km/h or more. Such points are particularly required where one high speed line splits into two or for crossovers on high speed lines. The slimmest inclination in use is 1:65 in France.

Points can be used on curved lines. Two cases can be distinguished: when both the diverging tracks curve in the **same** direction, and when they curve in **opposite** directions. In the second case, if the arrangement is largely symmetrical and if the radii of curvature are similar, they may be called wye points on account of their shape, Y. A typical application is in gravity marshalling yards between the hump and the classification tracks (chapter 12.1).

In figure 1, it can be seen that between the wing rail and the frog core the tread of the train wheel (or surface on which it runs) is interrupted. This increases the dynamic strain that impacts upon the frog and decreases passenger comfort when the rolling-stock runs over points. With higher movement speeds and lower turnout angles, this effect increases sharply. To enable a smooth running at high velocity, points with an uninterrupted tread are to be used. These points use a movable frog, requiring an additional point machine in the frog area. The movable frog is adjusted tightly against the appropriate side verge of the wing rail, creating thereby a continuous tread for the train wheels.

6.1.3   Other Solutions for Connection of Tracks

The summation of all the points placed at one station entrance is called the station throat. For compact design of the pointwork in a station throat with a minimum use of land area, the elements described below can be applied (figure 6.4).

A **diamond crossing** (figure 6.4, a) is used for the crossing of two tracks and represents a construction consisting of two sharp and two blunt crossing frogs. Diamond crossings can have movable frogs or not, mainly depending on the crossing angle and the required speed. In both cases, diamond crossings have to be route interlocked to prevent train collisions.

A **slip crossing** (figure 6.4, c) combines the functions of a diamond crossing and points. There are single slip crossings as well as double ones. A double slip crossing has eight blades, two sharp and two blunt frogs and four guard rails. The two pairs of blades at the same end of the slip crossing are switched together by one point machine. Slip crossings are somewhat difficult in their construction and are therefore used restrictively by many railways, but are a compact solution at stations whose area is restricted.

Double points, also called three-way-points (figure 6.4, b), permit the branching of a single track into three. It contains the two pairs of blades, five guidance rails and three frogs. The advantage of this solution is the compactness in comparison with two simple sets of points, but disadvantages are difficulties in the construction, problems pertaining to track circuits and the

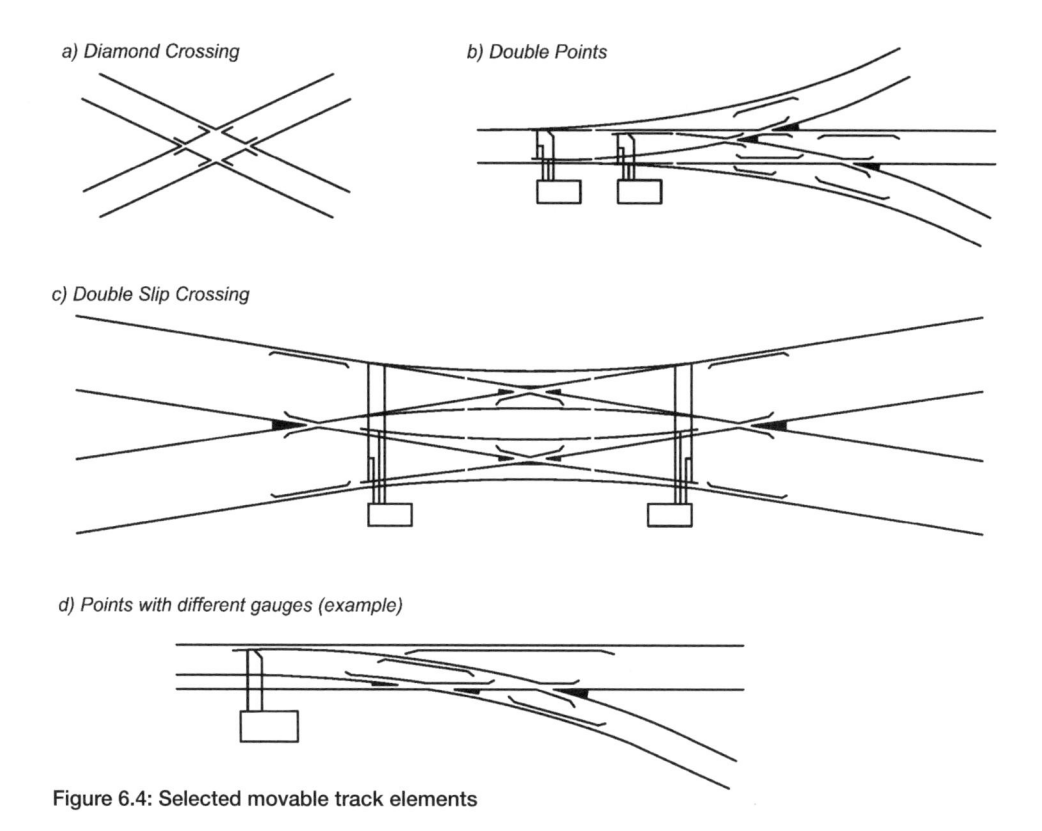

6.1 Kinds of Movable Track Elements and their Geometry

a) Diamond Crossing

b) Double Points

c) Double Slip Crossing

d) Points with different gauges (example)

Figure 6.4: Selected movable track elements

unsmooth running of trains due to the presence of three frogs. Therefore, their use is widely avoided.

Where different gauges (e.g. standard and narrow gauge, which frequently occurs in Europe) meet, **points with different gauges** can be required. Depending on the local arrangements, they can become rather complex in construction. A simple example is shown in figure 6.4, d.

At entrances to locomotive depots, in case it is necessary to distribute the rolling-stock from one or a few tracks to a large number of depot tracks, **turntables** can be used. They represent a girder with a track on which a locomotive is placed to enable it to rotate in a horizontal plane. Another purpose of turntables is to turn locomotives with a single cab.

A **traverser** is a similar device as a turntable. Here the movement is not a rotation, but a translation. The traverser is a very compact device to connect parallel tracks, with the disadvantage that the moving unit has to stop and that its length is limited. Traversers are mainly used in depots.

6.1.4 Arrangements of Several Movable Track Elements

In the following, typical forms of arrangements of several movable track elements are described, which are used for a compact track layout (figure 6.5).

A **crossover** (figure 6.5, a) represents an arrangement of two simple sets of points. It is the standard solution to connect two parallel tracks. The point positions in the routes are interdependent and both provide flank protection for each other (chapter 4.3.5). Therefore in many

6 Movable Track Elements

a) Crossover

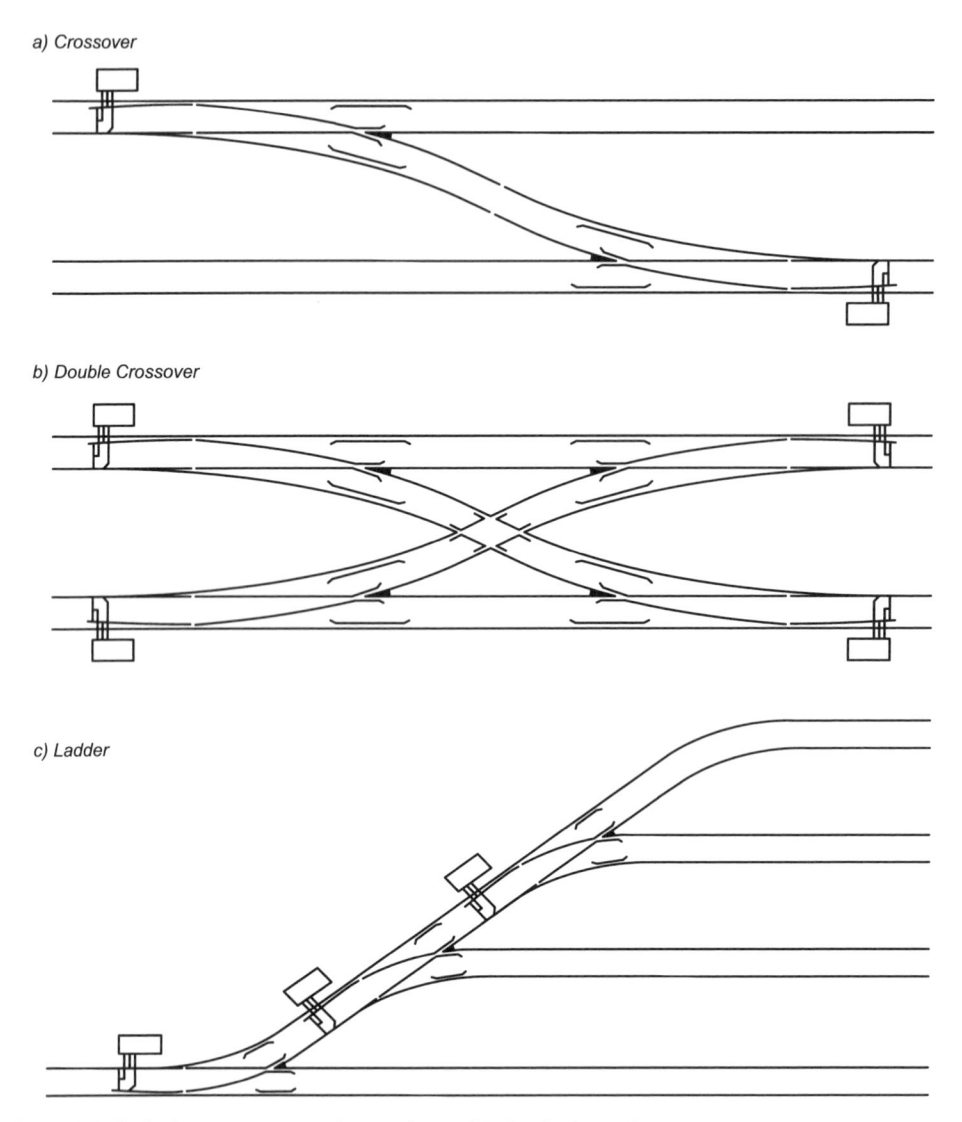

b) Double Crossover

c) Ladder

Figure 6.5: Typical arrangements of several movable track elements

countries these points are coupled in their operation and control to ensure the safety of train movements and reduce efforts in terms of facilities and cable maintenance.

A **double crossover**, also known as a **scissors crossing**, (figure 6.5, b) emerges when two crossovers meet together. It is a more compact solution than two crossovers, but requires an additional diamond crossing. As with the simple crossover, it enables independent movements along the straight tracks, but transfers from each side of those tracks to the parallel track exclude other movements.

A **ladder** (figure 6.5, c) is an arrangement of different sets of points with the same point angle to split up one track into three or more parallel tracks. The points are places in a line which is inclined by point angle towards the parallel tracks. Ladders are mainly used for the points in the station throat and yard necks.

6.1.5 Derailing Devices

The purpose of derailing devices is to prevent accidents caused by unintended movements of rolling stock by stopping these vehicles. Reasons for such unintended movements can be, among others:
- rolling away of parked vehicles from sidings, freight yard, warehouses or others due to error of staff or technical defect;
- rolling back of vehicles after a train separation in a section with rising gradient;
- non-compliance of signals or other instructions by drivers.

Such special derailing devices are double and single catch points and derailers. Application of these devices is aimed to stop rolling-stock units forcibly by derailing them.

The double **catch points** represent a part of a simple set of points consisting of the blades area and partly the connecting rails (figure 6.6, a). The simple **catch points** are simpler in their construction, although not less effective to prevent a spontaneous exit of rolling stock onto the

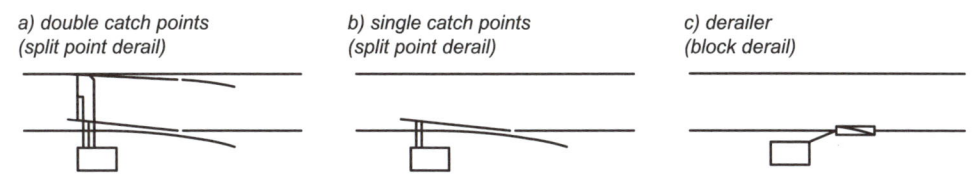

a) double catch points (split point derail)   *b) single catch points (split point derail)*   *c) derailer (block derail)*

Figure 6.6: Derailing devices (terms used in USA in brackets)

main tracks (figure 6.6, b). In their normal position the catch points are switched in the direction of dump, i.e. in the event of a spontaneous movement the vehicle will be derailed and come to a halt. To enable the passage of a train or shunting movement, the catch points are switched to the passing position and after the moving unit has passed, they are returned to their initial position. Under interlocking conditions, when the catch points are equipped with an electric point machine, the return operation is carried out automatically after route release, i.e. the preferred position of the catch points is the dump position.

The **derailer** is somewhat different in the principle of its operation (figure 6.6, c, figure 6.7). This device does not disrupt the track, for it represents a bar of a special profile placed from above onto the rail head. The protrusion at the bar provides for the derailing of the rolling stock in a defined direction. In its normal position the derailer is placed upon the rail and can be moved aside to enable a movement. Many railways do not permit the use of derailers in main tracks, but only in secondary tracks.

A comfortable alternative to derailing devices are so-called trap points. **Trap points** are complete simple points with a short track section attached to the diverging track and with a buffer stop or a sand drag at its end. Normally, the diverging track is never intended to be driven onto by a moving unit, but serves for protection only. Trap points are preferably used where vehicles occupied with passengers or with dangerous goods or moving at relatively high speed are expected to be stopped. They are more expensive in installation and require more space.

Figure 6.7: Derailer (Spain)

6.2 Safety Requirements at Movable Track Elements

The safety requirements at movable track elements can be summarised as follows:
1. Prevention of collision with other vehicles.
2. Prevention of derailment at interruptions of the rails or collision with parts of the movable track element.
3. Prevention of derailment by excessive speed in the diverging branch.

Safety requirement (1) is met by route locking functions (chapter 4) in connection with track clear detection (chapter 5). Particular problems in this context occur with the occupation of that track of a set of points or crossing which is not part of the running path for a certain movement. These problems and solutions for them are described in chapter 6.3.

Safety requirement (2) is met by the following functions in the mechanism of the movable track elements:
- Switching of the movable track element, i.e. the moving of its movable parts to the required position. Even if the points come to standstill in the intermediate position by error, they can be switched to an end position again.
- Locking of the movable parts in the required end position. Locking means in particular that one point blade must be fitted to the stock rail with low tolerance to ensure a continuous guideway, whereas the other point blade must be hold away from the stock rail (with higher tolerance) to ensure safe passing of the wheel flanges.
- Supervision to ensure that the movable track element has reached the required end position and remains there. This feedback information is a precondition for permitting a movement over the movable track element.

An additional requirement particularly in areas with frequent shunting is the trailability of points, which is required by many railways generally or in certain situations. This means that if a rolling stock unit erroneously passes the points from the trailing direction which does not correspond with the current end position of the points, it must be able to proceed over the points without derailment or damage by forcing the movable parts to the required position. However, the trailing of the points must be detected and usually further use of the points be prevented until inspection.

The units which fulfil these functions are described in chapters 6.4 and 6.5.

Safety requirement (3) is fulfilled by selection of the proper signal aspect (chapters 4 and 7), occasionally in combination with speed supervision (chapter 8).

6.3 Track Clear Detection at Points and Crossings

At the fouling point (chapter 6.1.2.1), the distance between the tracks varies between the railways and with the turnout radius; usually it is around 3.50 or 4.00 m. When technical track clear detection is used, the insulated rail joints (IRJ) or axle counters are placed at a certain distance from the fouling point to take into consideration the overhang of the wagon over the last axle (figure 8a) (Kononov/Lykov/Nikitin 2003; Sapožnikov et al. 2006). For example, on Russian railways this distance is 3.50 m. In Germany, an additional protective length against stretching of the standing train is provided and this distance increased to 6.00 m.

In case it is impossible to comply with the dimensional requirements (e.g. the points 1, 3, 5 and 7 of the ladder arrangement in figure 8b, the adjacent track section on the other branch track

6.4 Point Machines

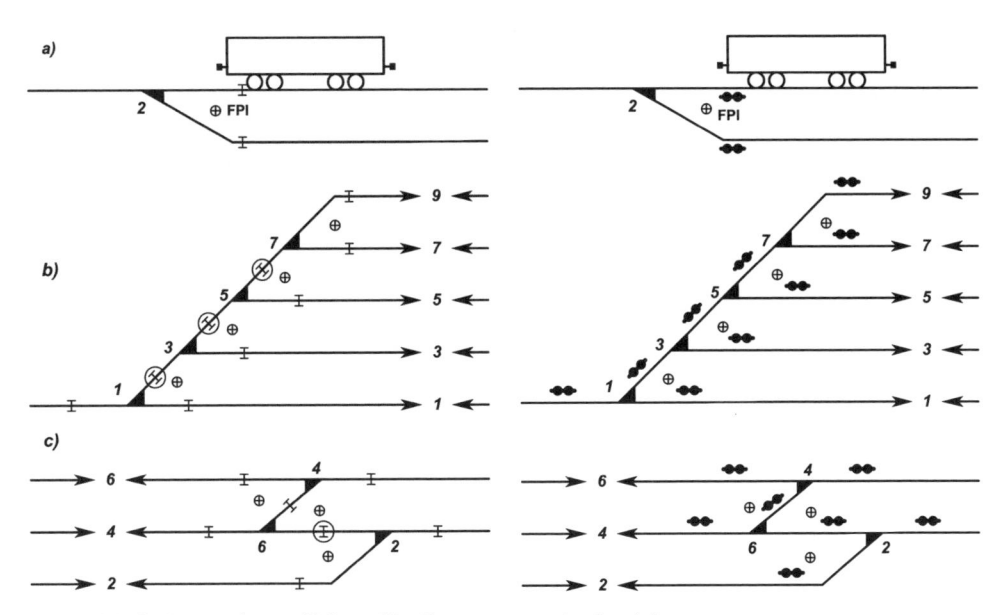

Figure 6.8: Safety against collision with other movements at points

must also be proved clear to permit the movement. In these cases the IRJ are called oversize IRJ.

In contrast, for the points of a crossover this rule does not apply if the points always give flank protection to each other, as the section between the points is short enough that vehicles are assumed not to stand there. For example, points 6 in figure 8c are dual protective points (chapter 4.3.5.4), giving flank protection to points 2 and 4. But in this particular case, it is coupled with points 4, giving flank protection preferably to points 4 and is therefore not locked for the diverging route over points 2. Therefore, the oversize rule applies for the IRJ between points 2 and 6, but not the one between points 4 and 6.

6.4 Point Machines

6.4.1 Overview

There are three functions associated with moving points:
- **switching the points** – applying a force to the blades to move them
- **locking the points** – holding them in position after the drive force is switched off
- **supervision of points** – reporting the actual current position of the blades (to the signaller or signalling system).

When points were first motorised, these functions were provided by three separate pieces of hardware. Nowadays combined machines are usually used, which perform all three functions.

The switching of points, locking of the blades and supervision of their end positions are carried out by the point operating gear (the point driving units) which may vary by the energy they use as well as by the means of locking the blades, trailability, commutation in the motive and control circuitry and also by the time required to switch a set of points from one end position to the other.

6 Movable Track Elements

The following switching technologies can be distinguished according to the types of energy they consume and the form this energy is transported and transformed:

- **Manual local** (figure 6.9). The points are switched locally by the physical strength of a person. The advantage of this solution is its simplicity in installation, the disadvantage the high manpower effort in operation and long time required due to the need of walking to each set of points. Manual local point operation is widely used in shunting areas and areas with low traffic, where points are switched by vehicle staff to relieve the signaller from this work. Checking of point position can be made by key locks. Here, a key can only be removed if the points are in the respective end position.

Figure 6.9: Manual point drive with key locks (Germany)

- **Manual remote.** The point levers are centralised and force is transmitted by wires or rods. This is the typical solution in mechanical interlocking systems and is used to a decreasing extent. The person does not need to walk to each set of points, but the possible length of the wires/rods and therefore the area of responsibility of one person is small, requiring high manpower efforts.

Figure 6.10: Electro-pneumatic point machine (USA) (photo: *Pachl*)

- **Electro-pneumatic** (figure 6.10). The blades are switched by compressed air energy. Supervision is usually solved electrically. The advantages of these point driving units are their small dimensions and simplicity of construction. One disadvantage is the necessity to install compressors and an air pipe network. Besides, the compressed air used as a working body requires to meet some stringent requirements pertaining to the acceptable levels of moisture composition, since it may bring about corrosion of pipelines (under high temperature conditions), and glaciation of choke joints (under low temperature conditions). Modern forms, until today, are frequently used in hump areas.
- **Electro-magnetic.** Electric power is transformed into mechanical power by solenoid magnets. Such point machines have relatively small dimensions if the weight of blades is low and are rather simple to operate and maintain. As the size and weight of the blades increases, the dimensions of material and energy intensity of such units grow sharply.
- **Electro-mechanical** (figure 6.11). Electric power is transformed into mechanical by means of AC or DC motors and transmitted to the blades by mechanical gear. Usually electric power is also used to provide supervision and control. Advantages are the high efficiency of electric motors, higher reliability, the opportunity of using the same cable lines for supervision and control and the stability of characteristics of electric current. For these reasons electro-mechanical

point machines are among the most widely used modern point machines.
- **Electro-hydraulic** (figure 6.12). Modern forms of electro-hydraulic point machines work similarly to the electro-mechanical ones, with the exception that the transfer of the rotation of the motor to the movement of the point drawbar is solved by hydraulics. The motor operates pumps which pump oil into a cylinder and move a piston there. Electro-hydraulic point machines of this type are frequently used in Western Europe and, together with electro-mechanical machines, are the most widely used modern point machines.

Figure 6.11: Electro-mechanical point machine (Poland, manufacturer: *Bombardier*)

The most frequently used modern types are the electro-mechanical and the electro-hydraulic point machine. In the following, they will be called Electric Point Machines (EPM). In some new systems, they are integrated into the sleeper to facilitate track maintenance work with big machines.

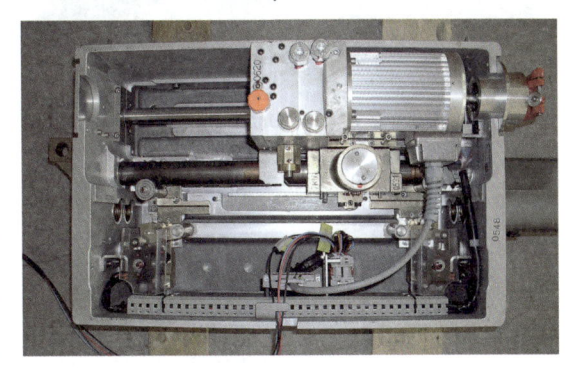

Figure 6.12: Electro-hydraulic point machine (manufacturer: *MKE*)

6.4.2 Electric Point Machines

Due to the high importance and complexity of EPM, they are described in more detailed here. The EPM may vary according to:
- the time needed to switch the points. For normal interlocked points, the switching time is between two and seven seconds. For special applications such as humps, rapidly operating points with a switching time down to half a second are frequently used (chapter 12.3.4).
- the location in relation to the points. EPM are usually on one side of the track, but they can also be between the rails, or even inside sleepers. The side of installation of an external EPM can be determined by the available space between the tracks, by convenience of maintenance and conditions for laying of cables and air pipes for removal of snow. Usually an EPM is installed on the outer side of a double track or in wide spaces between tracks.
- the working point of the force onto the blades: Usually there is one working point at the end of the blade. But for points which are designed for relatively high speed (high turnout radius), often several working points are required, either connected with the same EPM (figure 6.13, right) or with several coordinated EPM's (figure 6.13, left). The difficulty is to adjust the switching lengths at each working point exactly to the required bending line of the blade. If a movable frog is required, at least one additional working point is necessary here.
- the centralisation of operation: EPM can be operated centrally from an interlocking station or locally. The local control in Europe is only applied on secondary lines and in shunting ar-

6 Movable Track Elements

Figure 6.13: Point machines with several working points (photos: *Siemens*)

eas, in the USA also on main lines. Here, the switching of the points can either be initiated automatically by the approaching moving unit (usually from the trailing end) or by the driver, e.g. by a pushbutton on the trackside or by an infra-red sender in the locomotive. Simple interlocking functions can be provided in these systems, described in chapter 9.4.10.
– construction of the reduction gear. In case of mechanical transmission, mostly tooth, worm or screw gear arrangements are applied.
– the type of treatment of trailing of points. Points can be made as trailable, equipped with a device that provides specified resistance against the movements of the throw bar initiated from outside the EPM. This prevents the ruin of the EPM if trailing of points happens. The other type is non-trailable points that will become damaged when being trailed.
– the types of locking. This can be internal locking with the locking mechanism inside the point machine or external locking with the locking mechanism being located in the track. In most cases, but not generally, internal locking mechanisms are not trailable, whereas external locking mechanisms are. The use of one or other of the solutions differs mainly by country. In central Europe, external locking mechanisms predominate, whereas in Russia, Britain and the USA internal ones are mostly used.
– the way of commutation of operational and supervision circuits. The contact mechanisms are applied using mechanical contacts or contactless.

To meet the requirements of switching, locking and supervising, an EPM has three operational modes:

1. Working. the mode whereby points are being switched.
2. Supervising. the mode ensuring that one blade is fitted tightly against the stock rail, while the other is kept in a safe distance from it. A particular requirement is resistance against dynamic impact: The EPM and point fittings must be able to endure a dynamic impact induced upon them by the moving rolling stock.
3. Trailed. This mode is active if the points have been forced open by an impact made by rolling stock whereby the supervising mode becomes violated. There is no possibility for transfer into working mode (if the EPM is not trailable) or it requires special actions to bring it into working mode (if the EPM is trailable).

In spite of the above mentioned differences and of some differences in the construction details, the working principles of most EPMs are similar.

6.4 Point Machines

The generalised block diagram is shown in figure 6.14:

- In an electro-mechanical point machine, electric power is transformed into mechanical by means of an AC or DC electric motor M. The motor rotation is spread on to the reduction gear R meant to strengthen the angular momentum and to reduce the rotary speed of the motor. The motor is connected with reduction gear via branch sleeve which allows an insignificant radial displacement of shafts while retaining a parallel position of their axes. To protect the motor from overloads, e.g. if the blades do not reach their end position due to an obstacle, and to ensure the braking of the revolving parts of the EPM after the end of switching the points, a friction gear is inserted into the gear. The rotating movement is transferred into the progressive motion of the **throw bar** TB in the last cascade of the reduction gear.
- In an electro-hydraulic point machine, the electric motor M rotates the pump of the hydraulic gear R. This pump pumps oil from one cylinder into another and causes a relative movement between the cylinder and a piston. Either the cylinder or the piston is mechanically connected with the **throw bar** TB, the other is fixed.

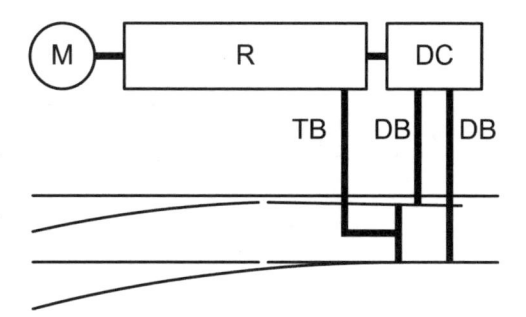

Figure 6.14: EPM block diagram (R is a mechanical gear in electro-mechanical and a hydraulic gear in electro-hydraulic point machine)

In both forms, the throw bar impacts upon the blades of the points through the point drive rod. The **detection contacts** DC provide checking of point positions and commutate the electric controlling circuits. Obtaining the checking signal about of point end position is only possible if the position of the **detection bars** DB conforms to that of the throw bar.

An important factor for exchangeability of EPM's of different manufacturers is compatibility of two kinds:

- Electrical compatibility of the EPM in the operation and supervision circuitry. An example for a standard is the German four-wire point circuitry (chapter 6.6.3.1).
- Mechanical compatibility at the interface between EPM and point drawbar, regarding mechanical connections, switching length and others.

Often compatibility is provided in one country, but not internationally.

For degraded mode operation and for maintenance, EPM shall enable the possibility of switching the points by the hand crank. During hand cranking, electrical movement must be prevented for safety.

The EPM influences directly the safety of train movements, since is checks the actual position of a set of points. The idea of supervision of point position is to verifiy the conformity between the detection bars and the throw bar. In order to check point position, one checking drawbar is attached to each blade. These drawbars are connected with the detection bars which move inside the EPM.

6.4.3 Supervision of Point Position on the Example of SP-6

To illustrate the principle of proving the point position, let us examine the checking block of the Russian EPM SP-6 (Reznikov 1985, Sapožnikov et al. 2008). According to the classification given in chapters 6.4.1 and 6.4.2, this EPM is electro-mechanical with possibility of using

6 Movable Track Elements

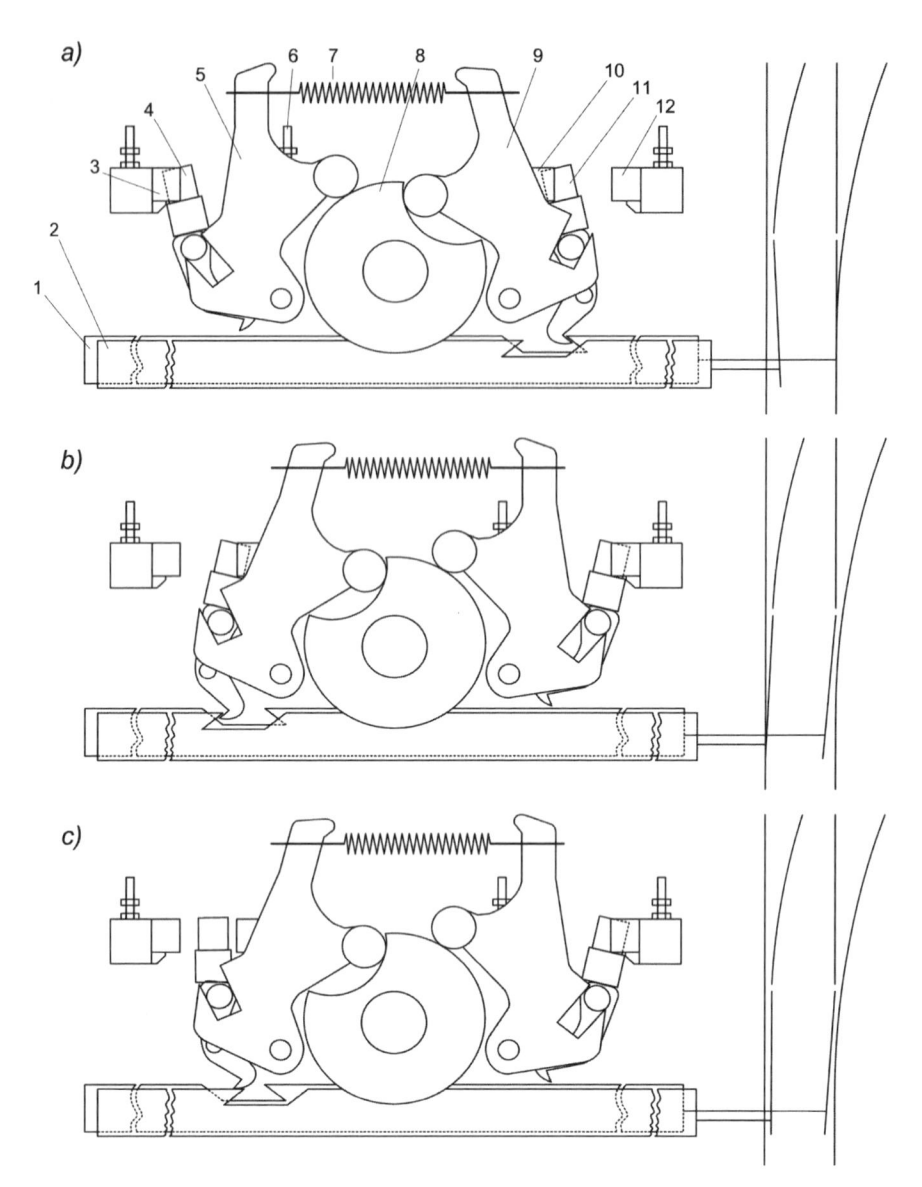

Figure 6.15: Checking of point position in SP-6

either AC or DC motors, with electrical supervision using mechanical contacts. The tooth gear transmission is designed to be not trailable with internal locking. EPM is designed to be installed on one side of the track.

The movements of the throw bar for the fixed distance and confirmation that the detection bars are in this position are verified by the switching levers 5 and 9, and jointly with them connecting levers 4 and 11 (figure 6.15). When the throw bar is located in its end position, the roller of one of the switching levers sinks down into a notch of the collar 8 mounted upon the main drive shaft of the reduction gear. The checking scheme is commutated with the detection contacts 6, 10. These contacts are closed by the connecting levers. The closing is possible if the beak-shaped end of the connecting lever is dropped into in the superimposed notches of the detection bars 1 and 2.

Figure 6.15a shows the state of details of an EPM checking mechanism when the blades are located in their end position, with the right blade fitted to the stock rail and the left blade free. Herewith the operational contacts 3 (controlling a control circuit of an EPM) are connected with the connecting lever 4, while the checking contacts 10 – with the connecting lever 11.

When the points are switched to the opposite position, the main drive shaft of the reduction gear rotates in a clockwise direction. Firstly, the roller of the switching lever 9 rolls onto the surface of the collar 8. That results in the following successive movements: the connecting lever 11 moves aside, disconnecting the checking contacts 10, and the operational contacts 12 become connected. Henceforth the throw bar begins moving and the blades of the points do the same together with it. This, consequently, provokes the movements of the detection bars 1 and 2. At the moment of the final movement of the throw bar, the notch of the collar of the main drive shaft becomes positioned under the roller of the switching lever 5, which causes it to move to the right under the influence of the spring 7. That results in the connecting lever 4 disconnecting the work contacts 3. If all elements of the EPM and the points are in working order, and therefore the blades and detection bars have moved to their end positions, the beak-shaped end of the connecting lever 4 drops into the superimposed notches of the detection bars. Owning to that, the checking contacts 6 (figure 6.15b) become connected.

They will not be connected, however, if at least one of the detection bars does not move for the specified distance, e.g. as a result of a breakage. In this case its notch will not be positioned under the beak-shaped end of the connecting lever and it will be propped against the surface of the detection bar (figure 6.15c). Neither the contacts 4 nor 6 will be connected.

In case of trailing of the points, the collar 8 and the main drive shaft do not revolve, but the detection bars do move. The beak-shaped end of the connecting lever is pushed to the surface of the detection bar by the splayed edge of the notch of this bar. In that case the connecting lever occupies an intermediate position disconnecting the checking contacts. At that time, the connecting and switching levers of the other blade do not change their position, and the operational contacts remains to be connected.

6.5 Point Locking Mechanisms

Under dynamic impact from passing rail vehicles, blades should be locked. As was already pointed out earlier, their locking can be external or internal. Besides, locking of the blades in the end position can be either form fitted (not trailable) or force fitted (trailable). In the following, some examples for locking mechanisms are described.

6.5.1 External Locking Mechanism: Clamp Lock

In these countries which use external locking mechanisms, there is a large variety of locks. However, the by far most widely used solution is the clamp lock. The clamp lock (figure 6.16) is trailable. Besides the clamp lock, in recent years modern optimised external locking mechanisms have been developed by different manufacturers, which are optimised for low friction and are therefore used for points which shall be switched very frequently. On high speed lines, a problem of trailable points can be the danger of unintended switching by dynamic impact. Therefore, special locks are often applied.

The drive rod is fixed to the point machine via the throwbar, but not to the blades. Instead, the blades are mounted to special lock arms. When the blade is unlocked (right blade in figure

6 Movable Track Elements

Figure 6.16: Clamp lock

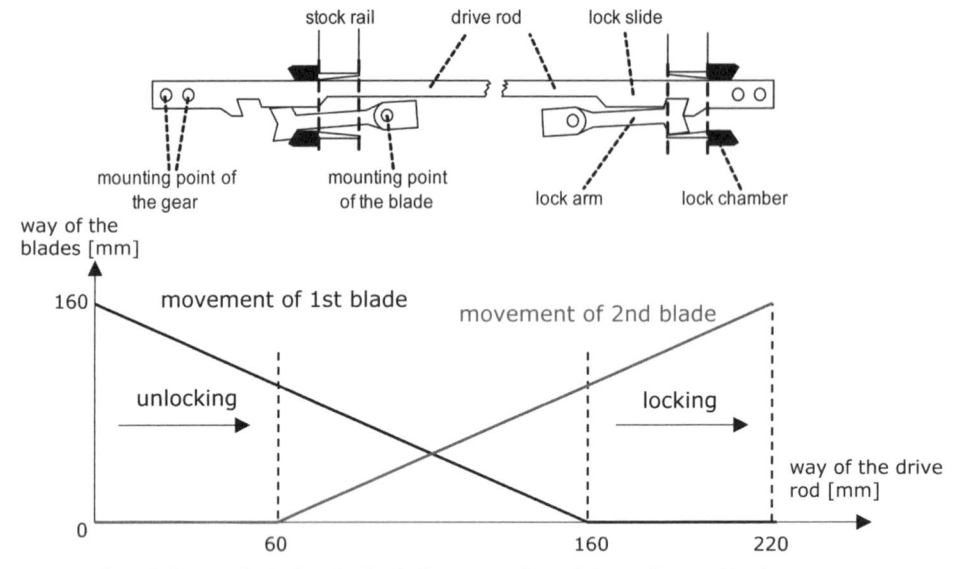

(Length figures refer to the situation in Germany and vary between the countries.)

6.16), the so-called 'swallow tail' of the lock arm is hold tight in the groove of the drive rod by the lock chamber and moves therefore together with the drive rod. In the locked position (left blade in figure 6.16), the lock arm is fixed between the lock slide and the locking piece and can therefore not move even if a strong force tries to push it away from the stock rail.

The switching process consists of three phases:

1. Unlocking phase. The open blade moves with the drive rod, whereas the closed blade is being unlocked.
2. Moving phase. This phase begins when the groove of the drive rod has reached the 'swallow tail' of the lock arm. Now both blades move with the drive rod.
3. Locking phase. This phase begins when the formerly open blade has reached the stock rail. Now this blade is being locked, whereas the formerly closed (now open) blade continues to move with the drive rod until the end position.

The trailing of points works as follows. When a force from the wheel of a rolling stock unit pushes the open blade towards the stock rail, the same process as in normal switching starts, with the difference that the force enters from the blade and not from the point machine. The open blade is pushed against the stock rail, unlocking the closed blade and then pulling it away from the stock rail to give the wheel flange the possibility to pass. When the formerly open blade has reached the stock rail, it remains in unlocked position, i.e. the third phase of the switching process (Locking) is skipped. Therefore no end position will be detected and no route can be set across the points. The slipping of the force-fitted coupling in the gear block prevents further switching of the points until technical inspection.

6.5.2 Internal Locking Mechanism

The Russian point machine SP-6 may serve as an example of an EPM with the internal locking. In contrast to the external clamp lock, the blades are rigidly connected with the drive rod.

6.5 Point Locking Mechanisms

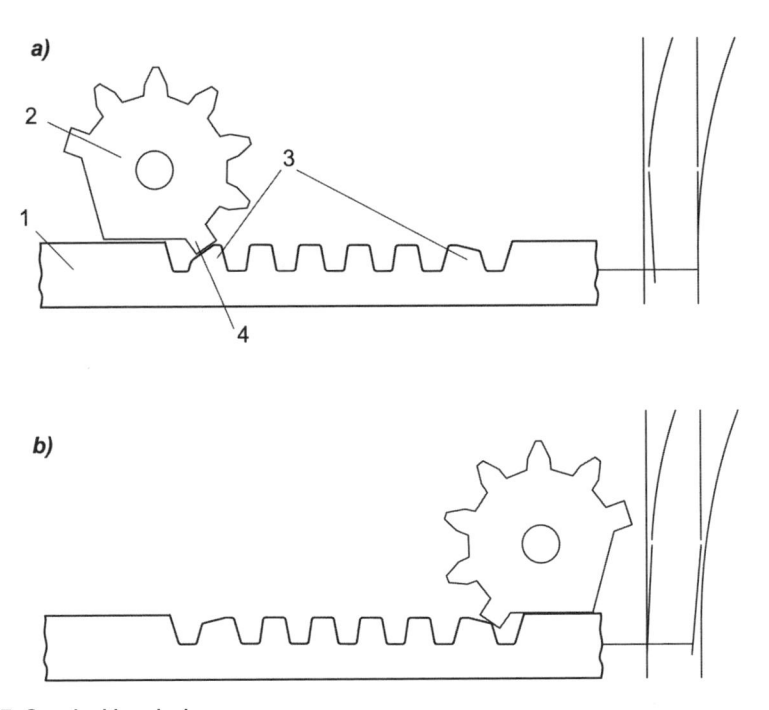

Figure 6.17: Cam locking device

The rotating motion of the motor is transformed into the progressive motion of the blades by the tooth gear with the cam locking device. This device consists of the cog-wheel of the main drive shaft 2 and the throw bar 1 (figure 6.17). The two outermost teeth 4 of the cog-wheel have a special form. The matching form is at the two outermost teeth of the throw bar 3.

When the electric motor is switched on, the main drive shaft begins to rotate together with its cog-wheel. At the beginning of rotation the crooked tooth 4 of the cog-wheel of the cam locking device unlocks the throw bar and begins to move it by pressing against it with its side edge. Then cog-wheel's normal teeth start to engage in a lock with the teeth of the throw bar bringing it to its other end position. At the end of the switching process the throw bar comes to a halt, while the cog-wheel continues to rotate, making yet some turn, which consequently results in the crooked tooth of the cog-wheel striking against the splayed tooth 3 of the throw bar. What happens as a result is the creation of an adjustable stop, precluding the throw bar from moving. Blades appear switched and locked against moving inside the track owing to rigid connection with the throw bar.

In case of trailing, the parts of cam locking device do not move. Therefore, in the EPM and in its fitting damage of various elements occurs. Often the damaged parts are the drive rod, front rod, detection contacts or bearings of main drive shaft. Because some damage cannot be found by visual survey, after trailing the EPM and fitting must be replaced.

6.5.3 Monitoring of Locking Mechanism

In a few examples, supervision of the locking mechanism is provided to detect displacement of the mechanism before it leads to point failure. As an example, a monitoring system for EPM internal locking mechanism on Shinkansen lines in Japan (Igarashi/Siomi 2006) is described.

6 Movable Track Elements

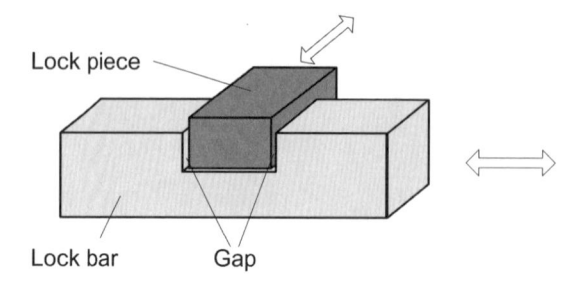

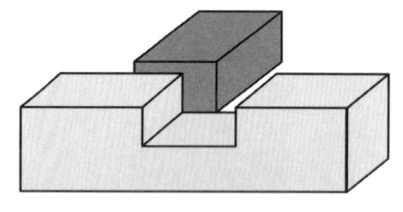

Figure 6.18: Relationship between lock bar and lock piece in Japanese example

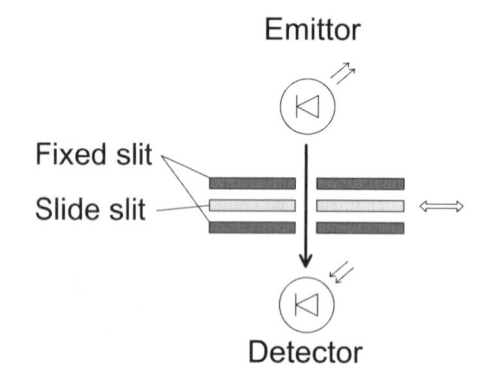

Figure 6.19: Structure of lock warp detector in Japanese example

Simplistically this mechanism consists of the lock bar and the lock piece (figure 6.18). The lock bar is connected to the blades by a drawbar and moves in a direct direction during each switching process. The lock piece moves in the orthogonal direction to the lock bar. After the blade has been switched, the lock piece is inserted into the notch comprised on the lock bar.

During operation, these mechanical parts can become displaced against each other. If this displacement becomes more than 1.5 mm, the lock piece can't be inserted into the notch of the lock bar. The blades are not locked and the EPM becomes disabled.

To detect displacement of the locking mechanism before it leads to disabling of the points, lock warp detectors are applied. The idea of the simplest lock warp detector is illustrated in figure 6.19. A fixed slit, a light emitter and a light detector are mounted to the EPM. A slide slit is fastened to the lock bar. While the displacement of the parts is in norm, the light from the emitter passes through the fixed and slide slits to the detector. When the displacement exceeds a defined degree, the slide slit shields the light and the displacement is thereby detected.

This simple detector allows detecting of malfunction, but it does not allow the precise measurement of the size of a gap. Therefore data from the lock warp detector cannot be used for detailed planning of lock adjustment works.

To eliminate this problem, an advanced monitoring system for EPM was created. A magnetic sensor is used as the measuring instrument for the gap between the lock bar and the lock piece (figure 6.18). This sensor continuously monitors lock mechanism conditions with high accuracy. This data is centrally stored and can be displayed to maintenance workers and operators to show lock displacement in real-time. This facilitates precise adjustment and planning of maintenance work.

6.5.4 Mechanical Key Lock

Mechanical key locks are used alone or in combination with other locking mechanisms like the clamp lock. They perform two different functions:
- As with the above mentioned mechanisms, locking of the points against dynamic mechanical impact on site.
- In contrast to the above, the interlocking of points and signals, other points etc.

Mechanical key locks are form-fitted and not trailable. They are basically used in three situations:
- Permanent interlocking dependence between manually operated points (figure 6.9).
- Temporary interlocking dependence between all kinds of points. e.g. during construction works.
- Temporary fixation of disturbed points in a defined end position.

In each country, there is a large variety of locks in use, each suited for one of these three situations. Whereas for permanent interlocking dependence, permanently mounted devices (figure 6.9) are used, key locking devised for the two temporary purposes are mounted flexibly in accordance with the situation.

In permanent installations, a simple set of points is normally equipped with two key locks: one for the plus and the other for the minus end position. The lock is mechanically connected with the respective blade by an additional bar to prove that it is safely fitted to the stock rail. Points can only be moved if both keys are present, and one key can only be removed if points are in the respective end position. In this case, this key can be locked into another set of points (for the logical possibilities, see section 4.2) or in the interlocking to transport the information that points are in the respective position.

The main advantages of these locking mechanisms are simplicity of the construction, while one of their disadvantages is the impossibility of remote control and a substantial amount of time required to establish a route. Manual key locks were the historically oldest form of interlocking. At present they are used mostly in shunting areas with low traffic, for temporary installations and in degraded mode operation in case of technical failures.

6.6 Circuitry of Point Operation and Control in Relay Technology

6.6.1 General Overview

The EPM control schemes are ones of the most important in railway signalling, as safety of train movement depends directly on their correct operation.

The three basic parts of the circuitry – managing, operational and supervision – can be distinguished in any control scheme. All these circuits are built in accordance with the performance specifications required for schemes with safety responsibility.

The purpose of the managing circuitry is intended to start EPM operation with checking the safety requirements. It provides the following functions:
- Switching points occupied by rolling stock must be impossible.
- Switching points locked in a route must be impossible.
- In case of fault of track clear detection, switching the points with auxiliary commands must be possible.
- The EPM must start from a short impulse irrespective of duration of pressing of the activation button.
- In case a set of points becomes occupied by a vehicle while moving, it must complete the operation already begun until reaching the end position.

- Any malfunction of circuit should be found out not later than the next switching of points.
- The switching must be reversible, which means that at any time during the switching process the buttons can be pushed again and the switching of points changes its direction.

The operational circuitry is designed to transfer electric power from an energy supply to the electromotor of the EPM. Depending on the location of the power supply for the operational circuit in the interlocking rooms or in the field, two types of operational circuits with central and local power supply can be distinguished. In similar manner, the circuits which are responsible for reversion of the EPM can be located centrally or locally.

The **operational circuitry** shall provide:
- double pole (multi pole) shutdown of the electric motor from the power supply when it is out of service, which eliminates a spontaneous point switching from an extraneous feed that can appear into the operational circuit in consequence of electric or electro-magnetic influence or connection between cable cores;
- automatic switching-off the motor by the detection contacts of the EPM when the points reach their end position;
- impossibility of the electric motor rotation from an energy source of the supervision circuitry, if supervision circuitry uses the same wires as the operational circuitry does;
- impossibility of operating the electric motor if the EPM is being switched manually by the hand crank or is open for maintenance work.

In the **control schemes** of EPM, electric motors of either DC or AC can be used, whose appropriate choice is determined by performance specifications. The motors must:
- have a sufficient starting torque to carry out the point operation, taking into consideration resistance (inertia) of the mass of transferable parts and the atmospheric conditions (hoarfrost, freezing, etc.);
- provide an opportunity of changing the direction of the motor shaft rotation (reversibility);
- be designed for remote control – their characteristics must not alter too abruptly if there are changes of voltage caused by a line drop.

As indicated above, points can be set in three different positions: two of these are end ones (called 'plus/minus', 'normal/reverse' or 'right/left') and another is intermediate. The EPM provides checking of these positions due to a specific construction and an algorithm of the detection contacts operation. The state of the detection contacts is constantly checked by the **supervision circuit**, in some systems during operation by the operation circuit itself.

The supervision circuity takes into account the following requirements:
- The circuit must be protected from mis-operation in the event of a short-circuit in a cable line or of connection between cable cores.
- Failure of any element must be detected immediately.
- Failure of non-safe elements (e. g. diodes, capacitors, polarised armature of supervisory relays, blowout of fuses) must lead to a fail-safe reaction.
- Depending on the accepted rules in the particular country, supervision of point position must not be influenced by repairing or maintenance of the EPM or must be switched off during these works.

6.6.2 Example with Type N Relays: Russian Five-Wire Point Circuitry

6.6.2.1 Overview

As an example of the EPM control scheme with the three-phase AC motor and management and supervision circuitry solved by type N (first class) relays (chapter 9.3.2.1), let us regard a

6.6 Circuitry of Point Operation and Control in Relay Technology

scheme that is widespread all over Russia's railroad network (Sapožnikov et al. 1997–II). It is called 'the five-wire' because of five cable wires installed from an interlocking tower to the EPM (figure 6.20).

6.6.2.2 Supervision

The EPM supervision scheme applies an AC supervision circuit with polar selectivity whose mode of functioning is based on the half-wave rectification of the AC. The supervision scheme's composition consists of the T4 transformer, the DC combined (chapter 9.3.2.4) relay OK, the rectifying unit comprised of the VD7 diode and the R2 resistor, an integrating circuit – the C1 capacitor, the R1 resistor – and the detection contacts.

Figure 6.21 illustrates the operational procedure of the supervision circuit. The AC arrives into the circuit from the secondary winding of the T4 transformer. When the points are located in their end position, one half-wavelength is bypassed by the VD7 diode due to lower resistance in that path, whereas the other moves through the OK relay coil. As may be seen from the oscillogram displayed in figure 6.21a, the relay appears under action of impulses of current of the identical polarity; therefore, the relay's mono-stable system activates, while the bi-stable one switches over into the position corresponding with this polarity. The point position is registered by the supervising relays PS (plus position), MS (minus position). A contact of the PPS relay is included in the scheme of PS and MS supervising relays to ensure that the actual position of the points equals the target position determined by the PPS (chapter 6.6.2.3) relay and that the contact of the bi-stable part of OK relay is not incidentally sealed.

During point switching the VD7 diode is shut off (figure 6.21b), and the OK relay is supplied with both half-waves and therefore deactivated. There is no supervision of the point position. In point's opposite end position, the corresponding detection contacts close. The polarity of connection of the VD7 diode is altered (figure 6.21c). The mono-stable system of the OK relay activates, and the polarized one switches over. The point position is registered by the MS supervising relay.

Thus, the T4 transformer is a source of energy for the supervision circuit, while the VD7 rectifying diode can be considered as the power source of the OK supervising relay. This achieves the requirement of power delivery to the supervising devices from the detection contacts side. Indeed, the OK relay is a DC relay and is not activated in the absence of the DC component formed by the VD7 diode. Consequently, in event of a short-circuit or abruption of the L1…L5 wires the supervision circuit is protected from a mis-operation to the dangerous side.

If the points are in their end position, electrical current circulates across all elements of the supervision circuit; therefore, any failure they may have is discovered immediately. The circuit does not interrupt if the lid of the EPM is open, because the lid contacts B1 and B2 are not included in it.

6.6.2.3 Initiation of Switching

The scheme's managing circuit is built upon two activating relays: the mono-stable relay NPS with two coils, each of them picking up the relay alternatively, and the bi-stable relay PPS located in the interlocking room. The NPS relay determines the opportunity of a point operation while the PPS relay defines the direction of point switching and performs memorization of the point's current position.

When a command to switch the points is issued (the SK point lever is rotated to the «–» position), the NPS relay picks up with the checking of safety requirements. The point switching is possible if the track circuit is not occupied (the front contact of the SP relay is closed) and the points are not locked in a route (the front contact of the Z relay is closed).

6 Movable Track Elements

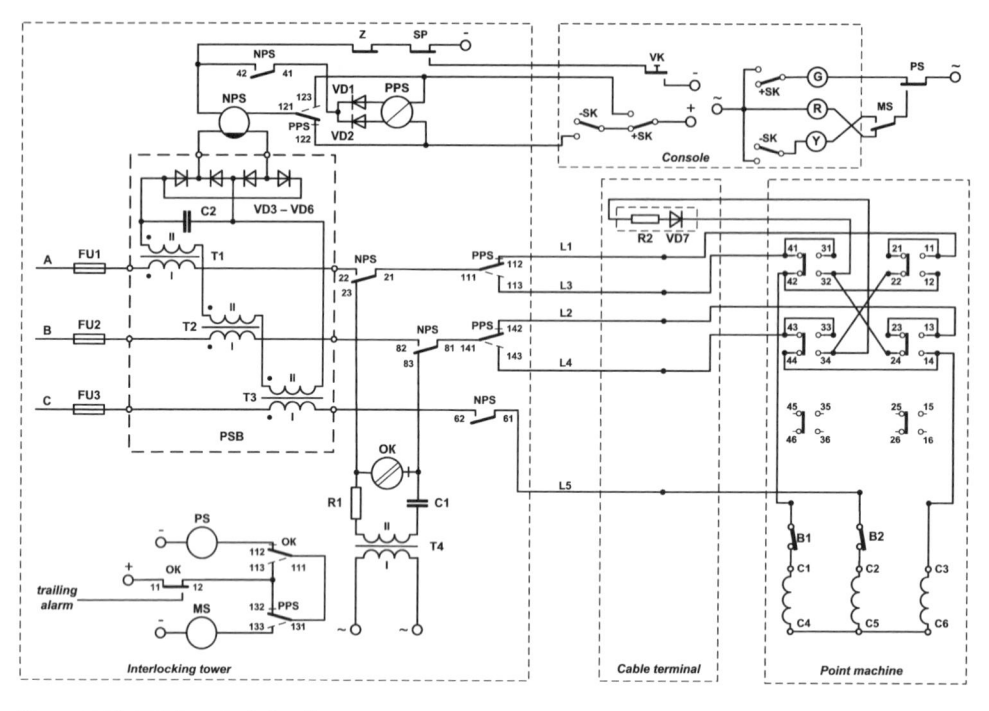

Figure 6.20: EPM control circuit

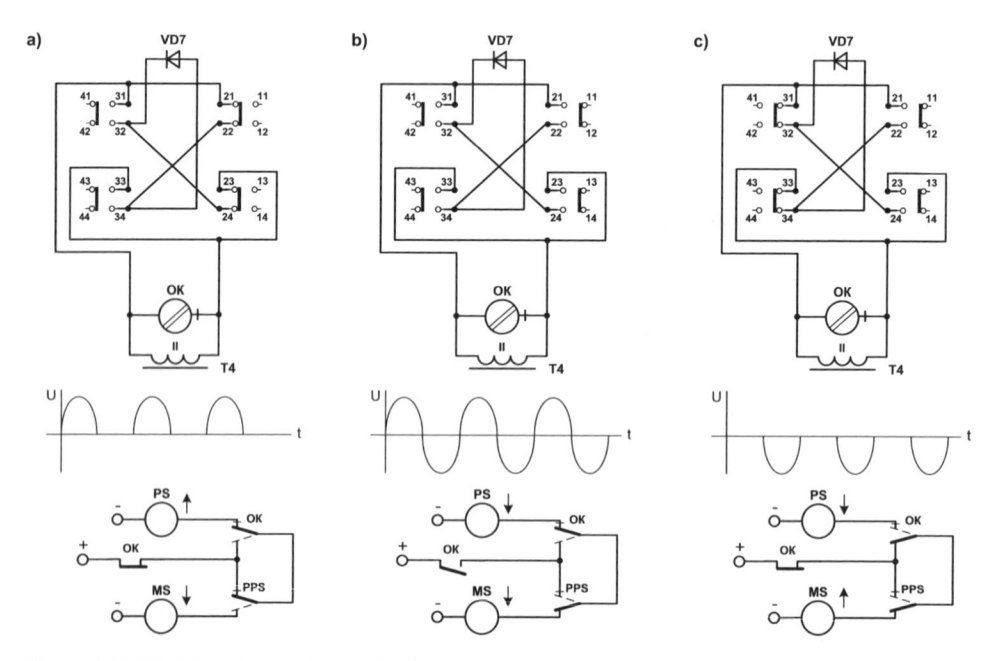

Figure 6.21: Working of supervision circuit

6.6 Circuitry of Point Operation and Control in Relay Technology

Back (break) contacts of the NPS relay switch off power of the supervision circuit, which causes disappearance of supervision of the end position (contacts 23, 83) and flashing of the red light on the console; they also plug in three phases of power supply to the operational circuit (contacts 22, 82, 62); and close the circuit of the PPS relay (contact 42). The electrical polarity of the PPS relay is now reversed in comparison with the previous interconnection, therefore this relay switches over. The 111 and 141 contacts of the PPS relay alter the order of the phases in the operational circuit and the motor begins to rotate.

The short duration of the activating influence on the managing circuit is provided by the 121 contact of the PPS relay, since it shuts off the circuit of the upper coil of the NPS relay. However, in normal operation the operational circuit retains NPS relay picked up during the whole time of the point switching by the lower coil.

6.6.2.4 Motor Operation

The operational circuit is comprised of the power supply source (the A, B, C phases), fuses FU1...FU3, line wires L1...L5, detection contacts (11-12, 13-14, 15-16 and 41-42, 43-44, 45-46) and lid contacts B1, B2. (When the EPM opens, B1 and B2 interrupt the operational circuit to prevent potential accidents and injuries to maintenance staff.) The phase supervising block (PSB) consists of the three transformers T1...T3 and the rectifier bridge collected on the VD3...VD6 diodes. It checks the actual closing of the operational circuit along all three phases and retains the NPS relay in picked up position with the motor current. The reason for that are effects of electro-magnetic saturation of the transformers and therefore its non-sinusoidal behaviour.

When the switching is complete, the detection contacts 41-42 and 43-44 open, the current through the motor stops; hence, the NPS relay drops down, providing with its contacts the tripolar disconnection of line wires from the power supply and connecting the supervision circuit to the power supply.

6.6.3 Example with Type C Relays: GS II DR (Germany)

6.6.3.1 Overview

In Germany, four-wire point circuitry is the interface used predominately between interlocking and EPM. It uses type C (second class) relays (chapter 9.3.2.1), resulting in simpler relays, but more complex circuitry than in the example in chapter 6.6.2. Whereas the four wires and the behaviour of the point machine is a standard for most applications in Germany, the circuitry inside the interlocking differs between the interlocking types, although the basic principles are similar. As an example, the point circuitry of GS II DR is described in detail. GS II DR is a relay interlocking type manufactured by *WSSB* in GDR and is still in operation to a large scale in East Germany. Adapted forms of GS II are also applied in other countries.

Figure 6.22 shows the managing circuitry (simplified) and figure 6.23 the operational and supervision circuitry. Table 6.1 explains the relays and other symbols. Unless stated otherwise, all relays are simple mono-stable DC relays (Kusche 1984, Arnold et al. 1987).

6.6.3.2 Supervision

Let us assume that the points are in plus position. The plus supervision relay is picked up and the supervision current passes all four wires between the interlocking and the EPM (figure 23) to prove all wires the point cable to be intact. The trailing supervision relay 1 does not switch although its first magnetic system is passed by the supervision current. The reason is that due to the plus supervision relay in series, the current is too weak to pick it up.

6 Movable Track Elements

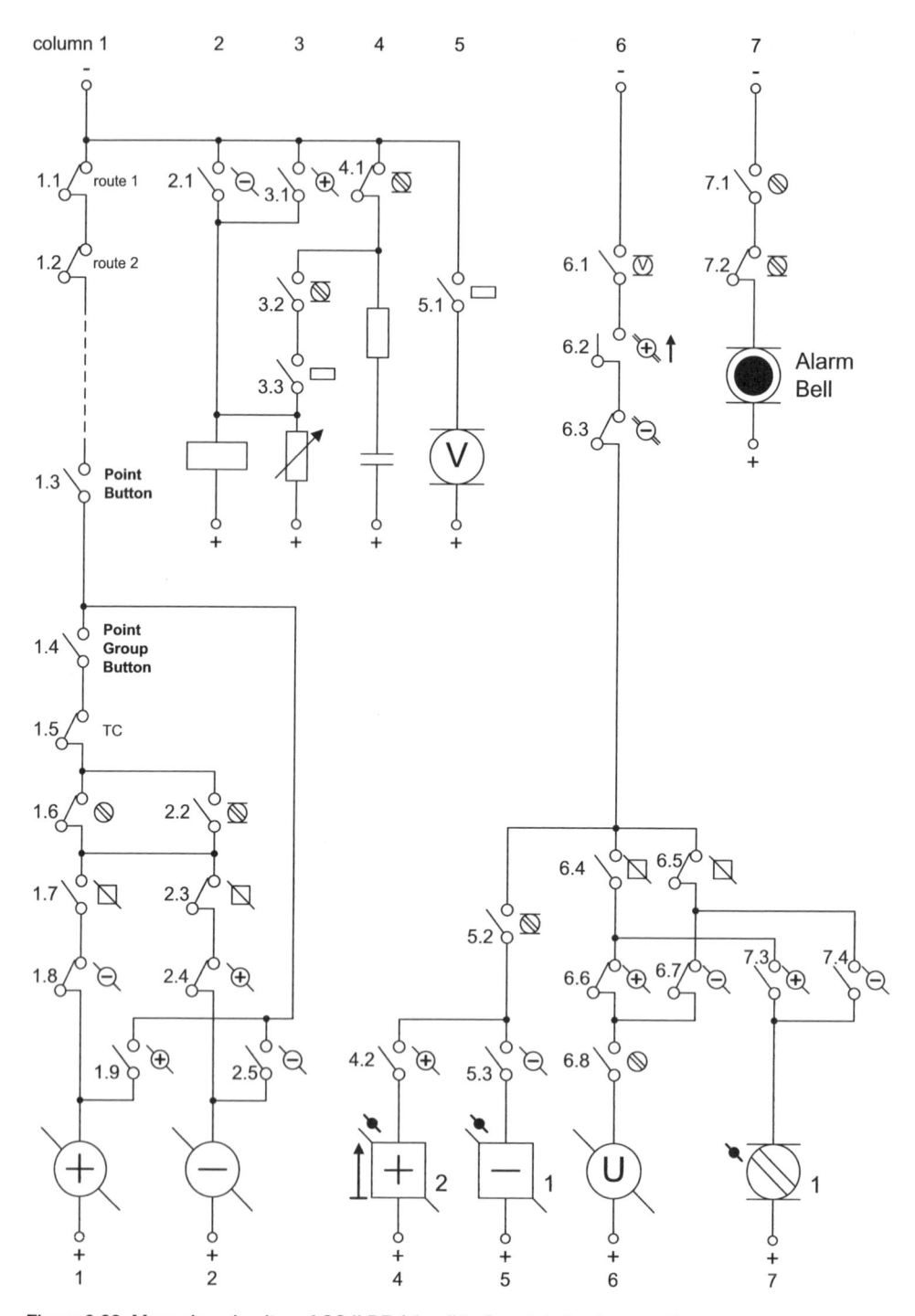

Figure 6.22: Managing circuitry of GS II DR (simplified); points in plus position

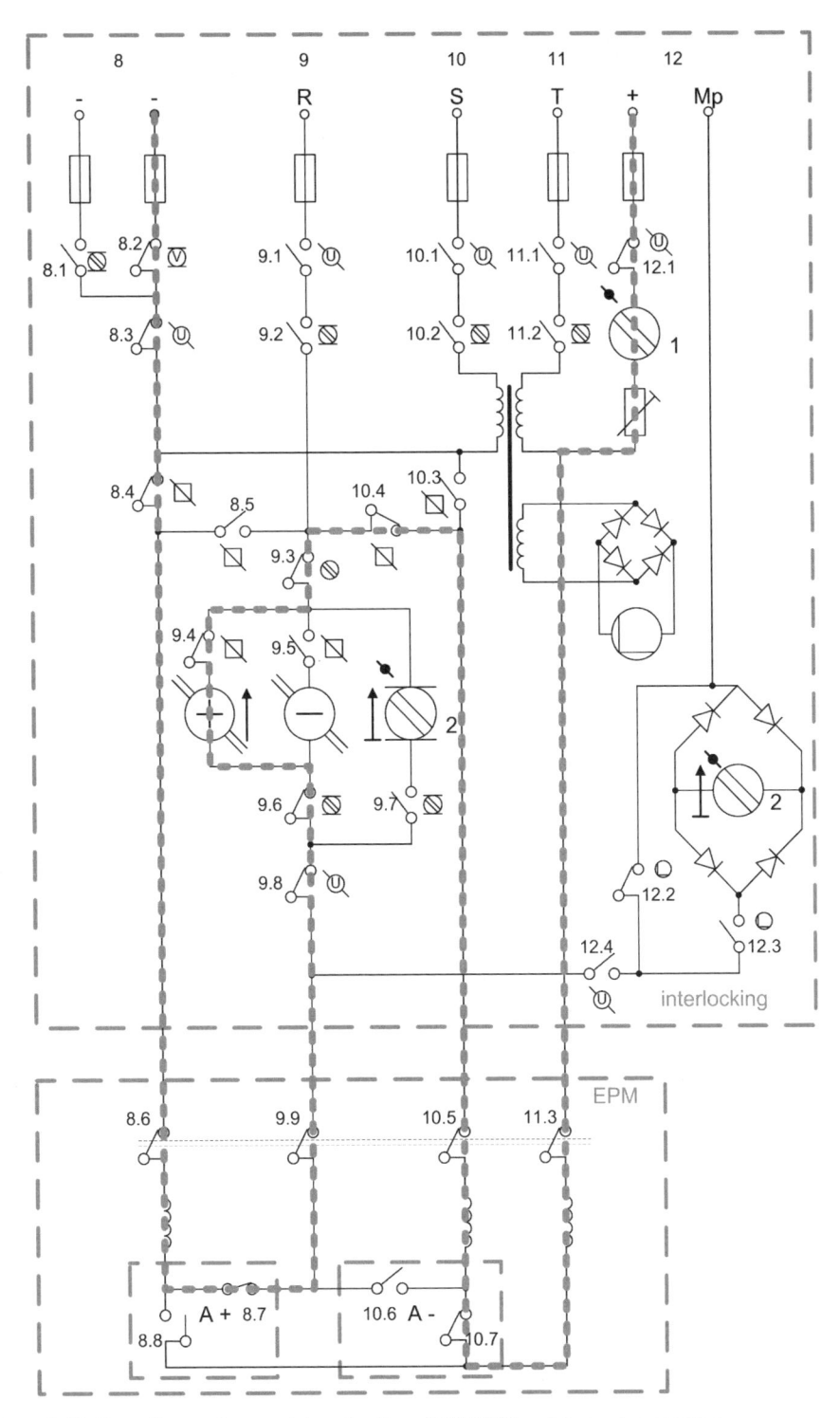

6.6 Circuitry of Point Operation and Control in Relay Technology

Figure 6.23: Operation and supervision circuitry of GS II DR; points in plus position

173

6 Movable Track Elements

Symbol	Name used in the Text	Explanation
R, S, T	feeding of operational circuit (three phases)	
Mp	zero point of the operational circuit	
(U)	current switching relay	This relay switches the power supply between operation and supervision.
(+)2 (+)1	target position relay	This relay (bistable, chapter 9.3.2.3) stores the target position on the points. In normal operation, this means the actual position if the points are in their end position or the position where it is moving to during the switching process.
(+)	plus setting relay	These relays initiate the switching process into the respective direction.
(−)	minus setting relay	
⊘	trailing supervision relay 1	These relays (both bistable, chapter 9.3.2.3) together detect the trailing of a points by a moving unit.
⊘	trailing supervision relay 2	
(+)	plus supervision relay	These relays supervise the respective end position if picked up.
(−)	minus supervision relay	
▭	auxiliary retarding relay	These relays together control the time delayed disconnection of the EPM if the end position is not achieved due to disturbance. This serves for protection of the EPM.
(V)	retarding relay	
(L)	auxiliary trigger relay	This relay switches coil 2 of the trailing supervision relay one in and out of the operation circuit and therefore helps to stop point operation after the points have reached the end position.
A+ A−	EPM contacts	These contacts are the equivalent to the detection contacts in chapters 6.4.3 and 6.6.2. They detect the end position of the point blades.
route1, route2, etc.	route contacts	These are contacts of the route circuitry whose purpose is to check if the points are locked in a route.
TC	track clear detection contacts	These are contacts of the track clear detection whose purpose is to check if the points are clear.
▪		attribute for a bistable relay
↑		coil of bistable relay held in up position by the bistable characteristics, although deenergised

Table 6.1: Relays and other Symbols of Point Circuitry of GS II DR (names liberally translated)

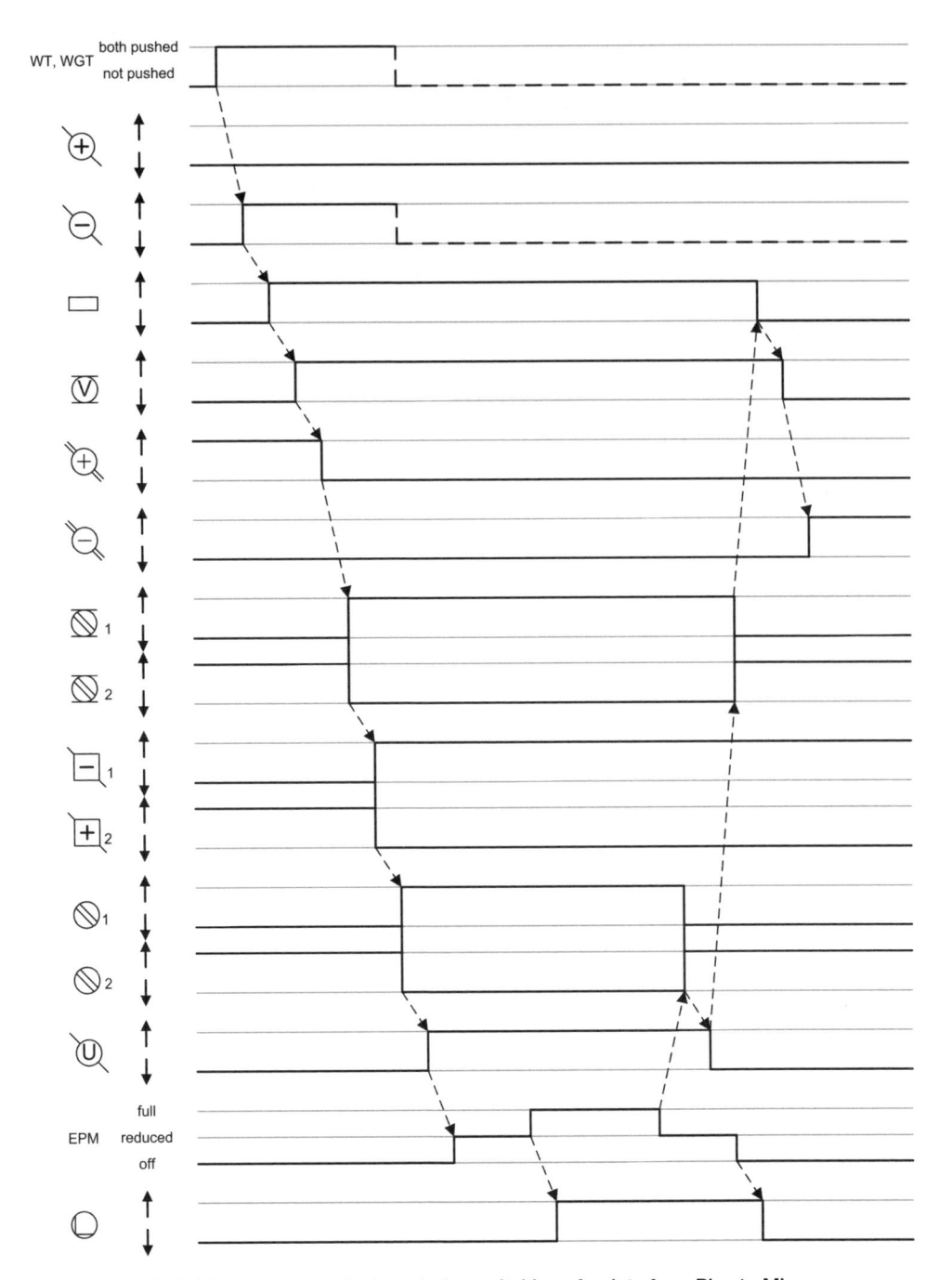

6.6 Circuitry of Point Operation and Control in Relay Technology

Figure 6.24: Switching sequence of relays during switching of points from Plus to Minus

6 Movable Track Elements

The minus position differs from the plus in the positions of the target position relay and of the EPM contacts. Therefore the minus supervision relay is picked up by current path over contacts 8.2, 8.3, 10.3, 10.5, 10.6, 9.9, 9.8, 9.6, 9.5, 9.3, 8.5, 8.6, 8.8, 11.3 and 12.1.

6.6.3.3 Normal Switching Process

Let us now assume that the points will be switched by an individual point operation action of the signaller. Automatic point setting, however, is provided in this interlocking type, but the related wires and contacts are not drawn in figure 6.22 for simplicity. According to German requirements, all relays must switch at least once during the normal switching process in order to detect failure of any relay. The sequence for switching from plus to minus is described in the following text and depicted in figure 6.24.

Individual point setting is initiated by simultaneous pushing of the point button related to the individual point and the master point button. The master point button serves for all points in the interlocking area and its purpose is to meet the requirement that, for prevention of unintended actions, in German interlocking always two buttons must be pushed to initiate an action. If the points are clear and not locked in any route, the minus setting relay picks up, initiating the switching process.

The minus setting relay activates the retarding circuit by closing contact 2.1. This circuit is adjusted to a time delay of approximately six seconds and disconnects the EPM for its protection if it doesn't reach the other end position within this time. The retarding relay disrupts the supervision current (figure 6.23, contacts 8.2) and therefore drops down the supervision relay.

Consequently, coil 1 of the trailing supervision relay 2 becomes energised by contacts 6.2 closing, causing the change of position of that relay. After this, coil 1 of the target position relay becomes energised by contacts 5.2 closing, which leads to a change of position of that relay. The target position of the points is now the minus. After that, trailing supervision relay one switches by closing of contacts 10.3 (current path via contacts 8.1, 8.3, 10.3, 10.5, 10.7, 11.3 and 12.1).

After switching of the target position relay (or later), the signaller can unhand the buttons, as contacts 5.3 and 7.4 do no longer need to be closed. This opens contacts 2.1 and therefore starts the unloading of the capacitor of the retarding circuitry.

After switching the target position relay and the trailing supervision relay 1, the current switching relay picks up via contacts 6.4 and 6.8 closed. It connects the operating current by contacts 9.1, 10.1 and 11.1 and disconnects the supervision current by contacts 8.3 and 12.1. In case of failure, the safe position of this relay is the dropped down (operation current off).

Now the motor starts to rotate, but due to unlinked phases with low performance (figure 6.25, left). The purpose for this smooth starting is to avoid jamming of the clamp lock (chapter 6.5.1) while unlocking, which could occur when too high force is applied. As in the transformer of the auxiliary trigger relay, both primary coils are passed by the same current in opposite directions, this relay does not pick up and therefore coil 2 of trailing supervision relay 1 is excluded from the circuit (contacts 12.2 and 12.3).

After the open point blade has moved a short distance and unlocking of the clamp lock (chapter 6.5.1) has started, the EPM plus contacts (8.7 and 8.8) switch, causing the linking of phases in star configuration and rotation of the point motor with full power (figure 25, right). The auxiliary trigger relay picks up to enable switching back of trailing supervision relay 1 later in the process (contacts 12.2 and 12.3).

When the point blades approach the new (minus) end position, the EPM minus contacts (10.6 and 10.7) change their positions. The phases of operational current are now unlinked, provid-

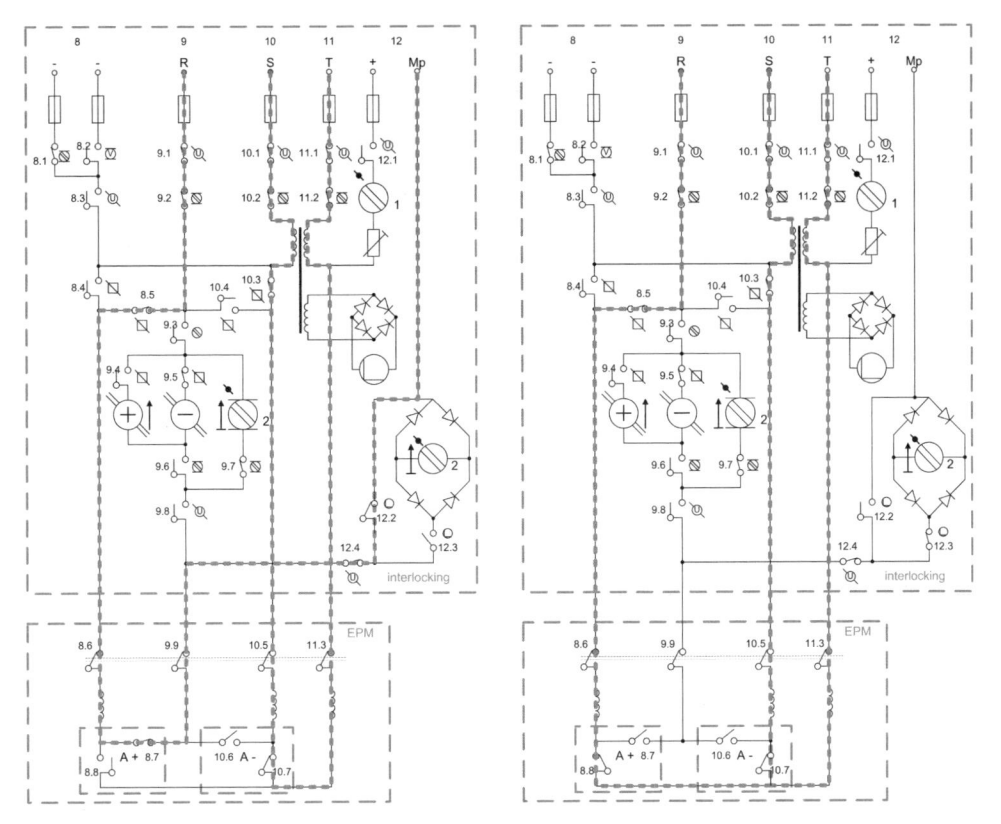

6.6 Circuitry of Point Operation and Control in Relay Technology

Figure 6.25: Initial operation of EPM with reduced power (left) and operation with full power (right)

ing for a smooth braking of the point machine. Thus, the trailing supervision relay 1 switches by its coil 2 energised. This drops down the current switching relay by opening contacts 6.8, which disconnects the operation current by opening contacts 9.1, 10.1 and 11.1 and connects the supervision current by closing 8.3 and 12.1. For a short time, the supervision current energises coil 2 of trailing supervision relay 2 via contacts 8.1, 8.3, 10.3, 10.5, 10.6, 9.9, 9.8, 9.7, 9.3, 8.5, 8.6, 8.8, 11.3 and 12.1. Trailing supervision relay 2 therefore changes its position, disconnects itself by opening contacts 8.1 and 9.7 and closes contacts 9.6 to energise the minus supervision relay few steps later. It also switches off the retarding circuitry by contacts 3.2 and 4.1. Now contacts 8.2 close and the supervision current energises the minus supervision relay (chapter 6.6.3.2).

6.6.3.4 Reversing of Switching Direction

Reversing of switching direction is explained on the example in that this occurs while the points are moving at full power from plus to minus (figure 6.25, right). However, reversing is also possible in other stages of point operation as well as after time-delayed disconnection of the EPM after not reaching the end position.

Pushing point and point master buttons energises the plus setting relay. This switches the target position relay by contacts 4.2, while contacts 6.1, 6.2, 5.2 and 4.2 are already closed before due to the normal switching process. By switching of contacts 8.4, 8.5, 10.3 and 10.4, the direction of rotation of the EPM changes.

6 Movable Track Elements

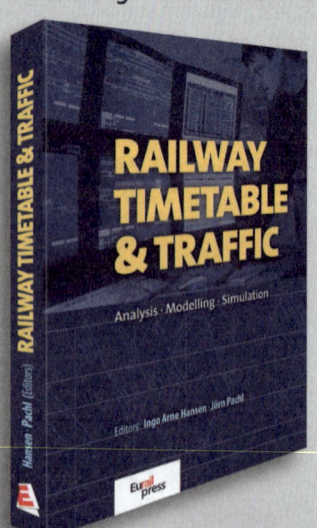

6.6.3.5 Trailing of the Points

Let us now assume that the points are trailed by a rail vehicle while being in plus position (figure 6.23). Moving the point blades switches the EPM plus contacts. Now the current path of the supervision current is via contacts 8.2, 8.3, 8.4, 8.6, 8.8, 11.3 and 12.1, which drops down the plus supervision relay. Trailing supervision relay 1 now switches by its coil 1 being energised by a higher current (compare chapter 6.6.3.2). This energises the trailing alarm bell by contacts 7.1 closed to attract the signaller's attention to the situation (figure 6.22). Trailing supervision is therefore based on the principle that this combination of positions of trailing supervision relays 1 and 2 does not occur in normal operation, although both relays also switch there to test their proper functioning.

Pushing point and point master buttons would now lead to no reaction, as contacts 1.6 and 2.2 are both open. Therefore, the points cannot be operated normally.

7 Signals
Gregor Theeg

7.1 Requirements and Basic Classification

The purpose of signals is to convey information and instructions to people, in this context mainly to the driver of a railway vehicle, but also to ground staff or to workers on the track. They are therefore an interface between the technical equipment and people. Examples of information to be communicated to the driver are:

- movement authorities (chapter 7.3.3)
- permitted speed (chapter 7.3.4)
- information about the direction of the route
- readiness for departure
- identification of vehicles and front/rear ends of trains
- position of points
- (in yards) signals related to actions on the hump
- commands for brake test
- signals for electric traction
- marking of mileage and particular positions (e. g. level crossing closure point) along the line

Signals have to meet the following general requirements:
- The driver must easily be able to recognise the signal for what it is.
- The driver must be able to understand the signal indication quickly.
- The information given by the signal must be unambiguous.
- The same information should always be given in the same way. The same signal aspects should be applicable for different cases.
- The driver must be able to memorise the information easily.
- The information shall be given in proper time, which means not too late, but also not too early, to prevent the driver from forgetting.
- Fail safe. In case of technical defect, the signal must never give a dangerously misleading indication. It can, however, show a more restrictive indication. This principle can be translated into practice in two ways:
  - Design of the signal system so that partial extinction always results in a more restrictive or an undefined signal aspect **(inherent fail-safety)**. The Dutch system is an example (chapter 7.5.5); others are used in parts of the North American network. Here, the advantage of this type of fail safe over reactive fail safe is that it protects not only against technical failure, but also against failure in perception by the driver. Another example in mechanical signalling is the use of upper-quadrant semaphores rather than lower-quadrant (chapters 7.5.1, 7.5.2).
  - Supervision of the signal and activation of a more restrictive aspect in case of failure **(reactive fail-safety)**. Most modern signal systems match into this category.
- Reliability: Technical defects should be rare.
- Economic efficiency which, for example, requires a low number of signal lamps.

In high speed traffic, a safe perception of trackside signals by the driver cannot be assumed due to the short time which is available for observing the signal. Therefore, railways which use high speeds have defined a limit of speed above which cab signals are obligatory to replace trackside signals. According to *UIC code 734*, railway lines can be distinguished into three classes according to speed and the related form of signalisation:

1. Conventional lines: up to ca. 160 km/h; signalling by trackside signals possible

7 Signals

2. Lines with speed: up to ca. 200 (220) km/h; trackside signals still possible and applied in some countries in Western Europe, but adaptations necessary (chapters 7.3.3.3 and 7.3.3.4)
3. High speed lines: over 200 (220) km/h; only cab signalling possible.

This limit for trackside signals is usually defined around 160 km/h, in some railways in Western Europe up to 220 km/h.

The speed restriction connected with movements on sight, e.g. for signalised entry into an occupied track, for shunting movements and in certain cases of degraded mode operation, varies between the railways and the situation. Usually it is between 15 and 30 km/h, but in special cases it can reach up to 100 km/h (e.g. open line sections in Saudi Arabia).

Signals can be distinguished by different criteria. One is the human sense reached by the signal. The practically important kinds are:
- **Optical** signals are suitable to give detailed information including the location of the signal.
- **Acoustic** signals are suitable to attract a person's attention independently, so they are suitable to warn of potentially dangerous situations.

The second criterion is the location of the signal (figure 7.1):
- **Trackside signals** are given from a position along the track.
- **Vehicle signals** are given from vehicles to persons outside the vehicles.
- **Cab signals** are given in the driver's cab to the driver.

Trackside signal:

Cab Signal:

Vehicle signal:

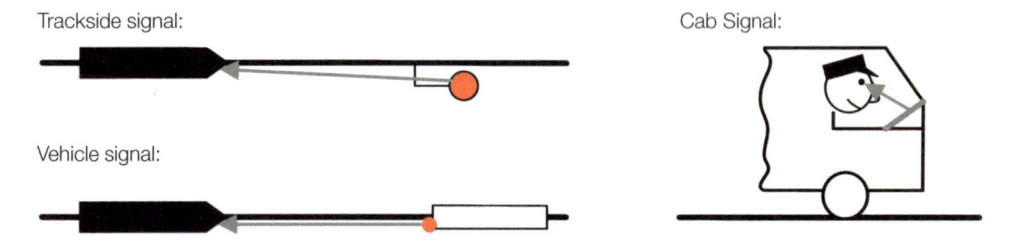

Figure 7.1: Classification of signals according to the place where given from

In some situations, cab and trackside signals are used simultaneously, but the amount of information which can be given differs. In some systems, the cab signalling gives less information than the trackside signal (e.g. ALSN in Russia, chapter 8.3.4.1), but often more (e.g. BACC in Italy, LZB in Germany and Spain, chapters 8.3.4.3, 8.3.6). If the signal indications do not conform, the priority of the one or the other is regulated differently: In countries of former USSR, the trackside signal has priority, but in German LZB the cab signal, for example.

By the technology of giving the signal, the main versions are (figure 7.2):
- **Hand signals** are given manually by a person.
- **Mechanical signals** are given by different position of objects.
- **Light signals** are given by different light arrangements and use of different colours.

Signals can be further distinguished:
- **Positive signals** give the information by the presence of an indication. This is the usual case today.
- **Negative signals** give the information by absence of an indication. Examples are an unlit light signal and mechanical signals standing edgewise.

7.2 Technical Characteristics of Trackside Signals

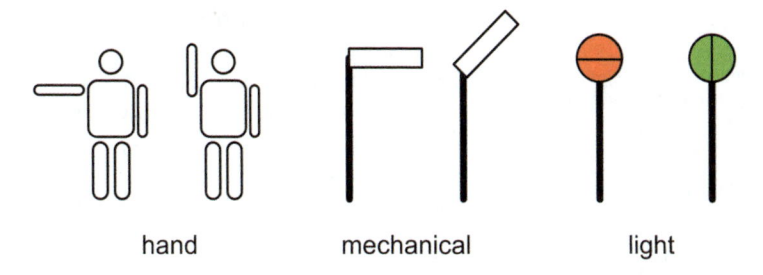

        hand   mechanical   light

Figure 7.2: Classification of trackside signals according to technology

Regarding the ability to switch between different aspects, the following forms can be distinguished:
- **Fixed signals** (e.g. signal boards) always show the same indication.
- **Switchable (multiple aspect) signals** can switch between different aspects.

As cab signalling is technically closely linked to train protection systems, it is described more detailed in chapter 8. As for the remaining, optical switchable trackside light signals are the most complex form. They shall be in the main focus of this chapter 7.

According to the formation of signal aspects, switchable optical trackside light signals can be distinguished as (figure 7.3):
- **Colour light signals** where signal aspects are distinguished using different colours.
- **Position light signals** where signal aspects are distinguished using different formations of lights of the same colour.
- **Colour position light signals** where signal aspects are distinguished using different formations of lights of different colours.

Colour light signals: Position light signals: Colour position light signals:

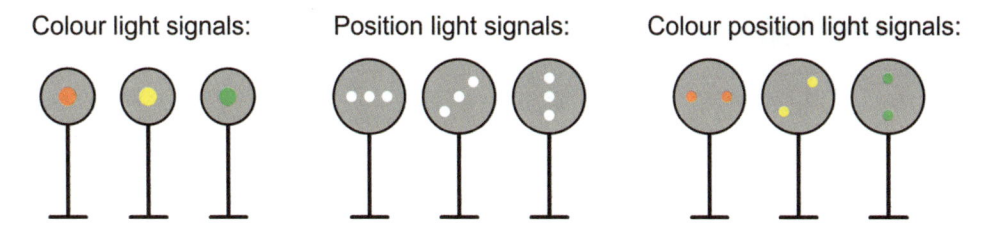

Figure 7.3: Examples for colour, position and colour position light signals

7.2 Technical Characteristics of Trackside Signals

7.2.1 Structure of Light Signals

The signal-giving device for switchable light signals – in short a light signal – can be mounted differently. The most common solution is on a high signal post directly beside the track it governs. The regular position right or left usually conforms to the normal direction of traffic (chapter 3.2.4). If there is insufficient space, signal heads may also be mounted on signal cantilevers or signal bridges. On some railways, signals may be constructed as dwarf signals. Particularly for shunting, in situations where all trains stop in front of the signal or where no other solution is possible or practical, dwarf signals are often used.

7 Signals

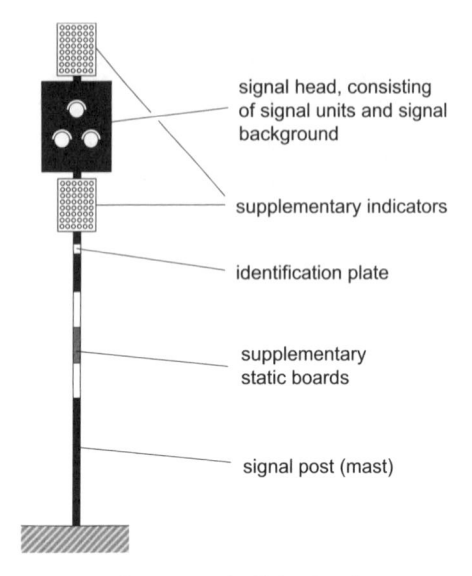

Figure 7.4: Structure of a light signal

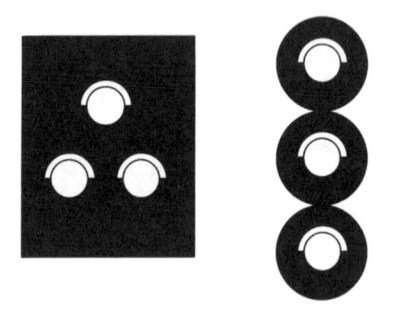

Figure 7.5: Arrangement of signal backgrounds

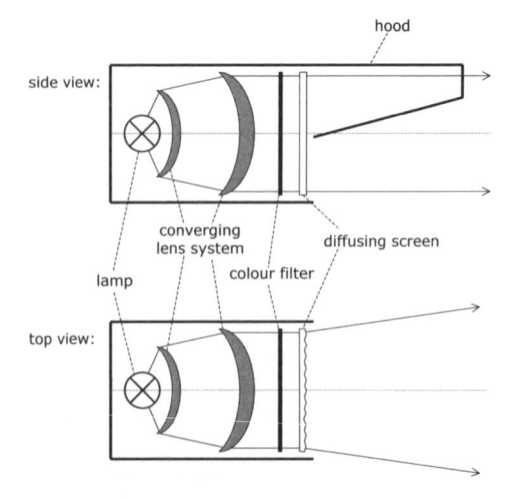

Figure 7.6: Structure of a signal unit

Figure 7.4 shows a typical structure of a light signal with high signal post. Either several signal units can be placed in front of the same background (practice in most European countries, figure 7.5 left), or each signal unit has its own background (dominating practice in North America, figure 7.5 right).

Light signals today are designed with bulb lamp or with light emitting diodes (LED). Figure 7.6 shows the typical structure of a **signal unit with bulb lamp**. The purpose of the converging lens(es) (usually one or two) is to gather as much light as possible from the lamp and to form a parallel beam of white light. Today, Fresnel (stepped) lenses are often used to reduce absorption of light. The light is filtered by the colour filter and diffused into horizontal direction by the diffusing screen. Often various diffusing screens can be selected to adapt the zone from which the signal is visible to the geometry of the approaching track.

The precise implementation varies. Often the colour filter function is included in the converging lens(es) by use of coloured lenses. Often no separate diffusing screen is provided, but a simple diffusing function is integrated into the converging lens system, meaning that this produces no exactly parallel beam.

To generate switchable signals in different colours, signals with a bulb lamp can be distinguished as follows (figure 7.7):
– **Multi-unit signals:** A separate lamp is used for each colour.
– **Searchlight signals:** Apertures are switched mechanically in front of a lamp which is always lit.
– In an intermediate form (e.g. used in Italy), several lamps are mounted in the same signal unit and the light from all lamps directed to the same exit by a lens, mirror and filter system.

An important difficulty in signals with bulb lamp is the prevention of **phantom lights**. These are effects caused by external light (usually sunlight) which imitates a lit signal

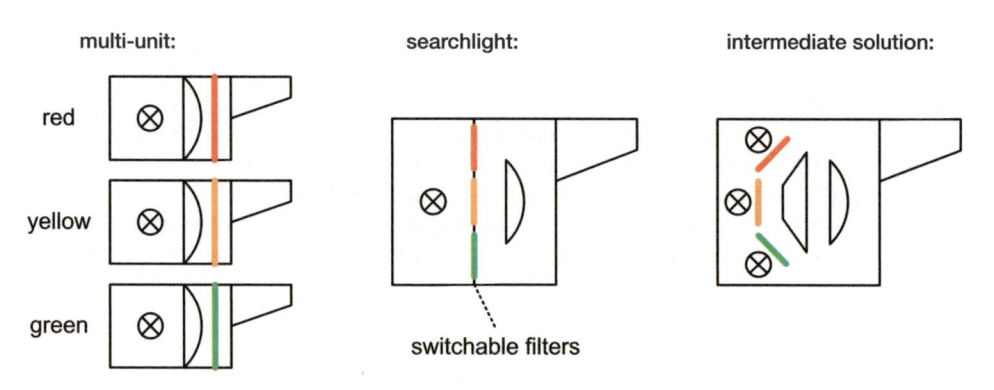

7.2 Technical Characteristics of Trackside Signals

Figure 7.7: Solutions to generate lights in different colours by bulb lamp signals (principles)

lamp. Either a bright image of the sun is focussed on the filament of the lamp, or sunlight is reflected on the surface of a lens or on a mirror (if this is used). To reduce this danger, a hood (figure 7.6) is fitted, and all interior surfaces in the signal unit are painted matt black. For the same reason, in multi-unit signals, mirrors are widely forbidden, although they can be used in searchlight signals in increase optical efficiency.

The advantages of searchlight signals over multi-unit signals are the lower number of lamps required, the impossibility of phantom lights and the more efficient optical system, whereas the disadvantage is mechanically moving parts for operation of the colour filters in an unsuitable environment, resulting in higher maintenance costs. Today, most railways prefer multi-unit signals.

In **LED signals**, which are coming increasingly in use, each spot light consists of a matrix of individual LEDs which emit light in exactly the required colour and the required direction. Therefore, no colour filters and lenses are usually necessary.

Technical solutions to form **subsidiary indicators** to display numbers, letters or geometrical forms are among others (figure 7.8):

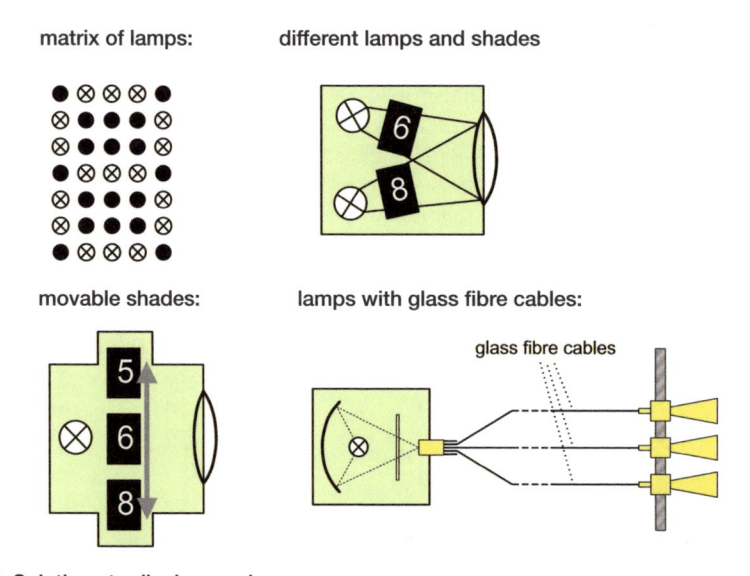

Figure 7.8: Solutions to display numbers

7 Signals

- A matrix of several lamps or LEDs, which are selectively switched on depending on the aspect to be displayed. This solution is expensive for signals with bulb lamps, but practicable for LED signals.
- Moving shades in front of one lamp, one shade for each aspect of the subsidiary signal. The disadvantage is increased maintenance efforts due to moving mechanical parts.
- One lamp with a shade in front of it for each aspect of the subsidiary signal.
- A sheave of glass fibres for each aspect, beginning at a lamp and splitting up the light to the desired form. This is the most modern, efficient and flexible solution for bulb lamp signals.

7.2.2 Optical Parameters

The (optical) **range of vision** of a signal lamp is determined by the luminous intensity of the lamp into the respective direction, the sensitiveness of the eye of the viewer and the absorption by the atmosphere, the latter depending on the weather conditions. The black signal background provides for a good contrast and therefore increases the range of vision. Designing the output of the lamps for the worst case (the densest possible fog) would be very disadvantageous under normal conditions due to the glare and high energy consumption of signal lamps. Therefore, their output is usually designed to give good results in hazy weather, but not for the densest fog. A typical requirement for the range of a main signal lamp is around 500 m under these conditions.

The **sighting distance**, in contrast to the range of vision, is also limited by obstacles between the signal and the viewer. However, obstruction of the signal for very short distances (e.g. by masts for electric power supply) is accepted. Railways have defined different regulations concerning the minimum sighting distance needed by the driver to perceive the signal aspect safely (chapter 3.4.1). The most typical are:

- A constant distance, independent from the speed of the approaching train (typically around 300 m)
- A constant time over which the signal has to be visible from the train approaching at maximum permitted speed (typically between 6 and 9 seconds)
- An AND-combination of the above two rules
- A distance value graded by the permitted line speed

If that minimum sighting distance cannot be achieved, several solutions are applied:

- Installation of repeating signal in rear of the signal
- Installation of a fix signal board in rear of the signal to attract the driver's special attention
- Repetition of the signal aspect in the driver's cab in several train protection systems

7.2.3 Retro-Reflection of Passive Signal Boards

Signal boards are usually unlit. To use the headlights of the train to make signal boards visible at night, use may be made of retro-reflection (figure 7.9). In contrast to disperse reflection (e.g.

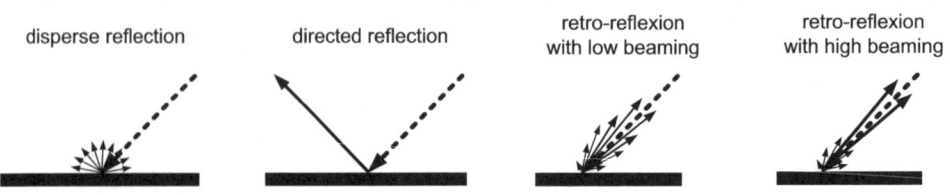

Figure 7.9: Kinds of reflection

7.2 Technical Characteristics of Trackside Signals

white wall) and directed reflection (e.g. mirror), retro-reflection is not a natural characteristic of a material surface, but is made artificially by multiple refection in spherical or prismatic structures. In the field of railways, in contrast to roads, retro-reflecting materials with high beaming are used. The reasons are that longer sighting distances to the signal are required for railways, while at the same time the front lights can be weaker on the vehicles.

7.2.4 Control and Supervision of Signal Lamps

7.2.4.1 Bulb Lamps

In contrast to other technical fields, incandescent lamps survived in railway signalling for a long time. The reason was the lack of alternatives for safe supervision: It can be assumed that light is emitted when and only when a current flows through the filament. To increase the life span of the lamp, some railways (particularly in North America) use approach lighting: The lamp is lit only if a train approaches. Other railways (particularly in Europe) operate the lamp at a lower current than that for which it is designed to achieve the same result.

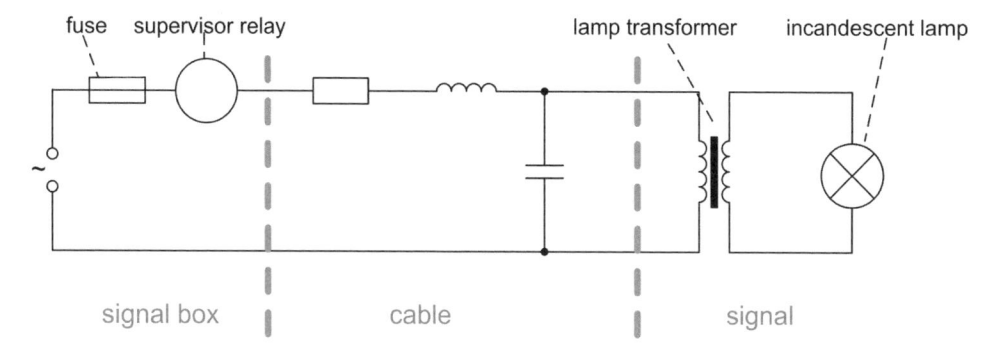

Figure 7.10: Lamp circuit for signal lamps

Figure 7.10 shows the typical circuitry for a signal lamp. Current is often transported to the signal at higher voltage to decrease losses of energy and transformed to lower voltage in the proximity of the lamp. The cable can be modelled as a resistor with an inductive component. Besides, the cable contains a capacitive leak impedance between the wires. A supervising unit in the interlocking, which is often a relay, supervises the presence of a current and, by this, proves the lamp lit. The following behaviour of the circuit must be ensured:

- In the case of open circuit (e.g. filament broken), the current via the leak impedance must not exceed a certain value which would prevent the supervising relay from dropping down. The longer the cable, the higher is the leak impedance and the stronger this current.
- In case of a short circuit, the current must be high enough to melt the fuse, which results in disruption of the circuit and dropping of the relay. The longer the cable, the weaker is the short circuit current.

The result of both conditions is the limitation of the length of the cable, depending on the type of cable. Typical values are between 5 and 10 km (figure 7.11, variant 1).

A modern solution, which is applied especially in electronic interlocking, is to transmit only information by an information cable between the signal box and each signal and to feed power directly at the signal by a ring line (figure 7.11, variant 2). The advantages of this solution are practically unlimited length of the distance between the signal box and the signal and lower material expenditure for energy cables.

7 Signals

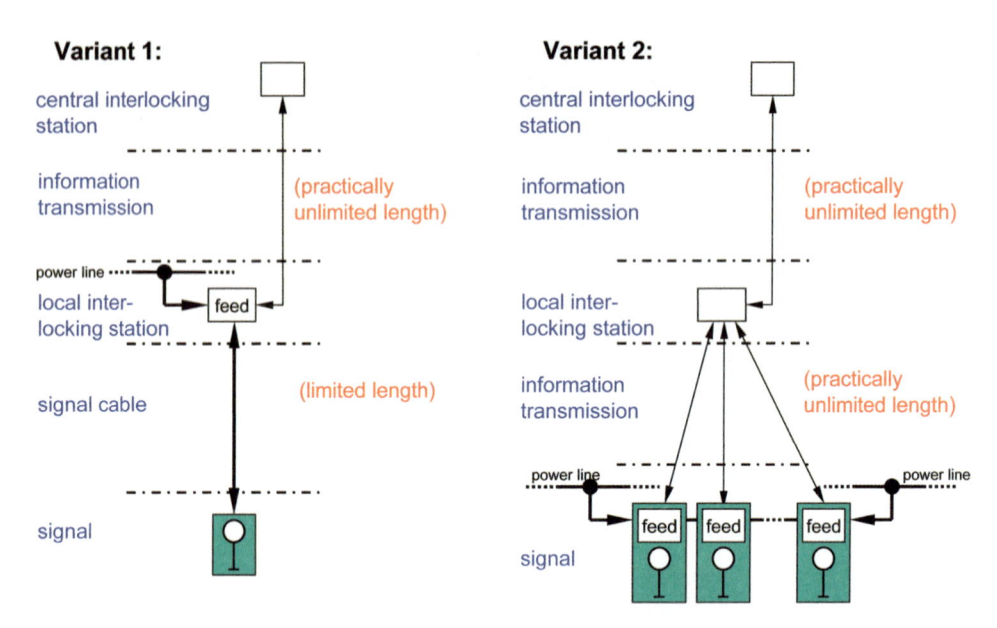

Figure 7.11: Power feed and control of signals in electronic interlocking

7.2.4.2 LED

In LED signals, the form of safety and supervision differs between the manufacturers. In contrast to bulb lamps, here the equivalence of current flowing and the lamp being alight cannot be assumed. LEDs degenerate gradually, which means that the light output slowly declines, but the consumed energy remains the same. Therefore the degeneration cannot be detected electrically.

However, in contrast to failure of bulb lamps, the degeneration of LEDs can be more or less predicted and the probability of complete failure of one LED within a defined time period is low. Therefore, the basic safety principle is that partial failure (a limited number of LEDs extinct and a predictable degeneration of the remaining LEDs) is tolerated and the light signal can still be considered as safely visible to the train driver. Risk analysis is used to calculate that the probability of system failure (light signal not safely visible) is below an extremely low, tolerable value. This principle is connected with fixed replacement intervals (e.g. 10 years), after which all LEDs have to be exchanged.

To exclude certain cases of complete failure, in several types of LED signals, the current flow over the LEDs is supervised, not necessarily individually, but also possible in chains. Besides, each LED is controlled by a separate driving unit, which means that failure of one LED or driving unit only effects this one LED.

Advantages of LED signals over incandescent lamp signals are:
- longer life
- easier maintenance due to calculable maintenance intervals
- immunity against phantom lights
- lower energy consumption

It can be expected that they will increasingly replace the incandescent lamps in future. However, the advantage of lower energy consumption is eliminated when LED signals are integrated into interlocking systems which were originally designed for the use of bulb lamps. Here, using special adaptation such as additional power consumers, LED signals are designed that way that they have the same electrical interface as bulb lamp signals and can be used in the conventional lamp circuit.

7.3 Principles of Signalling by Light Signals

7.3.1 Utilisation of Signal Colours

Unless using mechanical signals, train movements are usually signalised by colour light signals and partly by colour position light signals. In contrast with position light signals, the advantage of colour light signals is visibility over longer distances. For shunting movements, all these kinds of light signals are in wide use.

Historically, most railways used the following colours for signal flags and the night signals of mechanical signals (figure 7.12):
- Red for 'Stop'
- Green for 'Caution'. The meaning of this signal aspect was something between 'expect to Stop at the next signal' and 'Proceed slowly'
- White for 'Clear'

This use of colours conformed to the test results of the *Chappé* brothers, who tested the visibility of colours for optical telegraphs in 1792. However, the utilisation of the white light for 'Clear' causes significant safety problems. In modern times, there are many white lights near the railways, which might be mistaken for a signal showing 'Clear'. Phantom lights caused by sunlight are often white and also if the colour filter of a red or green light bursts, a 'Clear' aspect is displayed. Therefore, the white light today is only rarely found for permitting train movements. Instead, today *UIC code 732* defines the following colours:
- Red for 'Stop'
- Yellow (Orange) for 'Caution' (USA: 'Approach')
- Green for 'Clear'

The shades for these colours are defined exactly. These colours are in wide use today. Sweden is the only European country which still uses the white light for 'Clear' and the green light for 'Caution' (both flashing) at distant signals.

colour		visibility (Chappé)	main use of colours	
			historically	today
white	○		Clear	shunting and subsidiary signals
red	●		Stop	Stop
green	●		Caution	Clear
blue	●		-	Stop (for shunting) (some countries)
yellow/orange	●	not tested	-	Caution

Figure 7.12: Visibility of colours according to test results by *Chappé* brothers in 1792 and use of colours in railway signalling

7 Signals

Besides, white is often used to permit shunting movements and blue or violet to forbid them. Blue and violet are characterised by low visibility. Therefore, they are suitable for shunting movements, but not for trains running at higher speed.

Whereas green and yellow lights are also combined to indicate reduced speed, a principle among most railways is to use the red light for nothing else but Stop aspects (chapter 7.3.2). Exceptions are in North America and Italy, where red in combination with green and/or yellow is used for speed signalling. In other cases, red with an additional indicator or flashing red means to proceed slowly or on sight (without stopping before).

7.3.2 Stop Aspects

Most railways differentiate between two or more Stop aspects connected with different rules. The most common can be classified into these groups (figure 7.13):

– **Absolute Stop:** A signal showing absolute stop is only allowed to be passed with special permission (e.g. written instruction or auxiliary signal), where the signaller giving the permission takes responsibility.
– **Permissive Stop** ('Stop and Proceed'): After stopping, passage of the signal on sight with the driver taking responsibility is permitted.
– **Restrictive Permissive Stop:** The passage of the Stop signal with the driver being responsible is restricted to certain cases where a technical failure can be assumed. Indicators for such cases can be that the signal does not clear for a defined time and/or attempts to contact the signaller have failed.

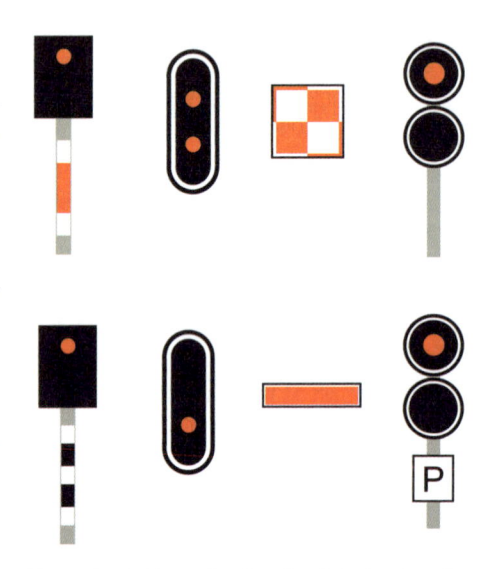

Figure 7.13: Examples for absolute (top) and permissive (bottom) Stop

The Absolute Stop exists in almost all railways and is mainly used for interlocking signals where movable track elements and opposing movements have to be protected. Besides, most railways use either the permissive Stop or the restrictive permissive Stop for block signals where only following movements shall be protected.

Besides the mentioned above, other Stop aspects are applied in several railways. One of these is the 'Advanced Stop' in France. Besides, some railways apply additional train-selective signal aspects to avoid heavy freight trains with low acceleration possibilities stopping in locations with rising gradient.

7.3.3 Signalling of Movement Authorities

7.3.3.1 Main and Distant Signals

Usually the locations limiting a movement authority (MA) are equipped with a main signal. These signals indicate if the train has to stop or is allowed to continue until the next main signal or similar block limiting point (chapter 4.3.2). The movement authority in itself does not include the order of departure for passenger trains, but it can contain an obligation to move on sight. In some countries, particularly the USA, different safety preconditions (e.g. points in

7.3 Principles of Signalling by Light Signals

correct position, opposing protection, following protection) are indicated by different signals, verbal permission or written instructions. In these cases, only the sum of all permitting indications is the MA.

A main signal can be either related to one track, or in some countries, especially in case of converging tracks in a station, to a group of tracks. The application cases for group signals (figure 7.14) differ between the railways. Generally they are likely to be applied for small stations on secondary lines, where all train movements for the same direction are signalised by the same signal, or for tracks used by freight train in medium sized and large stations. The tracks where the group signal is valid for are often marked by additional signal boards or by switchable indicators which relate the signal aspect to a certain track.

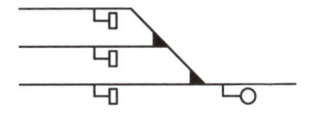

Figure 7.14: Group Exit Signal

With railways, in contrast to road traffic, the stopping distance is typically longer than the sighting distance of the signal. In most cases, the main signal has to be preceded by a **distant signal** (USA: 'approach signal') to enable the driver to decelerate in time. Distant signals are not necessary if the train approaches the main signal at low speed. Examples of such cases are:

– shunting signals
– secondary lines with low speed
– cases where the speed has been reduced by means of speed signalling before (e.g. chapter 7.5.7)

7.3.3.2 Two- and Three-Aspect-Signalling

In Two-Aspect-Signalling (figure 7.15) each signal is either a main or a distant signal and gives information about the state of one block section beginning at the main signal and ending at the next main signal.

When the distance between two consecutive main signals is reduced to almost the stopping distance of trains, Three-Aspect-Signalling (figure 7.15) makes sense. Here, the distant signal is locally combined with the main signal in rear. The signal can show three aspects: 'Stop', 'Caution' (one free section; expect Stop at the next signal) and 'Clear' (two or more free sections; no restrictions). Therefore, these signals are called combined signals.

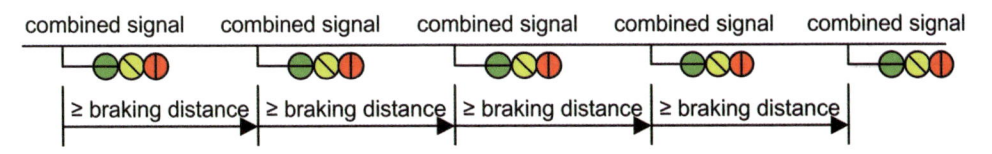

Figure 7.15: Two- and Three-Aspect-Signalling

7.3.3.3 Distance between Main and Distant Signal

The proper warning distance and hence the positioning of the distant signal is determined by the stopping distance of the train. However, the stopping distance varies, mainly depending on the following:

- the speed of the train before starting to brake
- the braking ratio of the train (chapter 3.4)
- the gradient of the line
- weather conditions

To bring the required braking distance of the train in line with the distance from distant to main signal in trackside signalling, two problems have to be solved: The different stopping distances of different train categories shall be adjusted with the fixed positions of the signals, and the distance between distant and main signal shall be adjusted to locally different speeds and gradients.

To adjust the stopping distance of the trains to the warning distance determined by trackside signals, often the permitted maximum speed of the train is graded by the braking ratio (chapter 3.4), where, for example, freight trains have a lower speed limit than passenger trains (figure 7.16).

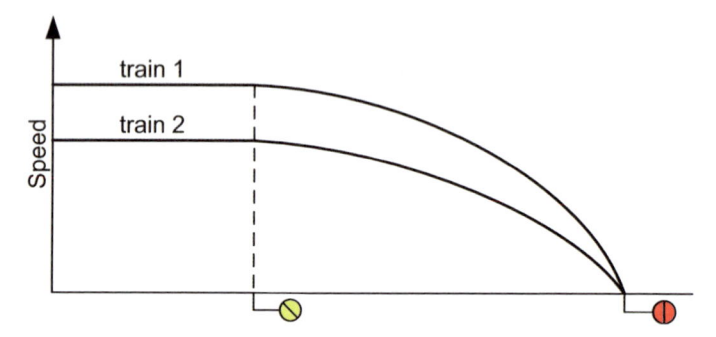

Figure 7.16: Relation between initial speed and braking deceleration under fix braking length

To adjust the warning distance to local parameters, the following solutions can be found in pure or mixed form among the railways (figure 7.17):

a) The warning distance is fixed and the driver has to adjust the braking process to the known warning distance.
b) The warning distance is varied with the stopping distance actually needed. Therefore, the

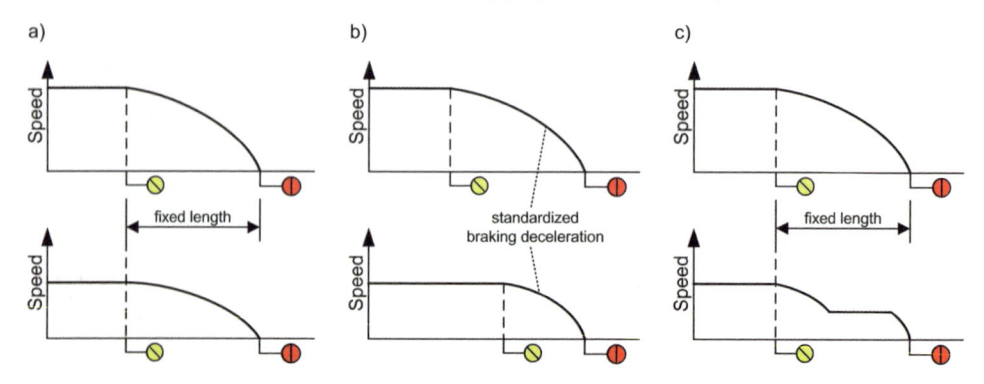

Figure 7.17: Adjustment of warning distance and braking process to local specialities

driver has to brake by a kind of standardised braking performance and can be sure to stop at the correct position.
c) The driver has to start braking with a defined braking performance until a defined low speed (e.g. 40 km/h) and then continue at this low speed until the main signal becomes visible. In the interests of service speed and line capacity, the distance to be run at low speed should be short.

Resulting from historical development, in each country there is a more or less uniform distance between distant and main signal of approximately 1000 to 1500 m on main lines in Europe and up to 3000 m in North America. With modern vehicles, this distance is sufficient for speed up to approximately 160 km/h to stop within this 'normal' warning distance.

7.3.3.4 Braking over more than one Distance between Signals

Cases can occur where the distance between two consecutive main signals is shorter than the stopping distance. Reasons can be:

- Due to high **capacity** requirements, the distance between main signals has to be reduced below the normal warning distance of approximately 1000 to 1500 m (chapter 7.3.3.3). However, by reducing the distance between signals to half of the braking distance, line capacity can be increased by a factor 1.3 to 1.5 only.
- Special situations in the **track layout** require a local reduction of distance between two main signals.
- **High speeds** over approximately 160 km/h shall be signalised by trackside signals, which makes the 'normal' warning distance (chapter 7.3.3.3) insufficient. This is done by some railways in Western Europe (e.g. France up to 220 km/h).
- Trains with very **different braking characteristics** have to be accommodated.

The following forms of signalling of such cases can be found among the railways in pure or mixed form (figure 7.18). Often different solutions will be used by the same railway. The suitability of the solutions for the above cases is different (table 7.1):

1. Speed Reduction in Steps

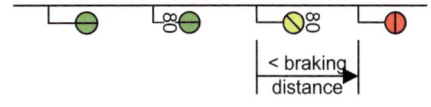

2. Preliminary Caution

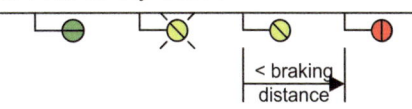

3a. Repetition of Caution

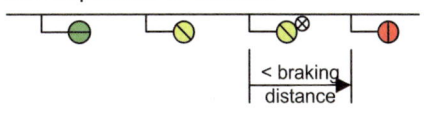

3b. Unlitting of Signal

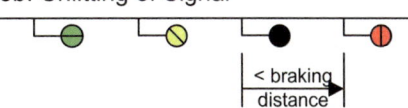

Figure 7.18: Solutions for short distances between signals

Solution No.	Situation			
	High capacity requirements	High speed	Special track layout	Different braking characteristics
1	Conditioned	Yes	Yes	No
2	No	Yes	Conditioned	Yes
3	Yes	No	Yes	No

Table 7.1: Suitability of different solutions for short distances between signals for reasons

7 Signals

1. The utilisation of speed signalling to reduce the permitted speed in two or more steps. Here the driver is indicated an exact speed not to exceed when passing each signal between the first speed warning and the stop signal.
2. Four-Aspect-Signalling with utilisation of the 'Preliminary Caution' aspect. Here, the signal showing 'Caution' is announced by 'Preliminary Caution' (USA: 'Advanced Approach') at the signal in rear. Some trains (those with high initial speed or low braking ratio) have to start the braking process at the 'Preliminary Caution' signal to be able to stop at the Stop signal, whereas others can consider the 'Preliminary Caution' as 'Clear'.
3. Extension of the warning distance over two sections between signals: 'Caution' is shown by a signal in proper stopping distance in rear of the stop signal, whereas the intermediate signal shows 'Repetition of Caution' (3a) or is switched off and marked by marker light (3b) or two consecutive caution aspects are used.

A special version of solution (1) is used in conventional Japanese railways (chapter 7.5.7), where no distant signals exist. The permitted speed from one main signal to the next is reduced in small steps. The difference of permitted speeds at each signal is small enough for the sighting distance of each signal to be the proper warning distance.

7.3.4 Signalling of Speed Reductions

7.3.4.1 Reasons for Speed Restriction

Various reasons can restrict the speed of trains:
- The maximum **speed of the train** is restricted by constructive parameters of the vehicles.
- The line speed is the speed the line is designed for regarding curve radii, easement curves, cants, visibility of signals and other parameters. The line speed is regulated in printed documents in the driver's cab, but also to be memorised by the driver and in some systems signalled by fixed boards.
- **Permanent speed restrictions** are locally restricted speeds below the line speed due to local conditions. These are regulated by printed documents on the driver's cab, but also to be memorised by the driver and indicated by fixed boards.
- **Temporary speed restrictions** can often be found at construction sites or as a result of defective track. They have to be signalled to the driver, usually by signal boards. Often more conspicuous or at least different types of signals are applied for permanent speed restrictions than for normal line speed.
- **Route determined speed restrictions** depend on characteristics of the current route, such as radii of diverging points, reduced overlaps, occupied tracks etc.
- **Speed restrictions in degraded mode operation** are connected with the utilisation of written instructions, auxiliary signals etc (chapter 7.4).

The **scheduled speed** is not safety related and can be varied. It is determined by the timetable as a recommendation to optimise railway operation, passenger comfort and energy consumption. The scheduled speed is usually indicated to the driver by printed timetables or comparable electronic indications in the cab and not by trackside signals. It can vary between train categories, e.g. for night passenger trains it is usually lower than for other passenger trains for comfort reasons.

For tilting trains, where applicable, higher line speeds and higher speeds at permanent speed restrictions can be achieved and have to be signalised to the driver either by additional information on signal boards or by cab signalling. For route determined speed restrictions, no increase can be achieved for tilting trains.

7.3 Principles of Signalling by Light Signals

From these different speeds, the driver has to choose the lowest for the current situation. To facilitate this, some railways integrate the signalisation of line speeds, permanent speed restrictions and in some cases even route determined speed restrictions.

7.3.4.2 Route and Speed Signalling

For route determined speed restrictions, in the first step the permitted speed has to be concluded from the route. In the second step, the driven speed profile has to be adjusted to that permitted. Two principles of signalling can be distinguished: the **route principle** and the **speed principle**.

The **route principle (route signalling)** is the historically older form, used by almost all railways until the beginning of the 20th century. It was already used in the middle of the 19th century before interlocking was introduced, where the position of the mechanical signal also served as an order to the levermen to set the points appropriately. Examples where the route principle is still used are the British Commonwealth, Spain, Norway, China, Western USA and Mexico. The driver is informed about the route the train is going to take and has to conclude the permitted speed using signal boards or his knowledge of the line.

During the 20th century, many railways, among these most European railways as well as those in Canada and Eastern USA, switched to the **speed principle (speed signalling)**. Here the interlocking concludes the permitted speed and signalises it to the driver as a direct order. Often the signal aspects of mechanical signals which were originally used for route signalling were now adapted to speed signalling: The aspect for 'diverging route' got the meaning 'speed reduction'. However, besides the speed information, the direction the train is going to take can be indicated as additional information to give the driver the possibility to prove if the correct route has been set.

The route principle has the advantage of giving the possibility to adapt the speed individually to each track element (figure 7.19), but it also has several disadvantages: The attention and knowledge of the driver is stressed to a larger extent, which makes mistakes in choosing the

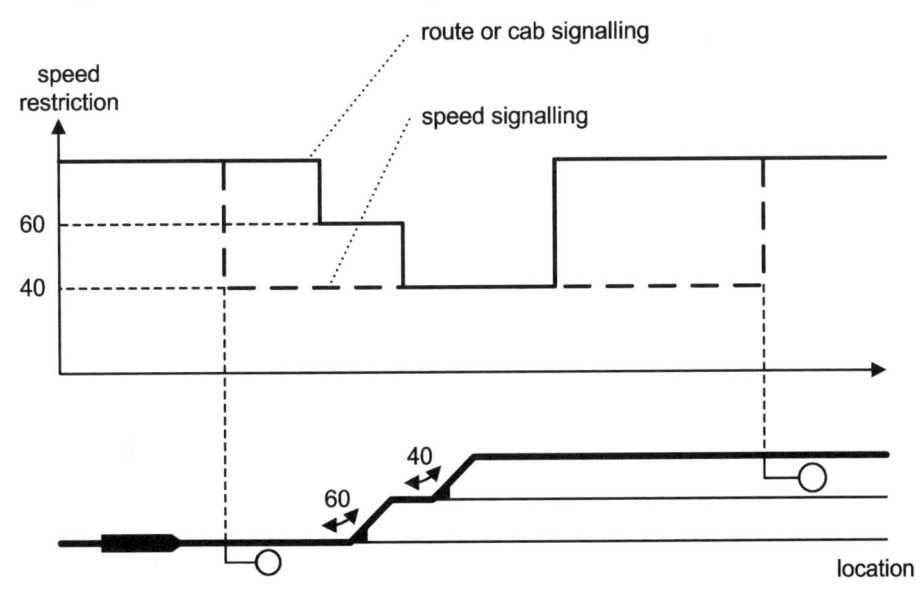

Figure 7.19: **Local validity of speed restrictions in route/cab and speed signalling**

speed more likely. This makes the allocation of train drivers to duties more difficult. Especially in large junctions with many routes, difficulties occur to signalise the routes clearly, particularly in track layouts with many different speeds. Besides, the adaptation of train protection systems to speed signals is easier than to route signals.

7.3.4.3 Speed Signal Aspects

A reduction of the permitted speed, in contrast to an increase, usually needs a previous indication to prepare by braking. Therefore, for a permanent, temporary or route determined speed restriction with two changes of the permitted speed, three basic signal indications are required (figure 7.20):

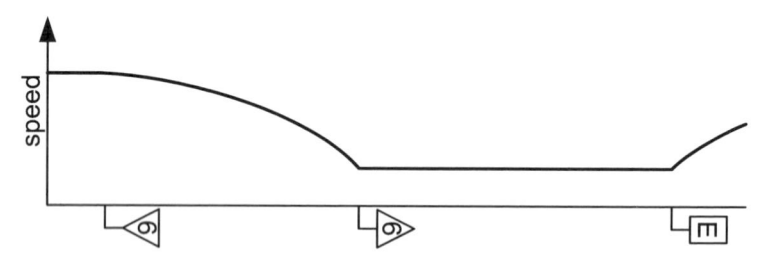

Figure 7.20: Basic speed signal indications

- **Speed Restriction Warning:** This signal shows the speed restriction and gives the order to start braking. In most signal systems, the permitted speed has to be specified, although in some signal systems, approximate values are accepted to reduce the range of different signal aspects. In case of slight speed reduction, such signals can even be unnecessary as the sighting distance of the restriction signal is longer than the stopping distance. For route determined speed reductions, the speed restriction warning and restriction signals are usually combined with the signals for movement authority (chapter 7.3.3); therefore a speed restriction warning signal shows a speed reduction valid from the next signal. The exact regulations concerning the details of the braking process connected with this aspect vary between the railways in a similar way as for distant signals showing 'Caution' (chapter 7.3.3).
- **Speed Restriction:** Upon passage of the signal, the respective speed must not be exceeded. Some signal systems repeat the exact value of the speed to the driver, others do not, as the driver has already been informed by the restriction warning signal. Some railways do even not use restriction signals in selected cases, which are replaced by the regulation to brake as soon as possible beginning at the restriction warning signal and then proceed at the indicated speed.
- **Speed Cancellation:** The driver has to be informed about the end of the speed restriction. Usually acceleration to line speed is allowed after the rear end of the train has passed the termination point. For permanent and temporary speed restrictions, typically an additional signal board is used giving the information 'end' or displaying the line speed. For route determined speed reductions, a second frequently applied solution is to define the termination point by regulations without additional signals. Typical termination points are described in chapter 4.3.2.3.

In cases where one speed restriction gives way directly to another, the cancellation aspect equals with the restriction aspect of the other speed limit. Restriction warning is only necessary in case the new speed restriction is lower than the previous.

7.3 Principles of Signalling by Light Signals

7.3.4.4 Forms of Signalisation of Different Speeds

Where speed signalling is applied, the following solutions to distinguish different speeds on the signal are the mostly used (figure 7.21):

- **Arrangement of colour lights** (colours, quantity and geometrical formation): A disadvantage of this principle is the low volume of possible speed aspects and the high number of lights the driver has to observe at the same signal. This principle is used in Central and Northern Europe to a decreasing extent, but still predominates in the USA.
- **Flashing lights:** Flashing and steady green and yellow lights are distinguished. Besides, distinction can be made between slow and fast flashing and between two lights alternately or simultaneously flashing.
- **Geometrical indicators:** Stripes in different colours or other different geometrical forms can be used to distinguish different speeds.
- **Numerical indicators.** Usually a number multiplied by 10 indicates the speed. With modern technology, these indications can be solved easily (chapter 7.2.1). The lower visibility compared with colour lights can be a problem, especially in cases where the numerical indicator degrades the 'Clear' aspect. However, this solution is increasingly used among railways in modern signal systems.

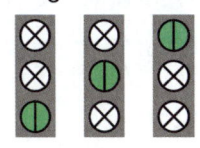

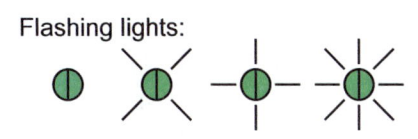

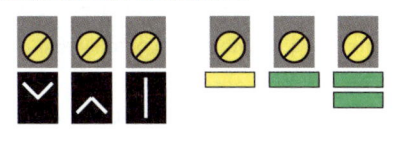

Arrangement of colour lights:

Flashing lights:

Geometrical indicators:

Numerical indicators:

Figure 7.21: Possibilities to signalise different speeds

7.3.5 Combination of Main and Distant Signals

Signal systems with speed signalling can be classified into three groups, concerning the form to combine the signal aspects of the main and the distant signal at the same place in three-aspect-signalling (figure 7.22) (Theeg/Maschek 2005):

- **Separate main and distant signals:** Both signals may or may not be mounted on the same post, but the signal aspects of the main and the distant signal are still separately displayed one above the other. In most cases, these signal systems are directly derived from the night light signals of the mechanical signals.
- **Combined Signals of 1st Grade:** In case the main signal must display a speed restriction, the aspects of the main and the distant signal are shown both above each other (figure 7.22, middle bottom). In contrast, if the main signal just shows 'clear', the signal aspects of the main and the distant signal are joined to a simplified aspect (in most cases the green for 'Clear' is not displayed).
- **Combined Signals of 2nd Grade:** In both cases, if the main signal indicates 'clear' and if it indicates 'speed restriction warning', the aspects are joined to a simplified aspect with fewer lights. Many newer systems use this principle to simplify the signal aspects to only one spot light plus speed indicators.

7 Signals

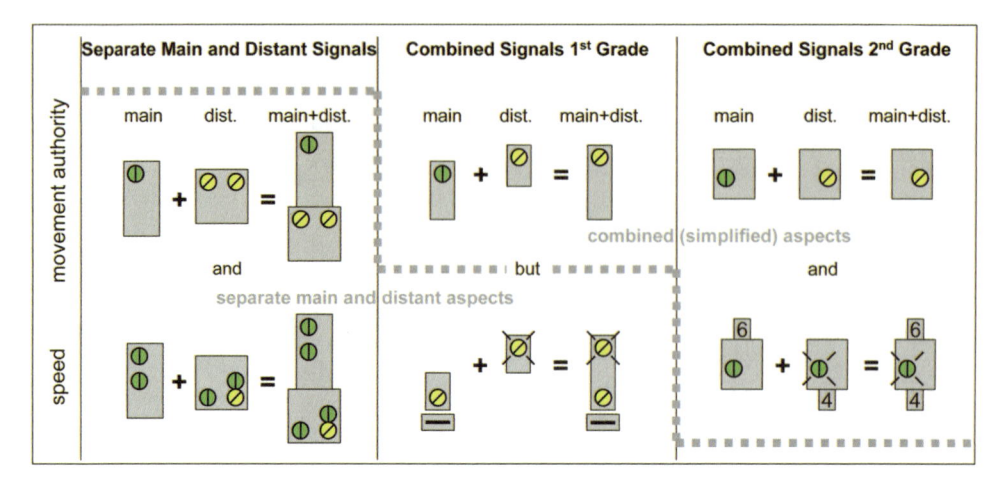

Figure 7.22: Combination of main and distant signals

Systems with combined signals can be distinguished by the amount of speed information given to the driver. If the speed information displayed at a main signal is valid until the next signal, from which a different speed restriction is valid, then the possibilities are (figure 7.23):

1. Full information about the permitted speed in this and the next section is displayed.
2. The speed for this section is always displayed, the speed for the next section only in case of a reduction when the driver needs this information to start braking in the proper time.
3. The signal is equipped with only one speed indicator which displays the lower of both speeds. When the speed restriction valid from the signal is not displayed, the driver has to remember the speed that has been announced at the signal in rear.

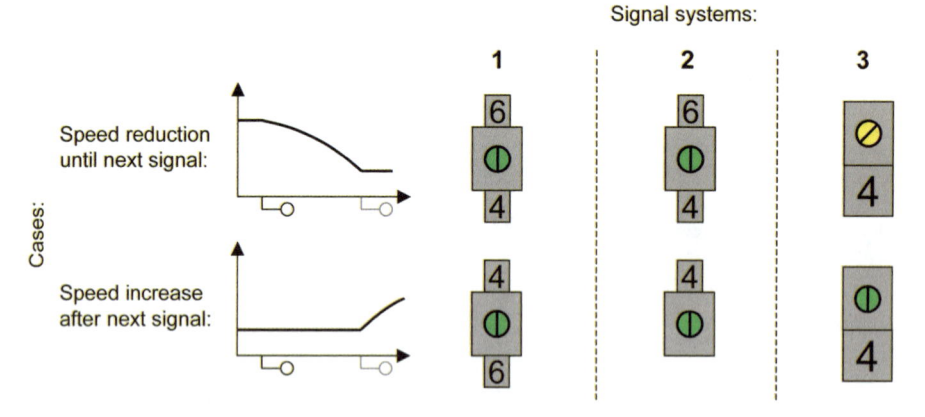

Figure 7.23: Different amount of information in speed signalling (fictive signal system examples)

7.3.6 Shunting Signals

Most railways distinguish between train and shunting movements (chapter 3.3) and signalise shunting movements by other, simpler signals. As shunting movements are slow, neither distant signals nor speed signalling for diverging routes are usually necessary. Besides, the required sighting distances are also low. This makes smaller signals, position light signals and other colours than for trains usable.

Shunting signals can either stand alone or be attached to a main signal as a subsidiary signal. In the latter case, the 'Proceed for shunting movements only' aspect is displayed together with the 'Stop' aspect of the main signal. Stand-alone shunting signals often have to be cleared or unlit if a train passes this signal to avoid confusion of the train driver.

Basically, the range of signal aspects for shunting movements has to include at least one 'Stop' and one 'Proceed' aspect. Besides, several railways distinguish between different 'Proceed' aspects, e.g. for movements into a free or occupied track, for straight or diverging movements etc. (chapter 4.3.10)

The following solutions for standalone shunting signals can be found:
- **Colour light signals:** The 'Proceed' aspect is usually represented by white, whereas the 'Stop' aspect is represented either by red, blue or violet.
- **Position light signals:** In most cases, the 'Proceed' aspect is represented by a vertical or diagonal formation and the 'Stop' aspect by horizontal white lights.
- **Colour position light signals and other mixed forms:** For example, 'Stop' and 'Proceed' are distinguished by colour, whereas different 'Proceed' aspects are distinguished by the formation of white lights.

7.4 Redundancy and Degraded Mode Operation

As in most cases bulbs are used, failures by destruction of the filament occur frequently. Most railways provide redundancy in the optical system to ensure high safety and reliability, by doubling either the lamp or the filament within the same lamp. Usually the extinction of the main lamp or filament is automatically reported by an alarm to the maintainer and maintenance actions must be initiated. For the red lights, redundancy is especially important for safety, whereas in case of failure of a permitting aspect, a more restrictive aspect can be automatically activated, which in most cases implies losses in travel time and line capacity.

Nevertheless, if a signal extinguishes, regulations apply which oblige the driver to assume the most restrictive aspect. Besides, often interlocking functions are implemented which actively set the signal in rear to stop to prevent danger.

Other frequent reasons for a signal not able to be cleared are technical failures in proving preconditions for clearing the signal, such as the clear track, point position etc. The principle of fail-safe requires that a technical failure has to lead to the safer condition, which means in most cases, the Stop aspect. Nevertheless, operation has to be maintained using methods of degraded mode operation (chapter 4.5.4). Such methods can be permissive driving (chapter 7.3.2) or authorisation by ground staff to the driver by
- **Hand signals** or **verbal permission**
- **Written instructions:** forms to be filled and signed by the signaller and handed over or dictated telephonically to the driver
- **Auxiliary signals** without any or with reduced interlocking functions, formed by additional signal lamps, to replace the written instruction, saving time for filling in and handing over the forms. The British call-on signal fulfils the functions of both, the auxiliary signal and the signalisation of regular movements on sight.

For the safety conditions related to the use of the auxiliary signal, see chapter 4.5.4.4.

7.5 Signal System Examples

In the following, some examples of signal systems are described. The examples are selected to represent a large variety of signalling solutions. Where appropriate (in combined signal systems with speed signalling), signal aspects are drawn in a table with the lines being the speed limits to be obeyed when passing the respective signal and the columns the announced speed for the next signal (figure 7.30, 7.31, 7.32, 7.34).

7.5.1 German Mechanical and 'H/V' Light Signals

The German mechanical signals (figure 7.24) are a good example to illustrate the historical development of light signals from mechanical signals. They are still used in several existing installations in Germany, though in declining numbers, and derived light signal systems are applied in several countries.

The signals are separate main and distant signals (chapter 7.3.5). The mechanical main signals are designed as semaphore signals with the upper semaphore arm to open and close the signal and the lower arm to restrict the speed. The signals are upper-quadrant, which means that the semaphore arm has to be raised, not lowered, to open the signal. For the main signal, three basic aspects are defined: 'Stop', 'Clear' and 'Proceed slowly' (40 km/h), the latter requiring a second signal arm. However, this second arm is not fail-safe, as in its absence or invisibility a more permissive aspect ('Clear') appears. A weak lamp behind colour filters below the semaphore arms provides signal indications at night.

The distant signals consist of a yellow disk. If the disk is presented to the driver, it announces a stop. If it stands edgewise (negative signal), it announces 'Proceed'. An additional arm below the disk indicates a speed restriction if standing diagonal. On some secondary lines with low speed,

Figure 7.24: German H/V signal system

only the fixed board is used instead of a switchable distant signal, which obliges the driver to expect Stop in all cases until he can see the main signal.

The H/V light signal system is directly derived from the night signs of the mechanical signals (figure 7.24). To be able to indicate different speeds, additional numerical speed indicators were introduced later.

More information on this example can be found in (DB Netz 2006).

Similar mechanical and light signals were used to a large extend among railways in Central and Eastern Europe. Today, the usage of these signal systems is decreasing, but they are still in use on several railways.

7.5.2 Belgian Mechanical Signals

The Belgian signal system from 1919 is a good example for three-aspect-signalling (chapter 7.3.3.2) by semaphores and the route principle solved with candelabra signals (figure 7.25). Similar signalling principles were also applied in other countries such as Britain.

The signals consist of maximum two semaphore arms, with the upper in red and the lower in yellow colour, also different in their geometrical shape. Four basic signal indications are defined: 'Stop', 'Caution', 'Expect slow movement' or 'Preliminary caution' and 'Clear' (figure 7.25). The signal system is designed such that most signals only need to be equipped with one semaphore arm at each post (block signals and separate main signals only with the red, separate distant signals only with the yellow arm). Only combined signals in interlocking areas which have to be able to show both 'Stop' and 'Expect slow movement' need two arms.

Signals protecting junctions are equipped with one post for each direction, with the semaphore arm at the respective post to be raised to permit movement into the respective track (candelabra signal). The highest post refers to the straight direction, and the others are situated to the right and the left in the correct order (route signalling, chapter 7.3.4.2). To indicate other speeds than 40 km/h for diverging movements, numeric signal boards are mounted below the respec-

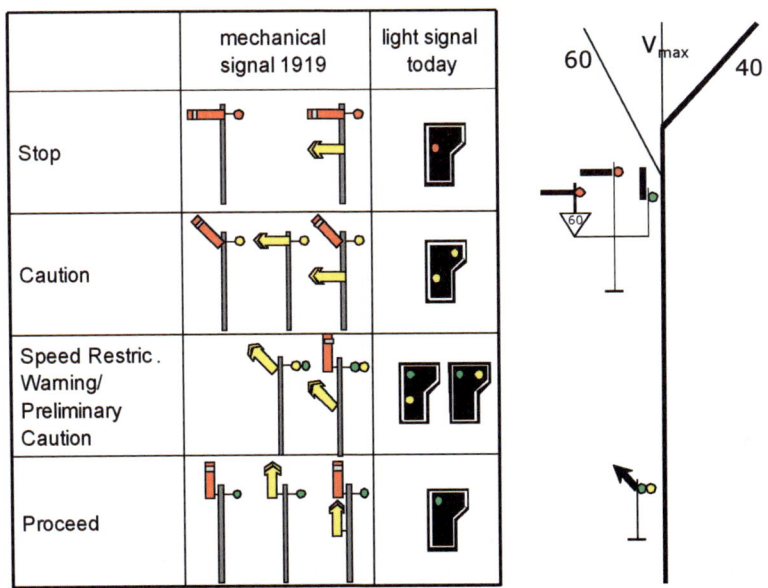

Figure 7.25: Belgian mechanical signal system from 1919 and modern light signals for comparison

7 Signals

tive semaphore arm. In the derived today's light signal system, speed signalling with numerical speed indicators is used.

More information on this example can be found in (Sasse 1941).

7.5.3 British Light Signals

The modern British signal system is a typical example of route principle by light signals. In a similar way it is widely used in the Commonwealth if light signals are used.

The movement authorities are signalised either by Three Aspect signalling (chapter 7.3.3.2) with the aspect sequence green → yellow → red or by Four Aspect signalling (chapter 7.3.3.4) with the sequence green → double yellow → yellow → red.

Signals protecting a junction are equipped with route indicators. For the straight route, no route indicator is displayed in Britain, but in South Africa it is, whereas for diverging movements up to three route indicators can be provided for the left and three for the right side (figure 7.26).

Figure 7.26: Lamp arrangement of British light signals (Four-Aspect-Signalling and maximum of route indicators)

The warning of speed restrictions in many situations is effected by pretending a Stop to the driver. Depending on the speed difference, the driver has to brake. The following signalling arrangements are used (figure 7.27):

− For low route speed (which means: high speed difference to the line speed), a stop at the signal protecting the junction is displayed to the approaching driver. When the train almost reaches this signal, it switches to 'Proceed' with route indicator **(Approach Control from Red)**.

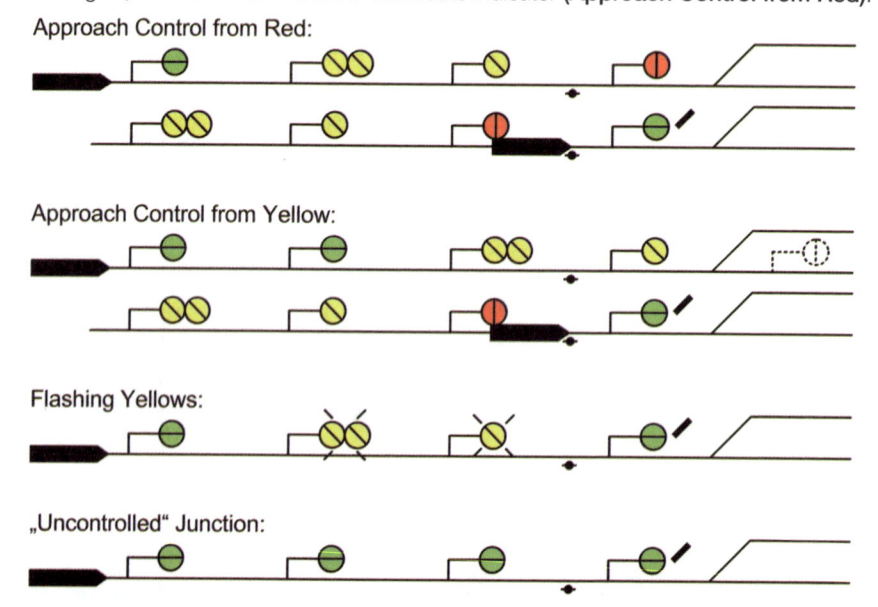

Figure 7.27: Warning of junctions in British signalling

7.5 Signal System Examples

- For medium route speed (approximately half of the line speed), a stop at the next signal beyond the junction is pretended to the driver before switching the signal which protects the junction to 'Proceed' with route indicator **(Approach Control from Yellow)**.
- In recent times, trains with high braking ratio are in operation. Therefore, a new hazard occurred in connection with Approach Control from Yellow: A driver of such a train, knowing about its braking capabilities, would start braking further beyond – and pass the junction at a too high a speed. Therefore, another method was invented to announce the junction unequivocally: **Flashing Yellows**.
- For fast routes where the diverging speed is only a little lower than the line speed, the sighting distance of the route indicator is regarded as sufficient warning distance, and no approach control is needed: so-called **Uncontrolled Junctions**.

A dynamic solution similar to Approach Control from Red is used to reduce train speed in case of a short overlap, so-called Delayed Yellow (figure 7.28). Here the route entrance signal is held at red and switches to yellow short before the train reaches it.

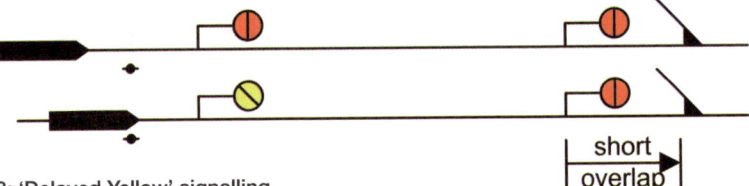

Figure 7.28: 'Delayed Yellow' signalling

More information on this example can be found in (RSSB 2004) and (Nock 1982).

7.5.4 OSŽD Signals

The *OSŽD* signal system is the result of efforts in the 1950s to provide a standard European signal system. In the former socialist countries in Central and Eastern Europe and Asia (member states of the *OSŽD* organisation), these attempts were partly successful, whereas in the other parts of Europe they mainly failed, apart from some standardised features of signal systems like the colours (chapter 7.3.1).

The system is based on the speed principle with combined signals of 1st grade. The signal aspect can be divided into the upper and the lower part (figure 7.29). The lower part gives orders

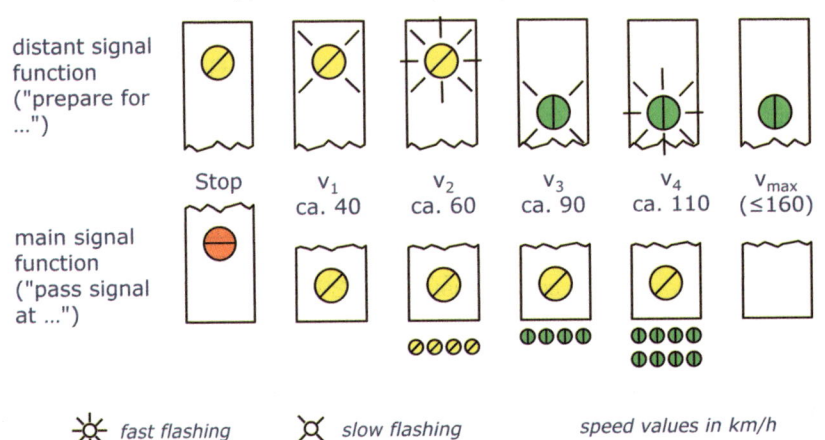

Figure 7.29: Principle of *OSŽD* signal system

7 Signals

about the section beginning at the signal (main signal function), whereas the upper part gives information on the next section (distant signal function) or the speed which is permitted after passing the points area (chapter 4.3.2.3).

The speed limit that is valid from the signal (main signal function) is specified by one or up to two green and yellow stripes. The speed to be prepared for (distant signal function) is displayed by flashing lights. If two flashing frequencies are used, they are made distinct by approximately the factor of two, so the aspects can be distinguished clearly by the driver.

The advantage of the unlit lower part of the signal meaning 'Proceed' is that block signals which do not protect any junction need to be equipped only with three lamps (one green, one yellow and one red).

The speed steps V1... V4 are variable and to be defined exactly in the signal regulations of each railway. Not all four speed levels need to be used and the exact speed can vary. This enables adaptation of the system to the point radiuses present differently in each country. Figure 7.30 shows the signal aspects in Czech Republic and Slovakia using all four speed steps.

More information of the OSŽD signal system can be found in (Piastowski 1960).

On the railways of former Soviet Union, a version of the OSŽD system with several specialities is applied. Some of these specialities were also adopted in PR China, where the OSŽD system is combined with route signalling principles imported from Britain in former times.

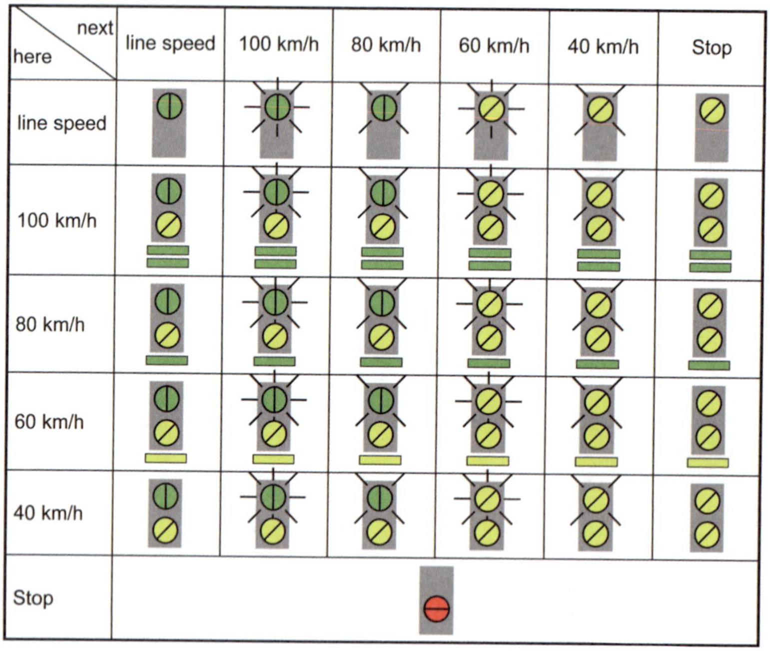

Figure 7.30: Signalling cases in Czech Republic and Slovakia (Česke Drahy 1998)

7.5.5 Modern Dutch Signal System

The modern Dutch signal system is a combined system of 2nd grade (chapter 7.3.5). It is the result of basic reorganisation of signalling principles in the 1950s. The most important ideas of the signal system are:

7.5 Signal System Examples

- The signal aspects are designed inherently fail-safe. That means, if the driver fails to perceive a part of the signal aspect or a part of the aspect extinguishes, always a more restrictive or an undefined signal aspect is shown.
- The signal indications are locally connected as closely as possible with the actions of the driver. Therefore, the 'Expect Stop' aspect got the meaning 'Apply the brakes, beginning at the signal, until you reach the speed 40 km/h. Then continue at 40 km/h until you see the next signal ahead'. Similar regulations are connected with the speed restriction warning aspect. Here the driver gets the order just in time when he has to carry it out.
- No distinction is made between train and shunting movements in signalling.

The most important signal aspects are (Bailey et al. 1995):

- Red: Stop
- Yellow: Caution (Reduce speed to 40 km/h, then continue at 40 km/h and be prepared to stop at a Stop signal.)
- Yellow + number: Speed restriction warning (Reduce speed to the indicated value.)
- Flashing green: Speed restriction (Proceed at 40 km/h.)
- Flashing green + number: Speed restriction (Proceed at the indicated speed.)
- Green: Clear

Although the ideas of the signal system do not match fully into the scheme, for comparison with other signal systems the signalling cases are shown in figure 7.31.

The signal system is inherently fail-safe. When the number is unlit by technical defect or the driver fails to perceive it, a more restrictive aspect is assumed: The driver brakes until 40 km/h ready to stop respectively continues at 40 km/h instead of a higher speed. If the red, yellow or

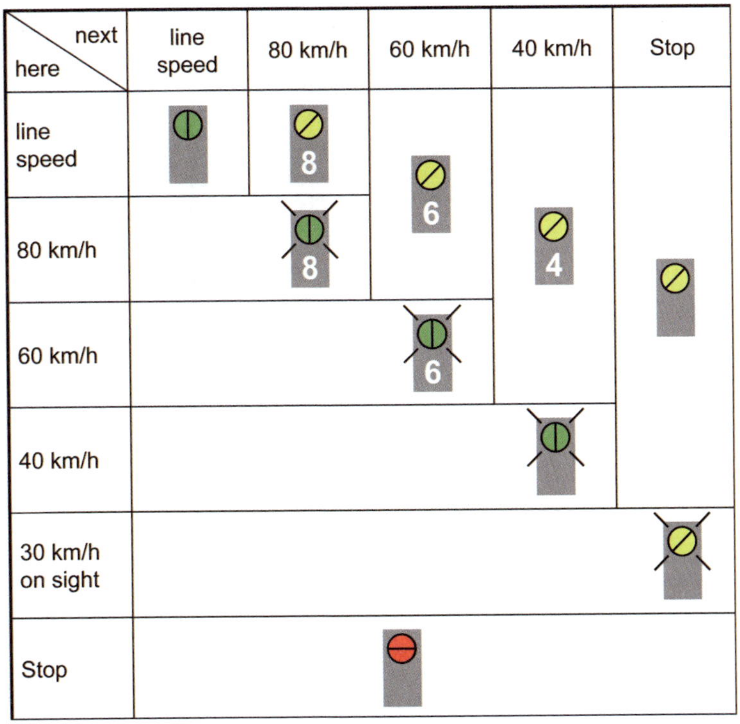

here \ next	line speed	80 km/h	60 km/h	40 km/h	Stop
line speed	🟢	🟡 8	🟡 6		
80 km/h		🟢 8		🟡 4	🟡
60 km/h			🟢 6		
40 km/h				🟢	
30 km/h on sight					🟡
Stop			🔴		

Figure 7.31: Signalling cases in modern Dutch signal system

green light is unlit, no or an undefined indication is shown, this obliges the driver to stop. And if the driver by mistake sees a flashing instead of a steady light (which, due to obstacles near the track, is more likely than the opposite case), the driver perceives a lower speed information. The flashing itself, however, has to be monitored technically.

7.5.6 German System 'Ks'

The German system Ks (figure 7.32), being introduced since the 1990s, is another combined system of 2nd grade with numerical speed indicators. The basic idea of the signal system is to separate the signalling of movement authorities and signalling of speed restrictions into two partial aspects. However, this would be particularly useful if in the same system signalling of different track-determined speed restrictions (line speed, permanent, temporary and route de- termined speed restrictions) were to be merged, which has not occurred in this signal system.

Movement authorities are signalised by the following three aspects:
- Red = Stop
- Yellow = Caution (one section free, expect Stop at the next signal)
- Green = Clear (two or more sections free)

Besides, the signals are equipped with two speed indicators, one for the restriction warning and one for the restriction aspect. In case a speed restriction warning is indicated, the green light flashes to attract driver's attention to the speed indicator. Figure 7.32 shows the example signalling cases for comparison with other systems.

More information of this example can be found in (DB Netz 2006).

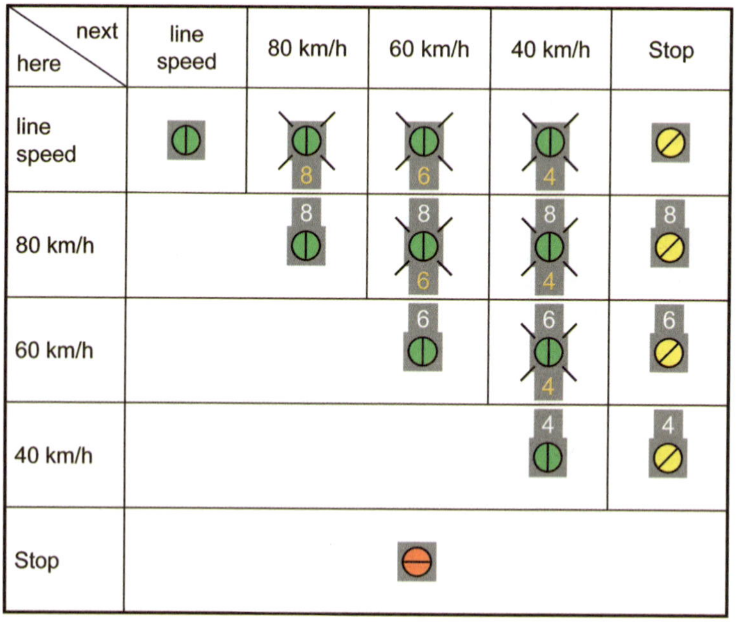

next here / line speed	line speed	80 km/h	60 km/h	40 km/h	Stop
line speed	⊙	⊙ 8	⊙ 6	⊙ 4	⊘
80 km/h		8 ⊙	8 ⊙ 6	8 ⊙ 4	8 ⊘
60 km/h			6 ⊙	6 ⊙ 4	6 ⊘
40 km/h				4 ⊙	4 ⊘
Stop			⊖		

Figure 7.32: Signalling cases in German system 'Ks'

7.5.7 Signal System on Japanese Commuter Lines

On several Japanese suburban lines (narrow gauge) with high capacity requirements, signals are situated very close together. No distant signals are used, but speed is reduced in small steps between two consecutive signals and the sighting distance is sufficient to reduce the speed. The speed is reduced cascade-shaped in several steps from the maximum speed of originally 130 km/h to Stop (figure 7.33). In the 1990s, an additional 'Proceed fast' signal aspect has been introduced to enable an increase of line speed up to 160 km/h on selected lines.

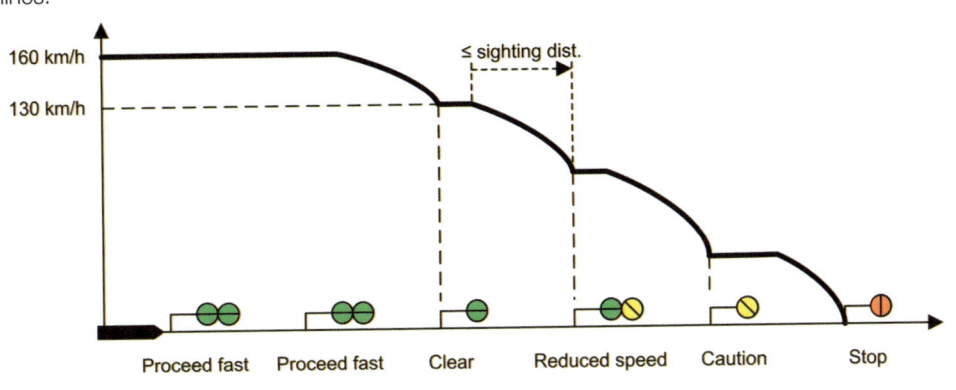

Figure 7.33: Japanese signal system for commuter lines

The speed steps of figure 7.33, added by another 'Restricted Speed' aspect (double yellow; 25 km/h) are also used to signalise reduced speeds required by diverging routes.

More information on this example can be found in (Ugajin et al. 1999).

7.5.8 NORAC Signals

NORAC (Northeast Operating Rules Advisory Committee) is a committee of different railway companies in the Northeast of the USA founded in the 1980s. It standardised operational rules and signal aspects for the participating companies without the necessity of resignalling. The result was the reduction to some fundamental signal indications, each expressed by different signal aspects. In addition, cab signals are used on some lines which are not described here.

The same strategy for harmonisation applied in Europe would be more difficult, as here the diversity of the operational rules behind the signal aspects is much higher than between the *NORAC* participants, although the diversity of signal aspects itself is lower.

NORAC operational regulations define following speeds and kinds of movement (1 mile = 1,609 km):

- Normal speed: Maximum speed of the line
- Limited speed: 45 mph for passenger and 40 mph for freight trains
- Medium speed: 30 mph
- Slow speed: 15 mph
- Restricted speed: Move on sight at a speed which permits stopping within half the sighting distance short of other rail vehicles, obstacles, movable track elements in the wrong position and signals. The speed is limited to 15 mph within and 20 mph outside interlocking limits. This is also the indication used for shunting.

7 Signals

- Permissive stop (stop and proceed at restricted speed)
- Stop

Based on these speed levels, the following 15 signal aspects are defined:
- 'Clear': Proceed at normal speed.
- 'Stop'
- 'Stop and Proceed'
- 'Restricting': Move at restricted speed, in interlocking areas until clearing all points.
- 'Approach': Caution; Prepare to stop at the next signal.
- 'Advanced Approach': Preliminary Caution; Prepare to stop at the second signal in advance.
- 'Limited/Medium/Slow Clear': Speed restriction; Proceed at limited/medium/slow speed until the train has cleared all points; then proceed at normal speed.
- 'Medium Approach Medium': Speed restriction; Proceed at medium speed until the next signal.
- 'Approach Limited/Medium/Slow': Speed restriction warning; Prepare to approach the next signal at limited/medium/slow speed.
- 'Medium/Slow Approach': Speed restriction + Caution; Drive at limited/medium/slow speed; expect stop at the next signal.

With the exception of 'Medium Approach Medium', no signal aspects are defined which imply

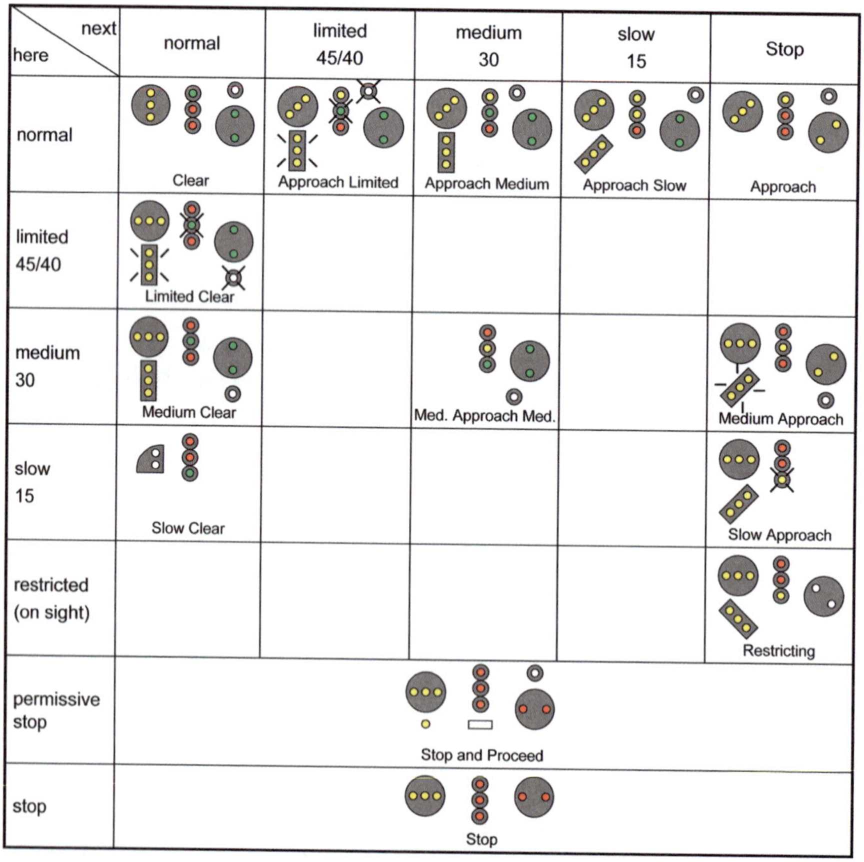

next here	normal	limited 45/40	medium 30	slow 15	Stop
normal	Clear	Approach Limited	Approach Medium	Approach Slow	Approach
limited 45/40	Limited Clear				
medium 30	Medium Clear		Med. Approach Med.		Medium Approach
slow 15	Slow Clear				Slow Approach
restricted (on sight)					Restricting
permissive stop			Stop and Proceed		
stop			Stop		

Figure 7.34: Signalling cases in *NORAC* signal system (selection of aspects)

7.5 Signal System Examples

a speed restriction and a speed restriction warning at the same signal. Other signal systems in the USA likewise in Europe provide a larger variety of such aspects. *NORAC* regulations, which were simplified to accommodate the signal aspects of different companies, assume that the range of vision of a main signal is sufficient to decelerate from medium speed to stop within it. Therefore, caution aspects usually imply that the driver does not need to reduce speed below medium until seeing the next signal.

Each signal indication can be expressed by between 2 and 12 different aspects. Figure 7.34 shows a selection, each of them existing in different variations. The advance approach aspect, which is not listed in figure 7.34, if existent, is always a variation of the Approach aspect with the yellow light(s) flashing.

The left example in each cell of figure 7.34 is a typical position light signal system. The lights on the upper (round) signal background express
- if the train can proceed at maximum speed (vertical),
- if it has to stop at the signal or pass it with reduced speed (horizontal) or
- if the train has to brake for a stop or a reduced speed (diagonal).

Speed restriction aspects are expressed by the stop aspect added by the lights of the lower signal background, whereas speed restriction warning aspects are similarly specified based on the caution ('Approach') aspect. Thus the signal system is inherently fail-safe on the extinction of a part of the lamps. The 'Slow Clear' aspect, when required in areas equipped with this signal system, can only be expressed by the dwarf signal.

The example in the middle (figure 7.34) uses a completely different principle of colour position light signals. It consists of three searchlight lamps. The number of red lights on top of the signal expresses the speed limit which is valid when passing the signal: No red light on top means normal speed, one red light means medium or limited speed, two red lights slow or restricted speed and three red lights Stop. Where necessary, limited speed is distinguished from medium, and slow from restricted by the non-red light flashing. The colours and positions of the non-red lights express what has to be expected at the next signal, that means what speed is valid after clearing the points:
- green means normal speed,
- yellow stop,
- two yellows or yellow above green means slow speed.

The purpose of the lower red lights is to fill up the empty spaces.

In the right example of figure 7.34, the signalling of movement authority is separated from the speed signalling like in the German Ks system (chapter 7.5.6). The movement authority (or the clear status of the block section if no complete MA is signalised, depending on the underlying rules) is signalised by red, yellow and green colour position light signals, whereas speed limits (except of 'Restricted Speed') are signalised by the position and flashing or not flashing of the additional white or yellow light.

More information on this example can be found in (NORAC 2003).

8 Train Protection
Gregor Theeg, Sergej Vlasenko

8.1 Requirements, Classification and Conditions for Application

8.1.1 General Overview

The human being is perhaps the weakest element in railway safety. To minimise the effects of mistakes by signallers, points and signals have been interlocked whenever possible from around 1870. It may thus be argued that the same approach is desirable for drivers.

Train protection systems to guard against driver error developed rather later to supervise the actions of the driver and, if necessary, to enforce safety. Normally, train protection systems protect only against errors, not wilful misconduct.

Many railways regard it desirable to supplement trackside signals with in-cab indications, or dispense with trackside signals altogether. Cab signalling functions, as they are technically strongly connected with train protection systems, are dealt with here rather than chapter 7.

Summarising, the functions of train protection/control systems can be classified into the three groups of:
– Cab signalling functions
– Supervision functions
– Intervention functions

While most modern systems supply all of these functions, many older systems still in use are less comprehensive.

For classification of train protection systems, the most used categories, especially in English speaking countries and Japan, are ATS (Automatic Train Stop), ATP (Automatic Train Protection) and ATC (Automatic Train Control). But as there is no common definition of the meanings of these terms, they are not used in this book for classification. Instead, another classification is developed in chapter 8.3.1.

8.1.2 Cab Signalling Functions

Cab signalling functions generally can be classified into the following groups:
– Non-selective warning signals
– Selective warning signals
– Visual repetition of trackside signals
– Continuous static speed information
– Dynamic speed information

Non-selective warning signals (mainly audible): Whenever the train passes a certain position, e.g. the location of a distant signal, a warning tone sounds to direct the driver's attention to the trackside signals, independently from the signal aspect. No information connection between the trackside signal and the train protection system needs to be provided for this function which is applied in old train protection systems.

Selective warning signals (again mainly audible): The audible signal is applied selectively in cases which imply restrictions for the driver. Usually, the cab signal is connected to signal aspects which require the start of a braking process, such as Caution (chapter 7.3.3.1) and Speed Restriction Warning (chapter 7.3.4.3).

8.1 Requirements, Classification and Conditions for Application

Visual repetition of trackside signals: The aspect of the trackside signal in advance (in some cases in rear), is repeated in the cab during the train's passage between two signals (the signal section), or while the train is within a defined partial section in the vicinity of the trackside signal. Under certain circumstances, this form of cab signalling can replace trackside signals, but in many cases it is used additionally: The cab signal is visible in any weather conditions, it gives the driver a positive reminder of the signal aspect, and in many cases gives information to the driver earlier than does the trackside signal. However, the cab signal does not provide any more information than the trackside signal and the driver is still responsible for estimating the braking requirement.

Continuous static speed information (figure 8.1): Not only are indications of trackside signals repeated, but the permitted speed under consideration of all restrictions is always displayed. In addition, speed restriction warning information can also be displayed, but the driver is still responsible for estimating the braking curve. In several systems, this form of cab signalling replaces trackside signals. In many modern systems, likewise in route signalling, static speed profiles can be adjusted individually to each track element instead of imposing one speed for the whole section between two trackside signals (figure 7.19).

Dynamic speed information (figure 8.1): Based on the static speed information, braking patterns are calculated on the train and/or in the trackside equipment. The technical system displays a guidance speed continuously, which must not be exceeded momentarily in order to comply with the next target speed, to the driver. For this function, information about the distance to the next braking target has to be present. This information can either be transmitted individually for each track section, or standardised by the uniform length of the sections. The latter case, due to its inflexibility, is suitable only for lines with almost uniform traffic, such as pure high speed lines or suburban railways.

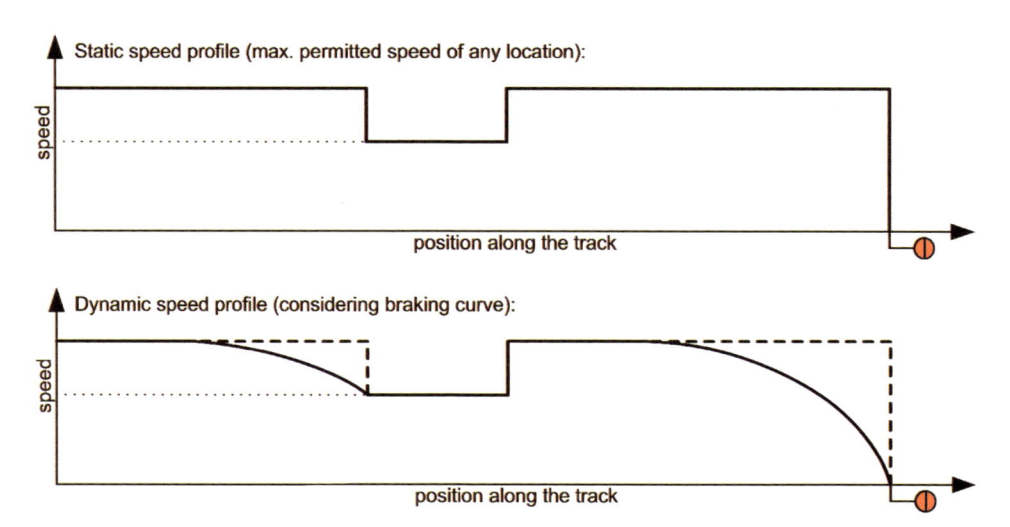

Figure 8.1: Static and dynamic speed profile

8.1.3 Supervision Functions

The following **supervision** functions can be found among the systems:
− Check on driver ability

8 Train Protection

- Check on driver attentiveness
- Train Stop function
- Braking supervision
- Compliance with speed limits

Check on driver ability. At regular intervals, independently from trackside information, the driver has to use an alertness device to guard against falling asleep or similarly, the so-called 'dead-man's handle'. The interval between depressing a handle, pushing a button, or whatever is required, can be either time or distance measured. As an additional feature in few train protection systems, the interval is shortened while the driver has to brake. Occasionally (e.g. USA), this device only needs to be handled if the driver did not undertake any other operation on the locomotive during a defined time interval.

Check on driver attentiveness. In certain situations, e.g. after passing a signal showing Caution, the driver has to acknowledge his attentiveness, e.g. by pushing a special button. Thus the danger from a driver failing to perceive a signal can be reduced significantly. However, many cases have occurred in which the driver pushed the button habitually and without braking, which is potentially highly dangerous.

Train Stop function. The passing of a red signal is detected, which results in an immediate emergency stop. A particular issue is permissive driving, driving on written instruction or on an auxiliary signal. This is enabled either by additional override handles in the driver's cab, by generally permitting the passage of the signal at very low speed or by a combination of these two methods. Except in cases with very low speeds, high braking performance or long overlap, this protective function in itself or in combination with attentiveness check is usually not sufficient to stop the train within the overlap.

As the above functions are not sufficient to bring the train to a halt before the point of conflict in most cases, modern systems provide **braking supervision**: When the train has to brake for a signal at danger or to comply with a speed restriction, the braking process is supervised continuously or at certain points. A problem here is different braking ratios for different trains, resulting in different braking curves (chapter 3.5). Different methods of braking supervision are used among the systems (figure 8.2):

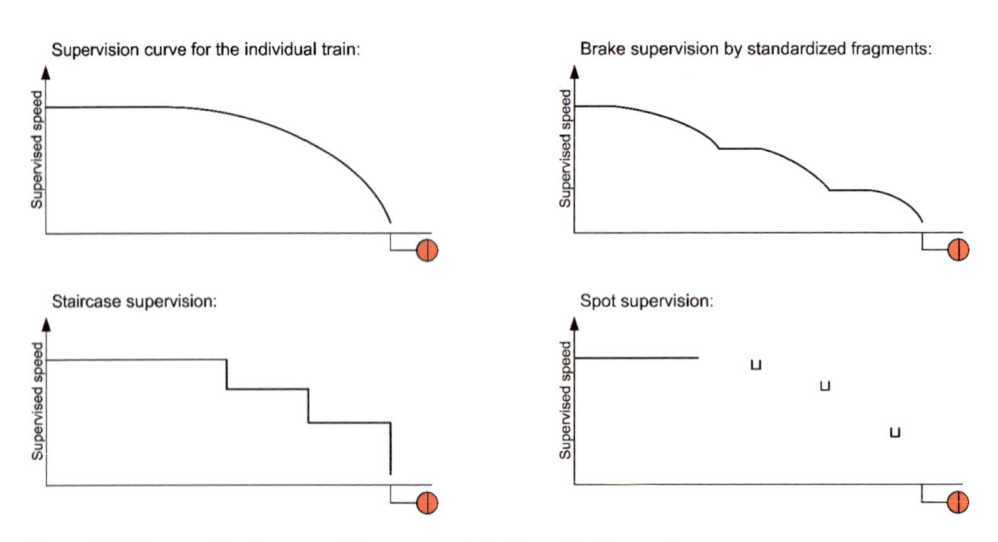

Figure 8.2: Forms of brake supervision curves in train protection systems

- The brake supervision pattern is calculated individually for the train and the track layout. One supervision curve is used for the whole braking process in rear of a Stop signal (one step brake control). Advanced systems with digital data transmission mainly use this method (chapters 8.3.5, 8.3.6).
- A stock of standardised fragments of brake patterns, differentiated by the speed level, proximity to the Stop signal and/or train category, is provided by the system. It is stored on the train computer or in the trackside control centre. Initiated by a trackside transmitter, the proper fragment is selected. This is the typical solution for systems with spot transmission and with low data volume (chapter 8.3.3).
- The supervision function has the shape of a staircase. This is the typical solution for systems with continuous transmission of signal aspects by coded track circuits (chapter 8.3.4), where the same data input is valid during the whole length of a track circuit.
- The speed is checked in form of multiple spots. The supervision speed decreases from one checkpoint to the next in approach to a Stop signal.
- Instead of checking the speed, the application of the brakes can be checked intermittently or continuously during the required braking process.

Some systems allow the driver to exit manually from the braking supervision if the signal has been upgraded, if this information cannot be transmitted automatically by the train protection system.

Compliance with speed limits: In addition to the supervision of the braking process, many systems provide checking of speed restrictions. These can be the maximum speed of the line, local speed restrictions, restrictions on the vehicles themselves and others.

For measuring the speed, two types of methods are used:
- Vehicle-based methods, which measure the speed by odometry, Doppler radar or others (chapter 5.2.6.3).
- Track-based methods which measure the time the train needs to travel a defined distance.

8.1.4 Intervention Functions

When the supervision functions detect a problem in the behaviour of the vehicle, **intervention** functions are activated. Most modern systems grade these. Possible levels of intervention are:

- The weakest is to warn the driver of a problem, mostly by an audible warning tone, and to demand correction.
- The next step applied on some railways is to switch off the traction power automatically.
- The next step is the service brake intervention.
- The strongest intervention function is the emergency brake intervention.

In case of exceeding speed restrictions or brake supervision patterns, some systems use different intervention measures consecutively. These will be applied according to different tolerance margins above the dynamic permitted speed (figure 8.3), and/or the duration of excess speed. Other systems only use one of these measures, mostly the emergency brake.

After passing a signal indicating 'Stop', the consequence in all systems which have this supervision function is an immediate emergency stop.

The brakes can be applied either until the train comes to a standstill, or until speed has been reduced below a safe limit. Many systems record problematic incidents.

8 Train Protection

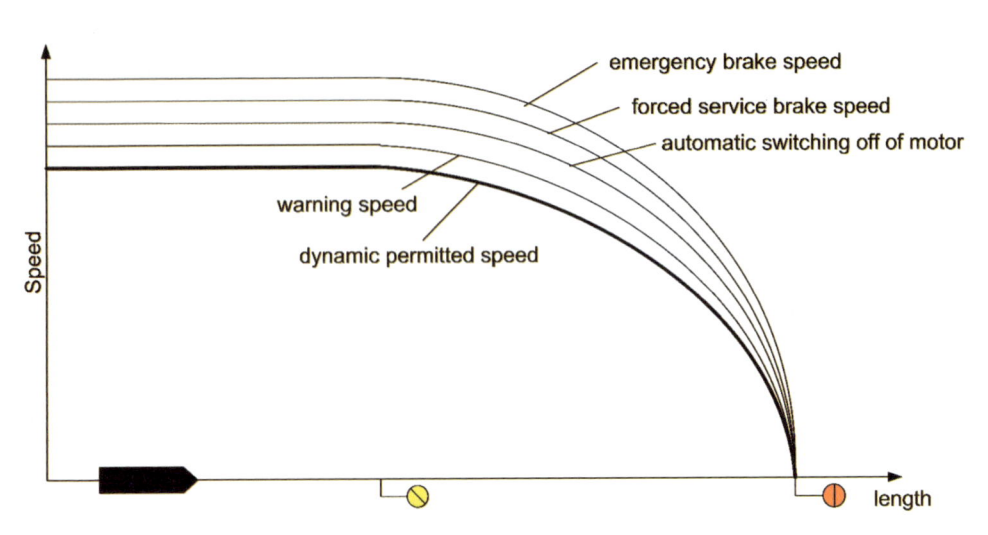

Figure 8.3: Speed limits for activation of intervention functions (maximum case)

8.1.5 Role in the Railway Operation Process

According to the role in railway operation, systems can be divided into
- **Auxiliary systems:** These systems provide additional safety, but do not substitute for the trackside signals. In most systems, if trackside and cab signal do not agree, the driver has to obey the trackside signal which is considered to be the safer signal there.
- **Independent systems with continuous guidance function:** These can replace the trackside signals, whereas trackside signals often are still present for operation of non-equipped trains and for degraded mode operation. If trackside and cab signals do not agree, the driver has to obey the latter. This can often give more detailed information (e.g. provide shorter block sections). These systems are especially required in high speed traffic, where the train driver is not able to perceive the aspect of the trackside signal safely, and in some metropolitan railways with high capacity requirements.

Most modern systems are designed fail safe, which means that a technical failure in transmission either leads to more restrictive commands or will be detected. Many older systems are not fail safe, which means that the driver must not rely on their correct function. Fail safe is therefore a precondition for using of information for cab signalling as a primary information source for the driver.

8.1.6 Automation of Train Operation

With complete dynamic speed profile present on the train, in principle, train operation could be automated. However, reasons which obstruct automation are mainly the lack of ability of automated systems to react to unpredicted situations such as obstacles on the track. Therefore, a further necessity for full automation of train operation is the continuous detection of external objects or their exclusion by barriers which cannot be passed either intentionally or unintentionally. This is very expensive on extended networks, but is practicable in some cases of metropolitan railways due to the limited extent of the network and the high density of traffic which makes the investment economically reasonable. Complete protection is never possible against very rare events, such as an object falling from a passing aircraft onto the railway.

Altogether, the following steps of automation can be distinguished:
- Manual driving without any automation: The driver is fully responsible for driving. This is the case without train protection systems.
- Manual driving with technical supervision: This is the case of a train protection system supervising the driver and enforcing safety in case of driver's error.
- Partially automatic operation: This is the case when some tasks of driving regulation are assigned to the driver and others to automatic systems. An example is ATC on Japanese high speed lines, where the driver is responsible for acceleration and platform stopping and the automatic system for safety related braking processes (chapter 8.3.4.4). Other examples are several modern systems with calculation of dynamic speed profile where the driver can select between manual and automatic driving (chapter 8.3.6).
- Automatic driving with human supervision: Here the train is normally driven automatically, but the driver watches the track and can take actions in case of danger or technical failure. Although this would be technically possible in many modern systems (chapter 8.3.6), it is rarely done for psychological reason: A driver whose only task in normal operation is watching the processes would not be able to act properly in emergency situations due to lack of attentiveness and driving practice (Yamanouchi 1979). This can be overcome by giving the driver some positive tasks, as implemented on the Victoria Line of London Underground in 1968.
- Full automation: In these systems no driver who would watch continuously the track is present on the train. However, in some cases a person who is normally in charge of other tasks (such as selling tickets) can take control if necessary. Fully automatic driving is currently applied on some single metropolitan lines (e.g. Paris, Lille, London, Vancouver and Copenhagen) and for special purposes such as airport shuttle trains.

8.2 Technical Solutions for Data Transmission

8.2.1 Overview over Forms of Transmission

The forms of transmission from the operational aspect can be distinguished into (figure 8.4):
- intermittent transmission including
  - spot transmission
  - interrupted linear transmission
- continuous transmission

In systems with continuous transmission there is – irrespective of possible short sections without connection such as radio holes – basically a continuous data link between track and train. However, data are usually transmitted by data telegrams in short time intervals. In systems with intermittent transmission, transmission is possible only at selected locations, determined by the trackside equipment. The technical solutions for continuous and interrupted linear transmission are similar and are therefore described together in chapter 8.2.3.

intermittent		continuous
spot	linear	
spot transmission	intermittent linear transmission	continuous transmission

Figure 8.4: Forms of transmission in train protection systems

8.2.2 Spot Transmission

8.2.2.1 Classification

For spot transmission there are various trackside devices which transfer the relevant data to the train. Classification is offered according to the following parameters:
- technical principle of transmission
- type of power supply
- uni- or bilateral transmission
- switchability of data content
- size of transmitted data
- redundancy
- longitudinal position along the track
- lateral position in the track

Intermittent transmission is effected by the following **technical principles**:
- Mechanical
- Galvanic
- Optical
- Inductive

The first two solutions were already developed in the 19th century, but are still used in older systems to a decreasing extent. The optical principle, after testing, did not gain relevant importance. The inductive principle is the basis of almost all modern systems. In chapters 8.2.2.2 to 8.2.2.5, the technical principles will be described more detailed.

Type of power supply: In most cases, trackside devices need energy for transfer of information. They receive this energy constantly from a trackside power source or short time from the passing train (e.g. transponder balises, chapter 8.3.5). There are also devices which transfer the information without power (e.g. a permanent magnet).

Uni- or bilateral transfer of information: In the majority of systems, data are only transferred from trackside to train. Some modern devices, e.g. balises, provide the possibility of bilateral transfer with connection with the control centre. This possibility, however, is used only in a minority of cases (chapter 8.3.5).

Switchability of data content: Transmitters can be classified into:
- fixed data transmitters, which always transmit the same information content, and
- switchable transmitters, whose information content can be switched by trackside input information, e.g. signal aspects.

Often both types of transmitters are used in the same train protection system.

Size of transmitted data: Older trackside devices can only transfer one bit (e.g. Indusi, chapter 8.3.3.1). If more bits are required, several such devices are installed. But modern types can transfer detailed data by the same trackside transmitter (chapter 8.3.5).

Redundancy: The information can be supplemented by redundancy, or even completely repeated during the transfer in order to detect and correct errors. Some systems use these possibilities, others do not.

Longitudinal position: Installation localities of devices along the track vary. To fulfil the attentiveness check function (chapter 8.1.3), a transmitter in proximity to the distant signal is required, whereas the trainstop function requires a transmitter in proximity to the main signal. For braking supervision, additional transmitters between these two points are applied in some systems, e.g. in PZB90 (chapter 8.3.3.1) and ATS-S (chapter 8.3.3.2), and for continuous speed profiles, information has to be upgraded in regular intervals along the whole line.

8.2 Technical Solutions for Data Transmission

Lateral position: Spot transmitters are installed mainly in the track, either centrally or in proximity to the right or the left rail. There are also cases with transmitters above rail level. As information depends on movement direction, the devices usually operate for one train direction only or transmit different information content for each direction. Their location in proximity to the left or to the right rail allows the train to detect the transmitter only for its direction. With transmitters located in the middle, directions have to be determined in another manner, e.g. by suppression for one direction or by the data contents containing direction information. Figure 8.5 shows an example for the positions of trainside antennas for various systems on the bottom of the locomotive.

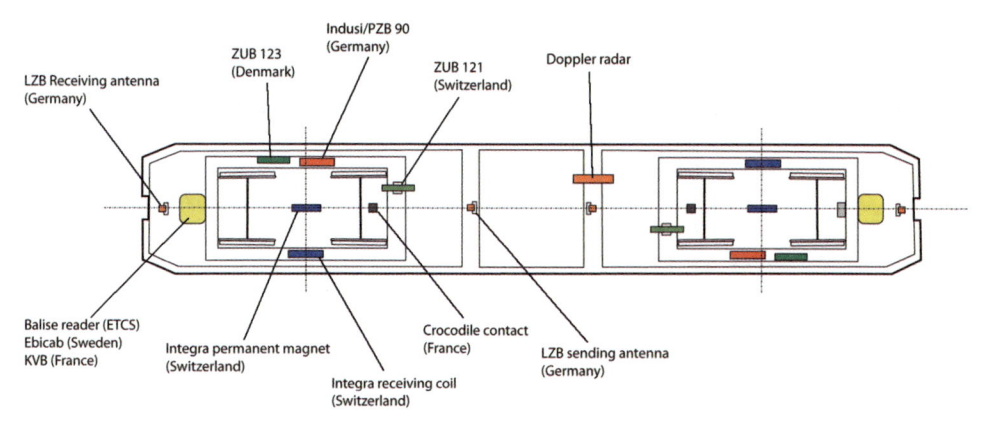

Figure 8.5: Positions of vehicle communication units for different train protection systems (graphic: *Pachl*)

8.2.2.2 Mechanical Principle

In mechanical systems, information is transmitted by a movable mechanical arm on the trackside which acts on a corresponding device on the train and therewith initiates reactions such as emergency stop upon passing a red signal. In the first systems a glass tube positioned on the roof of steam locomotives was connected to the pneumatic brake. Upon passing a signal showing stop, the trackside trip arm would break the glass tube and stop the train as a consequence (Theeg/Vincze 2007). An example which is still used is the Mechanical Trainstop (chapter 8.3.2.1).

8.2.2.3 Galvanic (Electrical) Principle

In the galvanic principle, information is transmitted by electric current between touching parts on the vehicle and on the trackside. A typical example is the French Crocodile (chapter 8.3.2.2).

The mechanical and galvanic principles were also used in combination. An example is the system introduced in 1905 in the USA (Barwell 1983). Here the presence of a trackside transmitter was indicated to the train mechanically, but the signalling information transmitted fail safe to the train by electric current.

8.2.2.4 Optical Principle

In the optical principle, special mirrors with varying reflecting properties were used as the trackside transmitters, whose position depended on signal aspects. A polarised beam was sent from the train, reflected by the mirror and came back to the train photodetector. Its spectrum con-

8 Train Protection

tained the information about the mirror position and therefore the signal aspect. Due to practical problems (e.g. keeping the mirrors clean), this principle did not gain much importance.

8.2.2.5 Inductive Principle

The inductive principle of spot transmission, generally spoken, is based on electromagnetic induction between parts on the trackside and on the train. According to the practically relevant transmitters, it can be broken up into the following most important types:
- resonant circuit method
- moved DC magnetic method
- transponder balise or beacon method

In the resonant circuit method, a permanently energised resonant circuit is on the locomotive. The track has a passive resonant circuit adjusted to the frequency of the train circuit and the information to be transmitted. If the leading vehicle passes above the passive resonant circuit, the coils between both interact with each other by change of inductance. Therefore, changes in the parameters of trainside resonant circuit (e.g. current, voltage, frequency) become measurable. Examples are described in chapter 8.3.3.

The moved DC magnetic method uses the effect of a current being induced in an electrical conductor which is moved in a magnetic field. The magnetic field can either be produced by permanent magnets (fixed data), by DC electric magnets (switchable) or by combination of both. Permanent and electric magnets can be used either together (with the one suppressing the magnetic field of the other, figure 8.6) or in sequence along the track to specify information. A disadvantage of this method is that its functioning is restricted to cases with a

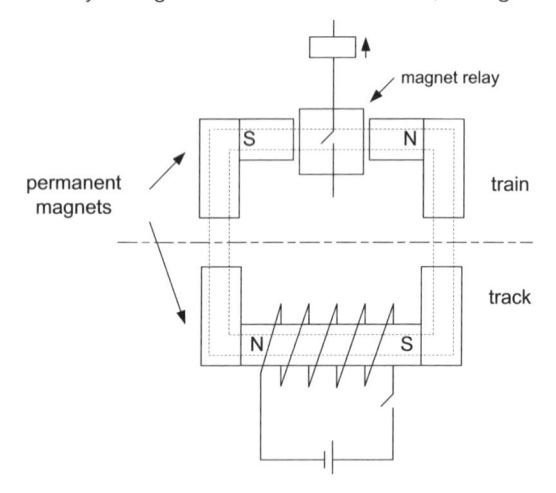

Figure 8.6: Permanent magnet with coil

certain minimum relative speed of the magnetic field and the conductor, which means a certain minimum speed (this minimum varies, but is lower than 10 km/h) of the train, so very slow moving vehicles get no information.

The most modern form of the spot inductive transmission is the transponder contact with short range radio beacons or balises. This contact operates at high frequency and allows transmission of complex telegrams. In most systems, the trackside equipment receives a power feed from the locomotive and gives out data of some hundreds of bits. Balises and beacons can carry either fixed or switchable data (chapter 8.2.2.1) or even provide bilateral transfer of information.

To perform safe data transmission, different solutions such as repetition of telegrams or checksums can be used. Furthermore, trackside transmitters are linked in most systems to detect of failure of one transmitter.

8.2.3 Linear Transmission

8.2.3.1 Classification

Linear transmission includes continuous as well as interrupted linear forms (chapter 8.2.1).

8.2 Technical Solutions for Data Transmission

Classification of linear transmission systems should take into account of the following basic criteria:
- technical principle of transmission
- uni- or bilateral transmission
- size of transmitted data
- length of transmission cycles
- reaction time on new information
- usage of technical additions
- centralisation of information generation and therefore possibility of operative change of the information
- expenses for equipment

Technical principle: The technical principle for all practically important linear transmitters is inductive. The information can be transferred through the following as the most important technical means, which are described more detailed in chapter 8.2.3.2:
- track circuit (figure 8.7, a)
- cable loop (figure 8.7, b)
- radio (figure 8.7, c)

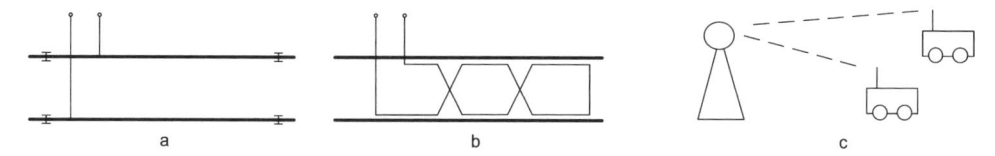

Figure 8.7: Technical devices for the linear transmission of information

Uni- or bilateral transmission: By bilateral data transfer, in addition to the train receiving data about movement authorities, permitted speed and others, it can transfer back the information about its current speed, position, condition of the brake system, train completeness etc. The cab signalling through track circuits provides only for unilateral data transmission; the loop and the radio can function in both directions.

The **size of transmitted data** depends on the channel capacity. Radio gives a wide frequency band for information interchange. Cable loops can transfer the information on frequencies up to 50–100 kHz. Channel capacity of track circuits is limited to 20 kHz. Thus the linear transmission media can transfer data from few bits (e.g. three signal aspects in ALSN, chapter 8.3.4.1) up to detailed movement authority and profile data like in ETCS (chapters 8.4.3, 8.4.5), depending on frequency band.

The **length of transmission cycle** is also important, particularly for transmission of urgent messages (e.g. emergency stop order). High train speeds demand short cycles (less than one second) to be possible, for low speeds cycles in several seconds can be acceptable.

Reaction time: The check of integrity of messages before decision-making is necessary. In some systems this results in actions being carried out only after repeated reception. It results in a delayed action. For example it can take up to three cycles or nearly five seconds in ALSN (chapter 8.3.4.1).

Technical additions: Track circuits in the area of isolated joints and radio in mountain areas can not always guarantee a full coverage with trackside transmission. In these cases some additional trackside antennas such as cable loops can become necessary.

8 Train Protection

Centralisation: Systems with continuous transmission can be centralised or dispersed. The centralised systems allow easy operative changing of information. Examples are the input of temporary speed restrictions, but also general emergency stop commands in case of natural disasters such as earthquakes.

An important economic factor is the **expense** of transmission equipment. The use of already existing track circuits does not demand significant investments. Cable loops demand relatively high additional efforts for cables, and their arrangement hinders track works. Transmission by radio between the train and the control centre is estimated to have a good perspective, although today it demands much investment.

8.2.3.2 Technical Principles of Transmission

When data are transmitted through **track circuits**, information can be coded by (chapter 5.3.4):
- constant signal frequency or
- modulated signal.

The historically first transmission system through track circuits used various frequencies, each frequency assigned to one particular information (system CSS in USA, chapter 8.3.4.2). But today, different forms of impulse, frequency or phase modulation are applied in most modern systems. Frequency and phase modulation allow higher volume of information than impulse modulation.

When **cable loops** are used, two types regarding the assembly are the most important in practical terms (figure 8.8):
- A cable loop, crossed periodically for positioning purposes: This form is used on some metropolitan railway systems as well as in the system LZB (chapter 8.3.6, figure 8.9)
- Cable with one conductor and return wire, stacked on the rail base. This type of loops is applied in ETCS L1 as infill loops (chapter 8.4.3, figure 8.10).

For **radio communication**, there is an European standard called GSM-R. Amongst other ap-

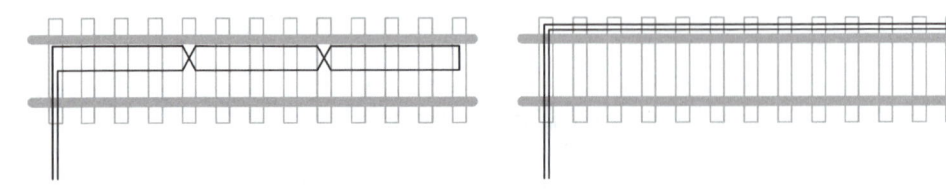

Figure 8.8: Forms of cable loops

Figure 8.9: Loop LZB

Figure 8.10: Euroloop

plications, it is used in ETCS (chapter 8.4.2). It is also applied outside Europe in various countries. Width of one channel is 200 kHz, carrier frequencies take zone 876–880 (Uplink) and 921–925 (Downlink) MHz. However, there are also other digital systems of radio communication which are being tested or used by railways for continuous data transmission.

8.3 Particular Systems

8.3.1 Classification of Systems

The train protection systems which are applied among the railways can be classified roughly into five groups according to their functions and the type of transmission (figure 8.11):
1. Systems with intermittent transmission and without braking supervision.
2. Systems with intermittent transmission at low data volume and with braking supervision.
3. Systems with continuous transmission of signal aspects by coded track circuits. As the boundary between systems with and without braking supervision is fluent, 3a and 3b are classified into the same group here.
4. Systems with intermittent transmission at high data volume and dynamic speed supervision.
5. Systems with continuous transmission at high data volume and dynamic speed supervision.

There is a tendency for the former systems the be the older ones and the latter the more advanced.

transmission \ functions	attentiveness check, trainstop function and others, but without brake supervision	with brake supervision in different form, but without dynamic speed profile	dynamic speed profile
intermittent	Group 1	Group 2	Group 4
continuous	(3a) Group 3 (3b)		Group 5

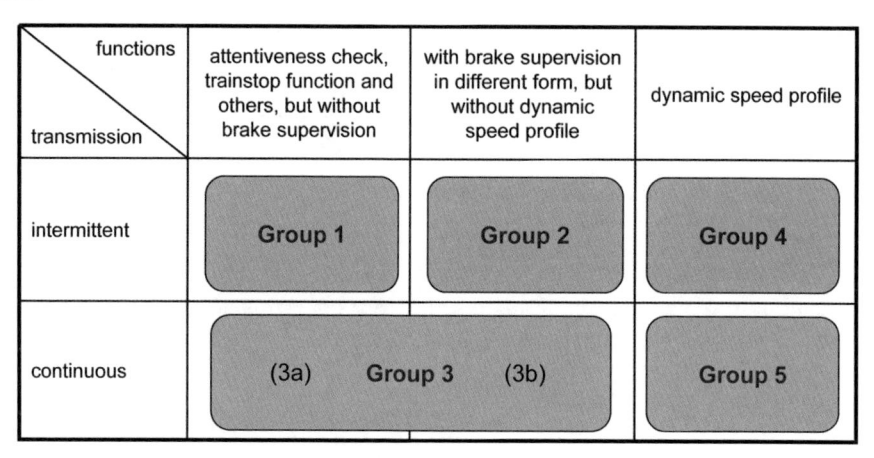

Figure 8.11: Classification of train protection systems

Most systems work independently, but there are some which supplement other train protection systems and can work only together with them. One example is the Russian SAUT (chapter 8.3.5.3) as a supplement to ALSN (chapter 8.3.4.1) with improved supervision functions. Another example is the German GNT (chapter 8.3.5.1) as a supplement to PZB 90 (chapter 8.3.3.1) to handle tilting trains for which speed limits are higher.

8.3.2 Group 1: Systems with Intermittent Transmission and without Braking Supervision

Systems without braking supervision (figure 8.11) have basically two supervision functions: They provide for an attentiveness check at the signals which can show 'Caution' (distant and

8 Train Protection

combined signals), and/or a Trainstop function (figure 8.12). The means of data transmission of these systems are simple, but different. Four examples will be described briefly: the mechanical Trainstop, the French Crocodile, the British AWS and the Swiss Signum.

The gain in safety resulting from the application of these train protection systems is limited. This insufficiency will be demonstrated using the example of the historic Japanese ATS-S system, which only provided a check of attentiveness at signals which can show 'Caution', independently from the current signal aspect: This system reduced the number of accidents due to stop signal violation only by half, and 98 % of the remaining accidents occurred after correct acknowledgement action by the driver (Kondo 1980). Even when it can be assumed that selective acknowledgement check (only when the signal actually shows 'Caution') and the additional trainstop function increase the safety, in most cases these systems are not sufficient for modern safety requirements. The trainstop function without brake supervision requires overlaps which are as long as the braking distance of the train. In most cases, these cannot be provided.

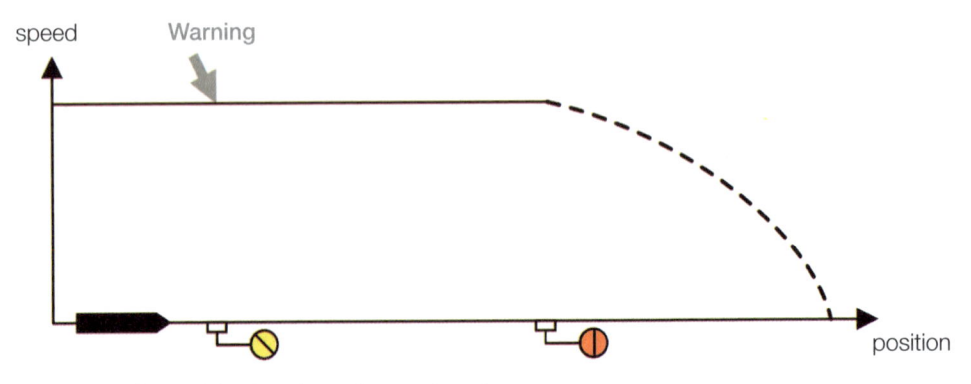

Figure 8.12: Supervision functions of group 1 systems

8.3.2.1 Mechanical Trainstop

The Mechanical Trainstop (figure 8.13) is a simple system with a mechanical contact at signals which can show 'Stop' (main and combined signals) and the complementary contact on vehicles. Depending on whether the signal shows a Stop or a Proceed aspect, the trackside contact is switched effective or ineffective. Passing an effective contact causes an immediate emergency stop. Besides the general safety disadvantages of systems of this group mentioned in chapter 8.3.2, these systems have several disadvantages:

- Movable mechanical parts cause wear and tear, which raise the maintenance costs.
- Special attention has to be given on the clearance profile of vehicles.

These disadvantages restricted the applicability of these systems to suburban and metropolitan railways with

Figure 8.13: Trackside device of Mechanical Trainstop on Berlin S-Bahn

uniform vehicles, low speed and high braking performance. Invented in Britain, various systems of this kind had been developed in past. Various mass rapid transit systems still use the Mechanical Trainstop, but the number is decreasing.

8.3.2.2 French Crocodile

The Crocodile, which is applied in France, Belgium and until few years ago Luxemburg, transmits the information about the aspect of a distant or combined signal by a galvanic contact between the rails (figure 8.14). The train equipment consists of an electric contact on the bottom of the locomotive which mechanically touches the trackside transmitter, a conducting connection between that contact and the wheels and a simple acting unit. Absence of a voltage between the rail and the trackside transmitter means that there is no restriction, 20 Volt DC depending on its polarity warns of a speed restriction or a stop. When passing a signal showing 'Caution' or 'Speed Restriction Warning', in older versions only an acoustic warning tone sounded in the driver's cab shortly before passing the signal. Newer versions also use an attentiveness test.

Figure 8.14: Trackside device of Crocodile (photo: *Dietmar Strobel*)

8.3.2.3 British AWS

AWS (Automatic Warning System) is an old British train protection system. The trackside transmission unit consists of two magnets between the rails in short sequence (figure 8.15), placed shortly before any signal which can show 'Caution'. The first is a permanent magnet with north pole on the upper side, which initiates a warning tone. The second is a DC electromagnet with south pole on the upper side which resets the train equipment in case the signal shows 'Proceed'. This is in accordance with the fail-safe-principle. In any other case (Caution or Stop), the train equipment has to be reset by the acknowledgement action of the driver, otherwise the train will be emergency braked. To remind the driver, a simple cab signal ('Sunflower', figure 8.16) is displayed. An important disadvantage is that the system does not distinguish between 'Caution' and 'Stop' aspects. For use on bidirectional tracks, the magnet has to be suppressed by an equally strong DC magnetic field with opposite polarity. More information on AWS can be found in (IRSE 1999).

Figure 8.15: AWS electro-inductor and permanent magnet, installed on a passenger line in England (photo: *David Stratton*)

8 Train Protection

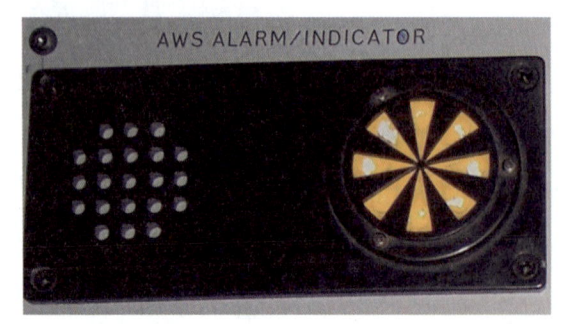

Figure 8.16: AWS Cab Signal

Figure 8.17: Trackside transmitters of Signum (middle and right coils) in Switzerland. The left device is of the additional system ZUB (chapter 8.3.5) (photo: *SBB*)

8.3.2.4 Swiss Integra Signum

Integra Signum is a Swiss train protection system which combines the trainstop and the attentiveness check function. The data transmission is inductive by direct current. The powered coil is positioned centrally at the bottom of the vehicle. When the vehicle passes above the central track coil, current is induced in the latter and induced back to the vehicle by the outer left-sided coils (figure 8.17, figure 8.18). A certain minimum speed of the train (approx. 5 km/h) is necessary for transmission. The trackside transponder can be switched ineffective by short-circuiting in case of allowing signal. By the positioning of the second pair of coils on the one or the other rail, directions are distinguished. The signal aspects 'Caution' and 'Stop' are distinguished by the polarity of the left coil. 'Caution' requires an acknowledgement action, whereas 'Stop' causes an immediate emergency stop.

More information on Signum can be found in (Oehler 1981).

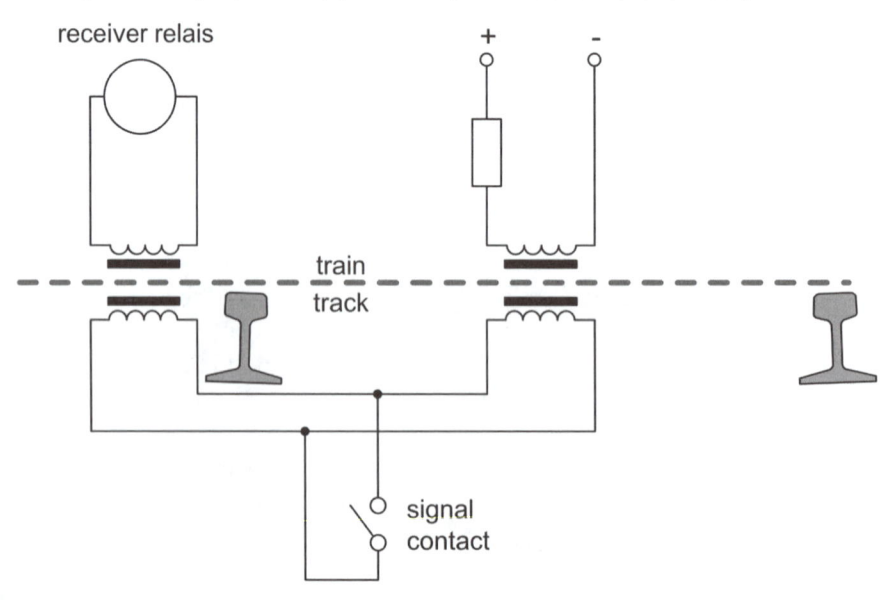

Figure 8.18: Data Transmission in Integra Signum

8.3.3 Group 2: Systems with Intermittent Transmission at Low Data Volume and with Braking Supervision

In these systems (figure 8.11), in addition to the attentiveness check and trainstop functions, the braking process is supervised in different forms, but without calculating a dynamic speed profile. For data transmission, resonant circuits are used in most cases. Each trackside resonant circuit is adjusted to a certain frequency out of a stock of several defined frequencies, with the frequency coding the information. The trackside resonant circuits can be switched effective or ineffective, or they can be switched between different active statuses (different frequencies), depending on signal aspects.

A disadvantage of many of these systems is that the ineffective (permitting) status cannot be distinguished from the absence of a trackside transmitter, which results in non-fail-safe behaviour of the system. Therefore, these systems are not suitable for cab signalling and have to work in the background as long is the driver is operating the train correctly. The driver must not be misled to rely on these systems.

Two examples, the German Indusi and the Japanese ATS-P, are described in detail in the following.

8.3.3.1 German Indusi/PZB 90

Indusi with its variations is the predominant train protection system in Germany, Austria and former Yugoslavia. It is used also in Romania and Turkey on few lines. The original Indusi, developed in the 1930s, had very simple functions and was continuously improved via the West German I60 and I60R and the East German PZ80 until PZB 90. The current version in Germany is PZB 90, to be described in the following (DB Netz 2001). During the development process, the complexity of the system increased, so that today's PZB 90, among the train protection systems, is unique for its complex supervision functions in combination with its simplicity in data transmission.

Figure 8.19: Indusi magnet

The trackside equipment consists of passive trackside magnets (figure 8.19) in the form of resonant circuits. These have an information connection to the trackside signals, and can be switched effective or non-effective. There are three types of trackside magnets, which are adjusted to the frequencies 500, 1000 and 2000 Hz.

The resonant circuits on the leading vehicle of the train permanently swing in these three frequencies. When the vehicle passes over a trackside magnet, the trackside magnet, if switched effective, inductively removes energy from the trainside resonant circuit with the respective frequency (figure 8.20). This loss of energy is evaluated by the on-board equipment. If the trackside magnet is switched non-effective, the resonant circuit on the vehicle is not influenced.

8 Train Protection

The trackside magnets are mounted to the right rail and the vehicle magnets above this position, distinguishing directions.

In approach to a signal showing 'Stop', the magnets are places as follows (figure 8.21):
- The 1000-Hz-magnet: at the distant signal, effective when the signal is at 'Caution' or at 'Speed Restriction Warning' of relatively low speed.
- The 500-Hz-magnet: 250 m in rear of the main signal, effective when the signal is at 'Stop'.
- The 2000-Hz-magnet: at the main signal, effective when the signal is at 'Stop'.

To represent approximately the different braking performance of trains, they are distinguished into three categories. The following descriptions (figure 8.21) refer to train category 'O', which are the fastest trains with permitted speed up to 160 km/h; most passenger trains belong to it. The other groups are distinguished mainly by the speed values of the supervision curves.

The permitted speed of the vehicles (max. 160 km/h plus a tolerance margin) is continuously supervised.

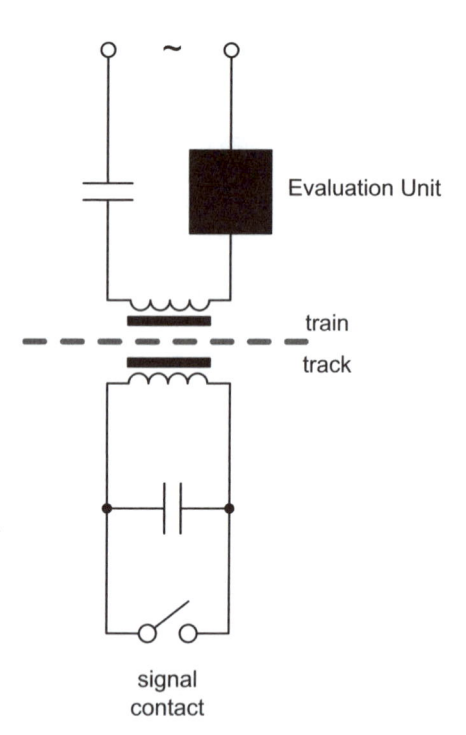

Figure 8.20: Data transmission in Indusi

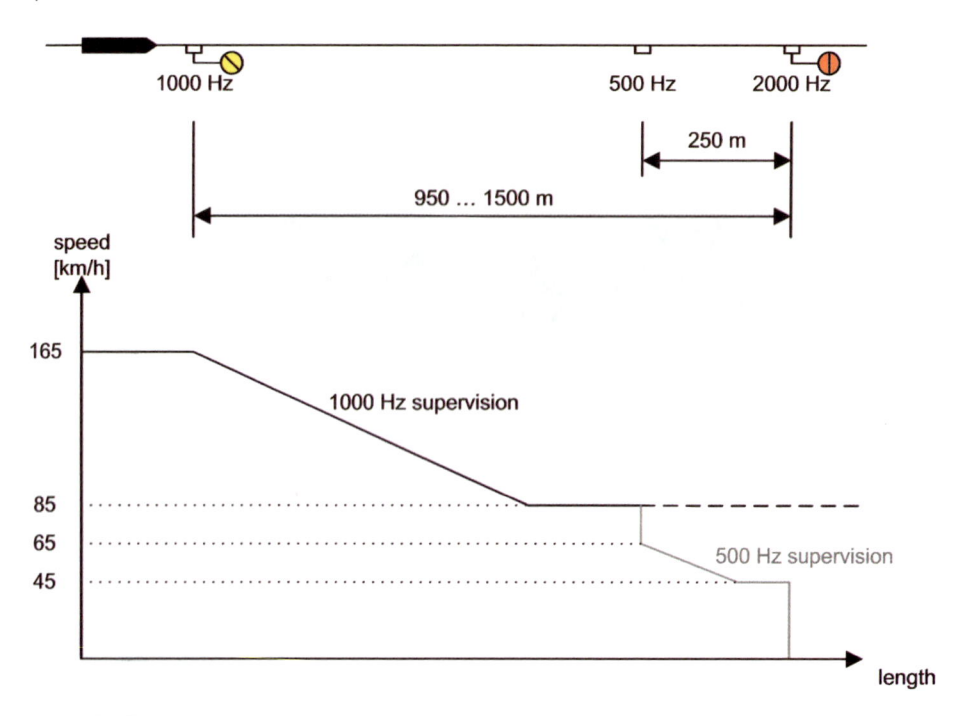

Figure 8.21: Braking supervision in PZB 90

Upon input from the 1000-Hz-magnet, the train driver has to push the acknowledgement button within four seconds. Then the supervision speed is steadily reduced from 165 to 85 km/h. As the distance from the distant to the related main signal can vary between 950 and 1500 m in Germany (normal value: 1000 m), the speed 85 km/h is supervised until 1250 m after the 1000-Hz-input (latest possible position of the 500-Hz-magnet). If within this distance no 500-Hz-input follows, the systems assumes the signal having cleared in the meantime and the supervision speed is released to the maximum of 165 km/h.

As this long speed supervision would be a great hindrance when the signal was cleared after passing the distant signal, the driver has the possibility of liberating himself from the 1000-Hz-supervision at the earliest 700 m after the 1000-Hz-input (earliest position of the 500-Hz-magnet) if he sees the 'Proceed' signal ahead.

Beginning at the 500-Hz-input, further reduction of speed from initially 65 to later 45 km/h is supervised. As the train is already close to the Stop signal, the driver has no possibility of liberation.

A 2000-Hz-input results in an immediate emergency stop. As speed supervision has been reduced to maximum 45 km/h before, this is sufficient to stop the train within the overlap in most cases. To pass a red signal in degraded mode operation, the cab is equipped with an override button.

Emergency brakes are applied until stop in case of failing to push the acknowledgement button, in case of exceeding the supervision speed, passing an active 2000-Hz-magnet and in several other cases of driver's error.

In case the train has stopped or gone very slowly between distant and main signal, it cannot be assumed that the overlap is still locked. Therefore, a more restrictive supervision becomes active in this case.

More information on PZB 90 can be found in (DB Netz 2001).

8.3.3.2 Japanese ATS-P

The system was developed in the 1970s and replaced the old ATS-S system on Japanese conventional lines. Whereas the old ATS-S system provided a check of attentiveness only, ATS-P provides advanced supervision functions including braking supervision which is even adjusted to downhill grades of the line.

Likewise in Indusi, data transmission is based on resonant circuits, but not loss of energy is measured, but the shift of frequency in the resonant circuit on the train, caused by the trackside resonant circuit. Normally, the resonant circuit on the train oscillates at the frequency 74 kHz. When passing over an active trackside resonant circuit, this frequency increases to one out of eight defined frequencies, depending on the trackside transmitter. The frequencies are filtered by a band pass filter on the train.

The trains are divided into seven categories, according to their braking performance.

The trackside coils are positioned in defined distance in rear of the Stop signal. Each input information initiates a related braking supervision (figure 8.22), differentiated by the train category. When the train reaches a new trackside transmitter, the previous supervision curve becomes ineffective. A fifth coil S_S with another frequency marks the position of the Stop signal and causes immediate emergency stop. Block signals have no S_S coil to enable permissive driving.

Other frequencies are used for following purposes:
- Braking towards a speed restriction of 50, 85 or 90 km/h. Different speed restrictions are identified by a sequence of two trackside coils with two frequencies.
- Cancellation of brake patterns and speed restrictions.

8 Train Protection

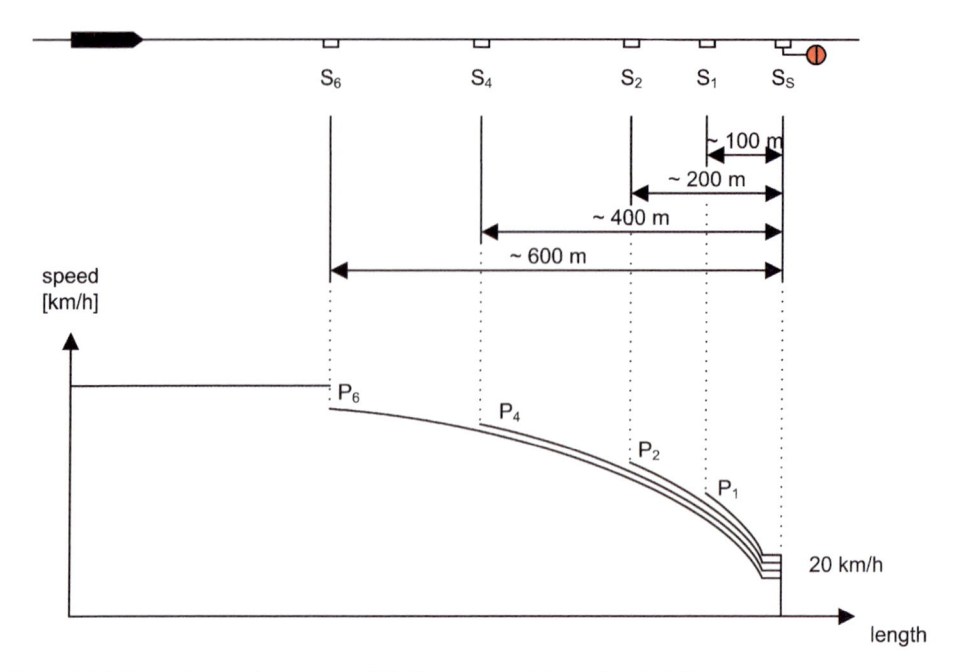

Figure 8.22: Supervision of Japanese ATS-P in approach to a signal at Stop

The system provides the possibility of adjustment to downhill slopes. In this case, the coils initiating the brake patterns are situated further in advance to compensate the lower deceleration (figure 8.23).

More information on ATS-P can be found in (Kondo 1980).

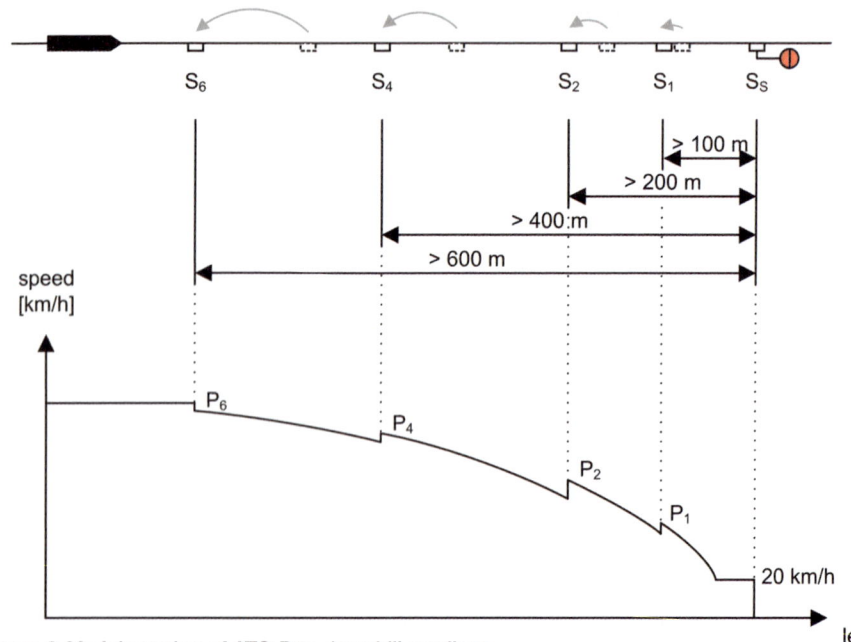

Figure 8.23: Adaptation of ATS-P to downhill gradient

8.3.3.3   Other Systems

Some other systems belonging to this group may be mentioned briefly.

The Polish **KHP** provides braking supervision not based on braking curves, but in the form of intermittently checking brake application during the required braking process. Data transmission is by resonant circuits and by interrupted coded track circuits with an effective length of about 700 m around each signal. During the length of these track circuits, the indication of the trackside signal is repeated in the cab. (Makowski 1992)

In the Spanish system **ASFA** (figure 8.24), the braking process is supervised intermittently in one position only: 300 m in rear of the signal at Stop, a speed of 35, 50 or 60 km/h, depending on the train category, must not be exceeded, otherwise the train will be emergency braked. Besides this speed, the line speed is checked, an attentiveness check is done at a signal at 'Caution' and a trainstop function provided at a red signal. Data are transmitted by resonant circuits with nine different frequencies. (European Commission 2004)

Figure 8.24: ASFA trackside transmitter

TPWS, which is applied on selected locations in Britain, is a supplement to AWS (chapter 8.3.2.3) and uses the same means of data transmission. For braking supervision in approach to a signal, one or more track-based speed traps are applied to check the speed in certain positions. Besides, the acknowledgement function at signals at 'Caution' and the trainstop function at Stop signals are provided. (European Commission 2004)

8.3.4   Group 3: Systems with Continuous Transmission of Signal Aspects by Coded Track Circuits

Systems in the third category (figure 8.11) transmit the aspect of the trackside signal ahead to the train through the rails. They were first applied in the USA, and then the idea spread to Russia and to different European countries, e.g. Italy, the Netherlands, Czech Republic, Slovakia and Hungary. Also the oldest high speed signal systems in Japan and France match this category.

The required track circuits are mostly also used for track clear detection (chapter 5.3) and transmission of block information (chapter 10.3.4). The signal aspect ahead is repeated in the cab, often in simplified form. The supervision functions reach from simple acknowledgement checks up to braking supervision with standardised fragments (figure 8.2). Basic advantages of systems of this group are the following:

8 Train Protection

1. In contrast to most systems of groups 1 and 2, these systems can be designed fail safe, so malfunction of the equipment leads to a more restrictive indication in the cab.
2. The train continuously receives the newest information in each position of the way. This prevents the driver from forgetting signal aspects and enables an immediate reaction of the system if signal aspects change.

However, an important safety-reducing disadvantage is that, unless the length of the track circuits is standardised or additional transmitters for length information are provided, calculation of an adjusted braking curve is not possible. To improve this, some systems with continuous transmission by coded track circuits are used together with intermittent transmission systems (e. g. System SAUT, chapter 8.3.5.3).

8.3.4.1 ALSN of Former Soviet Union

ALSN is used on almost 100 thousand kilometres in the countries of former Soviet Union or more than 10 percent of world railways. This system is installed on main lines but applied basically as additional equipment which supplements, and not replaces trackside signals in most cases. If there is a disagreement between trackside and cab signals, the driver has to obey the trackside signal.

The system was developed in the 1930s in the Soviet Union with use of experience of the first coded track circuits in the USA. In 1937 this system had received a medal at the International exhibition in Paris, but its introduction has been interrupted by war and has proceeded only after 1949 (Vlasenko 2006).

There are three codes displayed in the cab signal corresponding with the aspect of the trackside signal ahead. In case of three-aspect-signalling (chapter 7.3.3.2), these codes are (figure 8.25):

- red signal ahead (results in cab signal red-yellow)
- yellow signal ahead (results in cab signal yellow)
- green signal ahead (results in cab signal green)

The section beyond a signal at Stop is not coded, therefore the train will be emergency stopped (cab signal red). This is in accordance with the fail-safe-principle. Passage of a

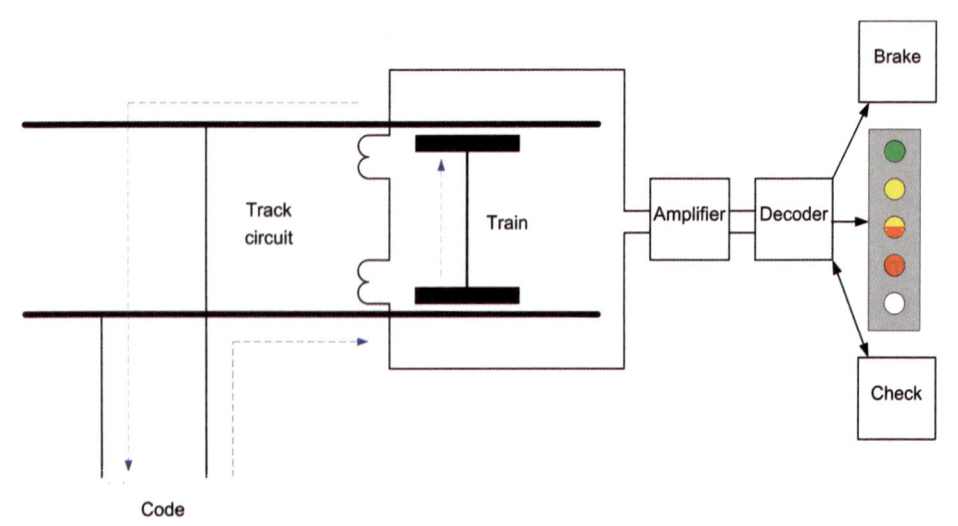

Figure 8.25: Code transmission from track circuits to the locomotive equipment

Stop signal can be authorised with driver's special action at the maximum speed 20 km/h (figure 8.26).

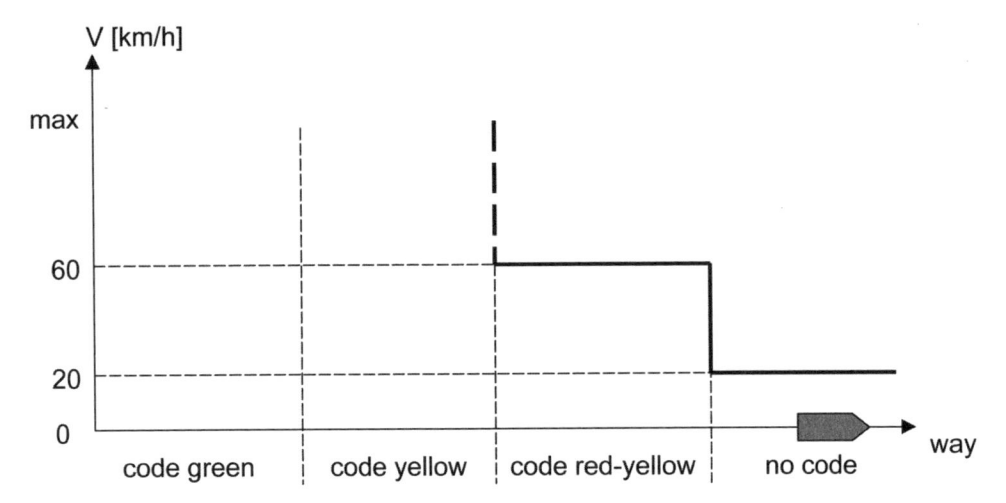

```
V [km/h]
  ▲
max│          │        │
   │          │        │
   │          │        │
 60├──────────┼────────┼────────┐
   │          │        │        │
   │          │        │        │
 20├──────────┼────────┼────────┼────────▶
   │          │        │        │
  0└──────────┴────────┴────────┴────────▶ way
  code green  code yellow  code red-yellow  no code
```

Figure 8.26: Braking supervision in ALSN (example)

As the number of signal aspects in and near stations is usually higher than three, there is a simple rule. If the train approximates a signal indicating the diverging route (the speed restriction for which can be between 25 and 80 km/h depending on the geometry of movable track elements (chapter 6.1.2.3)) or to a yellow signal (straight route, but only one block section is clear, the speed restriction after the signal is 60 km/h), the train receives the code yellow and driver has to select the proper speed according to the trackside signal. If two or more block sections are clear and the first section has straight route, the train receives the code green. If the following signal is red or auxiliary signal, the train receives the code red-yellow.

To distinguish the passage of a Stop signal from the entrance into a section without codes (e.g. secondary tracks in stations or secondary lines), there is the memory function of the previous code. No code after the code red-yellow is considered as passage of a red signal and the emergency brake commanded. No code after other codes results in a white cab signal (ALSN is switched off) and requires only periodic acknowledgement action.

The period of the code transmitted by the track circuit is 1.60 or 1.86 seconds (figure 8.27). Assignment of the two periods in the neighbouring track circuits is described in chapter 5.3.5. The carrier frequency depends on traction power supply and is 50 Hz or 25 Hz. As external influences with low frequency often occur, the reaction time of the system is three code periods. This means that the new cab signal becomes effective five seconds after a code change. This reaction time also applies if a train passed a Stop signal (Sapožnikov et al. 2006).

System ALSN itself (without the additional system SAUT, chapter 8.3.5.3) cannot transmit information about distance to the signal ahead. The distance between signals varies from 1000 to 2600 meters depending on local conditions. Therefore the supervision curve in system ALSN (without any additional transmission) is only staircase, but the frequency of driver's acknowledgement check by the codes yellow and red-yellow depends on train speed. Moreover, the change of a cab signal to lower speed is accompanied by a bell and the driver is obliged to confirm this by an acknowledgement. If the train receives the code red-yellow, its speed is limited to 60 km/h. The speed limit of 60 km/h upon passing a signal at 'Caution' may ap-

8 Train Protection

Codegenerator KPTŠ 5

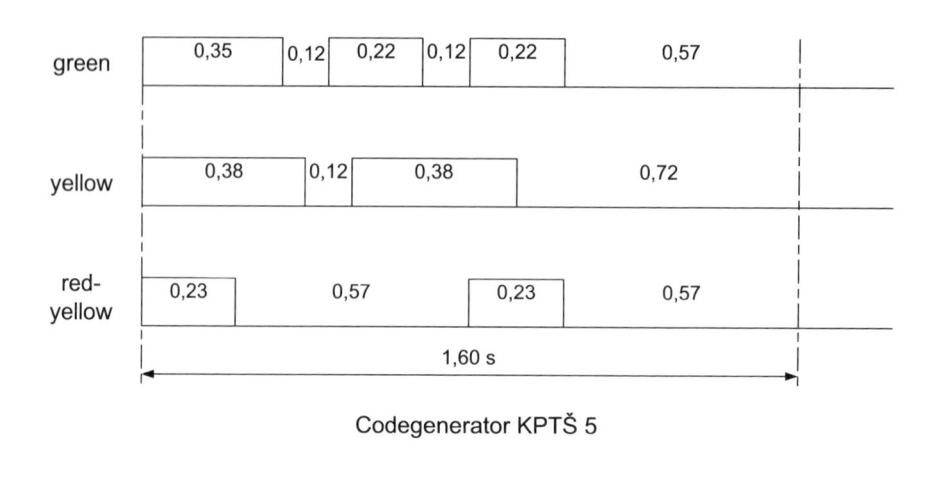

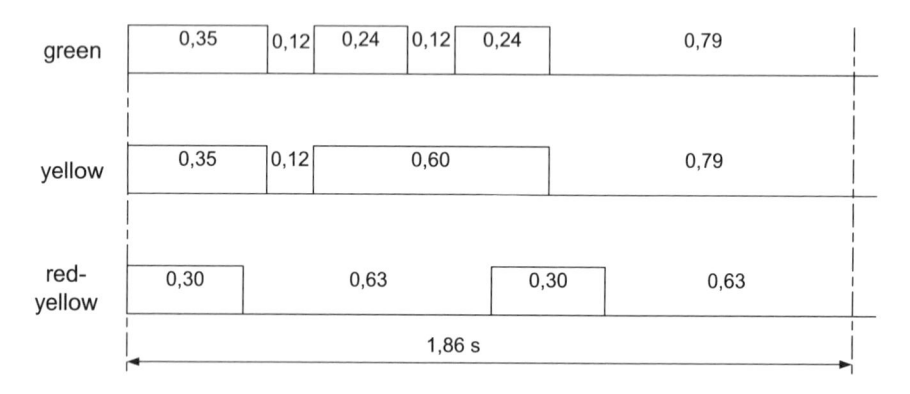

signal	segments (s)	total
green	0,35 — 0,12 — 0,22 — 0,12 — 0,22 — 0,57	
yellow	0,38 — 0,12 — 0,38 — 0,72	
red-yellow	0,23 — 0,57 — 0,23 — 0,57	1,60 s

Codegenerator KPTŠ 7

signal	segments (s)	total
green	0,35 — 0,12 — 0,24 — 0,12 — 0,24 — 0,79	
yellow	0,35 — 0,12 — 0,60 — 0,79	
red-yellow	0,30 — 0,63 — 0,30 — 0,63	1,86 s

Figure 8.27: Codes of ALSN

pear very low to readers in Western European countries, but is appropriate for generally lower speeds and long braking distances of heavy trains in this region. The maximum speed of trains on lines equipped with ALSN is 160 km/h.

According to new technical requirements, stations and open lines will have high frequency track circuits. This implies new functions for ALSN. When the track section is clear, the track circuit carries no code and serves for track clear detection only. Only in these sections where a train is detected or expected soon, the code is applied. On open lines, jointless track circuits are used whose areas of efficacy overlap by some tens of metres. When the train is in this overlapping area, the coding is shifted from one track circuit to the next. Therefore, only one track circuit carries the coding at the same time. But in stations, insulated rail joints are still applied and track circuits do not overlap. To transfer the coding upon entering a new track could cause failures due to inertia of the detectors and transmitters. Therefore in stations the codes are given in two track circuits at the same time: the one which is currently occupied and the one ahead. As soon as the train occupied a new section, the coding in the previous section is switched off. Likewise generally in ALSN, only these station tracks which are provided for non-stop train passage are coded.

8.3 Particular Systems

8.3.4.2 Other Systems for Conventional Traffic

Some countries in Europe and the USA use similar principles of cab signalling through track circuits. The system **LS** (figure 8.28), as an example, is used in Czech Republic and Slovakia. Its basic characteristics are the following (Mraz 1992, European Commission 2004):

- Four codes in accordance with speed restriction at the signal beyond are distinguished: 5.4 Hz – line speed, 3.6 Hz – 100 km/h (signal beyond shows 'Caution' or 'Speed Restriction Warning'), 1.8 Hz – 40 km/h (if the route is diverging, also used for higher branch speeds than 40 km/h), 0.9 Hz – Stop.
- The carrier frequency of the track circuits depends on traction power supply and is 50 Hz or 75 Hz.
- Block sections differ in length; therefore the speed supervision is staircase without calculated braking curves, but acknowledgement action is required if the actual speed is higher than the target speed. The passage of a red signal causes immediate emergency stop.

Figure 8.28: Code generator and cab signal in LS

The system **EVM** is used in Hungary and has the following principles (Mandola 1992, European Commission 2004):

- Seven codes: 1, 2, 3, 4 impulses (0,26-0,3 s) in the period means approaching a signal showing stop, 40, 80 km/h and max. speed. Four impulses with different impulse timing means 120 km/h, one impulse in two periods permits a speed of 15 km/h, continuous signal means clear line without speed information.
- Carrier frequency is 75 Hz.
- If speed reduction is required, the driver has to push the acknowledgement button at shorter intervals than normally. Besides, the brake application is checked.

System **ATB-EG** with coded track circuits is the older out of two train protection system in the Netherlands. Some characteristics are (Bailey et al. 1995, European Commission 2004):

- Five speed steps (40, 60, 80, 130 and 140 km/h) are used.
- Amplitude modulated speed codes on carrier frequency 75 Hz.
- If speed reduction is required, the system requires a periodic acknowledgment action every 20 seconds and checks application of brakes.
- The system does not distinguish between 40 km/h and stop.

In the USA, a cab signalling system with three aspects (**CSS**) is used since the 1920s. Its basis is the track circuits transmitting the signal frequencies 3 Hz (180 min^{-1}; unrestricted speed),

8 Train Protection

2 Hz (120 min-1; speed restriction ahead) and 1,25 Hz (75 min-1; Stop signal ahead) (Barwell 1983, wikipedia).

For higher speed (currently used up to 200 km/h) in Russia the system **ALS-EN** is designed. The phase modulated signal 174,38 Hz with a noise-suppressing code comes to the locomotive through track circuits. 48 cab signal aspects translate the data which show five block sections (about 10 km) and speed limitations beyond. The system ALS-EN on the line Moscow–St. Petersburg is not centralised: The equipment of each signal decodes the data, shapes the new information and sends a code to the next track circuit. (Sapožnikov at al. 2006)

On underground lines in the countries CIS the system 'Dnepr' is mainly used. Signal frequencies correspond to the velocity steps 40, 60, 70 and 80 km/h. Some underground railways have two frequencies as information about velocity beyond two following signals. The frequency band reaches from 75 up to 325 Hz. Cab signalling on new metro lines uses phase modulated signals (Dmitriev/Minin 1992).

8.3.4.3 Italian BACC for Conventional and High Speed Traffic

On Italian conventional lines and lines for increased speed up to 200 km/h a cab signalling system with four different aspects is applied. The signals on the carrier frequency 50 Hz are frequency modulated with frequencies 4.5 Hz (270 min-1; green signal ahead), 3 Hz (180 min-1; yellow signal ahead; reduce speed to 150 km/h), 2 Hz (120 min-1; speed restriction due to diverging route ahead) and 1.25 Hz (75 min-1; Stop signal ahead). As track circuits have almost equal length (1350 metres), it is possible to calculate and supervise the braking curve. A stop is therefore announced 2700 m in rear of the stop position.

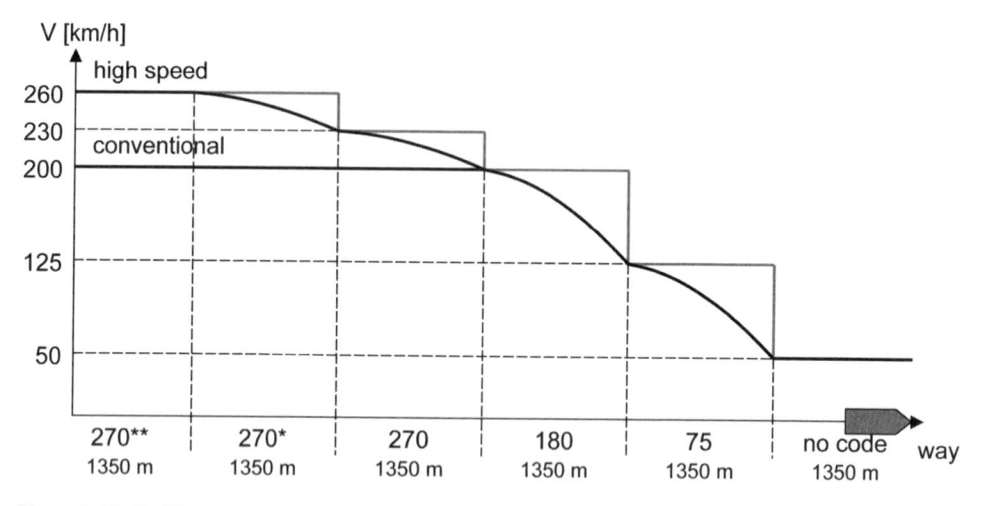

Figure 8.29: Braking supervision in BACC for conventional and high speed traffic

However, the braking distance is more than two track circuits if the train speed is higher than 200 km/h, therefore the system had been upgraded for the high speed line (figure 8.29). Also for diverging routes, additional higher speeds 100 and 130 km/h are to be used. Therefore, on high speed lines track circuits have an additional carrier frequency 178 Hz which in combination with the codes based on 50 Hz gives nine speed steps for high speed trains (table 1). The system is downwards compatible, which means that high speed trains can run on conventional lines and conventional trains on high speed lines using only the 50 Hz code and at speed

8.3 Particular Systems

Code name	with the carrier frequency 50 Hz	with the carrier frequency 178 Hz	Meaning (related to the end of the track circuit)	Permitted speed at the end of the TC [km/h] for trains with max braking ratio	Next trackside signal
270**	4.5 Hz	2 Hz	5400 m are clear	260	🟢
270*	4.5 Hz	1,25 Hz	4050 m are clear	230	
270	4.5 Hz	–	2700 m are clear	205	
180*	3 Hz	1,25 Hz	1350 m are clear, then speed restriction 100–130 km/h	155	🟡
180	3 Hz	–	1350 m are clear, then speed restriction 30–60 km/h or stop	125	
120**	2 Hz	3 Hz	speed restriction 130 km/h on the next track circuit	135	🔴🟡
120*	2 Hz	1,25 Hz	speed restriction 100 km/h on the next track circuit	105	
120	2 Hz	–	speed restriction 30–60 km/h on the next track circuit	65	
75	1.25 Hz	–	next signal is closed	50	🔴

Table 8.1: Cab and trackside signal aspects of BACC. The speed values vary between the trains

not higher than 200 km/h. This compatibility is necessary, as only one line (Rome–Florence) is equipped with BACC high speed system and trains continue onto the conventional network.

More information on BACC can be found in (Bianchi 1985).

8.3.4.4 Japanese ATC for High Speed Rail

The first Japanese high speed line opened in 1964, therefore the ATC system used there is the oldest cab signal and train protection system for high speed rail in the world. Today, on several lines the more modern so-called 'Digital ATC' is installed, which matches into group 5 (chapter 8.3.6).

The system is adapted to the requirements of a high speed network which is technically and operationally separated from the conventional network. The capacity and availability requirements are high and the required flexibility regarding different types of traffic and vehicles is low, which makes a limited number of signal aspects sufficient.

The trains are driven semi-automatically: The safety related braking for a signalled stop is controlled automatically, whereas not safety related processes like acceleration or platform stops are controlled under responsibility of the driver.

Information transmission from track to train is achieved by coded track circuits. Each block section is covered by two track circuits. All track circuits belonging to the same line have the same length, therefore the ends of the track circuits are used as fixed intermediate points for speed reduction. The result is a multi-step cascade braking curve (figure 8.30) in rear of a signalled stop. At higher speed levels, both track circuits of a block section transmit the same speed information. In contrast, in the last block section immediately in rear of the brake target both track circuits transmit different speed commands and stop the train in two steps. When the train has reduced its speed to 30 km/h, the driver can push an override button which ena-

8 Train Protection

bles him to continue at 30 km/h maximum speed for permissive driving. Otherwise, shortly in rear of the brake target point, an induction loop brings the train to a stop.

Figure 8.30 shows an example of the speed steps defined this way. The exact speed values vary between different Shinkansen lines and the time. In contrast to many other train protection systems, the staircase signalised speed is not a limit speed, but a target speed for regulation which is exceeded regularly. For diverging tracks in stations, an additional speed level of 70 km/h is defined.

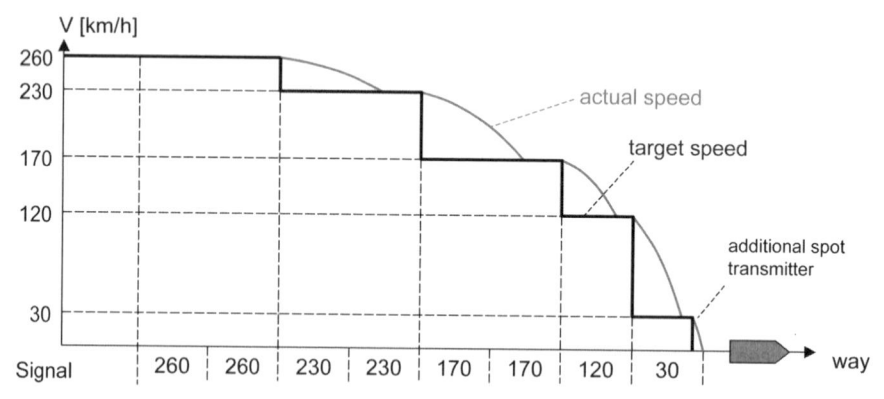

Figure 8.30: Brake control in Japanese ATC, example (Suwe 1988)

To avoid interference between neighbouring track circuits, the track circuits of the same double line use four different carrier frequencies: Thus in one track, the carrier frequency is alternately 720 Hz and 900 Hz and in the other track 840 Hz and 1020 Hz. The carrier frequency is frequency-modulated by a modulation frequency corresponding with a certain speed limit. Hereby, with only few different signal aspects, high speed traffic with high capacity is enabled.

More information on this system can be found in (Yamanouchi 1979, Suwe 1988).

8.3.4.5 French TVM 300

TVM 300 (figure 8.31) is installed on the older French high speed lines, the first of which opened in 1981. Like in Japan, the high speed lines are regularly used by high speed trains only, but these pass into the conventional railway network, too.

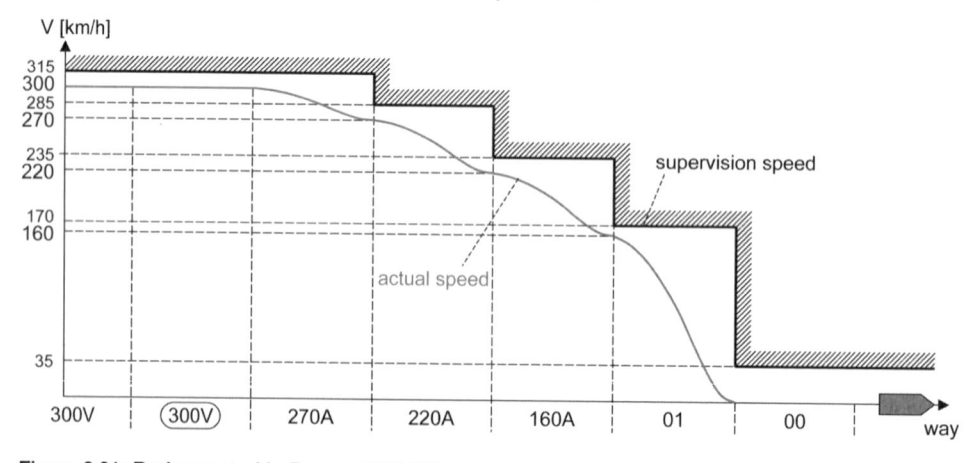

Figure 8.31: Brake control in French TVM 300

Regarding the data transmission and the control, TVM 300 has large similarities with the Japanese ATC. One main difference is that the driver regulates the braking process and is supervised by the technical system. The staircase speed information is a limit speed for supervision, and each block section is covered by exactly one track circuit. The length of the block sections is adapted to the gradient. Resulting from the staircase supervision pattern, the required overlap has the length of a complete block section.

8.3.5 Group 4: Systems with Intermittent Transmission at High Data Volume and Dynamic Speed Supervision

Systems of group 4 (figure 8.11) are modern systems for intermittent transmission. Due to the fail-safe behaviour and the possibility to supervise the complete dynamic speed profile, these systems are a safe solution up to high speeds if the efforts for continuous transmission are not considered as necessary. During the recent decades, many systems of this category have been developed. Although these systems are similar in their basic functional principles to a large extent, due to different data coding, the amount of detail information and the antennas, they are incompatible with each other.

The main trackside transmission media are:

– Transponder balises (figure 8.32) which work without trackside power supply by using energy sent from the vehicle unit to send data telegrams back to the vehicle. Figure 8.33 shows the principle of data transmission on the example of the system ZUB. In channel 1, the presence of a balise is detected by the principle of resonant circuit. Thereupon, channel 2 is switched on to send energy from the train to the balise. The balise uses this energy to transmit information back to the train (channel 3) (Fenner/Naumann 1998).
– Inductive loops with limited extension, which are usually powered from the trackside.
– Locally limited radio transmission devices.

Figure 8.32: Balise and trainside antenna of system ZUB (photo: *Siemens*)

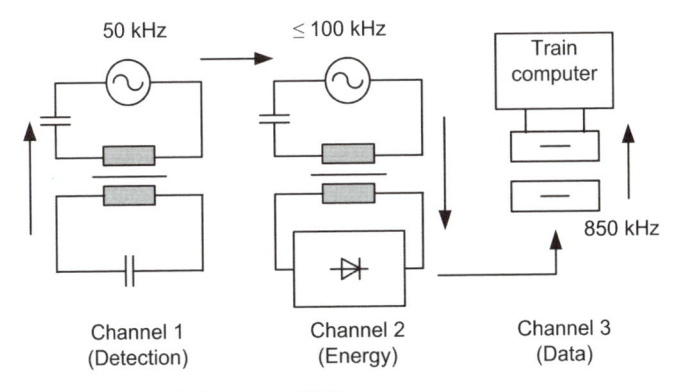

Figure 8.33: Transponder principle (example: ZUB)

According to the data contents, transmission media can be divided into:
- Static data transmission media whose information content is independent from trackside input such as signal aspects. They can transmit only static data such as line speed, gradient etc.
- Switchable transmission media whose information varies depending on the status of the trackside, especially signal aspects.

8.3.5.1 Systems With Static Line Data Stored in the Trackside

The majority of systems with intermittent transmission at high data volume and dynamic speed supervision store static line data in the trackside. These systems are suitable for large fleets and extended networks, which are typical for national/international railway networks. Examples are (European Commission 2004):
- Ebicab (Scandinavia, Portugal, Bulgaria)
- ATB-NG (Netherlands)
- TBL (Belgium)
- ZUB (Switzerland, Denmark)
- KVB (France)
- ETCS Level 1 (International)

Data are transmitted by transponder balises, loops and/or local radio units. The main transmission medium is balises in most of these systems. Inductive loops and radio units of limited length extension mainly serve for updating of signal aspects in approach to a main signal (Infill function).

Each data point consists of a single balise or a balise group transmitting data telegrams which contain line information. Transmission of detailed data can be achieved by using several 1-bit-transmitters in a line (French system KVB) or by each individual balise sending detailed data telegrams (most other systems).

Additional important information to be transmitted in most systems is the distance to the next data point, so-called linking information. If the train does not find the next data point at this distance, a technical failure can be concluded and protective measures activated. It follows that such systems are fail safe and are therefore suitable for detailed cab signalling.

The data telegrams include dynamic information about signal aspects and routes as well as static line information such as length information and gradient profiles. In most cases, systems supervise the complete dynamic speed profile continuously. Additionally, many systems provide continuous cab signalling function including the dynamic speed information which can replace the trackside signals. There is no principal speed restriction for the application of these systems, but in most countries these systems, as information is transmitted intermittently only, are restricted to conventional traffic up to around 160 km/h. In Sweden, even high speed traffic up to 250 km/h is guided by Ebicab without use of trackside signals.

Some other systems with balises are used for special purposes. An example is the system GNT in Germany which is used for tilting trains only. The system suppresses certain speed control applications of Indusi and supervises particular (higher) speed limits instead.

ETCS Level 1 as the most modern and the internationally most important example of this group is described in more detail in chapter 8.4.

8.3 Particular Systems

8.3.5.2 Systems with Static Track Data Stored on the Vehicle

Systems with static line data stored on the train are mainly applicable to metropolitan, suburban or local railway networks with a limited number of vehicles which are assigned to a separate network. The system ZSL 90, which is applied on different local railways in Switzerland, is an example. But these systems are also applicable on main railway networks where locomotives are assigned locally (partly in Russia).

Static track data are stored on the train and only dynamic data transmitted from trackside transmitters. In the example of ZSL 90, cable loops are applied only in station areas and, within these, continuously transmit dynamic data to the vehicle (Althaus 1994).

8.3.5.3 Example: SAUT

SAUT is an example of the combination of trackside and on train storage of line data. It is a relatively new system on Russian main lines as a supplement to ALSN. The combined system allows to know aspect of the signal ahead (owing to ALSN), and also distance to this signal (owing to SAUT) and therefore to calculate and supervise a brake curve. The trackside transmission media of SAUT are inductive loops with a defined length (figure 8.34). The system SAUT is in continuous redevelopment; there are currently three versions.

SAUT-U. The generator is installed in the beginning of each block section and transmits a high-frequency signal into the right rail. The length of the loop (section where this signal can be detected by the train) is proportional to distance to the following trackside signal. The information is registered thanks to a special locomotive antenna. But at entrance into station the distance to the following signal depends on a route. Therefore there are several loops at the home signal which are selectively powered depending on the route. Besides, an additional loop is installed one block section in rear of the station (at the distant signal) which transmits the exact speed information of the station home signal (because three ALSN-codes are not enough to code all speed steps; chapter 8.3.4).

SAUT-C (figure 8.35). The loops are only used in station areas, not on open lines. Distances, gradient and other

Figure 8.34: Several loops SAUT at the home signal

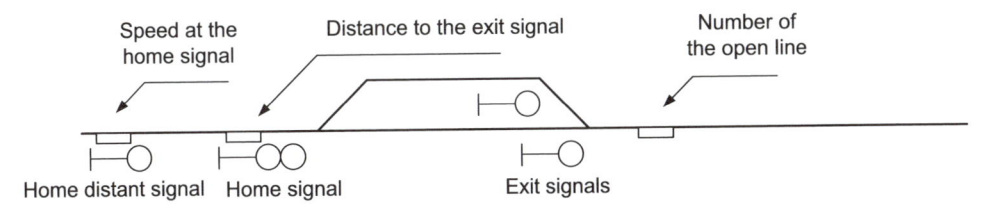

Figure 8.35: Data from loops to train in the system SAUT-C

8 Train Protection

track data of open lines are memorised in the locomotive computer; locomotives are assigned locally and move only inside the network of the local railway division. One loop at the exit of the station transmits a code number of the next line to the locomotive computer. The current locomotive position is determined by odometer. Every new block section corrects the odometer data. Section limits are determined by ALSN thanks to change of current in the track circuit: the current gradually increases at the approximation to the trackside signal, but drops sharply after passing the signal.

SAUT-CM is a further development of system SAUT-C. Loops are only located in station areas. The track data of the station are memorised in the locomotive computer. The code transmitted at the home signal contains the number of the route for the train which proceeds into the station. The length of the loop is no longer used for coding, therefore it is a uniform 10 m. Besides, the driver receives voice messages if the factual train speed is higher than allowed or the train approaches a Stop signal, a bridge, a tunnel or a level crossing.

For the future, the satellite positioning systems GPS and GLONASS are also intended to be used for definition of distance between the train and the trackside signal. Hereby, a precision of detection of two metres is reached thanks to reference stations. This solution is used in the development of the new system for locomotive safety KLUB-U (OAO RZD 2007).

8.3.6 Group 5: Systems with Continuous Transmission at High Data Volume and Dynamic Speed Supervision

8.3.6.1 System Structure and Data Transmission

The basic difference between systems of this category and systems belonging to the former category is continuous or quasi-continuous data link between track and train.

Among systems of this category, the following technical transmission media are applied:
- Coded track circuits: Examples are the Digital ATC (Watanabe et al. 1999) in Japan and TVM 430 (Guilloux 1990) which is applied on French and Belgian high speed lines and in the Channel Tunnel.

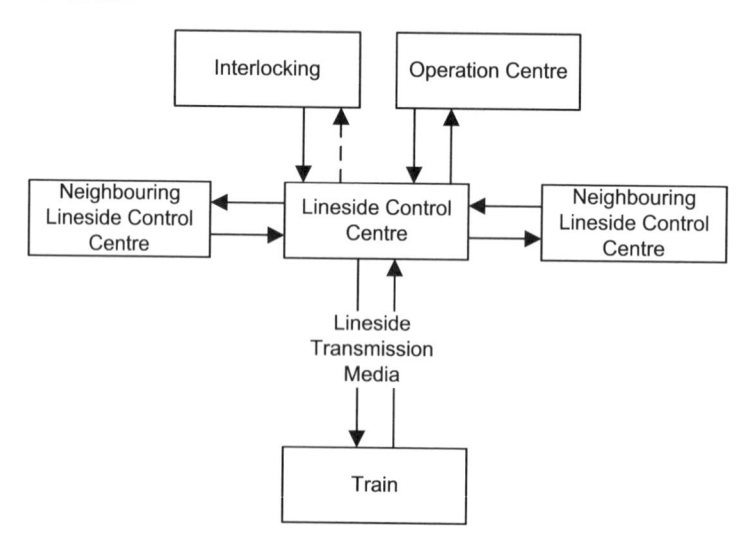

Figure 8.36: Information Structure of Systems with Continuous Transmission at High Data Volume (Wenzel 2006)

8.3 Particular Systems

- Cable loops: An example is LZB which is applied in Germany, Austria and Spain, mainly on high speed lines.
- Radio transmission: The most important example is ETCS Level 2/3.

Figure 8.36 shows the generalised basic structure of such systems. In contrast to most systems with intermittent transmission, information flow is centralised in most cases, using a lineside control centre. The lineside control centre and the train computer are in most cases vital redundant microprocessor systems. The functions of the components can differ in detail between the systems.

An important criterion to distinguish the systems is whether they are used as the only signal system on the respective lines or if they are used mixed with trackside signals. In the former case, system-inherent fallback levels are provided, or driving on sight is the only fallback level for degraded mode operation. In the latter case, the cab signal system often enables shorter block sections (figure 8.37) and therefore higher line capacity than the trackside signals.

Figure 8.37: Block marker on ETCS line (Photo: *SBB*)

The assignment of functions to the interlocking system or the train control system is basically defined as follows on lines for mixed traffic (LZB and partly in ETCS L2):

- The interlocking functions including track clear detection, which are needed for all movements on the line, are assigned to the interlocking system.
- The particular cab signalling and train protection functions, which are only applicable to equipped trains, are assigned to the train control system. In some cases, additional auxiliary functions for interlocking can also be carried out in the train control system, such as detecting the halt of the train for route release.

Resulting from this assignment of functions, route information has to be transmitted from the interlocking system to the trackside control centre (figure 8.36). For functions such as sending the information about the halt of a train to the interlocking system, a bidirectional data connection is necessary, otherwise a unidirectional connection suffices.

On pure high speed lines (systems TVM 430 in France and Digital ATC in Japan, partly also ETCS L2), this assignment of functions can vary. For example, the track clear detection outside stations can be carried out by the train control system.

8.3.6.2 Generation of Movement Authorities and Static Speed Profile

The main data required for the generation of the movement authorities and the static speed profiles are:

- **Static data** such as track topology, locations of points and signals, speed restrictions and others. These data are mostly stored in the trackside control centre. Alternatively, in closed networks where each train moves always on the same lines, they can be stored on the train. This is applied in Japanese Digital ATC, for example.
- **Dynamic data** such as routes, point positions, track occupancy and others. They change dynamically and are transmitted from the interlocking system to the lineside control centre or, in some cases for track occupancy, obtained by the train control system itself.

The movement authority and static speed profile is mostly generated in the lineside control centre and transmitted to the train, but also the train can contribute, e.g. in Japanese Digital ATC. Here the trackside control centre only transmits the number of accessible track sections, but the length of these sections is stored on the train together with other geographical data and the length of the movement authority calculated on the train.

8.3.6.3 Dynamic Speed Profile and Speed Control

To calculate the dynamic speed profile, the following data are the most important:
- Static speed information
- Braking-related characteristics of the line, especially the gradient, which are stored together with other geographical data in the trackside control centre or on the train
- Braking performance of the train which is stored on the train
- Momentary position of the train

The momentary position is detected roughly by reference positions and between two reference positions calculated train-based e.g. by odometry or Doppler radar. Reference positions are marked as follows:
- Digital ATC: The boundaries of the track circuits
- TVM 430: The boundaries of the track circuits
- LZB: Every 100 m the cable loop is crossed, detectable as a 180° shift of the electromagnetic field
- ETCS Level 2: In regular intervals, Eurobalises with fixed data are placed between the rails, transmitting absolute position information

From the data described above, the dynamic speed information can be generated train-based, track-based or mixed:
- In LZB as the historically oldest of these systems, the main limiting factor was the processing performance (and therefore the size and weight) of the train computer. Therefore, the brake curve is calculated in the trackside control centre and transmitted to the train as a code number of a standardised brake curve segment. This reduces the processes on the train computer, but increased the data transfer demand.
- In those systems based on coded track circuits or radio transmission, the main limiting factor is the data volume to be transmitted between track and train. Therefore, dynamic speed information is generated completely on the train. Especially in ETCS L2, an additional purpose of this assignment of functions is to define a clear interface in an interoperable system.

The static and dynamic speed information is displayed to the driver (figure 8.42). Using the dynamic speed profile, fully automatic train operation would be possible. However, this is done in only a few metropolitan railway systems (chapter 8.1.6).

8.4 ETCS

8.4.1 History + Motivation

In Europe there is currently a large variety of train protection systems (figure 8.38; UIC 2003). This is one of the main obstacles for international interoperability on the European continent. To eliminate this disadvantage, the European Train Control System (ETCS) was developed since the early 1990s as a future unified system for the continent and is still in redevelopment process. The main agencies in the development process are:
- the European Commission who initiated the process

8.4 ETCS

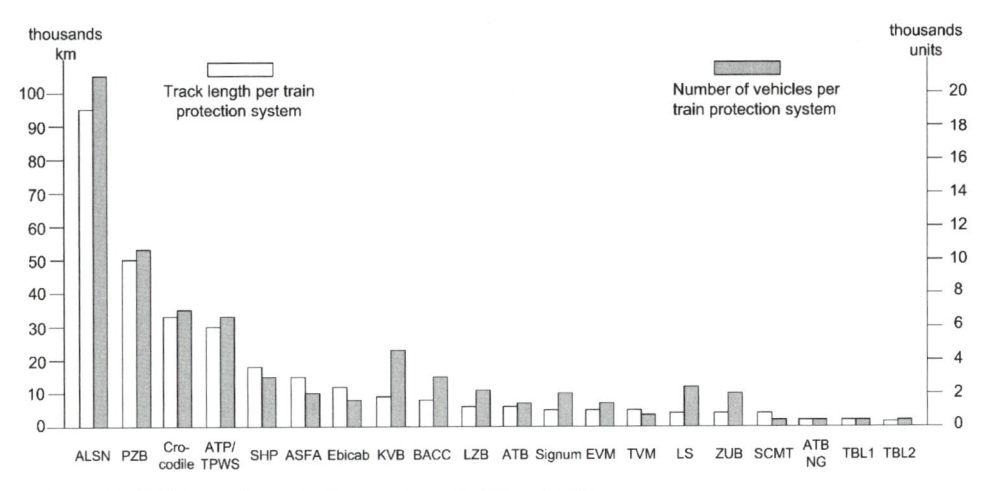

Figure 8.38: Existing train protection systems in EU and CIS

- UIC (International Union of Railways) who defined the functional requirements of the system
- Unisig (Consortium of the 6 (now 7) largest European signalling manufacturers) who specified the detailed technical solutions for the system

Beginning in the early 1990s, the specifications for ETCS are being continuously redeveloped, resulting in updated versions being introduced in intervals of few years (figure 8.39). Therefore, an important point is compatibility between the versions.

ETCS is currently being introduced on several railways in Europe. Problems in the introduction process are the high investment value in the existing national systems and the migration from the old national systems to ETCS requiring double-equipment of lines and/or vehicles for longer time.

Besides Europe, several countries outside Europe use ETCS on some lines. These are currently: Taiwan, South Korea, PR China, Saudi Arabia, Turkey, India, Australia and Mexico (Garstenauer/Appel 2007, Winter et al. 2009).

The following explanations on ETCS mainly base on (Unisig 2006) and (Unisig 2008).

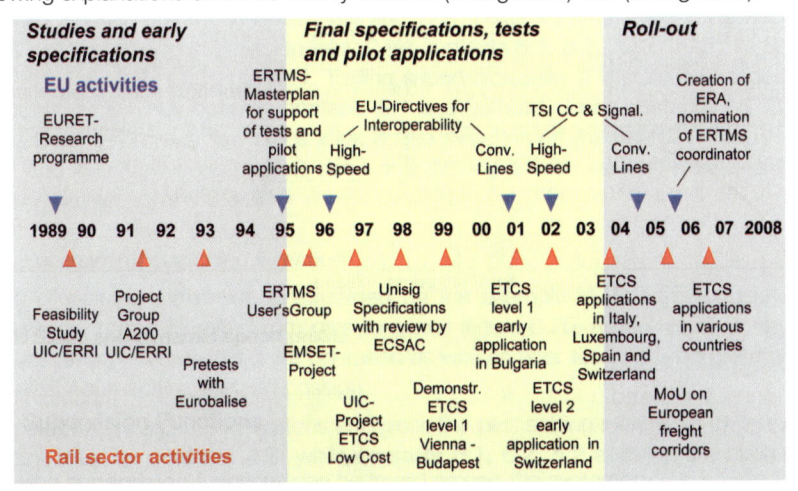

Figure 8.39: Historical Development of ETCS (Winter et al. 2009)

241

8 Train Protection

8.4.2 Application Levels and Technical Components

8.4.2.1 Application Levels

Regarding the equipment of the lines, ETCS specifications distinguish five application levels: the levels 0, 1, 2, 3 and STM.

The term 'ETCS **Level 0**' describes the situation where a vehicle which is equipped with ETCS moves in an unequipped area. The supervision functions are limited to supervision of a constant speed, which is the minimum of the maximum train speed and a general nationally defined speed limit for Level 0.

Level 1 (figure 8.40) matches into category 4 (chapter 8.3.5). The main transmission medium are transponder balises called 'Eurobalise' which transmit, among others, movement authorities and profile data to the train (which is not individually known) when passing above the balise. Balises can be fix data or switchable balises. The former store the data content in the balise itself (only static data), whereas for the latter, a Lineside Electronic Unit (LEU) selects the data according to input information (e.g. signal aspects). Balises are usually linked with each other, which means that most balise groups are announced by a previous balise group, enabling the detection of faulty balises by trainside distance measurement.

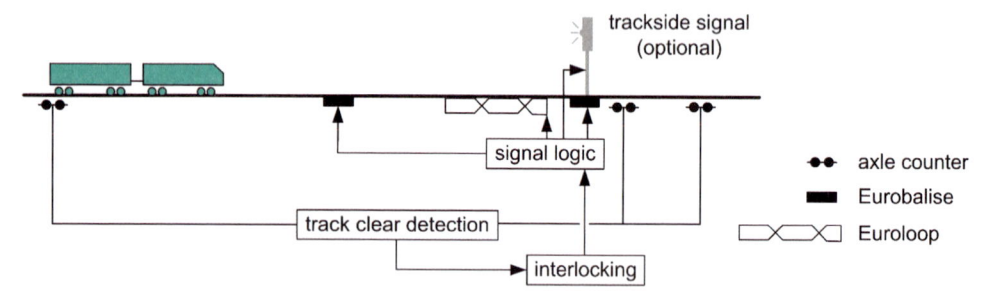

Figure 8.40: Data flow in ETCS Level 1

Besides the balises, linear infill devises can be used locally to transmit changes of signal aspects beyond. These are Euroloops (cable loops in the rail) or radio infill units. The third type of information between track and train besides the normal and infill is repositioning information. This is used in cases when in some signal systems a trackside signal doesn't know the exact path the train is going to take to specify the information gaps after the branching.

Based on the information received from the trackside and on the train data which include braking characteristics, the train computer calculates the dynamic speed limit which can be signalised to the driver in the cab signalling equipment and supervised.

As Level 1 provides continuous guidance functions by movement authority, trackside signals are optional, although used in most cases. Although there is no general limitation of speed, Level 1 is mainly used in conventional traffic for speed up to ~ 160 km/h.

If no infill devices are provided, trackside signals, at least in a simplified form, are practically necessary.

Levels 2 (figure 8.41) **and 3** (figure 8.42) can be classified into category 5 (chapter 8.3.6). Information can be continuously and bidirectionally transmitted by Euroradio, a radio standard based on GSM-R (chapter 8.2.3.2). The central trackside unit is the Radio Block Centre (RBC). It is responsible for a longer section of line, stores the static data and obtains dynamic data like signal and point positions from the interlocking stations in the area. In contrast to Level 1,

8.4 ETCS

Figure 8.41: Data flow in ETCS Level 2

Figure 8.42: Data flow in ETCS Level 3

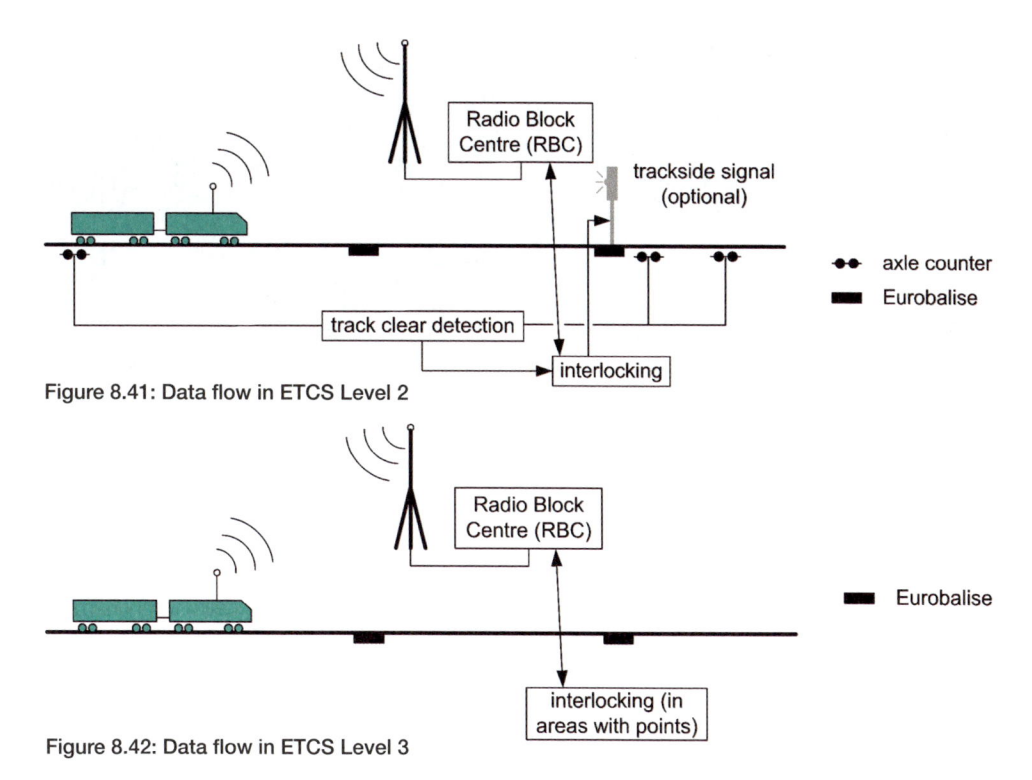

the trains are individually known in the RBC. The train requests new movement authorities in regular time intervals (usually every 60 seconds) or at particular events. Balises are only fixed data balises and serve, among others, for positioning of the trains.

A difference between Levels 2 and 3 is that in Level 2 ETCS only takes the responsibility for signal and train protection functions, whereas Level 3 also replaces the interlocking-based track clear detection by continuously checking train completeness on the train and transmitting this information to the RBC. Based on this, Level 3 uses the principle of moving block operation (chapter 3.4.2.2). Level 2 is being introduced on several lines, especially high speed lines. Level 3, in contrast, is gaining some importance in form of the Scandinavian 'ETCS Regional'.

Level STM (Specific Transmission Module) is designed for situations where a train which is equipped with ETCS moves on a line without ETCS, but with a national train protection system. This level, exactly spoken different levels, one for each national system, has been developed for the migration period. An additional module, the STM, is added to the onboard equipment to translate between the respective national system and ETCS (chapter 8.4.4).

8.4.2.2 Equipment

In the following, the main technical components of ETCS are summarised, beginning with the train equipment:

– There is one **Train Computer** on each vehicle from where a train can be controlled. It is a redundant fail-safe computer (European Vital Computer EVC) and conducts the ETCS-related coordination and calculation functions on the train. Tasks are the calculation of the dynamic speed profile, the storing of the train data, speed supervision, control of the operational modes and others.

8 Train Protection

Figure 8.43: Example of ETCS DMI (Winter et al. 2009)

- The **Driver-Machine-Interface (DMI)** consists of the cab signalling display (figure 8.43), emitters for acoustic signals and input devices such as for starting the mission, overriding an end of authority and input of train data. The cab signal displays the target distance to the next braking target, the speed which has to be obeyed there, the dynamically calculated current speed restriction and the actual speed of the train.
- The **Train Interface Unit (TIU)** serves for communication with the train equipment such as brakes.
- **Antennas** for Euroradio, Balises, and Euroloops communicate with the trackside.
- The **distance measurement equipment** (e.g. odometry, Doppler radar) calculates the speed and travelled distance.
- The **Juridical Recorder** stores several events for evaluation of critical incidents.
- One or several **Specific Transmission Modules (STM)** for communication with national train protection systems can be applied as an option.

The main trackside components of ETCS are:

- **Eurobalises** (figure 8.44) (used in Levels 1, 2 and 3) as spot transmitters are used in the form of single balises (so-called 'single balise group') or in groups of up to eight balises. They are either fixed data balises without any informational connection to other trackside devices or switchable balises with data connection, e.g. from signals, via a **Lineside Electronic Unit (LEU)**. A balise can transmit information for both directions of travel, but always distinguishes by direction. The directions are distinguished either by the internal numbering of balises within a balise group or by linking information from a balise group in rear.

Figure 8.44: Eurobalises of different manufacturers (Germany, *Siemens* and Italy, *Ansaldo*)

- **Radio Block Centre (RBC)** is the central trackside device in Levels 2 and 3. It is, among others, responsible for storing static line data, for obtaining dynamic data from the interlocking systems and for generating the movement authorities (MA) and profiles. The RBC communicates with trains by the **Euroradio** based on GSM-R (figure 8.45). The length extension of the area covered by one RBC varies by manufacturer and local situation an can be upto few hundreds of metres.

Figure 8.45: Trackside antenna of GSM-R (photo: *DB AG/ Hans-Joachim Kirsche*)

- **Euroloops** and **Radio Infill Units** (Level 1 only) are linear transmitters of limited extend for transmission of infill information related to a position in advance, such as upgrades of signal aspects.

8.4.3 Functional Concepts

8.4.3.1 Positioning of a Train

The train knows its position only in a linear form along the path. All positioning information is linear and relative to a reference balise group in rear. As an exception, the information of infill units in Level 1 is related to a balise group in advance.

Inexactness of the estimated location of the train has to be considered; therefore a confidence interval is calculated. The width of the confidence interval is narrow (although not zero) immediately after passing a reference balise group. It then increases linearly with the distance from the balise group due to the inexactness of train-based relative length measurement (figure 8.46). At each reference balise group the position is corrected. Depending on the usage of the position information, the following positions can be relevant (figure 8.47):

- the estimated position
- the maximum safe front end position
- the minimum safe front end position
- the maximum safe rear end position
- the minimum safe rear end position

In Levels 2 and 3, the train periodically or at certain events reports its position including the confidence interval, the train speed and direction to the RBC.

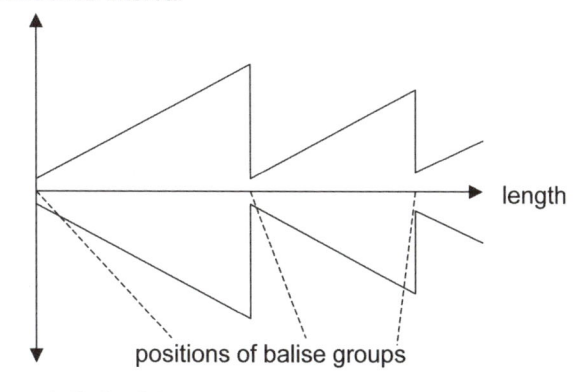

Figure 8.46: Confidence interval of ETCS length measurement in relation to balise positions

8 Train Protection

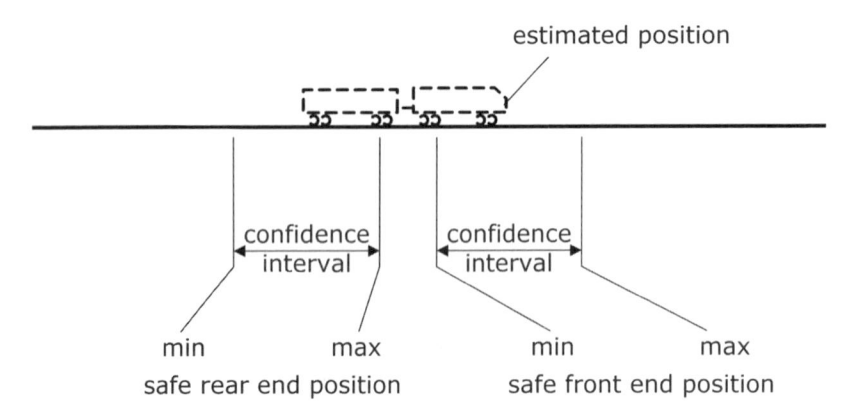

Figure 8.47: Confidence interval for the train position

8.4.3.2 Movement Authority (MA)

The **Movement Authority (MA)** is the permission for the train to pass a section of track with a defined length. The MA is transmitted from the trackside to the train: in Level 1 by balises and in Levels 2 and 3 by the RBC on request of the train. The MA can be split into sections and for each section a time-out value can be defined. Other time-out values can be defined for the target section and for the overlap. Each MA is valid until new MA information reaches the train or the time-out has expired.

Two purposes for the time-out can be identified:
- Conformity with the time delayed release of the last route section and the overlap when the train has a stop in the target section (chapter 4.3.3.6).
- In interlocking systems where routes can be released time-delayed by the signaller (chapter 4.5.4.3), to stop the train which might have no longer a safe route ahead.

The **End of Authority (EOA)** (figure 8.49) is the position where the train is obligated to stop with its max safe front end, comparable with a red signal. Beyond the EOA, a **Supervised Location (SvL)** (figure 8.49) can be defined and calculated as the target for the emergency braking curve. The SvL is:
- the end of the overlap if there is an overlap which has not yet expired.
- the danger point in advance of the EOA (e.g. fouling point of points) if there is no distinct overlap or it has expired.
- the EOA if neither an end of overlap nor a danger point is applicable.

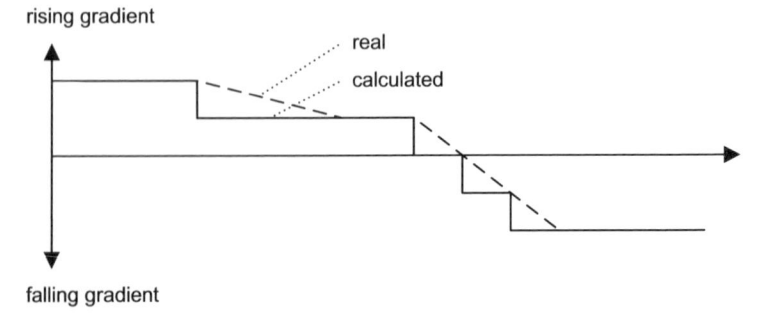

Figure 8.48: Definition of gradient profile

8.4 ETCS

Each movement authority is replaced by a new movement authority, which normally reaches to a position further beyond. There are some special situations of shortening a movement authority in Levels 2 and 3:

- **Co-operative Shortening** is used in cases where operational decisions require release of the route. The RBC sends a request for shortening to the train, whereupon the ETCS train equipment checks if the train can be stopped safely at the new EOA and accepts or rejects the new MA.
- **Emergency Stop Messages** are used in dangerous situations. Two kinds of emergency stop messages are distinguished:
    - When receiving an unconditional emergency stop messages, the train has to be tripped (stopped by emergency brake) immediately.
    - When receiving a conditional emergency stop message (which is always connected with a position), the train is only stopped if its min safe front end hasn't yet reached the stated position, otherwise it can ignore the message.

8.4.3.3 Profiles

Different characteristics related to the path of the train are transmitted as profile data. Profile data have the structure of a chain of values with the related beginning and ending positions and are transmitted from the trackside to the train. Examples for profiles are:

- The **Static Speed Profile (SSP)** includes the line speed, permanent speed restrictions and route related speed restrictions. It is required by the on-board unit to calculate the **Most Restrictive Speed Profile (MRSP)**. This is calculated as the minimum of all speed restrictions, including also temporary speed restrictions, the permitted speed of the train, mode related speed restrictions (chapter 8.4.4) and others.
- The **gradient profile** defines the gradient of the line and is needed by the train to calculate the braking curve (dynamic speed profile). For safety, in each segment, the minimum value of longitudinal inclination has to be considered (minus = falling; plus = rising) (figure 8.48).
- The **track condition profile** gives particular information related to electric traction, air tightness (required for air conditioning in passenger wagons), tunnels and bridges where stopping is forbidden, radio holes and others.

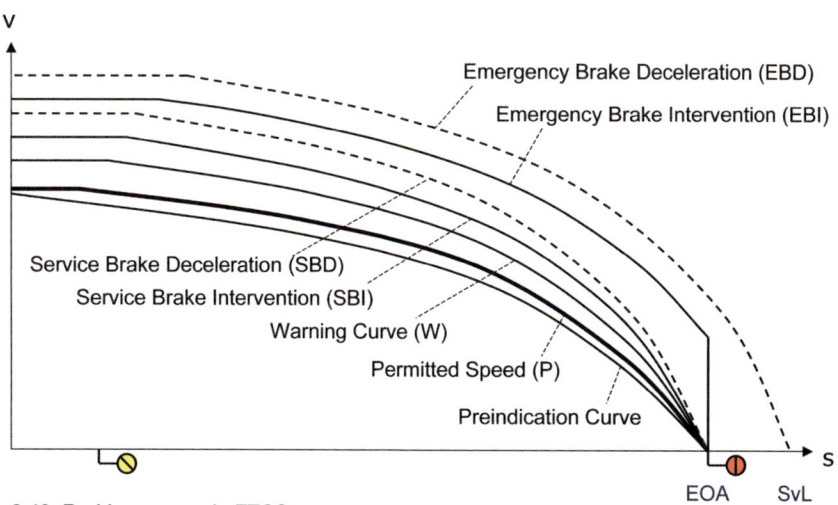

Figure 8.49: Braking curves in ETCS

- The **route suitability profile** defines criteria a train has to fulfil to be allowed to enter certain track sections regarding axle load, traction and others.

8.4.3.4 Supervision on the Train

The following explanations refer to the operation mode 'Full Supervision'. For specialities of other modes, see chapter 8.4.4.

For movements at constant speed, four different speed limits are relevant (from the lowest to the highest), marking a tolerance margin above the permitted speed (figure 8.49):
- The **Permitted Speed (P)** is displayed to the driver and shall not be exceeded.
- If the driver exceeds the **Warning Speed (W)**, a warning tone sounds.
- If the **Service Brake Intervention Speed (SBI)** is exceeded, service brakes are automatically applied. Considering the time which passes until full braking power is reached, the train will brake along or below the Service Brake Deceleration (SBD) curve.
- If the **Emergency Brake Intervention Speed (EBI)** is exceeded, emergency brakes are automatically applied, and the train brakes along or below the Emergency Brake Deceleration (EBD) curve.

The EBI is, depending on the speed, 5 to 15 km/h above the permitted speed. The SBI and W shall be defined that way that reaching of the EBI is avoided and that the driver has a real chance to react on the warning signals, considering reaction times.

In the braking process, an additional preindication curve is used to define the point where the driver firstly gets an announcement to start braking soon.

Whereas this and the curves P, W and SBI are calculated towards the EOA as braking target, EBI can be calculated towards the SvL (Supervised Location), which takes into consideration overlaps and danger points (figure 8.49).

With the SvL situated a longer distance beyond the EOA, this helps to avoid an important disadvantage: Although emergency brakes can usually effort higher braking deceleration than service brakes, only safe brakes can be considered for calculating the EBI which makes the EBI and EBD curves comparatively flat. With the other curves below the EBI, an unfavourable early start of the braking process would be required.

Besides supervision of the static and dynamic speed profile, additional functions are used to prevent unintended movements: Reverse Movement Protection, Standstill Supervision and Roll Away Protection.

Regarding the release of brakes, two kinds of brake interventions are defined:
- A **Train Trip** can be released only after the train has come to a standstill and after acknowledgement of the driver. This form is applied for exceeding the EOA (chapter 8.4.3.2), unintended movements and various technical failure cases.
- In contrast, after exceeding the speed, brakes can be released when reaching a safe speed.

8.4.3.5 Text Messages

Text messages for the train driver can be automatically generated or typed by the signaller or dispatcher. Besides the text itself, information about the time and location for displaying the message, restriction to certain levels or operation modes and the necessity of acknowledgement by the driver are transmitted.

8.4.3.6 Values for Variables

Predefined values of variables can be defined in one of three forms:

- **Fixed values** are general values which do not change from one application to the other. Examples are:
  - cycle for requesting a movement authority unless specific situations apply: 60 seconds
  - assumed adhesion value for slippery rails: 70 %
- **National values** are defined for a country or region and transmitted to the train as a complete data packet by balises. Examples are (with default value in brackets):
  - modification of adhesion factor by the driver (not allowed)
  - mode related speed restriction for shunting (30 km/h)
  - distance for brake application in roll away protection, reverse movement protection and standstill supervision (2 m)
- **Train data** are entered by the driver when starting the train. Examples are:
  - permitted speed of the train
  - braking characteristics
  - train length

8.4.4 Operation Modes

The mode describes the operational situation of the train. Altogether, 16 modes are defined in the System Requirements Specifications, Release 2.3.0, and two additional modes, LS and PS,

are being foreseen for release 3.0.0. In the following, the modes will be classified into six groups. As first groups, the modes for regular operation under supervision of ETCS are described:
- **Full Supervision (FS)** is the 'normal' case with the train fully supervised by ETCS.
- **Limited Supervision (LS)** is only defined for Level 1 and uses reduced functions. The optical cab signalling is deactivated and the only supervised speed is the EBI, therefore ETCS works in the background. An advantage of this mode is the possibility to start braking closer to the EOA (figure 8.49). This increases line capacity and, in existing systems, reduces the demand to change positions of trackside distant signals and therefore facilitates migration from national train protection systems to ETCS.
- In **On Sight (OS)** mode, the train proceeds on sight onto a track which can be occupied by vehicles or obstacles. This mode is selected by this restriction being received from the trackside. A mode related speed restriction is defined in national values (default value: 30 km/h).
- **Shunting (SH)** mode is used for shunting movements and can be selected by the driver or the trackside. Train data are deleted when entering SH mode, as a main purpose of shunting is assembling and splitting of trains. If selected by the driver, in Level 2 and 3 a permission of the RBC to shunt in the respective area is necessary. Besides the mode related speed restriction (national value, default 30 km/h), a list of balise groups which may be passed is defined to prevent the driver from exceeding the shunting limits.

The second group is a mode for degraded mode operation under responsibility of the driver:
- **Staff Responsible (SR)** mode is used to move the train under responsibility of the driver when no Movement Authority is available. One case is during and after an override when the driver passes a red signal e.g. on written instruction or permissively. Another situation particularly in Level 1 is after starting up the ETCS on-board equipment before reaching the first balise group. A movement on auxiliary signal is no override, but a movement authority is generated then, added by the related speed restrictions. In SR mode, the driver has to check track occupancy and positions of movable track elements in his own responsibility.

The third group are modes for prevention of hazards in critical situations:
- **Trip (TR)** mode is active during a trip. A train trip is only the emergency braking until stop, but not the braking to a safe speed in case of excessive speed (chapter 8.4.3.4).
- **Post Trip (PT)** mode is activated after acknowledgement action of the driver after a trip and permits the train to continue. The driver can select Start of Mission, Override or Shunting. When selecting start, the next mode is Staff Responsible in Level 1, whereas in Level 2/3 a MA is requested from the RBC.
- **Reversing (RV)** mode is used to move trains backwards in safety critical situations without the need to change cabs beforehand. It can only be selected by the driver in few especially marked locations. The only known application example is in long tunnels where in case of fire the danger of delayed or prevented evacuation by the driver changing cabs is higher than the danger by a train running blindly backwards in the tunnel.

The fourth group is modes related to the not leading locomotive in double traction, without further supervision functions:
- **Non Leading (NL)** mode is selected by the driver if the locomotive is separately controlled.
- **Sleeping (SL)** mode is selected automatically by train information if the locomotive is remotely controlled by the leading locomotive.
- **Passive Shunting (PS)** is used for a non-leading engine in a shunting consist.

The fifth group includes modes for movements on lines which are not equipped with ETCS.
- **STM European (SE)** mode is used in Level STM if the national system is able to provide movement authorities, speed and gradient profiles. The vehicle antennas of the national

train protection system transmit data to ETCS and ETCS signalises to the driver and supervises the movement, possibly with reduced functions.
- **STM National (SN)** mode is used in Level STM if the national system is not able to provide the data required for SE mode. In SN mode, the national system supervises the movement, but uses particular ETCS components (e.g. DMI, brakes and odometry).
- **Unfitted (UN)** mode is applied when the line is either equipped with a national system for which no STM is available, or is not equipped with any train protection system. This mode is only defined in Level 0. The only supervised speeds are temporary speed restrictions, train speed and a particular mode related speed (national value; default: 100 km/h).

The sixth group is modes related to different offline and disturbance statuses of the onboard equipment:
- **Stand By (SB)** mode is the initial mode after starting the onboard equipment. It includes self-test functions and supervision of the train to be stationary.
- **No Power (NP)** mode is automatically selected if the ETCS onboard equipment is not powered. The train is immediately tripped.
- **System Failure (SF)** mode is activated in case of a safety critical system failure of the ETCS onboard equipment. The train is immediately tripped.
- **Isolated (IS)** mode is active when the ETCS onboard equipment is isolated from the other onboard equipment to continue operation in case of the above described failures. Operation continues without supervision by ETCS.

8.4.5 Data Structure

The structure of data transmitted between track and train is described briefly here.

The smallest data unit is the **variable** which codes a single value.

Different variables are connected to a **data packet**. A packet has a header expressing the identification (type of packet), the length of the packet, the direction for which it is valid (only for packets from track to train), the scale for length variables (if applicable) and others. After the header, the packet can contain several more variables, depending on the type of packet Examples for packet sent from track to train are a movement authority, a speed or gradient profile, a temporary speed restriction, linking information to following balise groups or a mode transition order. Examples for packets sent from train to RBC are position information or mode transition information.

In communication from balise to train, a **data telegram** is the information content of one balise. In communication with RBC, no telegrams exist. The telegram size is standardised: Short telegrams have a length of 341 bits and long telegrams 1023 bits (830 user bits + safety attachment).

A **message** is a complete set of information in communication with balises or RBC. In communication with balises, the data telegrams transmitted by all balises belonging to the same balise group (1...8 balises) add to the message. A message contains several data packets and possibly additional single variables.

9 Interlocking Machines

*Gregor Theeg, Oleg Nasedkin, Ulrich Maschek, David Stratton,
Heinz Tillmanns, Thomas White, Giorgio Mongardi*

9.1 Classification

This chapter describes the technical equipment needed to fulfil the interlocking functions, called interlocking machines. Over the years, the technical progression was as follows:
- human or manual, without technical support
- mechanical
- (hydraulic/pneumatic)
- electric
- electronic

Basically, an interlocking system can be divided into three main functions:
- The **operation control level** includes the interface to the signaller and may include different non-vital functions of automatic operation control such as automatic train routeing etc.
- The **interlocking level** includes the vital functions to interlock signals, routes, movable track elements, block applications etc. with each other.
- The **element control level** includes functions of commanding, power and information transmission to and from the field elements, such as signals, movable track elements, track sections, level crossings etc.

Since a high degree of centralisation is applied, especially in electronic interlocking, the view on these functions has changed: Today the interlocking system can be operated from two different workplaces, one local and one central, which can be situated many kilometres apart from each other. The interlocking can be operated from either, although not from both at the same time. The operation can also be switched between different operation systems, which include automatic systems such as automatic train routeing. Therefore, considering the operation control level as an integral part of the interlocking loses its sense. Instead, the operation control level is often referred to as an external technical system. In the interlocking itself, a new functional level called operation interface level is required. This forms the interface to the external operation systems and validates input commands.

Theoretically, all combinations of the three levels (operation control, interlocking and element control) working with one of the above technical principles can be imagined, although not all of them make practical sense.

There are four almost pure forms of interlocking technologies (table 9.1):

	operational level	interlocking level	element control level
human 'interlocking'	–	human (to be reminded by the signaller)	human (walking between the elements)
mechanical interlocking	mechanical levers	mechanical (lever frame)	mechanical (wires or rods)
electric (relay) interlocking	electric buttons and illuminations	electric (relays)	electric
electronic interlocking	electronic (monitor, mouse or tablet, keyboard)	electronic (hardware/software)	electronic

Table 9.1: Basic interlocking technologies and technical application of the functions

- The human 'interlocking' is not a real interlocking, as no technical locks are provided. It is a situation in which the human in the form of the signaller or the shunting staff is fully responsible for checking the preconditions for clearing signals, switching movable track elements and for transmitting information to the field elements by walking between them. Historically, this is the oldest solution. For train movements it has been widely replaced by technical solutions for safety reasons, but for shunting movements this method is still widely used.
- In mechanical interlocking (chapter 9.2), the signaller operates mechanical levers which are interlocked with each other. Power and information transmission to the field elements is by wires or rods.
- In electric (relay) interlocking (chapter 9.3), the signaller operates buttons. The interlocking functions are in relay technology and the field elements are operated and controlled electrically. Especially in English speaking countries, historically also in Germany, the term 'electric interlocking' is used for an interlocking with electric element control, but mechanical interlocking functions. This use of this historic term, although inexact, originated when relay interlocking functions were not imaginable. To avoid confusion, the pure electric form of interlocking will not be referred to in this book as 'electric interlocking', but as 'relay interlocking'.
- The electronic interlocking (chapter 9.4) is a machine where all these functions are performed electronically by hardware and software. The interlocking logic is usually defined in the programmed software. Older electronic interlocking systems perform the element control in electric (relay) technology.

Besides these 'pure' forms, several hybrid forms exist because the shift from mechanical to electrical, as well as from electrical to electronic technology took place in several steps. Some of these forms are described in chapter 9.5. Besides, in several installations the operation of selected field elements is by other technical means than those for which the interlocking system was originally designed. Examples are electrical replacers for mechanical signals in mechanical interlocking and power operated mechanical signals in relay interlocking.

9.2 Mechanical Interlocking

9.2.1 Historical Development

In the first railways in the 19th century, the operation elements for points and signals were distributed in the field without interlocking between them. Around 1860, the first mechanical locking frames emerged in Britain. This concentrated the levers for element operation in a central place and provided mechanical interlocking functions between the elements. The idea of interlocking and centralisation of element operation was exported worldwide and different systems were developed by different manufacturers. These were adapted to the operational requirements of the respective countries. A great diversity of technologies was the result. In Germany around 1900 there were about 20 manufacturers, each of them producing its own type of interlocking. Later, concentration reduced the number of types in each country.

9.2.2 System Safety in Mechanical Interlocking

The safety of mechanical interlocking systems is ensured mainly by the strategy of **elimination of failures** (chapter 2.2.1.1). Mechanical components are dimensioned in a way that technical failures (e.g. the breaking of a lever or the loosening of a bolt) is (almost) impossible.

9 Interlocking Machines

9.2.3 Structure of Mechanical Interlocking Systems

Generally, a mechanical interlocking comprises the following:
- The mechanism and operation elements for operating the movable track elements and signals remotely, which represents the operational and the element control level in table 9.1.
- The interlocking between the elements, which forms the main part of the interlocking level in table 9.1.
- The communication between different signal boxes, which forms an additional part of the interlocking level in table 9.1 in some interlocking systems.

For the remote mechanical operation of field elements, basically two different technical solutions are applied: rigid connections (e.g. rods and pipes) and flexible connections (wires). Both form a connection between the field element and the operation lever in the signal box. The wires can be double wires, or single wires with a counterweight. Wires (figure 9.1) and rods are used differently by the railways. German railways used almost only wires, whereas in Britain and Russia wires and rods and in the USA pipes are/were used.

Figure 9.1: Wire connections with weight bar equipment for wire adjustment

In some railways, such as the German, each set of points is typically switched individually, whereas in other railways, such as the British and the French, the two points of a crossover are coupled and therefore set together by the same point lever (chapter 4.2.2). This means that the signaller needs the physical strength to switch both sets of points together. Besides the force needed to set the points, there is also the friction of the wire or rod system to be overcome by the muscle force of the signaller. This increases by distance and is higher for rigid connections than for wires. This effect limits the possible distance of field elements from the signal box. A second reason which limits the length is the stretching of wires by mechanical force and the outside temperature, and of the rods by temperature.

The interlocking within the same signal box is mainly carried out mechanically by moving parts which are interlocked with each other to prevent certain combinations of element positions. In some interlocking technologies, these mechanical locking registers are aided by electrical devices such as track circuits and wheel detectors for detecting the presence or passage of a train.

Communication between different signal boxes includes the line block system as well as communication in cases where different signal boxes contribute to the same route. In the oldest systems this was carried out without interlocking dependences. Here the signaller was safety responsible for sending the correct messages and carrying out the correct procedures after receiving a message.

9.2.4 Example: British Origin Mechanical Interlocking

9.2.4.1 Overview

In the 19th century, there were several manufacturers of mechanical interlocking in Britain, of which *Saxby & Farmer* was the most important. Their mechanical interlocking and its derivates were widely used throughout Britain, other Western European countries and the United States. Some of these machines still remain in service. In the USA, new installations were constructed as late as 1950. The first design was patented in England by *John Saxby* in 1856. With partner *John Stinson Farmer*, *Saxby & Farmer* became the world's first signalling manufacturer in 1860. There were other locking schemes, but by the 1880s, the tappet locking invented by *James Deakin* of *Stevens & Sons* in 1870 became the most widely used. By 1875, the Saxby & Farmer interlocking had become virtually the standard in the UK. Although the use of interlocking in England was already well established, the first installation in the US did not occur until 1870.

9.2.4.2 Geographical Location of Signal Boxes

In traditional British interlocking, stations and open lines are not distinguished from each other. Therefore, there is no hierarchical order between the signallers and signal boxes like in Germany (chapter 9.2.5.2). One signal box typically controls an area smaller than a station (e. g. a 'station throat'), normally limited by home signals on the entry from each track. The interlocking area is limited by the length of the rods for field element operation and by the number of operations a signaller can handle in a certain time.

9.2.4.3 Operation of Field Elements

Points are operated by levers (figure 9.2). The levers have two end positions: The straight position is 'normal' and the pulled position towards the signaller 'reverse'. The lever is fixed in each end position with a spring-loaded rod, the so-called catch rod at the back of the lever. The catch rod is connected with a catch handle near the top of the lever. When this handle is grasped, the catch rod moves up and the lever can be moved from its current end position. The catch handle does not return to the locking position until the full throw of the lever.

For transmission of the motion to the field elements in Britain, rigid rods dominate for points and single wires against a counterweight for signals. The reason is the greater distance of signals from the interlocking. In the USA, rigid steel pipes are used.

Figure 9.2: Levers in mechanical interlocking *Saxby&Farmer* (USA)

9 Interlocking Machines

9.2.4.4 Interlocking

The elements are interlocked in a locking box which is usually located below the signaller's floor. The method of interlocking most used is tappet interlocking, whose basics (only) are described in the following.

Each lever is connected with one or several tappets, which are flat bars of steel. The tappet is either connected with the lever itself or with the catch rod and moves, while switching, longitudinally within a tappet way (figure 9.3). The tappet carries notches which are adjusted to the form of the locks. The locks are placed in locking boxes. Each locking box can contain one or several lock channels, depending on interlocking type. Several locking boxes can be placed above each other in an interlocking system.

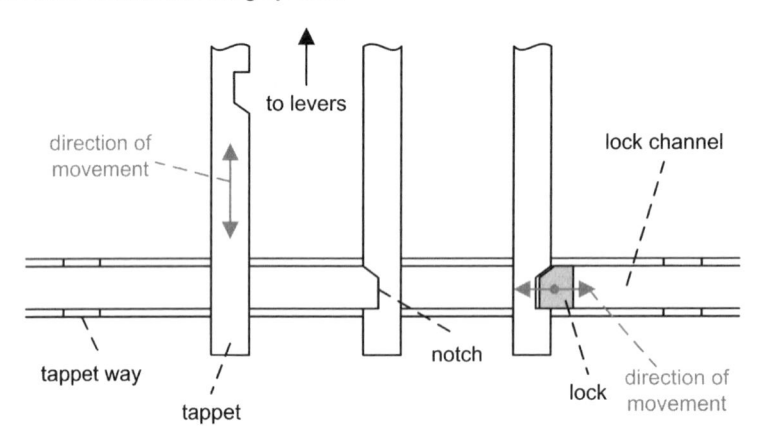

Figure 9.3: Principle of interlocking by tappets

In the following, some example solutions for locking functions will be described. Figure 9.4 shows a lock between two elements: When both levers are normal (A), either of them is free to be reversed. But as soon as one lever is reverse (B), the other is locked normal. Figure 9.5 shows another example: If lever 2 is normal, lever 1 cannot be reversed, and if lever 1 is reverse, lever 2 cannot be released to normal.

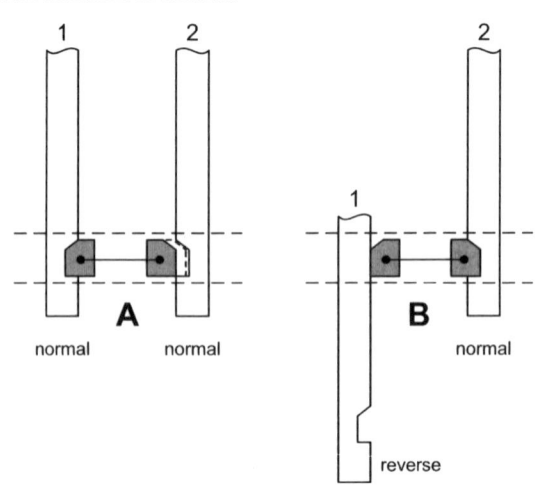

Figure 9.4: Example of tappet locking: 1 locks 2

256

9.2 Mechanical Interlocking

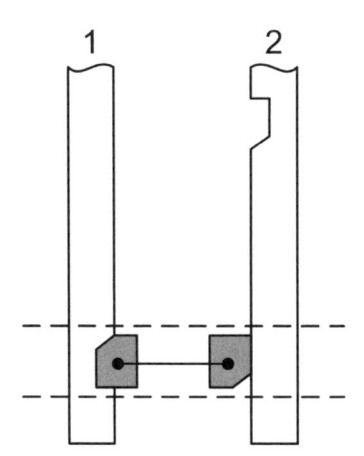

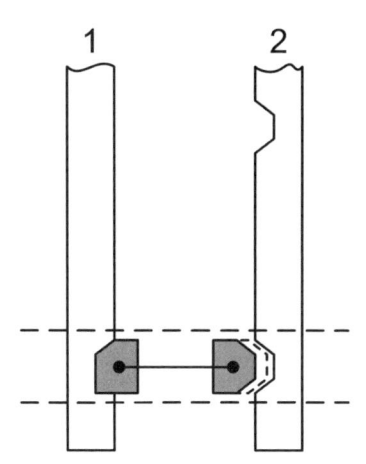

Figure 9.5: Example of tappet locking: 1 released by 2

Figure 9.6: Example of tappet locking: 1 locks 2 in either position

In the example of figure 9.6, lever 2 cannot be moved if lever 1 is reverse. This kind of locking is called '1 locks 2 in either position'.

In many cases, locking between two elements is not sufficient, but three or more elements have to be interlocked. An example of such 'conditional locking' (chapter 4.2.5) between three elements is shown in figure 9.7: The elements are signal 1 and points 2 and 3 here. If points 3 are reverse, signal 1 must lock points 2 reverse: That means that if point 3 and reverse and signal 1 open, the points 2 must be reverse. To enable this, tappet 2 is narrower than the normal width of a tappet, but moves in a normal tappet way (figure 9.7). Therefore, it can move sideways within the tappet way. If 3 is reverse, the lock between 1 and 2 is effective. But if 3 is normal, tappet 2 can move to side and both levers 1 and 2 can be moved freely. (However, signal 1 and points 2 need an additional lock '1 locks 2 in either position', see figure 9.6.)

In the oldest version, points were protected against being switched under a train by mechanical detector bars (chapter 5.2.3.1). Later, track circuit (chapter 5.3) detection was connected to mechanical locking by way of forced drop electric locks mounted on the interlocking machine and connected to the locking levers.

More information on this example can be found in (Such 1956).

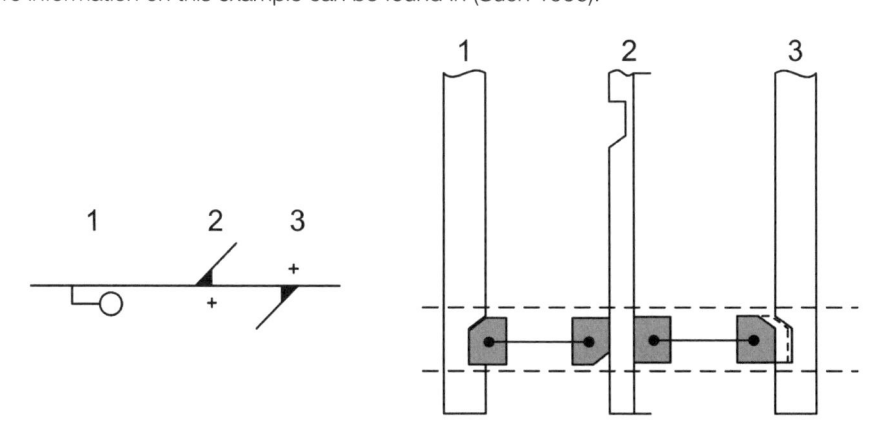

Figure 9.7: Example for tappet locking: Conditional locking 1 locks 2 reverse if 3 is reverse

257

9 Interlocking Machines

9.2.4.5 Interlocking between Different Signal Boxes

As each signal box works independently, there are no order and affirmation dependences like in Germany (chapter 9.2.5.5). Instead, each track between two interlocking areas, even if it is inside a station with many parallel tracks, is considered as a line for the block system.

However, cases occur where the same signal requires control from two different signal boxes (so-called slotted control), which is in effect an AND function. The most common case is a distant signal at the post of a main signal which is controlled from another signal box. By a combination of mechanical elements, the distant arm can move to the 'expect clear' position only if the main signal arm of the same signal and the signal ahead both show 'clear'.

Another case with dependence between different places is where industrial sidings branch from the line halfway between the interlockings. Upon telephonic request, the signaller of one interlocking gives an electrical release which allows the train driver or shunter to switch the points into the siding. This release is interlocked with other relevant signalling functions in the usual way.

9.2.5 Example: German Type 'Einheit'

9.2.5.1 Overview

Around 1910 in Germany, a unified form of mechanical interlocking was developed jointly by the *German Railways* and the interlocking manufacturers to replace around 20 different types used previously. In the following years, the interlocking type 'Einheit' (figure 9.8) became the most used in Germany and remained so until the introduction of relay interlocking.

Main technical characteristics of this form of mechanical interlocking are:

- In contrast to British forms, the elements in a route are not interlocked element by element, but a particular route drawbar (German: 'Fahrstraßenschubstange') interlocked with all route elements is used. This means that, as long as no route is set, points can be moved almost freely. This enables free shunting (unsignalled), which was typical for German railways at that time.
- Power is transmitted to all field elements by double wires.
- Safe communication between different signal boxes is by safe block instruments.

Figure 9.8: Operation elements of mechanical interlocking type 'Einheit'

9.2 Mechanical Interlocking

9.2.5.2 Geographical Location of Signal Boxes

According to the German operation concept, the 'Bahnhof' (for which no exact English translation exists, but which can be more or less translated by 'station', meaning the whole area and not just the platform) is regarded as the unit in operation and interlocking (chapter 3.1). The total length of a station is typically more than 1,000 m. On the other hand, the interlocking area controlled by one signal box is limited by the following factors:

- As no technical track clear detection is usually provided in German mechanical interlocking, the signaller has to observe the whole interlocking area assigned to the signal box visually from the window of the signal box.
- Before unblocking the open line section, end of train markers have to be observed. This is only possible with a signaller working at each station throat.
- The wire length of an element to be mechanically switched from the signal box is limited to approximately 400 m for points and 1,200 m for signals.
- As the grade of automation is low, the signaller has to carry out each step in route setting and release manually, which leads to limitations in the number of train movements a signaller can handle in a certain time.

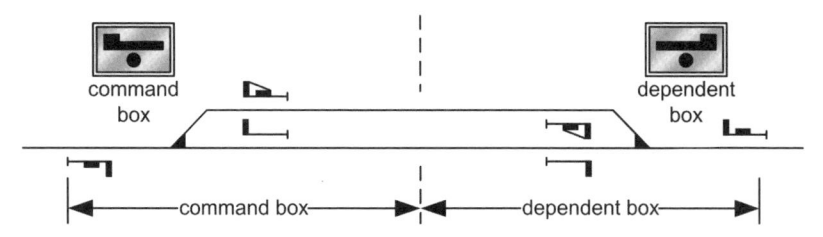

Figure 9.9: Signal boxes in a station by two signal boxes

For these reasons, a station is usually equipped with at least two signal boxes, or more in large stations with high traffic volumes, whereas for junctions one signal box is usually sufficient. There is a hierarchy between the signal boxes belonging to one station: One is the command signal box, staffed with the train director, whereas all others are dependent signal boxes, staffed with one leverman each. This affects the technical communication between the signal boxes: The leverman can only clear a signal for trains upon an order from the train director, whereas shunting movements can be controlled under the responsibility of the leverman.

9.2.5.3 Operation of Field Elements

In most cases, field elements are operated by characteristic levers which have to be turned through 180°. There are basically two types of levers: Point levers (figure 9.10) which serve also for derailers and additional point locks, and signal levers. The levers are equipped with a catch handle. The catch handle is connected with the point bar in the locking box. If the catch handle is grasped, the element can be moved and the respective element bar in the interlocking box is in the medium position. The catch handle can only be released in an end position of the lever, bringing the point bar to the upper or lower position. This gives information about the end position of the points to the interlocking box.

The lever is connected with the field element by a double wire, making a quasi-rigid connection between the lever and the field element. There are parts to guarantee the fail safe behaviour of the rigid connection.

If trailable points are used and they are trailed by a vehicle, the coupling in the point lever moves and prevents the points from being moved again by the signaller. It also makes a noise

9 Interlocking Machines

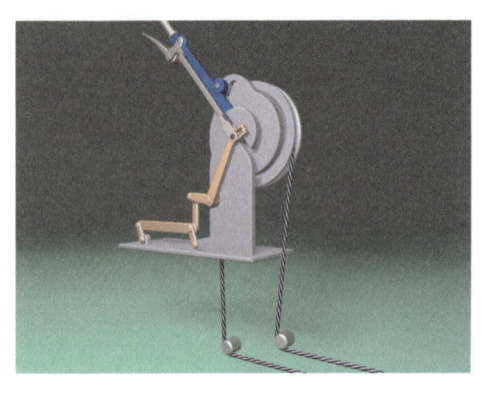

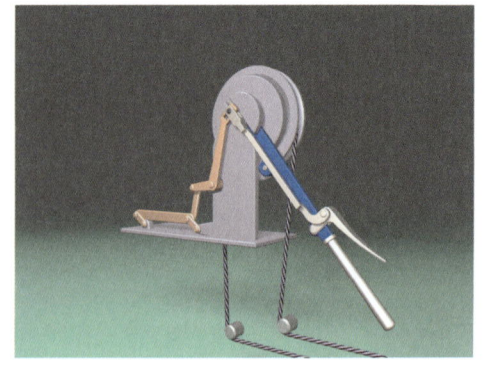

Figure 9.10: Point lever in German interlocking type 'Einheit' (illustration: *TU Dresden*)

during this action. The point bar in the locking box is then in the intermediate position, inhibiting all routes over these points.

9.2.5.4 Interlocking

In German mechanical interlocking, there are two steps when locking a route (chapter 4.3.8):
1. Mechanical (reversible) route locking
2. Electrical (irreversible) route locking

Whereas the mechanical route locking serves for the dependence between the signal and the movable track elements, the purpose of the electrical route locking is to maintain the route locked after returning the signal to stop and until the train has cleared the route (chapter 4.3.8). These two steps of route locking are logically still provided in current German relay and electronic interlocking systems, although the technical reason for the distinction (the impossibility of evaluating the passage of the train mechanically) has lost its importance in the meantime.

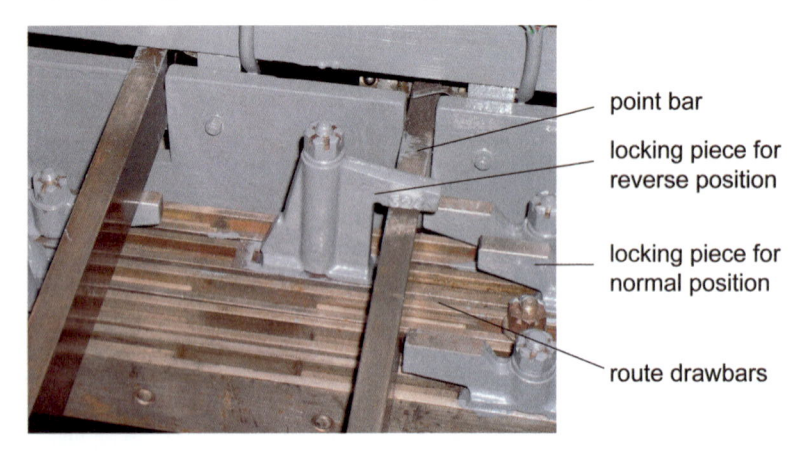

- point bar
- locking piece for reverse position
- locking piece for normal position
- route drawbars

Figure 9.11: Interlocking pieces of mechanical interlocking register type 'Einheit'

The mechanical (reversible) route locking is carried out by separate route drawbars in the interlocking box (figure 9.11, figure 9.12). These route drawbars cross the interlocking box in its whole length (usually several metres). From the normal position (no route locked), each route drawbar can be moved longitudinally into two directions by a route lever, locking two different routes. The route drawbar is interlocked with each movable track element belonging to the

9.2 Mechanical Interlocking

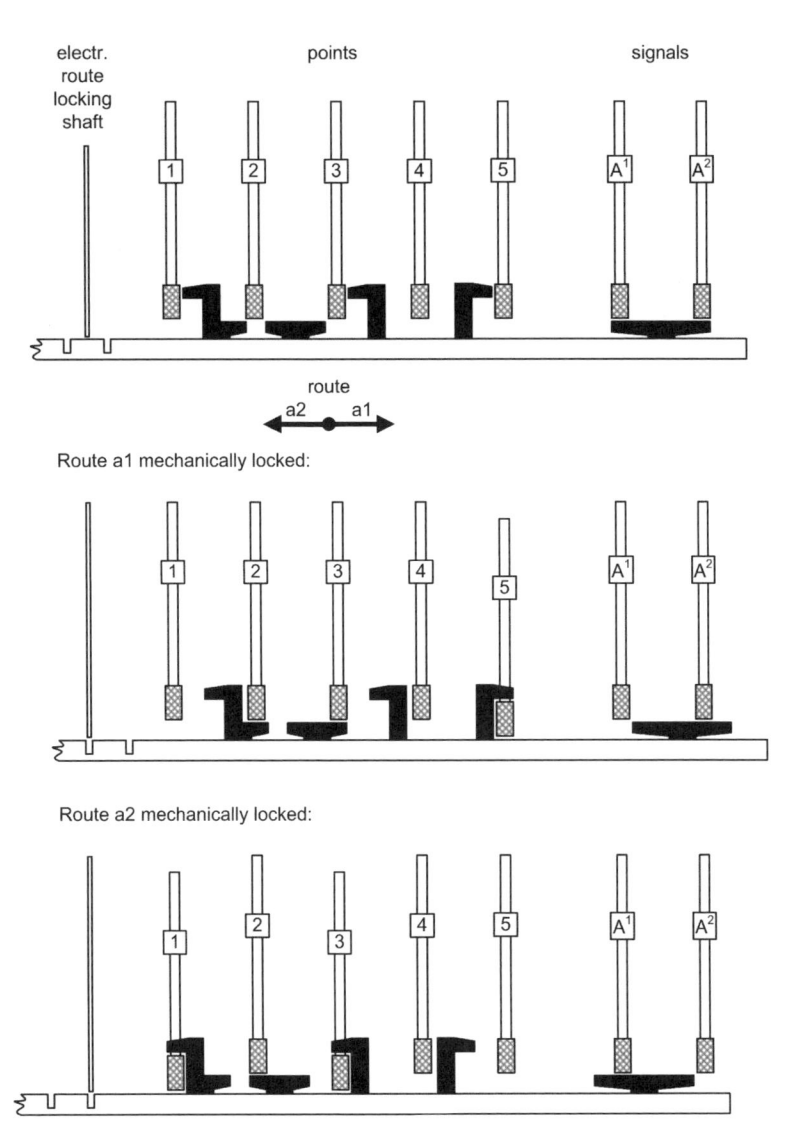

Figure 9.12: Working of mechanical interlocking register type 'Einheit'

route including flank protection and trailing overlap elements by locking pieces mounted on the route drawbar. The route can only be set if the respective point bar is in the proper end position, and then the point bar is held in that position for as long as the route is set. Interlocking between the route drawbar and the signal is carried out in a similar way.

The electrical (irreversible) route locking is carried out by electro-mechanical route locking instruments, which have some similarities with the block instruments used for the manual block of *Siemens & Halske* (chapter 10.3.3.2). By pushing the button on top of the instrument, a shaft enters into a notch of the route drawbar (figure 9.12, left) until this lock is released by a pulse from the train which has cleared the route. To detect the train, a combination of a wheel detector and a short track circuit is used to detect that the train has first entered and then cleared the route release position (chapter 5.2.3.2).

9 Interlocking Machines

9.2.5.5 Interlocking between Different Signal Boxes

Interlocking between different signal boxes can be carried out either between signal boxes belonging to the same station or between signal boxes belonging to different stations, junctions or block stations. Whereas the respective line block system for the latter case is described in chapter 10.3.3.2, the dependences between signal boxes of the same station are described in the following.

There are two reasons for interlocking between different signal boxes of the same stations:
- A leverman must not be able to clear a signal without the permission of the train director.
- For a route which leads into the interlocking area of a signal box other than the one controlling the route entrance signal, conditions in its interlocking area have to be proven. Therefore, permissions have to be exchanged between the signal boxes.

These procedures are solved technically by the same block instruments that are used for the manual block of *Siemens & Halske* (chapter 10.3.3.2). An electrical lock between the route drawbars connects the route drawbar of all signal boxes concerned that way. The route must be built up from the exit to the entry (figure 9.13). That means that this same route must be

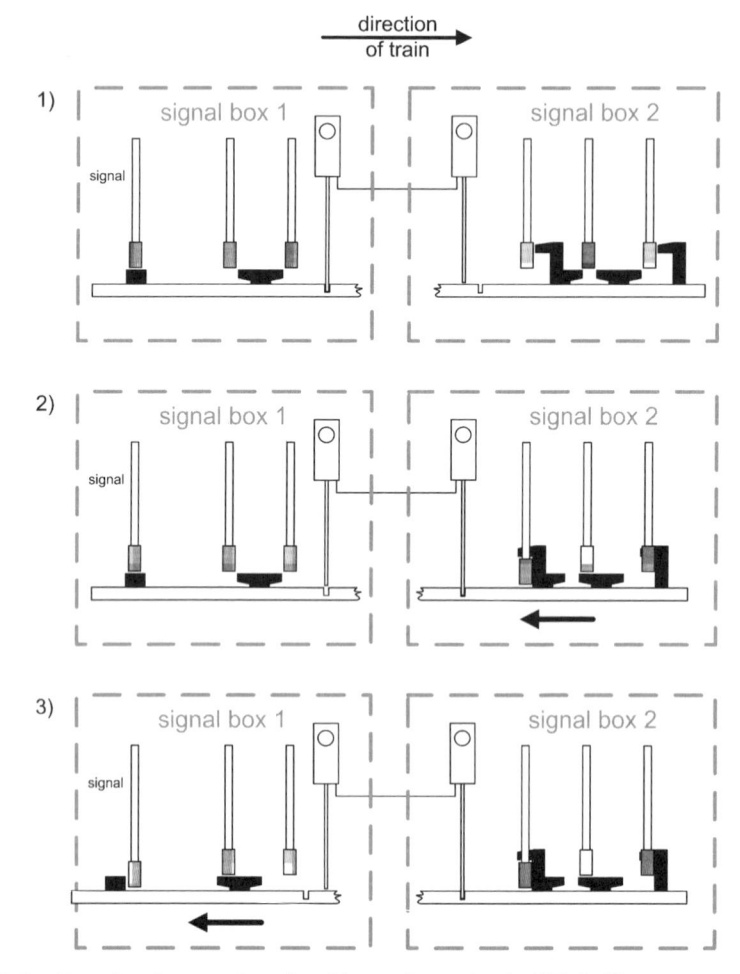

Figure 9.13: Locking of routes over two signal boxes in mechanical 'Einheit'

locked also in the areas controlled by signal boxes beyond. Thereby, the locking of a certain route has always to be initiated by the train director who gives the respective commands to the levermen.

More information on this example can be found in (Maschek/Lehne 2005).

9.3 Relay Interlocking

9.3.1 Historical Development

The transition from mechanical to electrical technology in interlocking was a slow process. As the first steps, the invention of the interlocked block instrument and of the track circuit around 1870 can be considered (chapter 3.1). Both devices provided functions which mechanical technology could not offer satisfactorily, and therefore supplement the mechanical interlocking technology.

The next step, beginning around 1900, was the development of systems with partly electrical and partly mechanical functions. Here, parts of the functions (usually the movement and supervision of field elements) were done electrically, whereas others (usually the interlocking functions) remained mechanical. (chapter 9.5.1)

Between the two World Wars, first so-called 'all-electric' signal boxes were developed as the first relay interlocking systems and installed in various countries. In the two decades after World War II, the relay interlocking was perfected and became the most widely used technology in the world. These still form the majority of existing installations.

A main focus of development in this period was modularisation of the relay installations, which resulted in the replacement of the tabular interlocking logic inherited from mechanical systems by a topological logic. This gives particular advantages in large stations.

9.3.2 System Safety in Relay Interlocking

9.3.2.1 Characteristics and Classification of Signal Relays

Safety in relay interlocking is based on the use of special safe signalling relays. The working principle of a relay is displayed in figure 9.14: An electric current energises a coil, which attracts an armature that opens and closes electrical contacts. In contrast to relays for non-safety purposes inside and outside the railway sector, signal relays need a number of special characteristics:

- They have a high operating threshold, so that the possibility of being erroneously energised by electromagnetic influence must be prevented.
- All contacts belonging to the same armature are rigidly joined together. This means that if the position of one is known, the positions of the others can also be known, with certainty.
- The contacts are made of materials that do not become welded together.
- The contacts of the relays are duplicated. The design of the relay ensures that should one contact become welded then the armature takes up an intermediate position with no contact closed.
- When the relay changes its position, all contacts which will be opened by this switching have to be open before any contact closes.

9 Interlocking Machines

Based on these requirements, the relays used in railway signalling and automation can be divided into three classes according to their safety characteristics:

1. Signal relays of **type N** (non-controlled), also called first-class-relays, can be used in safety related circuitries without monitoring of their dropping down. In particular, safe dropping down of the relay can be assumed due to the high mass of the anchor or by a sufficiently strong spring force. Contacts use materials (e.g. silver and carbon) which cannot form a weld which is strong enough to resist gravity or the spring pressure. Other characteristics of type N relays are high inductance, residual pin (to prevent complete closure of the air gap and therefore saturation of the magnetic circuit) and protective shroud (for protection against environmental impact).
2. Signal relays of **type C** (controlled), also called second-class-relays, are also suitable for safety related purposes, but their operation must be checked by circuitry. In particular, failure of the relay to drop down has to be considered, although the probability is much lower than for not picking up.
3. **Non-signal**-relays, also called **third-class**-relays, must not be used in safety related circuits.

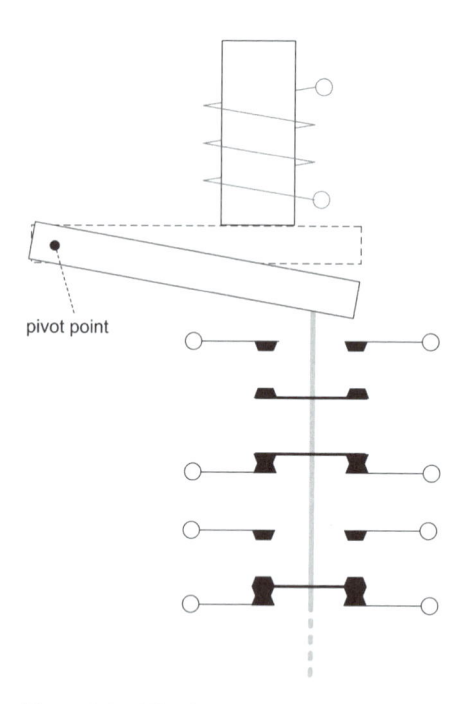

Figure 9.14: Working principle of a relay

The terms 'type N' and 'type C' are current official UIC terms, whereas the terms 'first/second/third class' are older terms which are still used in many countries.

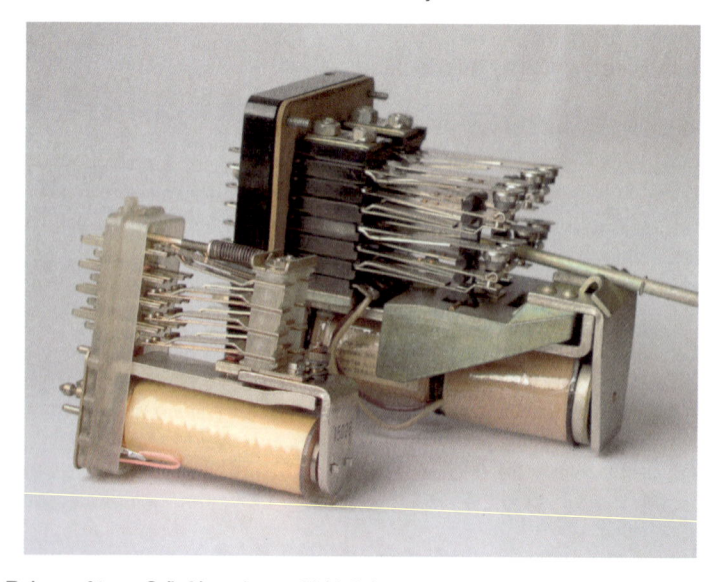

Figure 9.15: Relays of type C (left) and type N (right)

Third-class-relays are used for non safety-related automation functions in railways throughout the world. Their advantages are low costs and the usually faster acting of the relays. For safety functions, railways have different preferences regarding the use of type N or type C relays. Whereas type N relays are very expensive to manufacture and larger in size (figure 9.15), with the use of type C relays the problem is the much more complex circuitry. Type N relays are applied predominately in Western Europe, the USA and Russia, whereas type C relays are mainly used in Central European countries. Several functions such as locking the route irreversibly upon the approach of the train with the signal already being clear before (chapter 4.3.8) are only possible with type N relays.

Where type N relays are used, often safety functions and non-safety functions are separated in different blocks for economic reasons (example: chapter 9.3.6). Type N relays are only used for the safety functions in these cases, and third-class relays for the non-safety functions.

Regarding the stable positions, relays can take the following forms:
- Mono-stable relays have one stable position: When current is cut off, the relay always drops down. They are used for most applications.
- Bi-stable relays have two stable positions and are therefore applicable for storing a binary status of the interlocking system safely. Application examples are the locked status of a route and the required position of a set of points.

For safety reasons, the correct choice of contacts used in a circuit is important. Contacts can be divided into:
- **Closers**, also called **front contacts**: They are open when the relay is de-energized.
- **Openers**, also called **back contacts**: They are closed when the relay is de-energized.
- **Two way contacts**: In each end position, they close another circuit.

9.3.2.2 Forms of Mono-Stable Relays

The simplest form of mono-stable relays is the simple signalling relay as shown in figure 9.14.

Special forms of mono-stable relays are (IRSE 1999):
- **AC immune relays**: The relay operates with DC, but safely not with AC. This makes the relay immune against AC electromagnetic influences.
- **Biased relays**: The relay only operates if current passes through the coil in one direction, but not in the opposite direction.
- Relays fitted with contacts for **high voltages** and current. These are used for purposes such as control of point machines.
- **Double wound relays**: These contain two independent coils, either of which being able to operate the relay. These coils can have identical or different parameters (such as resistance, number of windings). They can be series-connected, parallel-connected or connected independently in different electric circuits. Coils with low resistance are applied to connect relays in series.
- **Slow pick up relays**: These relays are magnetically designed in a way that they do not pick up immediately when energised, but with a certain time delay.
- **Slow release relays**: These relays are magnetically designed in a way that they do not drop down immediately upon being de-energised, but with a certain time delay.
- **Vane relays**: These relays work similarly to an AC motor and therefore provide frequency and phase selectivity. Their main field of application is the track relay of track circuits. Therefore, this type of relay is described in detail in chapter 5.3.5.

9.3.2.3 Forms of Bi-Stable Relays

There are different technical forms of relays with two stable positions which remain in the position last set. These relays are used differently in the railways for safety and non-safety purposes:

- A **magnetic latching relay** incorporates a permanent magnet. The relay is picked up by energising the coil in the usual way. It is then held in the picked-up state by a permanent magnet, even when power is removed from the coil. The relay drops down when the coil is energised with reverse polarity (and remains dropped down when power is removed). Alternatively the relay may have a second coil for this purpose.
- A **toggle relay** (figure 9.16) contains two magnetic coils and one armature. When either coil is energised, the armature is attracted to that coil. The force of a spring then holds it in that position until such time as the other coil is energised.
- A **mechanically latching relay** (figure 9.17) has two coils and two armatures. The armatures interfere mechanically so that it is impossible for both to be dropped down at the same time. Energising one coil causes the relay to go into the respective position (or remain in that position). When power is removed, the relay stays in the position last set.

Figure 9.16: Toggle relay

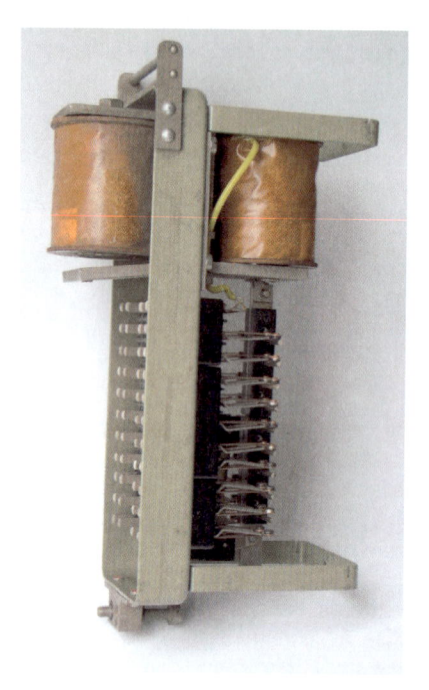

Figure 9.17: Mechanically latching relay

9.3.2.4 Combined Relays

Combined relays (figure 9.18) are relays with mono-stable and bi-stable systems in the same relay. Figure 9.18 shows an example. In this example, the mono-stable armature is energised by current of any polarity proceeding through the winding. The bi-stable armature switches if the polarity of the current is reverse to that of the last time it was energised. After being de-energised, the mono-stable system drops down, whereas the bi-stable one stays in its last position. An application example is the supervision relay in Russian point machine control scheme (chapter 6.6.2.2).

9.3 Relay Interlocking

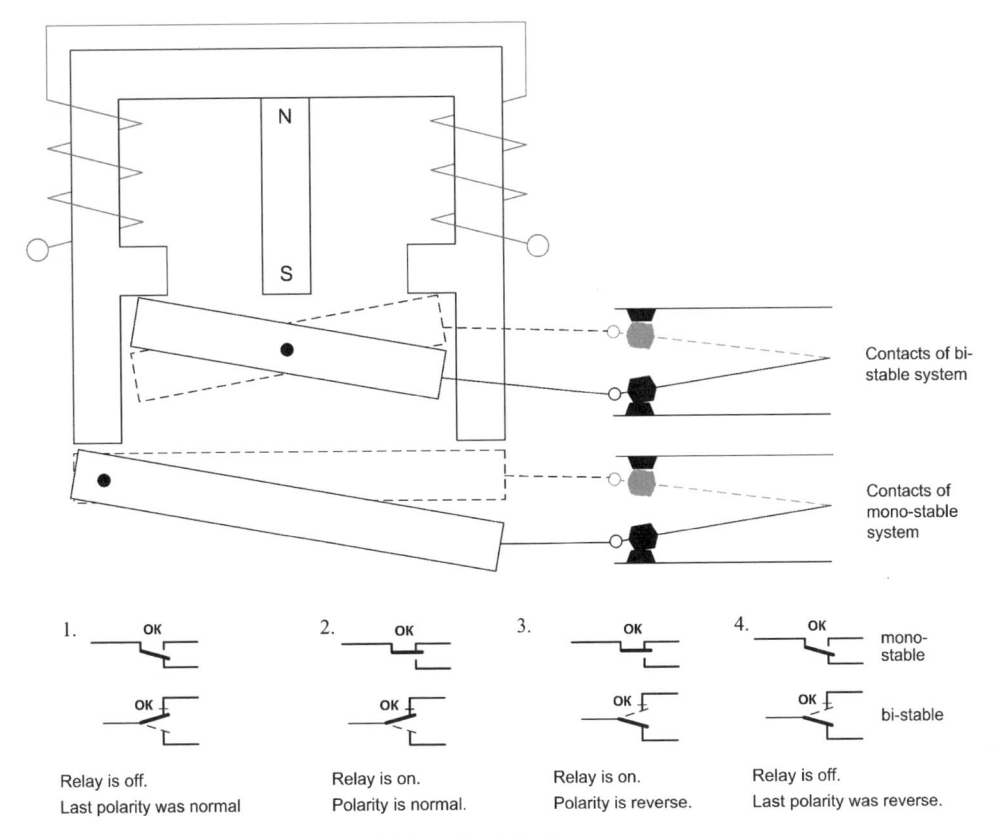

Figure 9.18: Combined relay (example) (graphic: *Lykov*)

9.3.3 Design of Relay Interlocking Systems

9.3.3.1 Structure

In relay interlocking, the interlocking, element control and operation control functions are realised by electrical circuits in relay technology. The relay racks (figure 9.19) are equipped with plug boards, in which the relays are installed. The wiring is held in the plug board, so replacing a relay does not require a change in the wiring.

Erroneous placing of a wrong relay to a wrong location could cause danger for railway operation. To prevent this, each type of relay and the matching plug board are usually equipped with a pin code which only permits the correct relay to be placed in the correct location.

There are two basic types of relay interlocking systems:
- The **free-wired** type, following the tabular principle of route formation (chapter 4.3.9): All relays are wired on site.
- The **topological block** type, following the topological principle of route formation (chapter 4.3.9): Standardised relay sets for each topological element are assembled and ready wired in the factory, and on site they are connected by standardised cables according to the track layout.

With the free-wired type, the wiring and checking work on site requires much effort. In the topological block type, the development of the interlocking type with all relay sets including those needed for certain central functions is relatively complex, but the wiring and checking on site as well as later changes are relatively simple. Therefore, in application (figure 4.37):

- the free-wired type is most suitable for small interlocking areas,
- the topological block type is most suitable for large interlocking areas.

9.3.3.2 Operation Functions

The operation is carried out by buttons which are to be pushed, pulled or turned. Whereas in Germany and North America, pushbuttons are preferred (hence the West-German name 'pushbutton interlocking', in Britain, France and Russia rotating switches are also used, especially for movable track elements. Pushbuttons are usually active as long as they are pushed, requiring the related switching processes to be initiated in this short time. Rotating switches, in contrast, have a medium normal position and one or two positions to the left and the right.

Figure 9.19: Relay racks

To illustrate the difference, a good example is the buttons/switches for point setting. With rotating switches, the medium position means that the points are able to be switched by automatic point setting, whereas the left and the right position immobilise the points in that position. Without automatic point setting, the switch has only two positions (figure 9.20). In push- and pullbutton technology, points are set to both directions by the same operation. The mobilisation and immobilisation of points for automatic point setting is often achieved by additional master (group) point buttons. There are often several such master buttons. Each is assigned a certain function and serves that same function for all related track elements.

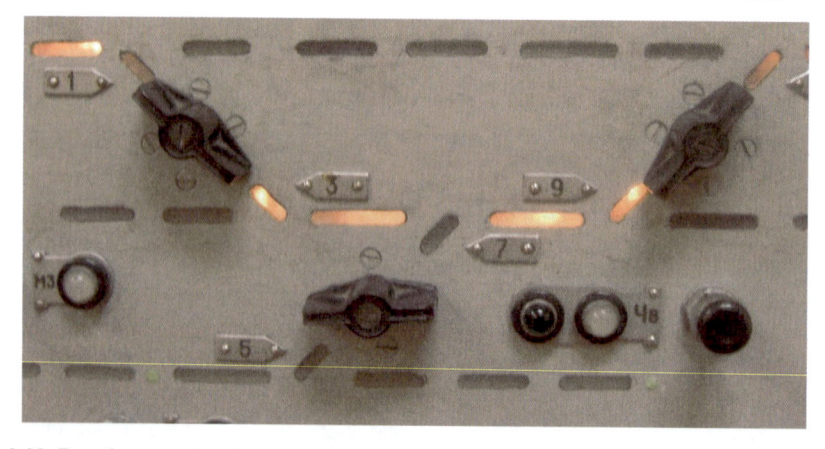

Figure 9.20: Rotating switches for points in relay interlocking (Russian example)

Where automatic point setting is applied, the most frequent operation to set a route is the entrance-exit (NX) operation. This means that a button/switch at the entrance location and another at the exit location have to be applied simultaneously or consecutively.

Without automatic point setting, each movable track element has to be brought to its position by a separate operation. To lock the route, there is either a separate button for each route, or one button serves for several routes with the same entrance or exit, and the route is selected by current point positions.

A frequent requirement is that at least two buttons always have to be applied for one command input (e.g. point and point master button) to avoid erroneous actions. The order of operations for a certain input command can be defined differently. These two buttons can be operated simultaneously (mainly in German influenced interlocking, chapter 4.1.1) or successively (mainly in British and North American influenced interlocking). Special functions such as selective overlaps, if provided, can be solved differently, e.g. by a separate exit button for each overlap.

Relay interlockings can be controlled locally or remotely. Due to the possibility of remote control of large areas from a CTC and the variability of operation surfaces, the operation control level of chapter 9.1 in some situations is not considered as an integral part of the interlocking, but as an external system. The operation control can be vital or non-vital, with the vital being the more frequent case. If a safety critical operation is to be recorded, this can be solved technically by sealed buttons and mechanical counters connected to the respective button.

9.3.3.3 Display Functions

Various status information is displayed to the signaller by control lamps which can be lit in different colours or unlit. Examples for such information are:
- occupation of track sections
- position of points
- status of routes
- status of line block
- status of central automation functions such as automatic point setting (on/off)

In early relay interlockings, control lamps and operation buttons were located without geographical reference. But soon those control lamps and buttons which are related to certain elements in the track layout were widely allocated topologically. Only such buttons and control lamps which refer to general statuses of the whole interlocking area are not located in the topology, but separately. This topological allocation facilitates the signaller's overview (figure 9.21).

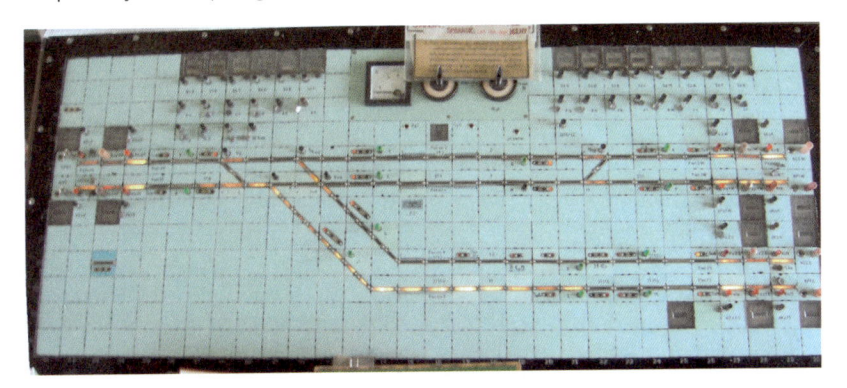

Figure 9.21: Topological operation table in relay interlocking (example from Poland)

9 Interlocking Machines

In large stations or in CTC, particularly where the length of the signaller's arms is not sufficient to push two buttons in different parts of the station simultaneously, large panoramic panels are often applied which serve for display purposes only. For command input, a simplified table with buttons, often without topological assignment, is installed at the signaller's workplace. Different workplaces can use the same panoramic panel.

9.3.4 Example: SGE 1958 (Britain)

9.3.4.1 Introduction

SGE 1958 is a frequently used British relay interlocking system. It is based on the tabular principle (free-wired type). SGE is the name of the manufacturer (*Siemens-General Electric Signal Company of London*), which was later absorbed by *GEC*. SGE 1958 is a typical example of British relay interlocking practice and was widely used on British railways until the advent of SSI (chapter 9.4.4) in the 1980s.

9.3.4.2 Operation

Routes are normally set by sequential entrance-exit-operation in the track layout panel: First a key is turned to select the route entrance and then a button pushed for the exit. If the entrance key is turned up, the train route will be selected, if it is turned down the shunting route.

To release the route, the key has to be turned back manually to its normal position. If a route which has already been approached is released before the train has passed it completely, a time delay applies (chapter 4.3.8.3).

Alternatively, points can be set individually. For this, three-position point keys outside the track layout are used: The key in centre position means that points can be set by entrance-exit operation. In left or right position the points are commanded to move to this position (providing they are free to do so) and locked there.

9.3.4.3 Route Interlocking Circuitry

The circuitry for route commanding is derived directly from the interlocking matrix. This is explained in the example of figure 9.22. Only the circuit for route No. 6, starting at signal 6 and reaching up to signal 4 is shown in figure 9.22. Similar circuits exist for each route.

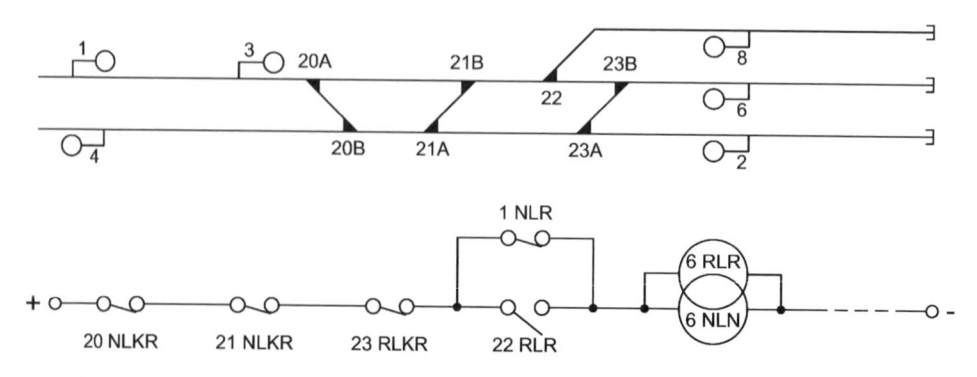

Figure 9.22: Example: Track layout and simplified route interlocking circuit for one route in SGE 1958

The relays NLKR (normal) and RLKR (reverse) of the point interlocking circuitry are picked up if the related points are in their respective positions or are able to move to that position. As can be read from the circuit in figure 22, route 6 can only be selected if this refers to points 20 (normal), 21 (normal) and 23 (reverse).

Besides, there is a conditional lock (compare chapter 4.2.5) between signals 1 and 6 and points 22: For reason of overlap beyond signal 3, signal 6 can only be opened if either signal 1 is closed (relay 1 NLR picked up) or points 22 are locked reverse (relay 22 RLR picked up).

The relay 6 RLR/NLR is a special form of electromagnetic latched relay (chapter 9.3.2.3) with two coils to be energised. It is bi-stable to hold the route information in case of loss of power. More information on SGE 1958 can be found in (Goldsbrough 1961).

9.3.5 Example: SpDrS60 (Germany)

9.3.5.1 Introduction

SpDrS60, manufactured by Siemens and developed around 1960, is the most frequently used relay interlocking type in Germany. Besides Germany, adapted versions are also applied in several other countries. SpDrS60 is a typical application example for the topological (geographical) principle (chapter 4.3.9) in relay interlocking: Relays belonging to the same topological element are grouped to a relay set. These relay sets are produced and tested in the factory and connected on site by standardised cables.

9.3.5.2 Relay Sets

There are three groups of relay sets (figure 9.23):
- **Topological sets** each represent one element of the track layout and are connected to each other by path cables representing the tracks between the elements.
- **Signal sets** are connected in a line to the route entrance/exit sets, representing the different indicators of a signal.
- **Central switching sets** are responsible for specific central functions in the whole interlocking area.

The following **topological sets** are the most important:
- **Point sets** include all relays for point moving, route locking, flank protection, supervision, occupancy detection, release and display functions related to a particular set of points. A slip crossing is represented by two point sets and a derailer by a point set without a track path cable at the branching end.

Figure 9.23: Relay set of SpDrS60

9 Interlocking Machines

- **Crossing sets** have similar functions as point sets without the moving and supervision.
- **Train entrance/exit sets** mark the entrance and exit positions of train routes and include several route-related functions.
- **Shunting entrance/exit sets** have the same functions for entrance/exit positions of shunting routes only.
- **Overlap sets** mark the end of overlaps.
- **Supervisory sets for intermediate sections** are applied whenever a track section has to be separately proved clear and this task cannot be assigned to any other topological element. They are needed rather seldom.
- Sets for different forms of **line block**.

Signal sets represent certain indicators of a signal. Their arrangement is closely related to the German H/V signal system (chapter 7.5.1). There are signal sets for main signals, distant signals, speed and route main and distant indicators, contraflow indicators, marker lights and others. All signal sets related to the same signal location are connected by signal path cables in a line, beginning at the respective train entrance/exit set. Shunting routes require no signal sets, as signal control and supervision is included in the shunting entrance/exit sets.

Central switching sets are responsible for central switching functions. They are present once in each interlocking area and contribute to many switching processes in the whole interlocking area. Tasks of the central switching sets are for example:

- Processing of information from master buttons for points, signals, routes, line block and others.
- Calculation of the speed limit for a route.
- Definition of priority when different track paths are possible between a route entrance and a route exit position selected (chapter 4.3.4).
- Storing of routes for automatic route setting functions (chapter 4.3.11).

Figure 9.24 shows a simple example of a set connection map.

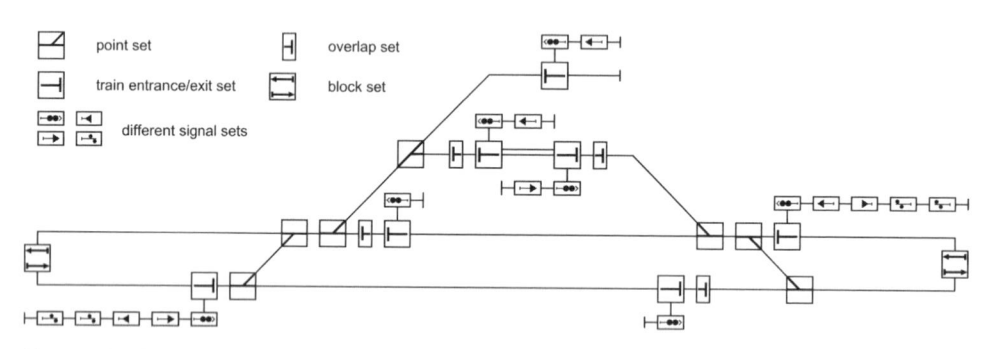

Figure 9.24: Set connection map of SpDrS60 (example)

9.3.5.3 Structure, Positioning and Wiring of the Relay Sets

Each relay set consists of several relays. Resistors, capacitors, inductors and diodes are added where necessary. The type C relay **K50** is used in the following versions:

- simple relay K50 (for most applications, figure 9.25)
- bi-stable relay K50
- relay K50 with contacts for heavy currents

9.3 Relay Interlocking

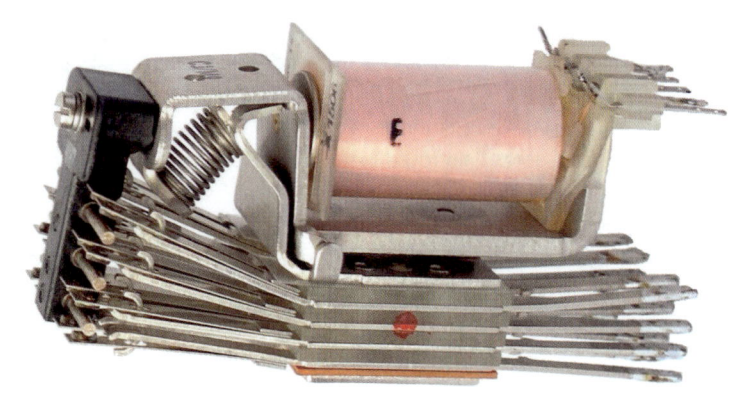

Figure 9.25: Relay K50

The relay sets are mounted in relay racks and are connected electrically by standardised cables. The following are the main types of cables:
- Track path cables with 20 wires each connect topological sets according to the topology.
- Loop cables connect all relay sets of the same type either in sequential or in parallel form with a central switching set.
- Cables for power supply.

In addition some relay sets take a configuration plug, which interconnects contacts within the set to give specific characteristics.

9.3.5.4 Procedures for Route Setting and Release

In the following, the procedures for route setting and release are described briefly and simplified at a functional level.

The first step after pushing the entrance and exit buttons simultaneously is searching for a route and testing it for availability. This process is described in chapter 9.3.5.5.

The next step is to **set the points** belonging to the running path and the overlap (without flank protection elements). For reason of power supply, the points have to be set sequentially. This is controlled by a sequential ring loop, with each point set retarding the current flow to the next set by a short time.

After this, the **points and crossings are locked** in the required position by switching the point locking relay in the negative relay set.

Then the **point and crossing sets are route locked**. In each point or crossing set the route locking relay switches. This is necessary to release each route element individually later, behind the train.

The next step is **flank protection search and supervision**. Each point or crossing relay set sends a flank protection search current to each path cable which is not in the running path.

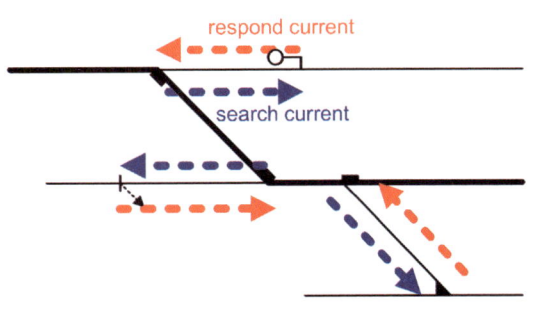

Figure 9.26: Flank protection search in SpDrS60

273

9 Interlocking Machines

When the search current reaches an element which can give flank protection, this element is set to the protecting position and locked and a respond current sent back to the searching element. (figure 9.26)

Now track **elements are proved clear** and if all elements are clear the **route is locked** irreversibly in the entrance/exit sets. This route locking function cannot be cancelled manually without special (registered) command (4.3.8.4).

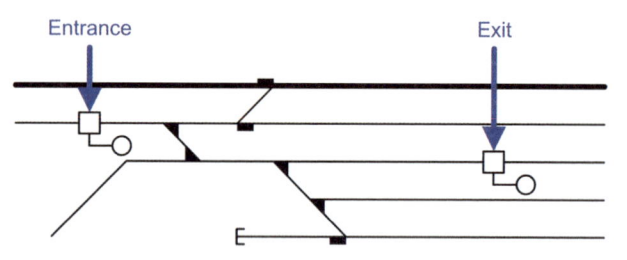

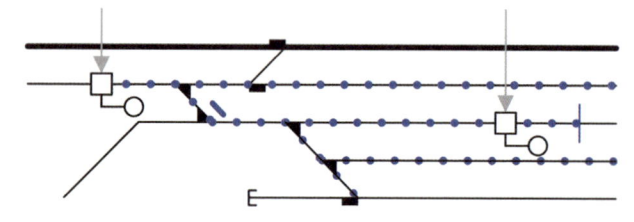

Now **signal selection** is carried out. Each point and overlap set in the route sends the information about its particular permitted speed to the signal selection set, a central switching set. This set calculates the minimum and gives it to the signal sets as permitted speed for the route. Then the main signal is cleared and after this the distant signal.

9.3.5.5 Route Search and Availability Testing

In this section, the processes of route search and availability testing are described in more detail, as they are a good means of understanding the topological principle.

The signaller pushes the entrance and exit buttons simultaneously. From the entrance set, a search current spreads into the related direction (figure 9.27). At this moment, the interlocking cannot distinguish which is the entrance and which the exit button, so the search current also spreads from the exit set but will not find a response.

Figure 9.28 shows the route search and availability testing circuitry in a point set. In each point set on the path the relay S1 or S2 picks up and current is newly fed in to continue along the route path. This is necessary to limit the electrical resistance in each circuit. In each point set passed by the search current, availability is tested as follows:

- If facing points are already locked in one position, e.g. as flank protection element for another route, the contact of relay L in figure 9.28 is open and search current can continue only in the available direction.
- If facing points are not locked for any route, the contact of relay L is closed and the search current spreads into both directions.
- If trailing points are locked in the opposite position, the contact of relay L is open and the search current blocked.

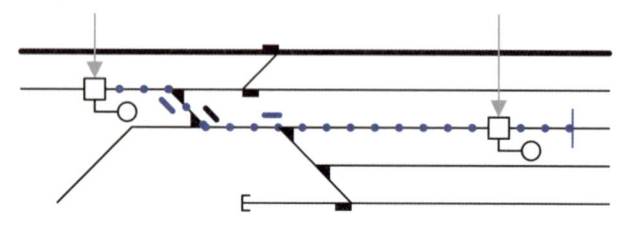

Figure 9.27: Route search and availability testing in SpDrS60

9.3 Relay Interlocking

- If trailing points are not locked or locked in the proper position, the search current can pass. Meanwhile, the points are marked in the required position by relay M+ or M- picking up. Relay M+ or M- remains picked up during the whole route search and availability testing process.

The entrance/exit and overlap sets laying in the path in opposite direction are responsible for the interlocking of opposing movements: If they are already being used by a route in the opposite direction, the search current is blocked and cannot continue along the path.

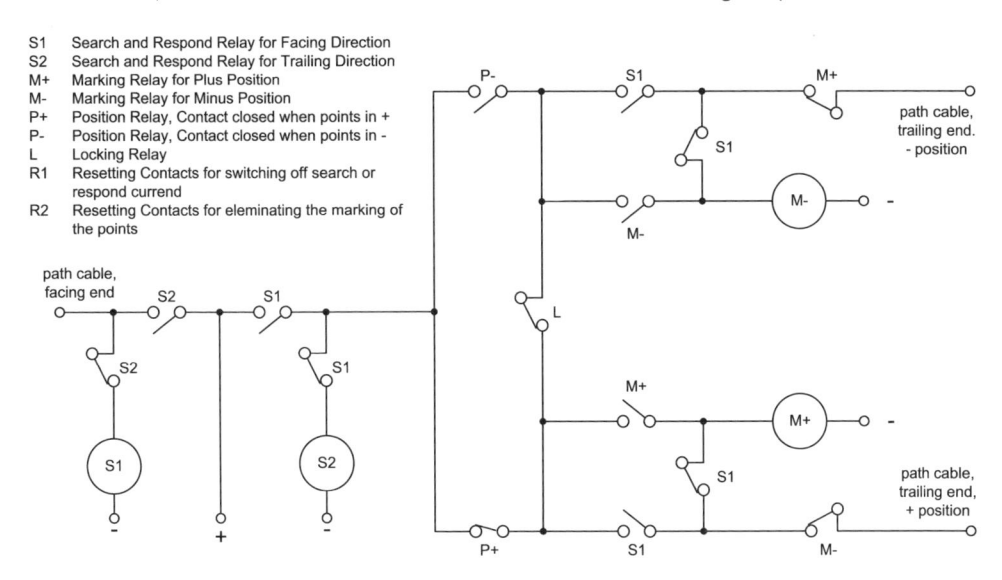

S1	Search and Respond Relay for Facing Direction
S2	Search and Respond Relay for Trailing Direction
M+	Marking Relay for Plus Position
M-	Marking Relay for Minus Position
P+	Position Relay, Contact closed when points in +
P-	Position Relay, Contact closed when points in -
L	Locking Relay
R1	Resetting Contacts for switching off search or respond currend
R2	Resetting Contacts for eliminating the marking of the points

Figure 9.28: Route search and availability testing circuitry in a point set in SpDrS60 (simplified and with adapted names and symbols); Points are in + position

When the search current reaches the exit set (marked by the related route button pushed), the overlap is also marked. At the overlap set, the current turns and flows back as respond current (figure 9.27). The respond current passes facing points of the route (which are reached by the respond current from the trailing end) in the same way as trailing points were passed by the search current before. Therefore, all facing points in the running path and overlap are now marked in the proper position by picking the relay M+ or M- up (figure 9.28).

As the trailing points have already been marked by the search current before, the respond current finds its way back to the entrance set. An unambiguous path is now defined by the marking of the points (M+/M- relays) and has been checked for availability, but track occupancy has not yet been checked.

9.3.5.6 Modifications of SpDrS60

In the 1970s, SpDrS600 as an improvement of SpDrS60 was developed, but was applied only to a minor extent. However, the functional logic of SpDrS600 formed the basis for the development of Simis electronic interlocking (chapter 9.4.6).

More information on SpDrS60 can be found in (Siemens 1978) and (Schmitz 1962).

9.3.6 Example: UBRI (Russia)

9.3.6.1 Introduction

In Russia, the most widespread interlocking system is **U**nit-**B**lock **R**elay **I**nterlocking (UBRI, figure 9.29), which also uses topological logic. Various modifications have been made since 1960, and it now controls more than 100,000 sets of points in the Russian railway network. The development of UBRI has involved experiences of engineering, implementation and operation of previous relay interlocking systems.

Figure 9.29: Unit-block relay interlocking

UBRI, likewise SpDrS60, is a highly modular interlocking system with the advantages of significantly reduced efforts for engineering, installation and testing.

9.3.6.2 Relay Sets

Each relay set of the system includes relays that realise functions of the corresponding track element: points, track section or signals. The connections between the relay sets are wired in accordance with the topology of the track layout (figure 9.31 a).

A special feature of UBRI is that separate relay sets are provided for the non-vital selection functions and the vital interlocking functions, both forming separate systems of topological circuitry. By this, faster acting of the non-vital functions can be achieved. The main functions of the non-vital route selection group of relays are:

– registration of operator's actions for route setting;
– determination of the route direction (odd/even) and category (train/shunting);
– determination of the path and commanding of point setting;
– checking of point setting along the path of the selected route.

The main functions of the interlocking group of relays are:

– checking of safety conditions along the path of the selected route;
– locking of the route;
– clearing of the signal with the checking of safety conditions;
– monitoring of a train passing via the route, releasing the route;
– manual releasing of the route.

There are 16 types of relay sets of the interlocking group. These relay sets are linked with each other by 8-wire connections according to the track layout. This group uses type N (first class) relays – relays of NMŠ type in old systems and RÉL type (figure 9.30) in newer systems.

9.3 Relay Interlocking

Figure 9.30: Relay RĖL

In the route selection group, there are seven types of relay sets. These are linked with each other by 4-wire connections according to the track layout. This group of relays consists of KDRŠ relays (fast acting coded relays), which are non-safety relays.

Figure 9.31 b shows an example of an UBRI system.

- Interlocking group
- Routes selection group

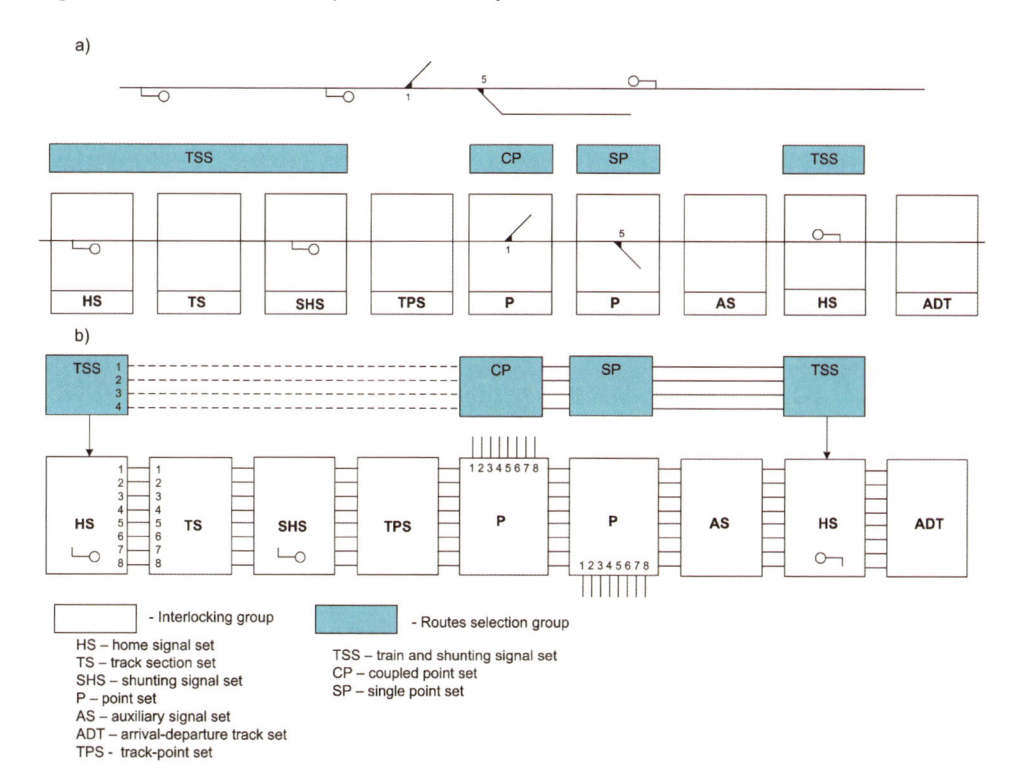

HS – home signal set
TS – track section set
SHS – shunting signal set
P – point set
AS – auxiliary signal set
ADT – arrival-departure track set
TPS - track-point set

TSS – train and shunting signal set
CP – coupled point set
SP – single point set

Figure 9.31: Partly relay set connection map of UBRI

277

9 Interlocking Machines

9.3.6.3 Non-Vital Route Selection Group

Route-selection group of relays consists of the following functional blocks (relay sets):
- **Signal sets** define entrances and exits of routes and can be subdivided in accordance with signal types (train/shunting) and function (entry signal, exit signal, intermediate signal).
- **Direction sets** define type (train/shunt) and direction of a route.
- **Point sets** define the required point positions in the route and command switching if the actual position does not conform with that required. Point sets can be subdivided according to the type of points (single set of points or two combined sets of points).

Each of the four wires connecting the relay sets can be assigned to a particular function:
1. The first wire (figure 9.31 b) connects these relays which are intended for fixation of the sequence of control button pressings. Any route is set up by sequential pushing of entrance and exit buttons. If the route should be set up along an alternative path, buttons for these deflection points are pushed after pushing the entrance button. Pushing the entrance button fixes the type of route (i.e. train or shunting route) and the direction of the movement. After pushing the exit button, the route is selected.
2. The second wire provides automatic joining of the elementary routes into a composite route.
3. The third wire provides task formation for point operation under control of the entrance relay set on the whole length of the composite route.
4. The fourth wire serves for checking the correct positions of points after their operation. After all points have been proved to be in their positions, the relays of the interlocking group become active.

Figure 9.32 shows the circuitry for the commanding of points. Each simple point set contains two relays for point commanding: PCR to command the points to plus position and MCR to command them to minus position. Accordingly, each relay set for coupled points contains three such relays: 1PCR, 2PCR and MCR. The circuit is formed in accordance with the track layout between the entrance and the exit. In case the track layout permits more than one path

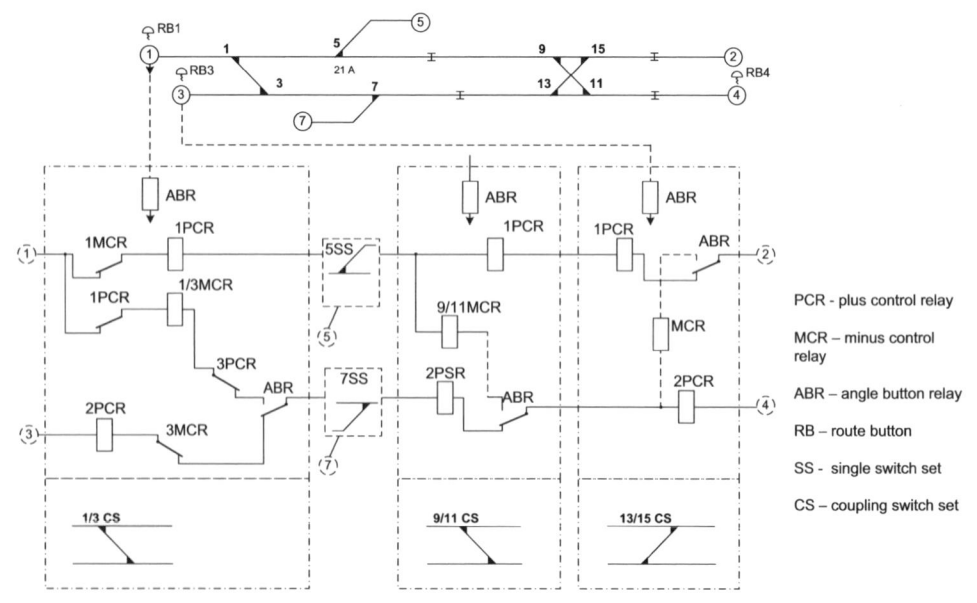

Figure 9.32: Circuitry for point commanding in UBRI

between entrance and exit, the preference variant of route (e.g. from entrance 1 via crossover 1/3 reverse instead of crossover 9/11 reverse in figure 9.32) is selected by contacts of the angle button relay ABR. When pushing the route entrance button, a corresponding angle relay in the point set picks up and electrical circuits for possible routes are formed. Then one out of the possible routes is selected by pushing the exit button.

9.3.6.4 Vital Interlocking Group

Vital **interlocking group of relays** consists of following main blocks (sets):
- **Track section sets** check the occupancy of main tracks and exclude opposing routes.
- **Track point sets** check the occupancy of a track section with points inside it and fulfil point locking.
- **Signal sets** control the signals and determine the signal aspect. These sets can be subdivided in accordance with the type of signal (entry signal, exit signal, intermediate signal).
- **Point sets** provide the checking of point position and are arranged according to the topology of a railway station.
- **Start sets** control the points and indicate the point position.

The relay sets of the interlocking group are connected by cables with eight wires in accordance with the track layout. The assignment of the wires to certain safety functions is as follows:

1. The first wire connects the section checking relays of route under the stipulation that all safety requirements are fulfilled. Such requirements include: proper position of the points, clear track sections, proper control of flank protection and oversized sections and absence of conflicting routes. The picking up of the relays locks the route sections and prepares the circuits of train movement control (wires 3, 4 and 5).
2. The second wire is used for clearing the signal by picking up the signal relay by means of corresponding control relays of the route sections. Simultaneously, locking of the route sections is checked to make the selection of conflicting routes impossible. Additionally, the fifth wire is used for selection of the signal aspects. Monitoring of signal lamps allows the detection of filament fusing, and switching the signal to protected state.

   The signal relay for a train signal is dropped down and the signal closed five seconds after a train entry into the track circuit in advance of the signal, or when any supervised condition is violated.

   The signal relay for a shunting signal is dropped down when the shunting unit has completely passed the signal.
3. Third, fourth and fifth wires are used by route relays for supervising train movement over the route sections and for releasing the sections after the train has passed.

   Each route section has two route relays; the first registers the train entry into the given section under the precondition that the previous section has been occupied before. The second route relay registers the clearing of the given section and entry of the train into the following section. If these two conditions are fulfilled, the section is released.

   As previously noted, the fifth wire is used for two different purposes:
   – when the route is being set - for aspect selection of the signal in accordance with the aspect of the signal ahead,
   – when the route is released – for switching on the second route relay of the section.
4. The sixth wire is used for the unlocking of sections when an unused route is manually cancelled. In this case, track occupancy is checked and protective time delays are taken into account in accordance with the occupation of the approach section. This approach section represents:
   – for an entry route – the track section between the distant and the home signals,

- for an exit route – the departure track,
- for the exit part of a through route – the section between home and exit signals,
- for a shunting route – the section immediately in rear of the signal.
5. The seventh and eighth wires are used for an indication of the state of the corresponding track sections (clear, occupied) on the operation table.

In addition to the connections with neighbouring relay sets according to the topology, the relay sets have standardised connections with common equipment (comparable with the central switching sets in SpDrS60, chapter 9.3.5.2) and individual control circuits.

One purpose of these connections is the emergency release of route sections in the case when they did not release automatically after the train passed. For this purpose, particular relays are provided in the relay sets of track-point and track sections. This action is registered and a time delay of three minutes applies.

9.3.6.5 Modifications of Unit-Block Relay Interlocking System

From the time of development (1960), UBRI was subject to various modifications.
1. Two variants of electronic blocks were designed in place of route selection relay sets:
   - with the use of transistor based logical elements OR – NOT (Rezekne station, 1967 year)
   - with the use of transistors and thyristors (Obukovo station, 1969 year).
2. An improved type of relay interlocking was developed in the 1970s to provide simplification of design processes of non-standard circuitries and new relay sets. As the result, in the 1990s, the system of industrially mounted relay interlocking was developed. This system is characterised by the use of new types of small-sized relays and an increased number of relay sets and wires in the connecting cables (up to 31) with easy handling and connections. New functions of this industrially mounted electrical interlocking include:
   - the possibility of releasing a route not only one section after the other, but alternatively as a whole,
   - the possibility of point locking along the path in the case of using auxiliary signal,
   - a new algorithm of train movement control,
   - the possibility of queuing of routes which cannot be set immediately due to route conflict (chapter 4.3.11).

But the tendency for broadening of functional possibilities has led to an overcomplicated circuitry, excessive expenditure for relays (about 100 relays for one set of points), and high costs of the interlocking systems. As a result, electronic systems have gained in importance.

Various modifications of relay-electronic systems have been used on Russian railways since the end of the 1990s. Any of them represents a combination of the vital interlocking group of Unit-Block Relay Interlocking (safety functions) and the electronic non-vital operational functions. This affected the workplace of a local or central signaller.

9.4 Electronic Interlocking

9.4.1 Historical Development

The first electronic interlocking systems were applied in the 1980s and have since been further developed. Various systems in different versions are offered by different manufacturers, each being applied in one particular or in different countries. Usually, each manufacturer offers various systems with different grade of complexity, size of the interlocking area and adjustment to

9.4 Electronic Interlocking

special national requirements etc. However, there are two standardised forms in national railways: SSI in Britain and SMILE in Japan.

The tendency in most advanced developments is to provide various functions for maximum safety also in degraded mode operation in case of element failure, which was not possible to this extent in relay interlocking systems.

In many application cases, components of different manufacturers and types are combined. For example, an operation control system of one manufacturer is used to control a relay or an electronic interlocking system of another manufacturer.

9.4.2 System Safety in Electronic Interlocking

In electronic interlocking, the interlocking functions are mainly defined in programmed software. Microelectronic technology, in comparison with mechanical and relay technology, has several unfavourable safety characteristics:
- Due to low voltages used, electronic components are highly sensitive to external influence.
- Electronic components, in contrast to relays, have no preferred direction of failure, but the way they fail cannot be predicted. This makes inherent fail-safe design impossible.
- Electronic components can change their characteristics over time.
- Due to high complexity, the prevention of systematic errors in manufacturing is difficult.
- Due to high complexity, checking processes and changes in the equipment are difficult.

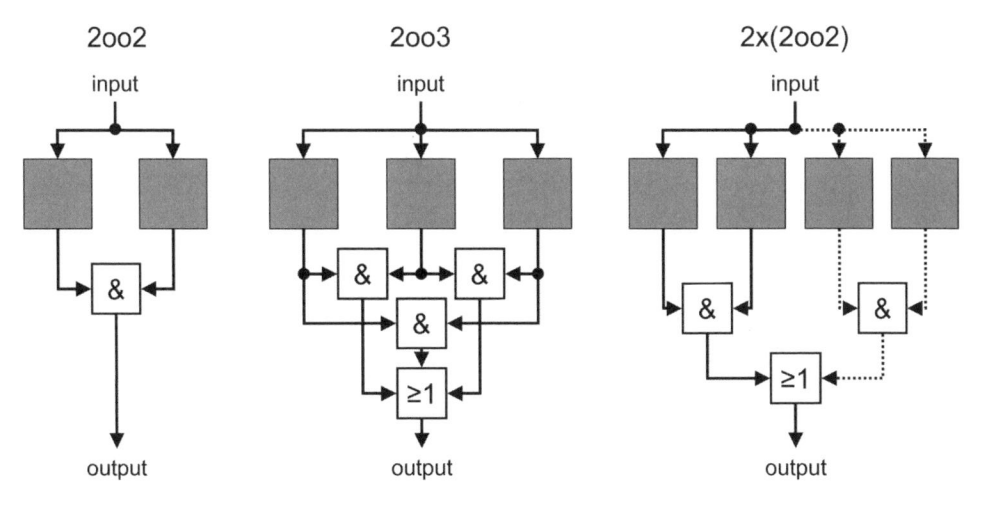

Figure 9.33: Redundancy in electronic systems

To overcome some of these deficits, hardware and software redundancy and diversity are used to a different extent. Hardware redundancy means that the same functions are processed in different hardware channels and the results compared. This mainly helps to exclude spontaneous errors of the electronic system. Hardware redundancy is used in almost all electronic interlocking systems.

The following system designs of redundant systems are applied:
- 2 out of 2 (2oo2) system: In these systems, the safety functions are processed in two independent channels and finally the results are safely compared. If both results are equal, the output is used, otherwise a fail-safe reaction leads the system into a safe (traffic-hindering) state.

9 Interlocking Machines

- 2 out of 3 (2oo3) system: The functions are processed in three independent channels. If a failure occurs in one channel, then that channel is isolated. The interlocking continues as a 2oo2 until the failure is repaired.
- 2*(2oo2) system: One redundant 2oo2 subsystem is the active and the other works in standby mode. If a failure occurs in the active subsystem, it is isolated and processing is continued by the standby subsystem.

Therefore, all these systems provide **safety redundancy**, whereas the latter two systems provide additional **availability redundancy** to maintain operation in case of an error and therefore reduce the probability of hindering states.

Diversity helps to exclude systematic errors in design. An alternative or supplement to diversity is a very strict checking process. Diversity can take different forms:

- Diverse hardware: Different hardware products are used in both channels
- Diverse operation systems, e.g. Windows and Linux
- Diverse software, e.g. for the definition of interlocking functions and track layout data

9.4.3 Structure of Electronic Interlocking Systems

9.4.3.1 Hardware Structure

Figure 9.34 shows the basic functional structure of an electronic interlocking. This does not mean that one functional block is necessarily identical with a certain hardware component, although in many systems it is.

The **operation control level** (chapter 9.1) is often provided by external remote control systems or by workplaces far away from the interlocking area. Therefore the operation control level is usually not considered as an integral part of the electronic interlocking system, but as a separate system. However, interfaces between the operation control systems and the interlocking itself have to be defined. These are occasionally named as **'operation interface level'**.

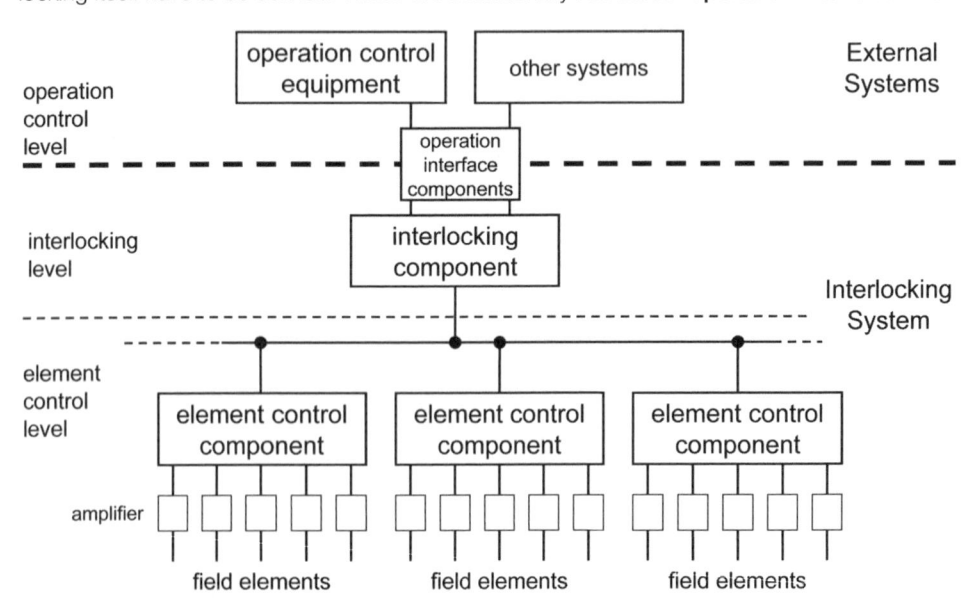

Figure 9.34: Functional structure of electronic interlocking

9.4 Electronic Interlocking

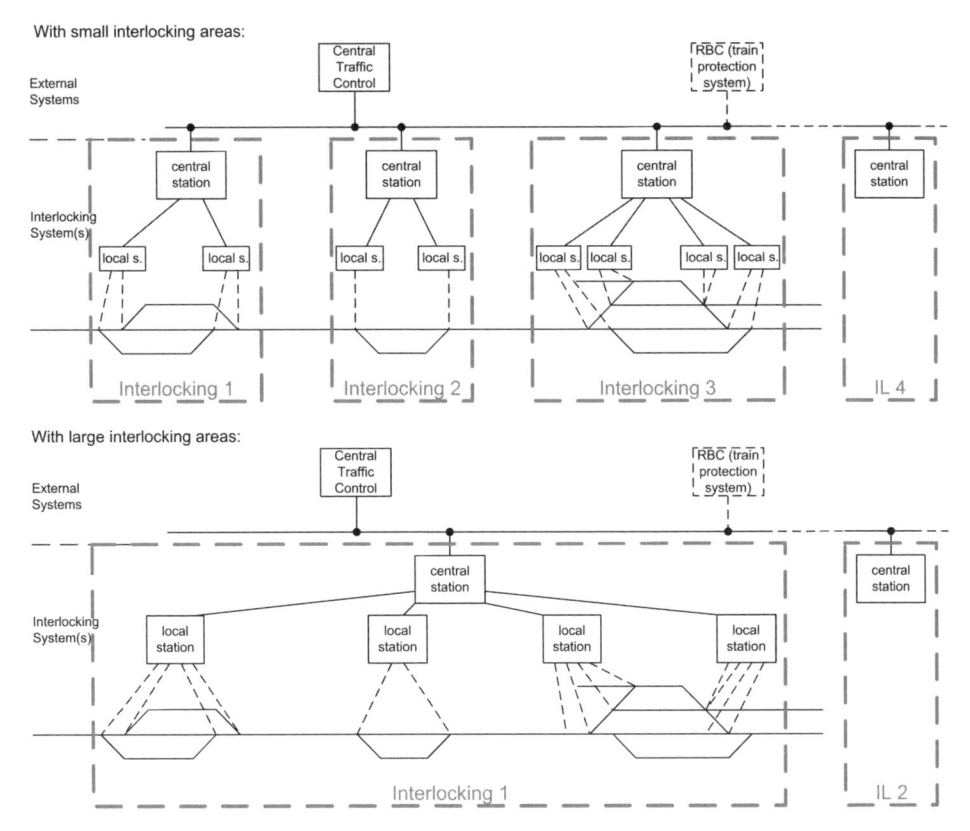

Figure 9.35: Geographical assignment of electronic interlocking

The **interlocking level** and the **element control level** (chapter 9.1) are integral part of the electronic interlocking system.

Train protection and train control systems (chapter 8) can interface with the element control level (often in systems which obtain information from the trackside signals) or with the interlocking level (often in systems with continuous transmission and centralised movement authority calculation).

Diagnostic functions provide checking of the state of components, failure detection and deactivation of faulty components. Diagnostic functions are necessary in all three levels. Often they are allocated in a separate hardware block.

According to the territorial distribution, most interlocking systems include central and local interlocking stations (figure 9.35). The central stations include parts of the interlocking level or the whole interlocking level, whereas the local stations include the element control level and, in some cases, the remaining parts of the interlocking level.

The size of the area of responsibility of the central and local interlocking stations varies much between the interlocking systems. In many systems, a central station is responsible for an area which covers usually one station and possibly the neighbouring open line section, whereas one local station is responsible for few (around 1-5) sets of points and/or signals. In other systems with very large interlocking areas, particularly Simis and L90 with their variants (chapter 9.4.6) and ACC Multistation (chapter 9.4.9), one central interlocking station can control a line section of up to around 100 km, with one local interlocking station being responsible for each (small) station.

9 Interlocking Machines

9.4.3.2 Software Structure

In the software structure, generally the following software components can be distinguished:
- The **system software** is responsible for the processing of the electronic components itself.
- The **interlocking software** defines the interlocking functions independently from the particular track layout. It is usually designed for a type or version of interlocking. Particularly in interlocking systems which are dedicated for international market, it can be further divided into:
  – Generic functions
  – Specific functions of the particular railway company
- The **location software** defines the track layout and all local specialities. It has to be designed particularly for each project.

Regarding the route formation, either the **tabular** or the **topological principle** is used (chapter 4.3.9). However, the difference between these principles is not as obvious as in relay interlocking and the advantages and disadvantages are slighter.

Due to highly complex software, in most systems commissioning, testing and later alterations cause additional difficulties. For the checking of software, particularly the location software, many railways and manufacturers use special simulators. These simulators replicate the behaviour of the field elements without the need to connect the interlocking to the real track layout. This reduces the obstructions to rail traffic during the construction, testing and commissioning phase of the interlocking.

9.4.3.3 Operation

Operation control is usually solved in a topological image. Input commands are given by keyboard, tablet (older systems) and/or mouse (newer systems). Often different monitors are used with different tasks, which can be more or less flexibly assigned to the particular monitors:
- **Area overview pictures** give an overview over a larger area (e.g. a big station or several stations). In some interlocking and operation control systems, the most frequent commands can be input in the area overview pictures, whereas in others it serves only for visualisation.
- **Zoom pictures** show details of each track element. Either all operations have to be input there or they are needed mainly for degraded mode operation or other less frequent operation actions.
- Alarm lists give out any possibly dangerous occurrences and actions. These lists can be given out by monitor and/or printer.
- Utilities like automatic train routeing are often managed in an additional monitor.

The grade of centralisation in operation control varies much between the railways (chapter 11). In decentralised operation control, typically one signaller is responsible for a certain station. In highly centralised operation, signallers for a large area are concentrated in one operation control centre. The latter gives high flexibility in adjusting the size of the area of responsibility of one signaller to the current workload (chapter 11.4), but causes additional difficulties in the case of technical problems when corrective action has to be carried out locally.

Operation control systems can be designed in different safety levels. In **vital (safe)** systems safety-related commands, e.g. in degraded mode operation, can be input. In **non-vital** systems, certain actions in degraded mode operation have to be achieved otherwise, e.g. by a train driver in the field. Also mixed systems with vital and non-vital functions are possible. As an example, figure 9.36 shows the local control panel, which can be optionally applied in ZSB2000 interlocking for input of track clear messages by a train driver.

Vital operation control systems require additional safety measures to ensure safe monitor pictures and safe processing of input commands. These methods are usually based on redundancy in the process. However, non-vital systems can also be equipped with availability redundancy.

9.4.3.4 Communications

Transmission between the geographically separated components is solved by electrical cables, optical cables or radio information, usually based on addressed data telegrams. Safety is provided by redundancy of the data telegrams. For availability, redundancy of the physical data lines on different paths is often also provided.

9.4.3.5 Power Supply

To ensure the availability of the system, back-up power supply is needed in most electronic interlocking systems (likewise in relay interlocking). The primary power source is mainly public networks. In case of temporary network failure, most interlocking systems provide autonomous emergency power supply by batteries and/or generators to ensure operation for a defined time.

Figure 9.36: Local control panel in ZSB2000 interlocking (photo: *Scheidt&Bachmann*)

9.4.4 SSI (Britain)

9.4.4.1 Origin and Application

Solid State Interlocking (SSI) originated as an open standard developed by *British Railways* in collaboration with *GEC* (now *ALSTOM*) and *Westinghouse*. Development started in the 1970s and the pilot scheme at Leamington Spa, a medium sized station in the English Midlands, entered service in 1985. The system was then adopted as standard for new work on British Railways.

With adaptations for other national requirements, SSI has become one of the most widely used interlocking systems, especially in Western Europe and countries of the British Commonwealth. Examples of adapted systems are:
- the Poste à Logique Paramétrisée (PLP) system in Belgium *(SNCB/NMBS)*
- the Poste d'Aiguillage Informatisée (PAI) system in France *(SNCF)*

SSI was one of the earliest electronic interlockings to be put into service. The experience gained during the process of validating the design fed directly into the CENELEC specifications (EN 50126 / 8 / 9) that were produced subsequently.

9.4.4.2 System Structure

SSI covers the interlocking and element control levels of figure 9.34. The interlocking level is performed by a centralised 2 out of 3 processor normally located in the signal box, whereas field elements are controlled by distributed object controllers at the trackside (figure 9.37). As a result, SSI can control station areas and block sections between stations without distinction, which means it uses route logic also for controlling the open line (chapter 10.4.4). This is in fact the norm in Britain.

9 Interlocking Machines

operation control level (non-vital): panel *or* electronic control system → PPM

interlocking level: IXL — DIA

element control level: DLM → ACM (axle counter), PM (points), SM (signal), UM (different field elements)

PPM	Panel Processor Module		ACM	Axle Counter Module
IXL	Interlocking Main Processor Module		PM	Points Module
DIA	Diagnostic Main Processor Module		SM	Signal Module
DLM	Data Link Module		UM	Universal Module

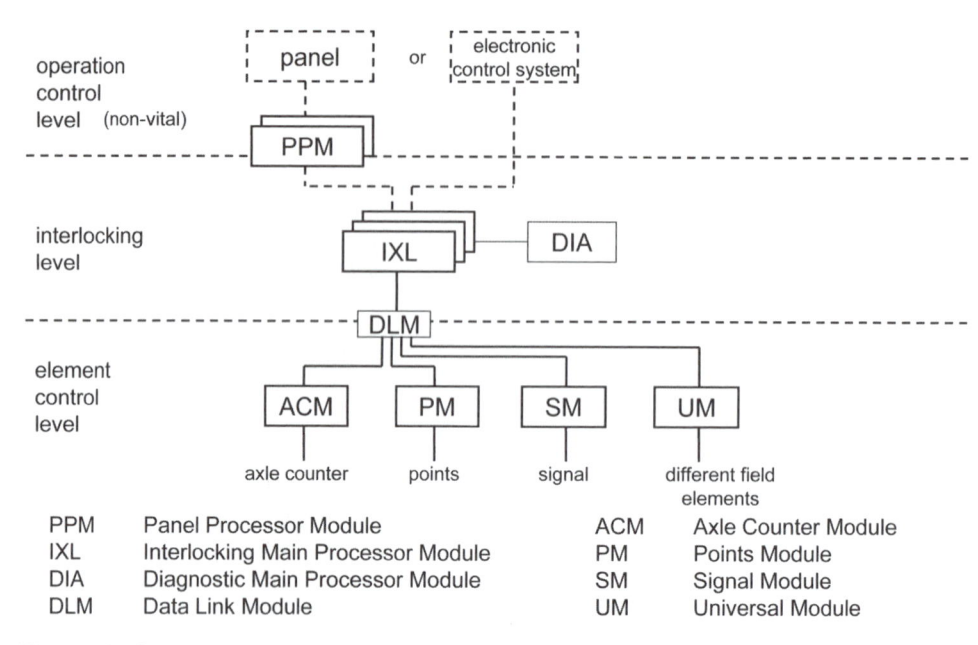

Figure 9.37: System structure of SSI

The original design of SSI was based on the Motorola 8502 microprocessor. The software is programmed onto EPROMs which are fitted to the cards in the factory. They cannot be accessed without breaking a physical seal.

The local object controllers called trackside functional modules (TFM, figure 9.38) are co-located with the elements which they control. Each one is a 2 out of 2 processor system controlling a group of vital outputs. They are used to drive functions such as signal lamps and point machines and monitor a group of vital inputs, such as train detection and point positions. The original trackside functional modules (TFM) were:

Figure 9.38: SSI trackside equipment, including Trackside Functional Modules and Data Link Modules

- Points module (PM) optimised for use with British Railways electro-pneumatic point machines. Each PM can drive 1 or 2 groups of points, with a 'group' meaning points which are always switched together, e.g. the ends of a crossover.
- Signal module (SM) optimised for British Railways main line and shunting signals. One signal module controls 1 or 2 signals.

The inputs of both types are also used for train detection and other functions. The TFM are by far the most numerous physical elements in a typical project. Therefore, they have to be low-cost as well as being robust for housing in lineside cabinets.

A separate terminal provides diagnostic and fault-finding facilities for the interlocking itself and for all the signalling equipment under its control (e.g. signal lamp failures, point failures). These include a printed record of all faults, and analysis of multiple fault conditions. When SSI was first introduced, this level of assistance for the signalling maintainer seemed quite revolutionary, though today it has of course become standard practice.

9.4.4.3 Software

The software is structured into two parts:
- The **fixed software**, which includes the system software of chapter 9.4.3.2, does not change from one station to another. It serves as an interpreter for the specific application data and carries out system functions including communications and management of redundancy.
- The **specific application data**, which includes the interlocking and location software of chapter 9.4.3.2, express all interlocking functions. They are written in data language designed specifically to be used by signal engineers. The internal structure of the language is organised in terms of signalling functions.

This software structure gives the system high flexibility, while requiring a disciplined approach to data preparation (e.g. use of approved data constructs). When adapting SSI for a new railway, the fixed software is not changed, as the specific application data express both the railway's signalling principles and the location data.

The SSI interlocking contains a memory called the 'state of the railway' which is updated continuously. It includes memories for every route request, signal, points, track circuit and route declared for the specific application, as well as various general-purpose functions (binary latches, timers). The SSI data language provides tests for reading individual items in this memory, and commands for writing to them. For example:

P101 cdn   might be a test for 101 points controlled and detected in the Normal position
S23 set y   might be a command to light the Yellow aspect of 23 signal.

Other statements are used to relate trackside functions to telegram bits in the trackside data link system.

Logical constructs are built up from these tests and commands using IF…THEN…ELSE in the usual way, e.g.:

if TAB c , TAD c , (TAE c or TBC c) , P101 dn then L1234 l else L1234 f \	If AB, AD and either AE or BC track circuits are clear AND 101 points are detected in the Normal position, then set the function L1234 to 'Locked', otherwise set it to 'Free'

The data for a specific application contains declarations of all the field elements (signals, points, track circuits etc.) and of internal signalling functions associated with them such as routes and sub-routes (route partitioned by train detection sections etc.). It also contains instructions for:

9 Interlocking Machines

- Updating the states of objects in the memory using the information reported by the TFMs
- Processing inputs from the signaller (route requests, point controls)
- Sending commands to the TFMs to drive the external objects
- If required, instructions for processing inputs from an NX Panel (to derive panel requests from button and switch operations), and outputs to it (basically lamp drives)

The diagnostic system also uses data in order to give 'intelligent' fault messages (e.g. 'Signal X lamp Y failure' rather than simply 'error on message A byte B bit C.' This data is derived automatically from the other information by the application design tools.

Data is also created to permit simulation in the design office of the whole of the trackside part of the signalling installation. This data too is generated automatically.

Examples of SSI data language: Stages in Setting route R12(1M)

In response to a route request received from the signaller, the following commands are to be carried out. (Obviously this only shows the main features, to illustrate the overall principle):

*QR12(1M) if	Conditions for accepting request for route R12(M): IF
R12(1M) a	the route is available
P101 cnf	the points 101 are in the correct position OR are free to be moved to that position
UBA f , UCA f ...	opposing sub-routes are not locked
then	Actions to be taken if the route is available: THEN
R12(1M) s	lock the route
UBA l , UCA l ...	lock all the sub-routes in it
P101 cn	command the points to the required position

The above is actually written

*QR12(1M)   if R12(1M) a , P101 cnf , UBA f , UCA f , ...
            then R12(1M) s , UBA l , UCA l , ... , P101 cn
            \ .

Example of SSI Data Language: Signal clearing controls

if R12(1M) s	If the route is set
S12 set stick	In effect tests that the train has not yet passed through the route
TAB c , TAC c ...	Track circuits proved clear
P101 cdn	Points proved in the correct position

Example of SSI Data Language: Signal aspect controls

The selection of signal aspects as a result from the aspect of the signal ahead and the route is handled as follows:

S12 set rip	Command to light the route indicator (if any)
G s 7 \	This statement identifies the telegram bit to set if the Red aspect is to be displayed
S14 seq 4 , G s 7654	This statement identifies the signal ahead (signal 14), specifies a 4-aspect sequence (red, yellow, double yellow, green) and identifies the telegram bit(s) to set if a Clear aspect is to be displayed

9.4.4.4 Operation

SSI uses a non-safe operation control level. The interface with the **operation control level** can be one of two kinds (IRSE 1980):
- The operational level can send route requests to the interlocking.
- The operational level can send route entrances and exits to the interlocking, and the interlocking includes route selection logic to determine which route should be set.

The former is applied with electronic operation control systems, whereas the latter mode permits compatibility with British-type entrance/exit (NX) panels which are used also for relay interlocking. Although this interface is non-safety, it is duplicated for high availability.

9.4.4.5 Communications

The interlocking and the trackside functional modules are connected by dedicated vital data links which can use special copper cables, or channels in a standard telecommunication system. Safety is assured on these links by multiple levels of coding, with superimposed Manchester II and Hamming codes. The links are duplicated for availability, and can follow different routes along the railway line where the geography permits for added protection.

The central interlocking polls each trackside functional module in turn in a fixed cycle, thus ensuring that all output and input functions are kept up-to-date.

More information on this example can be found in (IRSE 1991).

9.4.5 SMILE (Japan)

9.4.5.1 Origin and Application

The Japanese SMILE (**S**afe **M**ultiprocessor for **I**nterlocking **E**quipment) interlocking is a joint venture development of three companies (*Nippon*, *Daydo* and *Kyosan*) as a standard electronic interlocking system for *Japan Rail (JR)*. SMILE exists in two versions: the original SMILE for big stations and Micro-SMILE for small stations. The following explanations concentrate mainly on the version SMILE. The first SMILE interlocking went into operation in 1985.

9.4.5.2 System Structure

In SMILE, vital and the non-vital functions are separated in different modules. The fail-safe microprocessor (FSM) as one component performs all safety critical functions, which is facilitated by the fact that no safety is required for operation control in this interlocking system.

The FSM works with hardware redundancy (2oo3) with identical software in all channels (figure 9.39). The channels work synchronously. Each bit which is transmitted between the CPU and the ROM/RAM is compared and, in case of deviation, corrected by the MVR (Majority Voting Restorer). Due to this high checking frequency, errors are corrected very quickly, reducing the probability of dangerous double errors. To detect errors, the FSC (Fail-Safe Comparator) compares the status of the bus before and after the MVR. A certain number of errors is tolerated, but in case of excessive number of errors, the MCC (Mode Control Circuit) switches off the respective channel.

Within the interlocking, all components (up to 14) are connected by the SMILE-Bus (figure 9.40). Besides the fail-safe microprocessor, these are several non-safe components such as interface components to the dispatching centre, the traffic supervision microprocessor, the schedule planning microprocessor and others. These components use a common memory.

9 Interlocking Machines

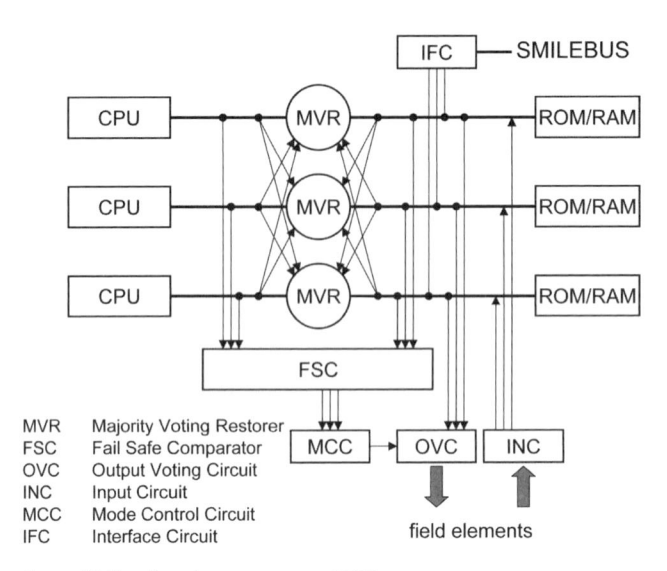

Figure 9.39: Structure of fail-safe microprocessor FSM

MVR Majority Voting Restorer
FSC Fail Safe Comparator
OVC Output Voting Circuit
INC Input Circuit
MCC Mode Control Circuit
IFC Interface Circuit

Some of the non-safe components are equipped with availability redundancy (hot stand-by).

One interlocking is typically responsible for the area of one station.

The SMILE-Net connects local external components such as local operation workplaces and passenger information, but also remotely controlled interlockings (figure 9.40). The only interlocking component which communicates directly with the SMILE-Net is the Schedule Planning Microprocessor. All other components communicate via a particular interface between SMILE-Bus and SMILE-Net.

SMM System Monitoring Microprocessor
CIM CTC Interface Microprocessor
SPM Schedule Planning Microprocessor
TSM Traffic Supervision Microprocessor

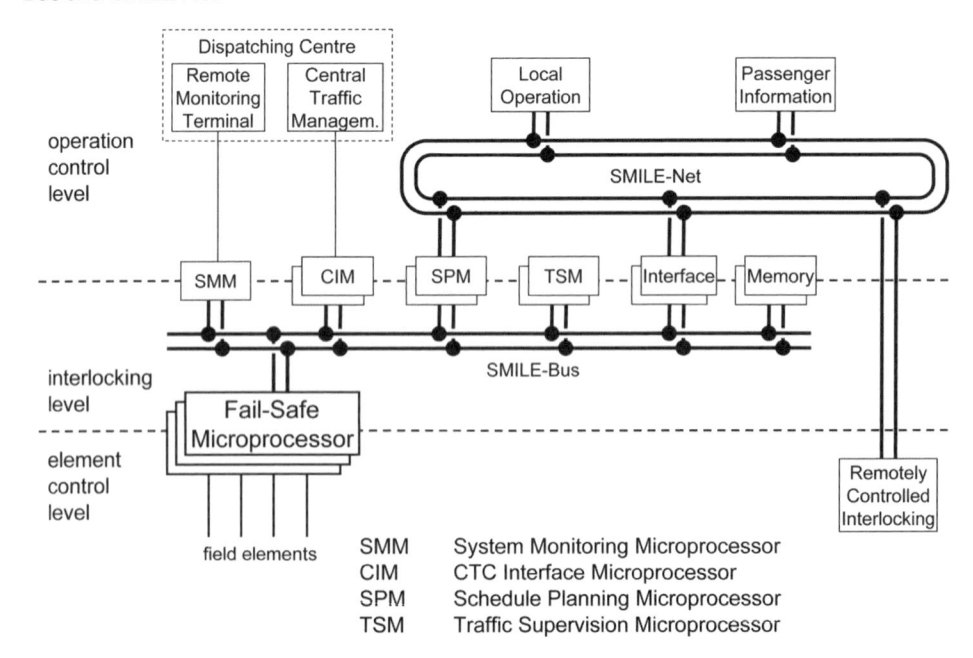

Figure 9.40: System structure of SMILE

290

9.4 Electronic Interlocking

9.4.5.3 Software

Both versions, SMILE and Mirco-SMILE, use the same software. The software is divided into many function modules. The local software of the particular station is structured according to the topological principle (chapter 4.3.9).

9.4.5.4 Operation

The operation control workplaces are non-safe. Operation can be done from the CTC or locally. In the first installations, an operator console was used, but probably today this has changed to monitor and mouse. The track layout of the station is displayed in a geographical image, similar to other electronic interlocking systems.

9.4.5.5 Communications

The interlocking components communicate via two redundant buses. The SMILE-Net for communication is a redundant fibre-optic cable which connects the components in circular manner.

More information on this example can be found in (Akita 1985), (Akita/Nakamura/Wanatabe 1987) and (Maschek 1996).

9.4.6 Simis and L90 with Derivates (German origin)

9.4.6.1 Origin and Application

In contrast to Britain and Japan, Germany has no standardised electronic interlocking system manufactured by various companies, but the particular systems of each manufacturer. Nevertheless, as the systems were designed for the same requirements of DB, there are large similarities in the functions and in the operation interface.

The main manufacturers are *Siemens* and *Thales* (formerly *Alcatel SEL*). Besides Germany, the products of German manufacturers are used worldwide, especially in Central and Eastern Europe, the Iberian Peninsula, the Middle East and China.

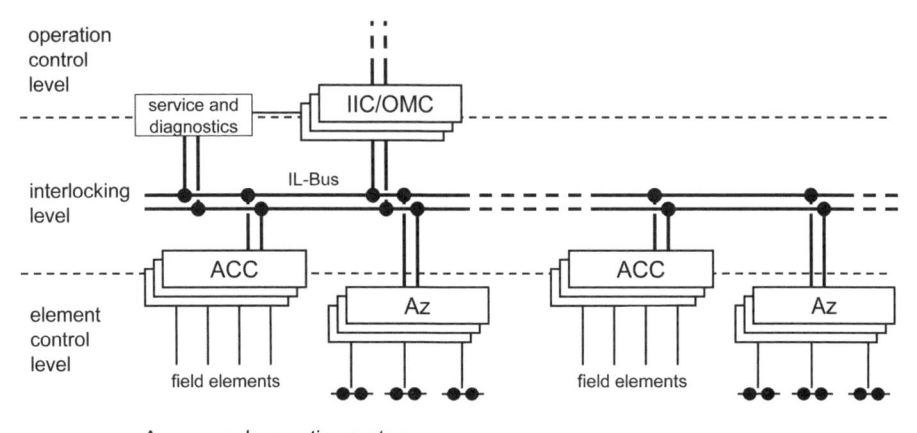

Az   axle counting system
ACC   Area Control Component
IIC   Interlocking and Interface Component
OMC   Overhead Management Component

Figure 9.41: System structure of Simis-W

9 Interlocking Machines

The basic type of *Siemens* is Simis, out of which different variants have been developed. Simis firstly went into operation in the 1980s. The currently most important version on the international market is Simis W, with different adaptations for different national requirements.

Thales offers basically two systems: the older system L90 (Locktrac 6111), used in Germany and other countries, and the newer L90 5 (LockTrac 6151), used in different countries mainly outside Germany. L90 firstly went into operation in 1989 in Neufahrn i.NB. Also here, different adaptations have been developed for different national requirements and operational situation.

9.4.6.2 System Structure

In L90 and L90 5, the modules are divided by the functional levels (figure 9.34) into Interlocking Module (IM) and Field Element Controllers (FEC) (figure 9.43). In Simis (example: Simis-W, figure 9.41), in contrast, the hardware distribution is different: IIC/OMC (figure 9.42) and ACC are responsible for the interlocking functions cooperatively, with IIC/OMC being comparable with the central switching sets of a topological relay interlocking (chapter 9.3.5.2).

Figure 9.42: Part of Simis interlocking (Żywiec, Poland)

The territorial structure is that of high centralisation: A central interlocking station (including IIC/OMC in Simis and IM in L90) can be responsible for a portion of single or double line of about 50 to 100 km, whereas the area of a local interlocking station is approximately the size of a medium size station.

If in Germany the electronic interlocking is included into an operation control centre, (chapter 11.5.2), the central interlocking station is called the 'sub-centre', with the CTC even one level above.

SCM	Safe Communication Module
IM	Interlocking Module
FEC	Field Element Controller
AzLM	Axle Counting Module

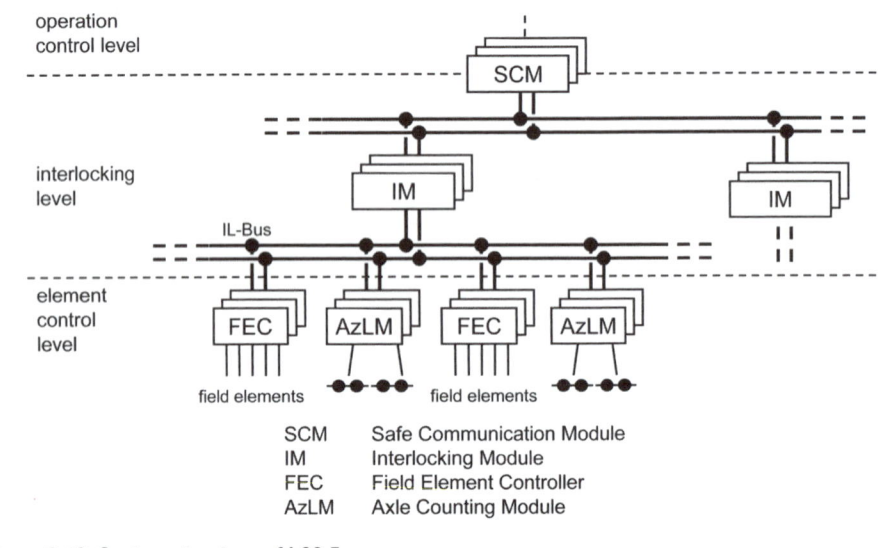

Figure 9.43: System structure of L90 5

Both systems work with hardware redundancy, but without diversity. The hardware redundancy is carried out mainly in 2 out of 3, in older versions and reduced versions for secondary lines also 2 out of 2 systems.

9.4.6.3 Software

In both interlocking systems, the software is distinguished into the three parts described in chapter 9.4.3.2 (with different, product-specific names).

Simis-W and most other versions of Simis use the topological principle, whereas L90 and L90 5 use the tabular principle. But as in L90 and L90 5 short partial routes are used, there is no operationally significant difference in the route release process.

9.4.6.4 Operation

In Germany, a standardised MMI system is used with two versions manufactured by Siemens and Thales which are similar, but differ in some details. Normal operation, unless an automatic train routeing (ATR) system is used, is done by consecutive clicking the entrance and the exit signal. Other commands such as individual point setting and safety-related commands in degraded mode operation, as well as alternative routes and alternative overlaps, have to be input by selection in a context menu related to the respective element. The MMI can perform vital functions with registration of safety commands.

In the international context, also other MMI systems are applied in connection with the described interlocking systems.

9.4.6.5 Communications

In Simis-W, the components are connected by the vital interlocking bus. Data are transmitted by standardised data telegrams. The telegrams include a safety attachment. The interlocking bus is doubled for availability redundancy. In case of failure of one channel, communication is done by the remaining channel with the data telegrams duplicated and inverted.

Optionally data can be transmitted remotely between parts of the interlocking bus by copper or fibre optical cables. Proprietary or public data lines can be used. Data have to be converted by special modems in the sections of the bus. To detect errors, checking telegrams are sent in regular cycles.

In L90, serial data connections are used between the components. The data lines are doubled for reliability redundancy, each telegram is sent on both lines. The safety of the data telegrams is performed by a combination of code redundancy and duplication/inversion of telegrams. Optionally data can be transmitted remotely between the interlocking module and the element control modules of L90 via IP network. Proprietary or public data lines can be used, preferably via doubled physical connections. A safety attachment will be used, and additionally when using public lines, data is encrypted/decrypted by special modems at both ends of the connections.

In L90 5, IP connections are used between the components. The physical connections are doubled for reliability redundancy. Data telegrams are supplemented with safety attachments. In regular cycles life sign telegrams are sent to assure reliable connections.

More information on both examples can be found in (Maschek 1996) and about Simis in (Schubath/Grotheer 2002).

9.4.7 Ebilock

9.4.7.1 Origin and Application

Microprocessor-based interlocking system of Ebilock type was developed by the Swedish company *Ericsson*. The first, Ebilock-750, was installed at Sweden's Gothenburg station in 1978. For control of field elements, only relay object controllers were used in this system. The following generation became the Ebilock-850, first implemented at Hallsberg station. Object controllers of Ebilock-850 system were constructed on the basis of electronic technology and were used to control point machines, signal lamps, balises, etc.

The implementation of Ebilock-950 began in 1996. Its distinguishing feature comprises the decentralised location of equipment, so that object controllers can be placed in direct proximity of the objects to be controlled. The system has been installed in Korea, Latvia, Portugal, Poland, Lithuania, Russia and other countries.

9.4.7.2 System Structure

The structure of Ebilock-950 is depicted in figure 9.44. Ebilock-950 is available in two alternative versions for centralised and decentralised equipment locations respectively. In the centralised version, the Interlocking Processing Unit (IPU), which realises logical interdependencies between objects, and the field element controllers (System of Object Controllers SOC) are both located in the signal tower. The decentralised version assumes the IPU location in the tower and the SOC distributed in immediate proximity to the controlled objects. The object controllers system serves as the basis for direct supervision and control of objects on the railway station.

One set of IPU can control up to 150 logical objects (actual objects of railway station in the computer program), up to 1,000 objects of IPU (points, signals, relay coils, contacts of relays), which corresponds approximately to railway stations with 40-60 sets of points. The number of objects controlled can be increased by using more IPUs.

The IPU is designed as a 2*(2oo2) system. There are two IPU (figure 9.45) and each of them has two processing units FSPUA and FSPUB. Both units work synchronously and concurrently and process system and interlocking functions in two diversity software channels A and B. In turn, both IPU computers work synchronously and independently, one in on-line mode and another in standby mode. The standby computer does not affect the operations of the on-line

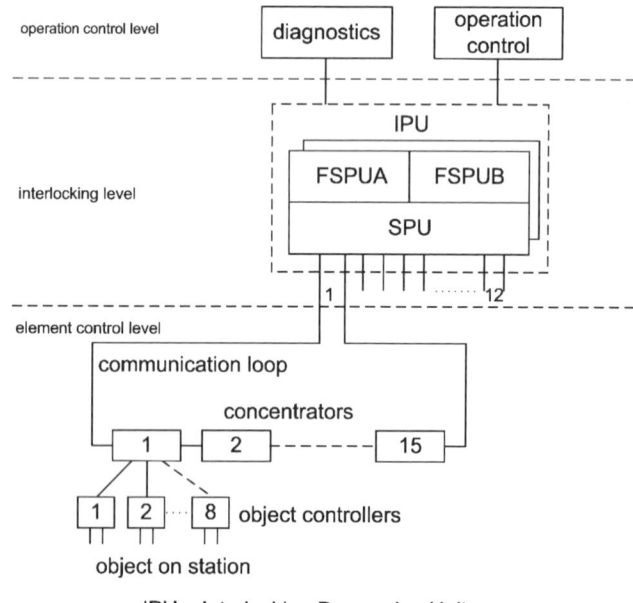

IPU – Interlocking Processing Unit
FSPU – Fail Safe Processing Unit
SPU – Service Processing Unit

Figure 9.44: Structure of Ebilock-950

Figure 9.45: IPU of Ebilock

computer, but continuously receives all actual information concerning the state of on-line computer. Thus, in the case of failure of the on-line computer, standby one has all the information needed to undertake the processing.

The Service (or communication) Processing Unit (SPU) fulfils all the asynchronous functions, for example input-output operations of commands and data. The device works under the control of the UNIX compatible real time operational system DNIX.

Every IPU module has its own communication subsystem which is intended for communication with concentrators and workplace of the signaller.

For computation process safety, the following measures were envisaged:
– Software diversity in the A and B channels
– In every computation cycle the following actions are carried out:
    – cross-comparison of input, intermediate and output data,
    – checking of the software integrity,
    – checking of the dynamics and relevance of the data processed,
    – checking of time parameters of the program and consequence of program modules fulfilment,
    – checking of program memory and main memory.

Object controllers supervise and control specific field equipment. Object controllers receive orders, which are transmitted by the Central Interlocking System (IPU) through the concentrators (COM), and transform the orders into electrical signals. Similarly, the signals received from field equipment are transformed into telegrams containing information about the states of the equipment, and transmitted to the IPU. Figure 9.46 shows the basic structure of the system of object controllers.

The kernel of the object controller represents a board of supervising and control (CCM). The board checks states of contacts: four channels for safety control of the contacts (track circuits), six non-vital channels for data input and two non-vital channels for data output. Software of the board is formed in accordance with functional destination of the object controller.

9.4.7.3 Software

Figure 9.47 shows the basic software structure of Ebilock-950. The main program components fulfil the following functions:

9 Interlocking Machines

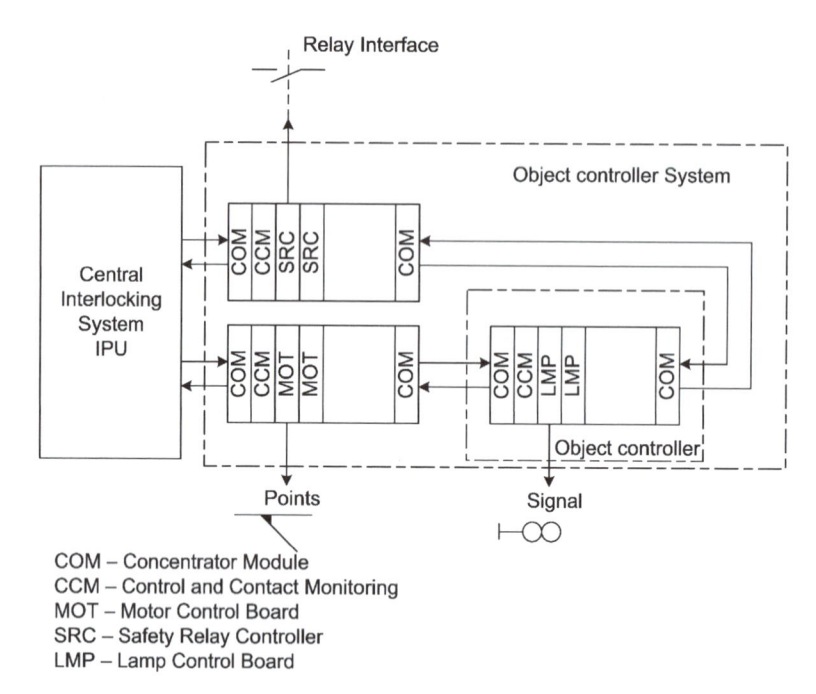

COM – Concentrator Module
CCM – Control and Contact Monitoring
MOT – Motor Control Board
SRC – Safety Relay Controller
LMP – Lamp Control Board

Figure 9.46: Object controller subsystem

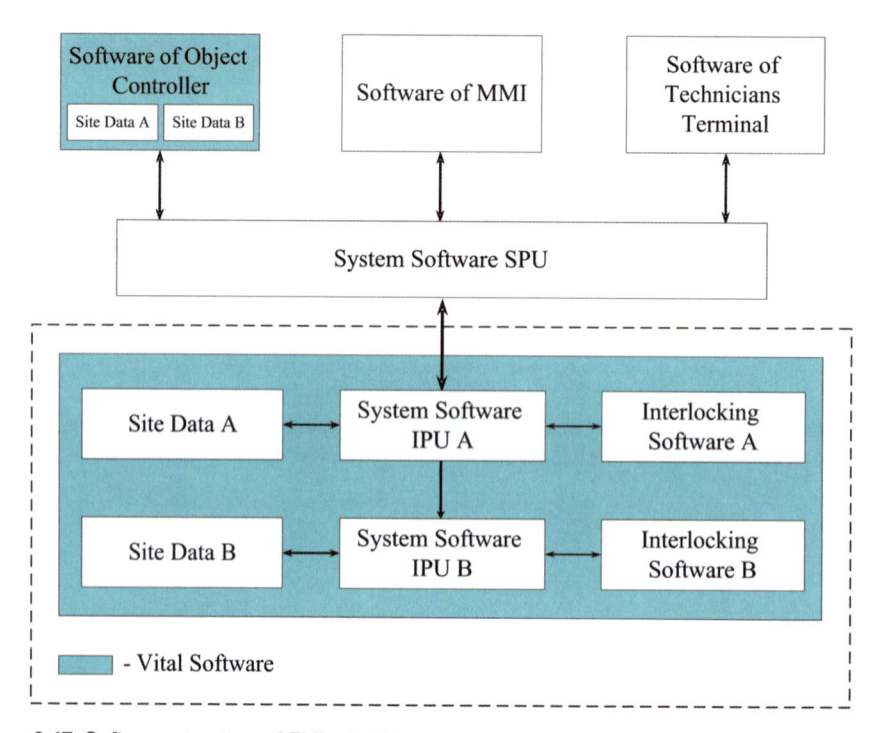

Figure 9.47: Software structure of Ebilock-950.

9.4 Electronic Interlocking

- SPU system software handles with communications between IPU and external subsystems such as object controllers, MMI, etc. SPU is running under DNIX operation system. No safety requirements are set for the SPU software.
- IPU A and IPU B system software fulfils input and output functions in communications with field objects and executes operations of interlocking logic. This software (and the following software components) is safe due to diversity (two software channels). The IPU A and IPU B software is developed in ANSI-C environment.
- Software of interlocking logic fulfils the interlocking functions and produces MMI control commands. It is developed by means of dedicated programming language – Sternol.
- Site data A and B includes data such as the identification and topology of field objects, configurations of the object controllers etc.
- The software of the object controllers fulfils the control functions for signals, points and other field elements including the safe communications with the control circuitries.

9.4.7.4 Operation

The MMI of Ebilock-950 system is similar to most other modern electronic systems with use of mouse and a geographical image (figure 9.48). The operation surface is non-vital.

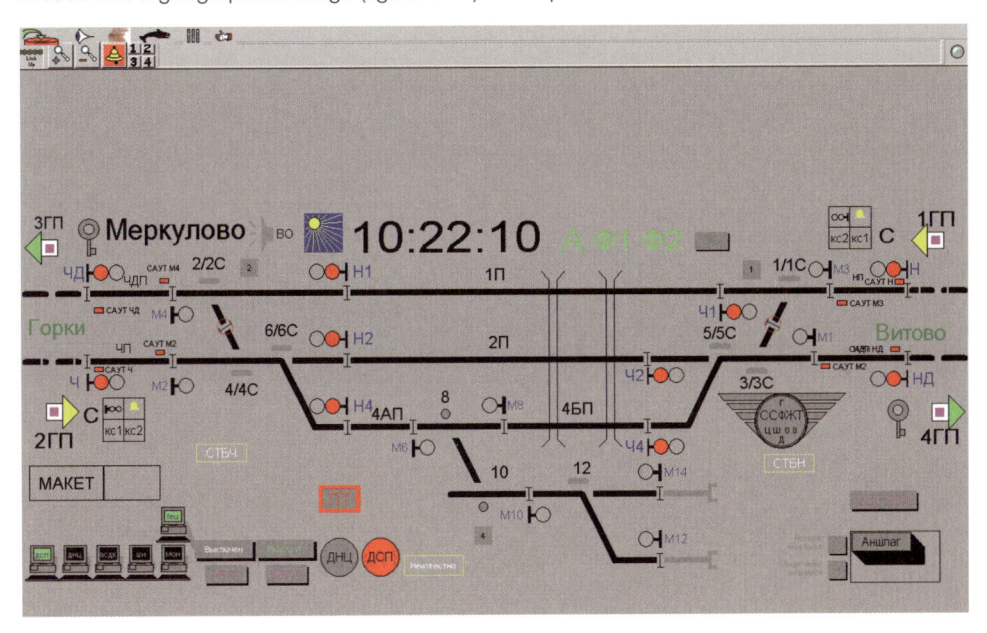

Figure 9.48: Topological image in Ebilock-950 (example from Russia)

9.4.7.5 Communications

Communication means in Ebilock-950 system can be divided into those within the interlocking cabin and those with object controllers. Modules of the interlocking cabin are connected with Ethernet protocol. Blocks of transmitted data are validated by means of checksums (without error correction).

Communication between IPU and object controllers is provided by means of HDLC protocol, on the physical layer – V.24 protocol.

9 Interlocking Machines

9.4.8 EC-EM (Russia)

9.4.8.1 Origin and Application

The first Russian microprocessor-based interlocking system EC-E was put into operations on Šossejnaja station near St. Petersburg in 1997. The computerised control system of UVC PC 1001 with tripled redundancy, developed for safety processes of nuclear engineering, became the technical foundation for realisation of the system functions.

During the time that followed, microprocessor-based interlocking development proceeded towards the creation of dedicated microprocessor systems. The system of EC-EM became one such system. Its basis is a computerised control system, developed by the *Radioavionika* company, and algorithms and programs of the EC-E system (from Šossejnaja station).

For the first implementation of the EC-EM system, New-Peterhoff station was chosen. At the following implementations on stations Nazija and Žicharevo equipped with centralised automatic block systems, EC-EM algorithms were extended for realising automatic block system logic. One interlocking usually covers the area of one station.

9.4.8.2 System Structure

The technical basis of EC-EM is made up of dedicated computer control unit DCU-RA. The functional structure of EC-EM is represented in figure 9.49.

The CPU (Central Processor Unit) is dedicated to the realisation of route interlocking algorithms and is a 2oo3 majority redundant computer. When in operation, the CPU interacts with the signaller's workplaces and communication blocks (CB). The CPU sends the actual data

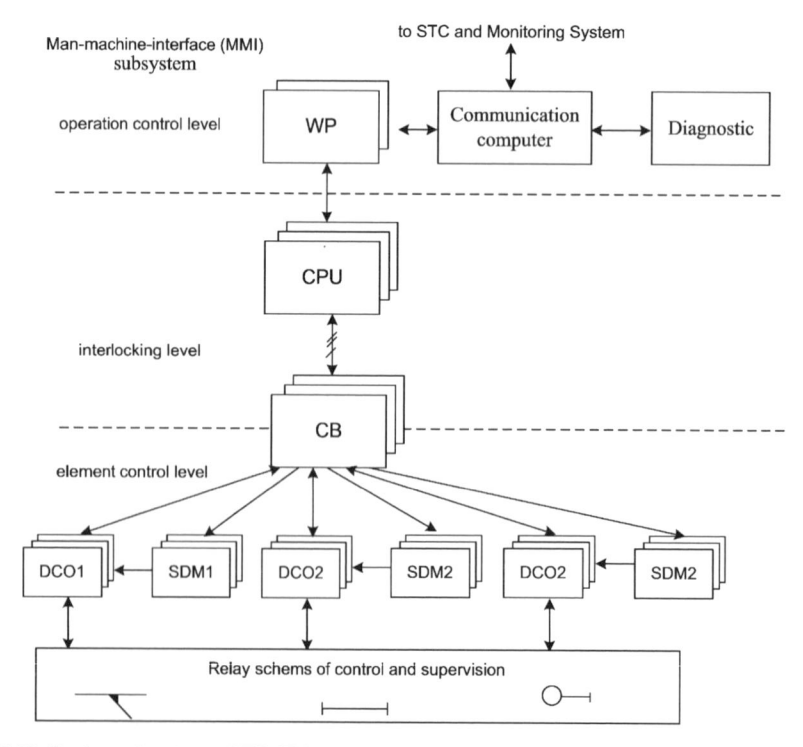

Figure 9.49: System structure of EC-EM

regarding the states of control objects to the workplace (WP), and the workplace sends corresponding commands to the CPU. On the basis of information received from communication blocks (CB), algorithms of central dependencies are calculated and output data are formed. The CPU is produced on the basis of an 80586 processor with the clock frequency 133 MHz.

The Communication Block (CB) provides data exchange with the CPU and programmed control of facilities, intended for communication with field objects (DCO), and modules of safety disconnection (SDM). The Communication block is also made of 2oo3 majority redundant computers. During every work cycle, CB fulfils:
- The checking of the DCO
- The testing of the internal resources of the CB;
- The software-based processing of test results, mutual data exchange between different CB's, transfer of diagnostic data to the CPU.

A relay interface with field elements is realised in the EC-EM system, thus relay circuitries of direct control and supervision of points and signals were retained. Correspondingly, the DCO consists of modules for data input and output. Input DCO modules provide processing of 48 discrete safety inputs from the relays. Output DCO modules provide processing of 56 discrete safety outputs to the relays. One movable track element requires 2 discrete safe inputs and 2 outputs, whereas a signal requires 8 inputs and 5 outputs.

The SDM represents a dedicated power supply unit for feeding of data output modules and provides immediate power interruption in case of failure according to the Fail Safe Principle. For its work, it is necessary for at least two (out of three) computing channels to be correct.

In order to check input data and processing results, an interchange of data is carried out between the CPU channels and their 'straightening' is fulfilled, too. The resulting data are used for the computation of central dependencies. Information is exchanged for comparison before it is transferred to the CB.

Failure proofing and safety of DCU operations are provided by:
1) triple redundancy of the equipment and communication lines (2oo3),
2) continuous checking and cyclic testing and following comparison of computation channels,
3) usage of fail-safe technology in blocks with high demands on safety.

9.4.8.3 Software

Software of the EC-EM system includes system software (SS DCU) and technological software (TS DCU). The latter consists of interlocking logic and application data. SS provides application user programs operations on the DCU hardware. Main cycle of the application programs is equal to one second and consists of:
- majority sampling of DCO input data,
- call of the leading function of application programs,
- majority sampling of output data and transfer of the data to DCO.

In each cycle, an inter-channel comparison of the dynamic part of the database (DPDB) of applications programs is provided. During several cycles, the comparisons of codes and fixed database part in RAM are fulfilled also.

SS DCU provides diagnostics and continuous testing of CPU hardware. In case of finding any fault, it reacts by repetition of the respective operation or withdrawal of the faulty channel. SS DCU provides operations of the CPU in modes of 'ring', 'chain' and 'pair' transparently for application programs.

SS DCU provides data interchange through RS-485 interface with DCO, with operator's workplace and with mobile workplace of maintenance staff.

9 Interlocking Machines

TS DCU is built up on functional principles. Such an approach assumes a realisation of interlocking functions of the system, for example:
- Setting the routes and clearing the signals on the basis of checking safety conditions
- Aspect selection of the signals
- Releasing of track sections after a train has passed

Each of the above mentioned technological functions has its own operational array. The array stores requests which are formed on the base of signaller's input commands. The presence in the array of such requests initialises the execution of the corresponding technological program. All technological programs interact with each other through the database. The database includes full information concerning the field equipment and its localisation.

9.4.8.4 Operation

The status of elements is displayed on the monitor as a topological image of the whole station, without distinction between area overview and zoom images. The input of commands is handled by keyboard and mouse in the topological image.

There are three operational modes envisaged in the EC-EM system:
- **Basic operational mode:** This mode is used when all subsystems and elements are working correctly and checking of all safety conditions is provided.
- **Auxiliary operational mode:** This mode is entered in the case of partial failure of field equipment. The signaller takes responsibility for these safety conditions which cannot be checked by interlocking.
- **Emergency operational mode:** This mode is entered in the case of failure of the processing unit. It provides direct control of movable track elements and signals without checking of safety conditions. To enable this, the system is equipped with directly wired emergency control panel.

9.4.8.5 Communications

The EC-EM system uses a relay interface and communicates with field equipment (points and signals) through electric cables. System electronic equipment is concentrated in the interlocking cabin. For interaction between the central processing unit and communication blocks, three-channel communication line RS-422 is used; for interaction between the central processing unit and the workplace doubled RS-422 interface is provided. Validity checking of the transmitted information is organised by means of check sums and periodical data exchange and comparison.

9.4.9 ACC (Italy)

9.4.9.1 Origin and Application

ACC (Apparato Centrale con Calcolatore) is an electronic interlocking system manufactured by the Italian company *Ansaldo STS*. It has been developed in Italy, based on the needs and requirements of the *Italian Railways (RFI)*, and later adapted in other versions to other countries' customers (e.g. Great Britain, Brazil, China, India). The first trial implementations were established around 1990. The original version has been applied for single station use, whereas a newer 'Multistation' version was developed later to control several stations along a line or within a junction area.

9.4 Electronic Interlocking

9.4.9.2 System Structure

Figure 9.50 shows the basic system structure of ACC. The interlocking functions are performed in the Central Interlocking Unit (CIU). It is designed in redundant hardware 2oo3 for safety and availability. The field element controllers (FEC) control and supervise the field elements in a defined area. They are designed as a 2oo2 safety electronic system, with a fully static or relay based interface to the field element. They communicate with the CIU via the Area Controllers (AC), which are 2x(2oo2) safety redundant electronic systems.

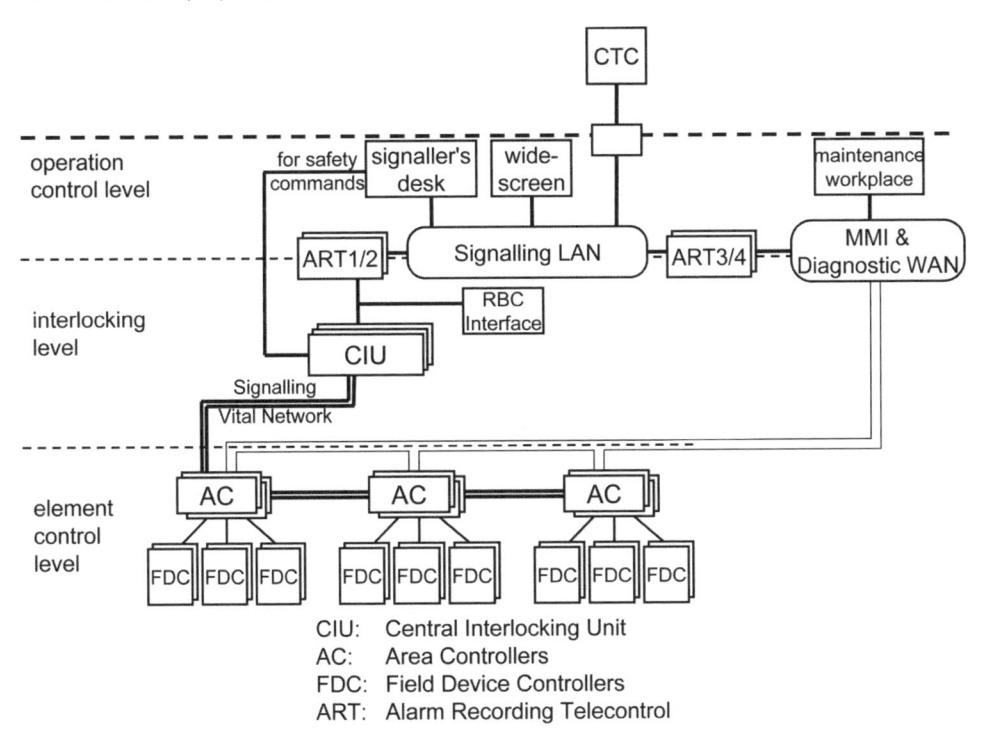

CIU: Central Interlocking Unit
AC: Area Controllers
FDC: Field Device Controllers
ART: Alarm Recording Telecontrol

Figure 9.50: System structure of ACC

ART (Alarms Recordings Telecontrol) subsystem is responsible for non-vital diagnostic and communication functions. The main purposes of ART1/2 are operation interface, recording and alarm functions. It is connected to the signaller's workplace by the non-vital Signalling LAN. CTC's can also be connected to the Signalling LAN via a particular interface. ART3/4 performs interface functions with the diagnostic workplace in the Control Centre and manages a Wide Area Network, which realises the connection to distributed MMI positions, for both the signaller's and the maintainer's functions.

Geographically, the CIU, the ART system and the operator interface are situated in the Central Post (central interlocking station), whereas the AC and FEC are in the Peripheral Post. The control area of a Peripheral Post is limited by the maximum cable length.

In the 'Mulistation' version, there are several peripheral (or local) locations, each of them accommodating several Peripheral Posts. The Peripheral Posts are connected with the CIU by a Signalling Vital Network. Besides, each peripheral location can accommodate signaller and maintainer MMI, all linked with each other and to the main MMI server by means of a devoted back-bone communication network.

9.4.9.3 Software

The software in CIU is divided into three levels in accordance with chapter 9.4.3.2. These are system software, application database (interlocking software) and geographical database (location software).

9.4.9.4 Operation

Traditionally, one station is controlled from one or more operator's workplaces, which are situated in the Central Post. Multistation version provides a platform for centralised operation control and through the MMI WAN makes MMI positions available in each peripheral installation.

Normal command input can be done by mouse, the status of the site being displayed on a topological image on monitor and/or wide-screen panel. The monitors provide area overview and zoom images. Safety of the image is ensured by its generation in two different programs, with a hardware vital watch-dog. For safety related commands, a particular procedure is defined, that in older versions required the use of a 'vital keyboard'. In modern versions, also procedures without vital keyboard by procedural measures are possible.

9.4.9.5 Communications

Communication between the CIU and Peripheral Posts (signalling vital network) as well as between CIU and ART1/2 is performed by synchronous serial fibre optic link. The links are doubled for availability redundancy.

Communication within the ART system is performed by the Signalling Local Area Network (LAN), that is a standard ethernet duplicated link.

Communications between the Central ART and Signalling LAN and the distributed MMI is realised by means of the MMI and Diagnostic Wide Area Network, an hyper ring Ethernet duplicated backbone.

More information on this example can be found in (Ansaldo 2008).

9.4.10 Local-electrical Operated Point Switches (LOPS)

LOPS were firstly developed for industrial railways and the system has spread, in the meantime, to several railways. They represent a decentralised form of electronic interlocking for shunting areas, with simplified interlocking functions. LOPS are designed as modules and can be used for single points as well as for large shunting areas. The advantages of LOPS, which increase the efficiency of shunting, are:
- In contrast to the usual electronic interlocking, the installation costs are much lower, and the points can be set by train drivers or shunting staff. Thus ground staff for signalling are not required.
- In contrast to manually operated points with key locks, switching of the points can be done much faster (e. g. without leaving the driving cab) and maintenance costs are reduced.

Essential components of a LOPS system are:
- Points with point machines
- Point signals to indicate the end position to the driver
- Decentralised electronic control equipment with simple interlocking functions
- Track clear detection
- Operation equipment.

9.5 Hybrid Technologies

The interlocking functions prevent the switching of points which are currently occupied, or up to which another shunting movement is approaching. They can also provide for simple interlocking functions between points (chapter 4.2). Points are often designed with a preferred position, to which they return whenever they are not being used. Spring points (chapter 6.1.2.2) are also often used.

Operation can be either by switching each set of points individually, or by entrance-exit operation. In the latter case, the entrance is determined by the current position of the vehicle, whereas the exit has to be selected by the driver or shunting staff.

For command input, either pushbuttons in different height (for a person on the trackside, on the step of a wagon or in the driver's cab) or even complete panels with different pushbuttons for different directions (figure 9.51) are used (German solution). Or points can be commanded by a handheld using radio transmission (North American solution).

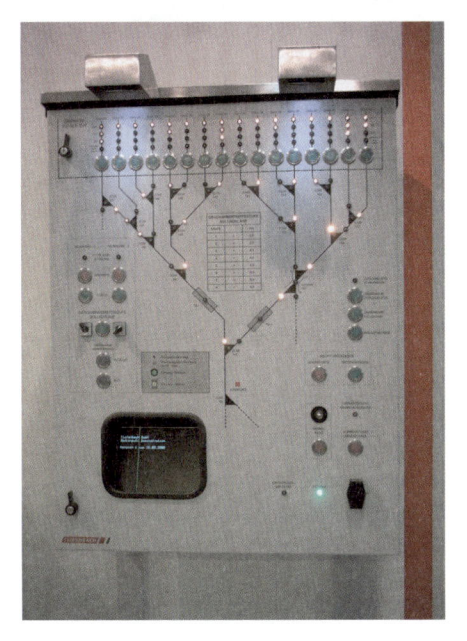

Figure 9.51: Local Operation Panel for LOPS (manufacturer: *Tiefenbach*)

9.5 Hybrid Technologies

9.5.1 Hybrid Mechanical and Electrical/Pneumatic/Hydraulic Forms

In such hybrid interlockings, the interlocking functions typically remains mechanical, but the control of field elements is by electricity, pneumatics or hydraulics. In the US, the first power interlocking with the pneumatic control of field elements was installed in 1876. From 1882, George Westinghouse's company *Union Switch and Signal* installed the first interlocking with hydraulics (using water in summer and a non-freezing solution in winter) and hydro-pneumatic control of field elements. Development of the electro-pneumatic switch valve in 1891 was followed immediately by installations of electro-pneumatic interlockings. The first interlocking with electric control of field elements was installed in Central Europe by Siemens in 1898. Figure 9.52 shows the later version E43.

9 Interlocking Machines

Figure 9.52: Electro-mechanical interlocking E43

The early advent of track circuit detection for all interlockings in the US also generated another hybrid. This hybrid was a conventional mechanical interlocking machine with electrical contacts instead of mechanical connections for signal operation. This concept was later expanded to electrical contacts instead of mechanical connections for some or all of the points in an interlocking. The interface between the electric and mechanical portions of the interlocking was found in electric locking of the mechanical switch locking levers and sometimes in the addition of point circuit controllers for proof of point position.

At the time of their invention, these interlocking technologies were named electric interlocking, pneumatic interlocking or hydraulic interlocking as appropriate, with power interlocking or power frame as a summarised term. In English-speaking countries, this nomenclature survived the introduction of relay ('fully electric') interlocking and is still used today. The term electro-mechanical interlocking in Germany refers to the above described solutions with mechanical and some electrical interlocking functions, but electric control of field elements. In the USA, this refers to a mixed form with the interlocking functions in mechanics and the element control in mechanics for some and by electricity for other field elements.

The main advantage of these hybrid forms over mechanical interlocking is the extended range of element control, as the signaller does not need to use his muscles to expend the force to operate the elements against the friction in the points itself and in the wire or rod system. Another advantage is in the size of the interlocking machine. The levers and locking elements could be much smaller than those of a mechanical interlocking. Thus, the interlocking tower could be smaller for a given track arrangement than was possible with mechanical interlocking. The practical importance of these advantages was the following:

– In countries where track circuits were already used for track clear detection (e.g. USA), the introduction of such interlocking helped to enlarge the area which could be operated from one signal box. Thus they could reduce the need for staff and were widely used.
– In countries where no track circuits were then used (e.g. Germany), the only important advantage was that one signaller could operate more routes in the same time, but the interlocking area of a signal box remained limited by the signaller having to prove visually that all tracks under his responsibility were clear. As the technology was more expensive in installation, electro-mechanical interlocking was limited to large nodes with dense rail traffic. Hydraulic and pneumatic forms gained only very little importance in Europe.

9.5.2 Hybrid Relay and Electronic Forms

Another frequently used hybrid form is relay-electronic interlocking with interlocking function in relay technology and operation control functions in vital or non-vital electronic technology. Such systems have been used since the 1970s in different countries. Several interlocking types were designed in this form originally, whereas in other cases this solution is the result of integrating relay signal boxes into electronic operation control centres.

These hybrids to some extent combine the advantages of both, the relay and the electronic technology:

- The relay technology at the interlocking level enables good safety behaviour with limited effort and the option for easy alterations. As the interlocking functions are comparatively simple, the relay technology is able to manage these functions with acceptable effort.
- The electronic technology in the operation control level enables flexible operation functions. Especially in systems with non-vital man-machine-interface, the difficulties of electronics for safe processing are not crucial, and these parts can be manufactured rather cheap.

In many cases, these hybrid forms offer solutions which are economic in installation and flexible in operation. Interlocking systems of this type are used in many countries, such as France, Poland and Russia.

The North American concept of Centralised Traffic Control, developed in 1927, has developed towards a similar hybrid. CTC involves a central interlocking machine operating a number of remote control interlockings distributed along a railway line. The technological breakthrough was the development of a way to send all control and indication functions through a single pair of wires by using a signal of long and short pulses. The control machine would encode the desired interlocking control movements when the operator pushed the start button. The selected interlocking would respond, decode the message, and operate the interlocking control circuits in the field. The control machine would continuously poll all the interlockings for status. Any change in status (e.g. signal or point position changes, change in track circuit status), would be encoded by the interlocking and transmitted to the control machine.

In order to use this technology, it was necessary for the control machine to be non-vital. This arrangement made transition to an electronic/relay interlocking combination relatively easy. The early installations of the 1980s used a solid state control machine employing a relay code emulator to transmit and receive codes. Electronic interlocking is now well established and current installations involve electronic non-vital control machines, data transmission by microwave, radio, or fibre-optic cable, and electronic interlockings.

10 Line Block Systems
Sergej Vlasenko, Gregor Theeg, Ulrich Maschek

10.1 Classification

This chapter describes the technical systems for safety on open lines. The logical principles on which these systems are based are described in chapter 4.4. In contrast to that, the main focus of chapter 10 is to describe the technical solutions for block systems.

Table 10.1 classifies the systems according to the following criteria:
- Technical or non-technical safety system: The main focus of chapter 10 (and the whole book) is technical systems. The non-technical systems are described briefly in chapter 4.4.1.3 and safety overlays for these systems in chapter 10.2.
- Centralised or decentralised block system. In decentralised systems (chapter 10.3), the block information of each block station is processed locally at the respective block station and exchanged with neighbouring block stations. (Note that station, junctions etc. do also count as block stations in the context of their functions for safety on adjacent line sections.) In centralised systems (chapter 10.4), the information of several block station is processed centrally, which requires longer information connections to the block signals and track clear detection.

centralisation	Safety provided by:	
	persons obeying rules	technical systems
decentralised	e.g. telephone block; chapter 4.4.1.3	chapter 10.3
centralised	e.g. Track Warrant Control (TWC), Direct Traffic Control (DTC); chapters 4.4.1.3, 10.2	chapter 10.4

Table 10.1: Classification of systems for safety on open lines regarding centralisation and technical equipment

Chapter 10.5 also describes briefly first experiences with moving block systems.

10.2 Safety Overlays for Staff Responsible Safety Systems

Before the introduction of safe block systems, non signal-controlled operation was used with responsibility for safety in the hands of staff. However, technical systems were used occasionally, which were not safe systems in themselves, but gave some support to reduce the risk of human errors. Later, block systems were often developed out of these systems. Therefore, first these technical safety overlay systems are described briefly here.

In Europe, traditionally decentralised systems with safety responsibility in the hands of persons were used (telephone block), whereas in North American centralised systems of dispatcher-controlled unsignalled operation DTC (Direct Traffic Control) and TWC (Track Warrant Control) dominated (chapters 4.4.1.3, 3.4.3). After World War II, centralised systems also came in use in several European countries for secondary lines (e.g. ZLB: 'Zugleitbetrieb' in Germany) to reduce costs for decentralised operations in each station or for signalling equipment (figure 10.1). In all these systems, safety is basically in the responsibility of people.

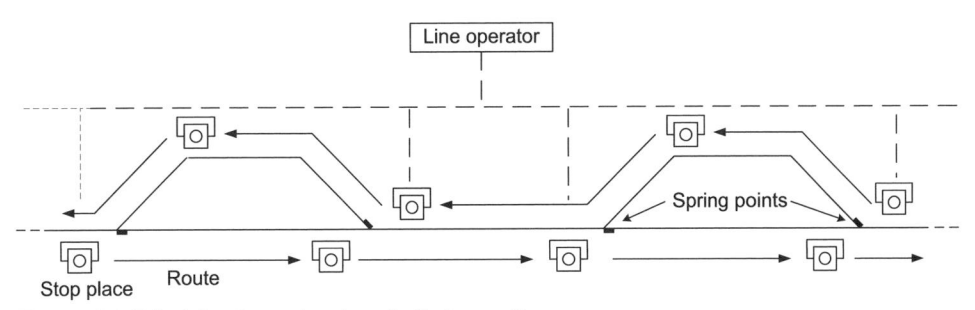

Figure 10.1: Principle of non signal-controlled operation

To increase safety nevertheless, several systems were developed with additional safety overlays for following and opposing protection. In contrast to a signalling system, these overlay systems do not guarantee safety on their own, but only help the operator and the drivers to reduce the probability or the consequences of human errors. Safety responsibility in normal operation basically remains at the dispatcher, who issues the movement authorities verbally after the driver has sent him the train position. Examples of such safety overlay systems are:

- The train itself recognises another train in its proximity and gives a warning signal to the driver or initiates automatic emergency braking if two trains come too close. For detection of the relative position between two trains, direct radio communication between trains or satellite positioning based methods can be used.
- The occupation of open line sections is detected track-based and a train protection system stops a train that tries to enter an occupied section. This may be either because of driver error or because of impropriate movement authority given by the dispatcher.
- The dispatcher has to reserve a path for the train in a non-safe interlocking-like system before issuing the Movement Authority verbally. Reservation of sections for different trains at the same time is refused, and trains trying to enter a non-reserved section are emergency braked automatically. But in contrast to an interlocking system, these locking functions work in the background and the Movement Authority to the driver is issued only verbally.

Most of the systems where the safety responsibility is with people do not provide for point control by the dispatcher. Therefore points are operated locally by the train driver (generally in North America) or spring points (chapter 6.1.2.2) are used.

10.3 Decentralised Block Systems

10.3.1 Overview

Figure 10.2 shows a classification of decentralised block systems by different criteria, excluding some special cases. As described in chapter 4.4.3, block systems can firstly be distinguished into token and tokenless block systems.

Further criteria to distinguish tokenless block systems are:

- The continuity of unblocking (chapter 4.4.3). The line can only be unblocked either singularly in one moment after clearing (chapter 10.3.3) or continuously during the whole time the section is clear (chapter 10.3.4).
- The grade of automation. manual, semi-automatic and automatic systems can be distinguished. In manual systems, all blocking processes are done manually. In semi-automatic systems some blocking processes (e. g. the blocking after a train has entered) are done automatically, whereas others (e. g. unblocking after the train has completely cleared) are done

10 Line Block Systems

method to ensure clear status of the block section	token block	tokenless block		
	presence of a physical element on the train	singular unblocking after movement		continuous unblocking
		by operator	by technical system	
technical track clear detection on open line	not necessary		yes (track-based detection)	
level of automation	manual	manual	semi-automatic	automatic
technical systems (examples)	(electric) token block	manual block	relay block	automated relay block; COMBAT	automatic block
described in	10.3.2	10.3.3		10.3.4

Figure 10.2: Classification of decentralised block systems

manually. In automatic systems, all processes including the unblocking (but not necessarily the change of direction) are done automatically. Automatic block requires to eliminate the necessity of visual observation of end of train before unblocking. Solutions can be continuous track clear detection (chapters 5.3, 5.4) or an end of train detection system (chapter 5.2.7).

However, the criteria 'continuous unblocking' and 'automatic' are not identical. There are also systems which use track clear detection and work fully automatically, but transmit the unblocking information only singularly to the previous block station. The clear status of the track is only checked upon clearing of the section by a train and later (unauthorised) entry of rolling stock will not be evaluated by the block system. An example of a system which exists in both versions, the semi-automatic and the automatic, is RB II 60 (10.3.3.3).

Systems using end of train detection instead of track clear detection can also work automatically, but transmit the unblocking information only upon clearing.

10.3.2 Token Block Systems

10.3.2.1 Overview

Token block systems originate from Britain. The authority to occupy a certain track section is issued to a certain train by the ownership of a physical element. This element is traditionally a staff, but can also be a disk, a paper with a defined text written on it or others. Occasionally also a person (a pilotman) who has to be present on the train to permit entry, fulfils the same function. The functional processes in token block systems are described in chapter 4.4.4, whereas this chapter 10 deals with the technical solutions. One-Train-Staff Systems and Train, Staff and Ticket Systems are still used in several countries on secondary lines, but also in degraded mode operation in case of failure of technical block systems. In some countries the token includes the permission to drive back e. g. after maintenance work on open lines.

The token block system can also be supplemented by a simple train protection system. For example, the driver places the token into a device on the locomotive. With this information, the trainstop intervention is suppressed, which would otherwise occur upon passing a trackside resonant circuit at the exit of the station.

The following chapter 10.3.2.2 deals with Electric Token Block. Radio Electronic Token Block (RETB), a centralised system which simulates the exchange of tokens electronically, is discussed chapter 10.4.3.

10.3.2.2 Electric Token Block

In Electric Token Block (see also chapter 4.4.4), several tokens exist for each block section. The tokens are interlocked in stationary token instruments (figure 10.3) on both ends of the respective block section to ensure that only one token can be out of the instrument at the same time. The tokens belonging to the same block sections are identical, but those of neighbouring sections vary physically. This prevents the 'wrong token' problems (principle of key and keyhole). The locking of the tokens is undertaken electrically. Different geometrical forms of tokens and the related instruments are described detailed in (Doswell 1957).

Electric token block, likewise the other forms of token block, originated from Britain and had a large historical distribution in different countries. Today the system is little used. Figure 10.4 shows the decline in usage of these systems on an example.

Figure 10.3: Stationary instruments of electric token block

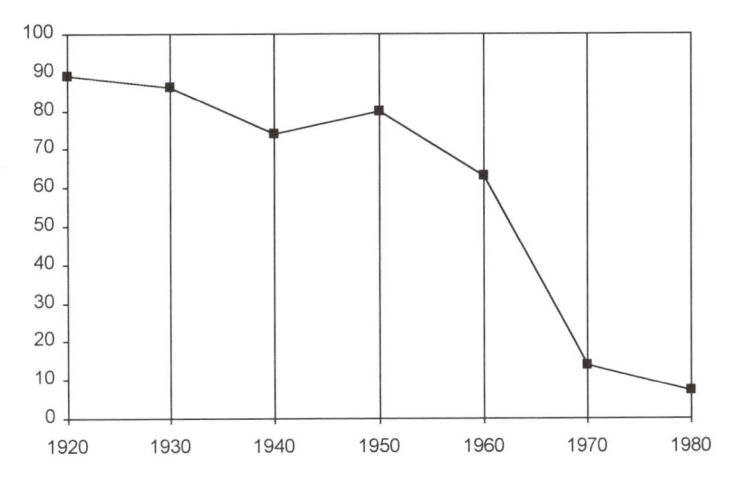

Figure 10.4: Percentage of electric token block on open lines in Soviet Union (information source: Railways Museum St. Petersburg)

10.3.3 Systems with Singular Unblocking upon Clearing

10.3.3.1 Overview

Systems of this category are tokenless systems where the unblocking information (setting the status of the track section to 'clear') is only transmitted punctually after the train has cleared the section. If later a moving unit enters the section unauthorised, the block system will not recognise this occupation, unless a particular 'alarm' status is defined.

10 Line Block Systems

The non-technical ancestor of these block forms is the telephone block. All messages (offering of trains, report of departure and report of arrival) are done telephonically. A strictly regulated wording is used and messages have to be repeated to avoid misunderstandings.

Technical block systems, according to the grade of automation, can be divided into:
- **Manual Block Systems.** All actions (acceptance, blocking and unblocking) are done manually, with the power for information transmission provided by the hands of the signaller.
- **Semi-Automatic Block Systems.** According to the definitions in the particular countries, this includes a wide variety of arrangements, for example:
  - All blocking processes have to be initiated manually, but the power is provided technically. The nomenclature differs between the countries. For example, in Germany these forms are also called 'manual'.
  - The blocking occurs automatically upon occupying the block section or while setting a route onto the open line section, but the unblocking is initiated by a person. The reason is that the end of train has to be checked visually before unblocking.
- **Automatic Block Systems.** Besides blocking, unblocking is also done automatically upon clearing. This requires continuous technical track clear or end of train detection. Only the acceptance of trains, unless automatic train routeing systems are used, is to be done by persons, as it depends on operational decisions. However, on double lines there is usually a normal direction of traffic. Therefore the change of direction is necessary only relatively seldom or even technically impossible there.

The historically oldest form of block instruments were used in Britain in the 19th century (figure 10.5) for double lines. Both ends of the block section are equipped with those instruments which simultaneously show 'line clear', 'line blocked' (meaning: reserved for a certain train) and 'train on line' by a needle in the respective sector. One pair of block instruments serves for each direction. The block instruments are operated by the signaller at the exit of the block section and identical information is displayed at the entrance. Those messages which originate from the entrance have to be transmitted to the exit by an exchange of bell signals. However, these block instruments in most cases were not technically interlocked with the signals and therefore served only for remembering the status (IRSE 1999).

Figure 10.5: Historical British block instruments for double line

The oldest interlocked block instrument was invented in Germany in 1872 (chapter 3.1). The practically most important block form which uses these instruments is the manual block with three block instruments, manufactured by *Siemens & Halske* (chapter 10.3.3.2). It was once much used, but now only to a limited extent in countries of Central Europe. The number of working installations is rapidly declining.

With the introduction of relay technology, a large diversity of relay block systems was developed in different countries, mainly in Europe including Russia. Their operation became partly or even fully automated. For the development of these systems, an important requirement was often the compatibility with the block systems of neighbouring older (mostly mechanical) signal boxes (chapter 10.3.3.3).

10.3 Decentralised Block Systems

10.3.3.2 Manual Block Siemens & Halske

Block Instrument

In 1872, *Carl Frischen* patented the interlocked block instrument, in German called 'Blockfeld', which became the basis of the manual block in several countries (figure 10.6).

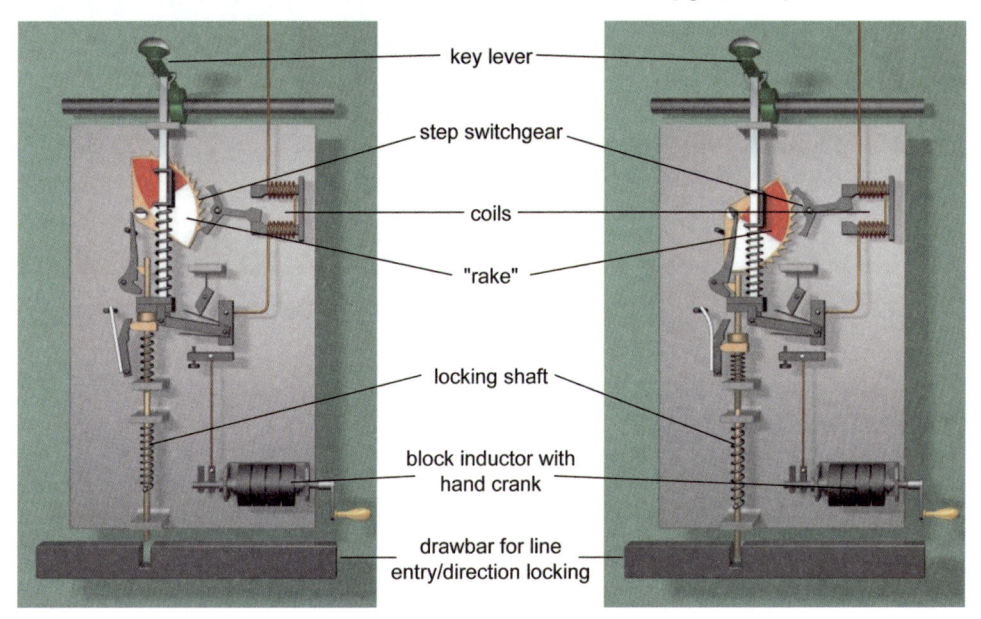

Figure 10.6: Block instrument (left: unblocked, right: blocked) (simplified graphic: *TU Dresden*)

The block instrument is blocked by pushing the key lever on top of it. This action pushes the locking shaft down, where it remains due to an internal lock inside the block instrument. The locking shaft is interlocked mechanically with a particular drawbar for line entrance/direction locking. This bar mechanically locks all signals onto the respective line in stop position. While pushing the key lever, the signaller turns a hand crank belonging to the inductor. This induces a block current (AC with low frequency) which moves a step switchgear by one tooth in each period, together 12 teeth, and makes the red sector of the 'rake' visible to the signaller. The induced current flows to the corresponding block instrument at the other end of the block section.

When a blocked block instrument is reached by the block current induced in the opposite block inductor, the 'rake' is moved back towards the unblocked position and opens the mechanical lock.

Summarised, this means that during blocking process, one block instrument changes from unblocked to blocked operated by the signaller, whereas the other changes from blocked to unblocked, operated by the block current.

Arrangement of Block Instruments

In the most used version with placed direction, each block instrument has exactly one partner, which is a corresponding block instrument at the other end of the block section concerned. There are three block instruments for each end of the block section (figure 10.7) with different function, but the same technology (compare chapters 4.4.5.1, 4.4.5.3):
- Entrance Instrument (German: Anfangsfeld): This instrument is blocked after a train has entered the line section and locks all signals leading onto the respective line section in Stop. The partner

10 Line Block Systems

of the entrance instrument is the exit instrument of the neighbouring block station.
- Exit Instrument (German: Endfeld): This instrument is in unblocked position when a train is expected; in all other cases it is in blocked position. It doesn't lock any signals in its own block station. The partner is the entrance instrument of the neighbouring block station.
- Direction Instrument (German: Erlaubnisfeld): This instrument is blocked when the respective train sequence station doesn't have the permission to send trains to the line section. It locks the same signals as the entrance instrument. The partner is the direction instrument of the neighbouring train sequence station.

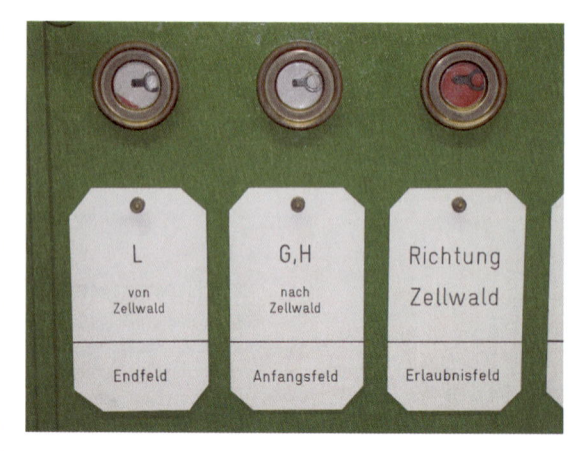

Figure 10.7: Block instruments for a single line section

On a single line, each block station is equipped with entrance and exit instruments for each direction. Only train sequence stations where the sequence of trains can be changed (e. g. stations, junctions) own direction instruments (figure 10.8), but each block station, by a contact of its entrance instrument, disrupts the wire dedicated to exchange of direction automatically if the block section ahead is occupied. On double lines, usually no direction instruments are provided, as signalized traffic is only possible on the right track in this old technology.

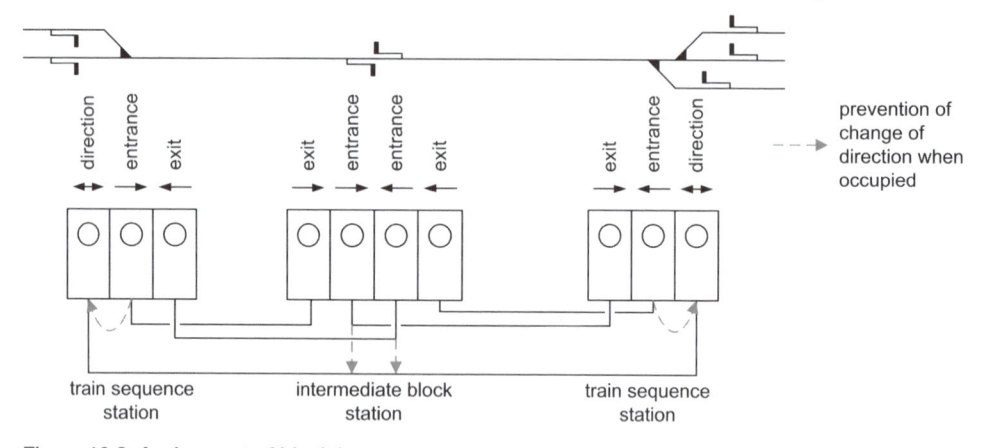

Figure 10.8: Assignment of block instruments on a single line

Process of Block Working

In the following, the processes in the block system are described (compare chapter 4.4.5.3).

The system works with placed **direction**. Changing the direction is only possible when the section between neighbouring train sequence stations is clear and no signal leading onto this section is open. Only the station currently owning the direction can change it by blocking the own direction instrument, which unblocks the corresponding direction instrument.

With clearing the signal, a **signal repetition lock** is activated mechanically and safely by the signal lever. After releasing the signal to stop, it will still lock all signals leading onto the same

10.3 Decentralised Block Systems

line section including the operated signal itself in closed position. Therefore, no dangerous situation can occur if the signaller forgets to block the line afterwards.

After the train has entered the section and after releasing the signal to stop, the signaller blocks the line by blocking the entrance instrument. As the entrance block instrument fulfils the same locking functions as the signal repetition lock, its purpose is mainly to transport the information to the block station in advance.

To prevent erroneous **unblocking** of the line section when the train is still there, the train is involved in the process. For this purpose, a combined detector consisting of a rail contact (chapter 5.2.2) and a short track circuit (chapter 5.2.3.2) is applied in the end of the block section. This combined detector and its evaluation circuitry enable the exit instrument to be blocked when three conditions are fulfilled in this sequence:

1. The signal must have been open.
2. The train must have passed the rail contact.
3. The train must have cleared the short track circuit.

This detector does not replace the observation of the end of train markers by ground staff, as it doesn't detect wagons lost on the line section.

More information on this example can be found in (Maschek/Lehne 2005).

Variants

This type of block has been adapted to different countries with different principles of block working (chapter 4.4). For example, in Switzerland a version adapted to neutral direction is applied (Oehler 1981). Here the dependences between the block instruments in the corresponding block stations are more complex, as the same block instrument corresponds with different partners.

10.3.3.3 Relay Block RB II 60

Overview

The relay block RB II 60, designed in East Germany by *WSSB*, is a good example of a technical migration process. With the appearance of relay interlocking, the need for a block interface to neighbouring mechanical interlockings arose. The relay block was so successful that even today it still serves as an interface between electronic interlockings of different manufacturers. The relay block RB II 60 is made upon two principles:
- the idea that, except of three in manual block, only one locking element is sufficient and
- the adaption of the circuitry to compatibility with manual block.

Relay block RB II 60 is used in the semi-automatic as well as in the automatic version:
- In the semi-automatic version, the line is blocked automatically, but the signaller has to unblock it by pushing a button after observing the end of train marker.
- In the automatic version, also unblocking is done automatically thanks to technical track clear detection on the line.

Block Relay and Block Current

To enable communication also with neighbouring stations which are equipped with manual block (chapter 10.3.3.2), a generator is used to produce a similar current as generated by the hand crank and a special block relay to evaluate this current (approximately 60 V and 12 Hz).

The block relay, which is used by RB II 60 and other block forms, consists of basically three parts (figure 10.9):

10 Line Block Systems

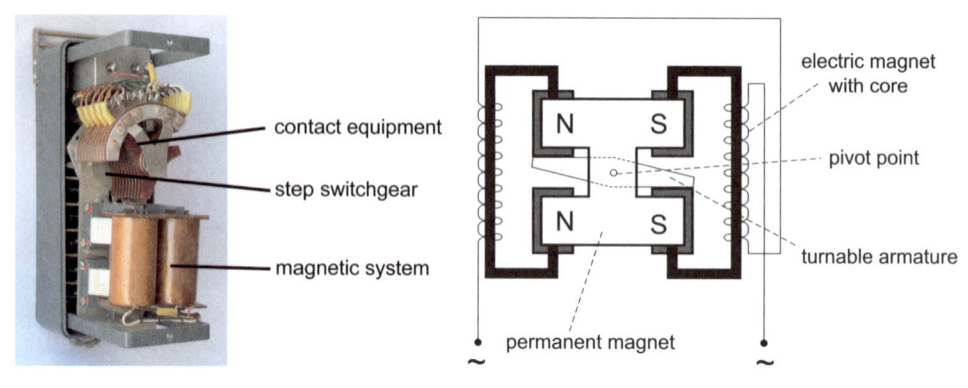

Figure 10.9: Block relay of RB II 60

Figure 10.10: Magnetic system of RB II 60

- The magnetic system (figure 10.10) contains a turnable armature which switches in each half wave of the block current. In absence of a block current, it is fixed in its position by the permanent magnet.
- The step switchgear contains a cogwheel with 72 teeth. A full revolution of the cogwheel is performed after three blocking and unblocking processes.
- The contact equipment contains several isolated contact segments with three arms each. These are rigidly connected with the cogwheel of the step switchgear.

Block Logic

Figure 10.11 shows the schematic block logic of RB II 60. The three block instruments of manual block (chapter 10.3.2.2) could be reduced to only one block instrument for following reasons:
- As both, the entrance and the direction instruments, lock the same signals, they can be summarised to one block relay.
- As the exit instrument doesn't lock any signal in the own interlocking, it can be replaced by a simple relay.

The information connection with the neighbouring block station is the same as in manual block, therefore either a station with manual block or another one with relay block can be the neighbouring. The information from the block relay is directed either to the direction or to the entry lines, depending on the position of the signal repetition lock, which means whether the signal has shown 'proceed' before (and therefore a train entered the block section) or not.

More information on this example can be found in (Kusche 1984) and (Maschek/Lehne 2005).

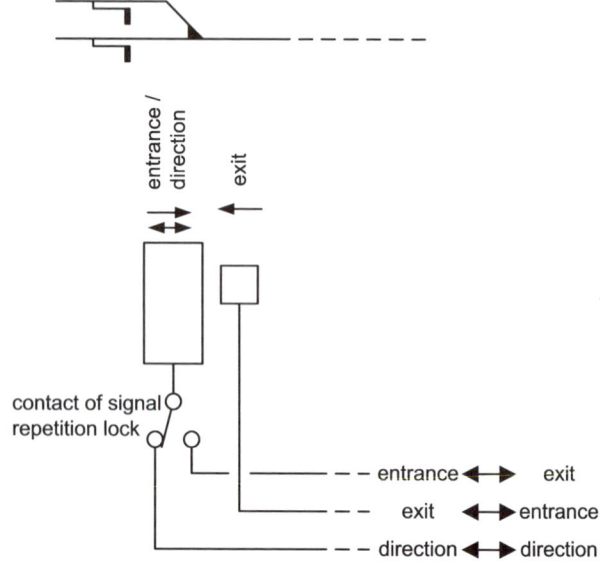

Figure 10.11: Block circuit of RB II 60 (simplified)

10.3 Decentralised Block Systems

10.3.3.4 Relay Block RPB GTSS

This block is applied in countries of the former Soviet Union, mainly on single lines without intermediate block stations between the train sequence stations. The system works with the principle of neutral direction (chapter 4.4.5.2). The working algorithm consists of the consecutive transfer of three messages:

1. Permission from the receiving station: train may be sent,
2. Blocking from the departure station before its exit signal has been cleared,
3. Unblocking from the arrival station after the train has come there and the end of train been seen.

After unblocking the relay circuitry comes back to the initial condition. Now each station can give permission for the next train upon telephonic request (neutral direction).

For transfer of these messages, there are two wires which can be used also for telephone connection (figure 10.12). In order to distinguish permission and unblocking messages, they are transferred with different polarity.

The exit route from the station A can be set only after receipt of the permission from the station B. Permission relay P of station B sends positive polarity, control relay C of station A picks up, but the current is too weak to pick up relay O of station B as well. After typical station operations the exit signal is cleared. Its clearing is accompanied by transfer of the blocking message through contacts of signal relay S from the station A to the station B where the occupation relay O (line is blocked) picks up and remains in up position by a stick circuit. The arrival of the train to station B is technically detected (usually through sequential occupation and clearing of station track circuits). Then the signaller checks the completeness of the train and gives confirmation of its arrival through the acknowledgement relay R. Control relay C of station A, which is a bi-stable relay, switches due to contra polarity.

This system can also be applied with block stations between stations. In this case, the two-wire line is disrupted, and in the block station the similar devices are established for both

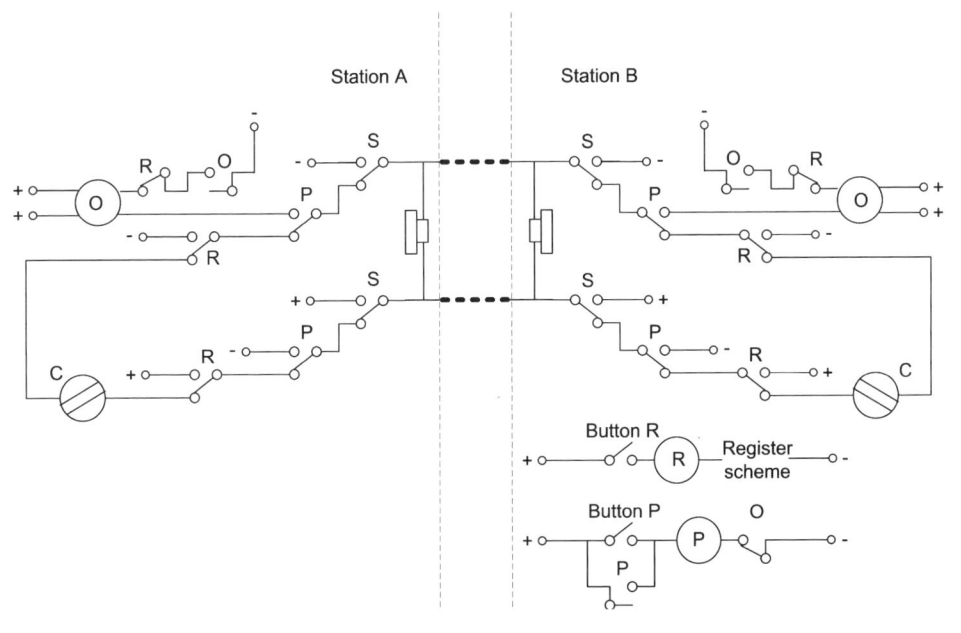

Figure 10.12: Circuit of RPB GTSS

sides. The difference is that the signaller of the block station cannot give, but can only transfer the permission.

This system can be applied on double lines, too. In this case stations are connected by two two-wire cables, each of them being used for one direction. Therefore the circuits are not symmetric. The permission is not required as each track is used for one direction only.

10.3.3.5 Japanese Electronic Block for Secondary Lines (COMBAT)

Beginning in the 1980s, Japanese railways introduced an electronic block for secondary lines with low traffic which is operated partly automatically and partly by the driver, but without operation ground staff. Together with the related detection system (chapter 5.2.6.2) it got the name COMBAT (computer and microwave balise-aided train control system). Each train is individually known by the block system and detection is done by ID number at fixed locations (chapter 5.2.6.2). The system is therefore only applicable to closed networks. The working principles are as follows:

- Each train carries an on-board unit for communication. This unit contains a unique identification number of the train.
- When the train is ready to depart, the driver pushes a starting button. The request including the train ID is sent by short-range radio or optical wireless transmitter to the station controller and from there via cable to the station controller of the neighbouring (receiving) station (neutral direction).
- After an affirmative answer, the signal can be cleared.
- Train departure is detected by track circuits, which are present in the station, but not on open line. The 'train on line' message is sent to the receiving station. The home signal of the receiving station is time controlled and clears a defined time after train departure.
- The arriving train sends its ID to the station equipment. Trackside train completeness check is not necessary due to the characteristics of the train. The receiving station then unblocks the line.

In each station a 2 out of 2 fail-safe computer is used for information processing. This computer also fulfils interlocking functions inside the station. The communication between stations is done via cable.

More information on this example can be found in (Sasaki 1986) and (Sasaki/Wakabayashi 1989).

10.3.4 Systems with Continuous Unblocking

10.3.4.1 Overview

The ancestors of systems of this group were introduced in the end of 19th century upon the invention of the track circuit in the USA. The entrance to a block section permanently receives the occupation information from the track circuit. This means that, in contrast to the systems described in chapter 10.3.3, an unauthorised occupation will be detected.

Train movements in these block systems can either be signalled by trackside signals or by cab signals or by both simultaneously. In the latter case, in case of disagreement between trackside and cab signal, different regulations are defined about which has priority (chapter 7.1). Most block systems of this group work with permissive block signals (chapter 7.3.2), which provides for a high capacity (however, at low speed) even in case of failure of equipment.

The basic specifications of some widely used block systems are compared in table 10.2.

10.3 Decentralised Block Systems

Country	Name of block system (or train protection system which includes block function)	Electrical supply for traction	Length of block section	Detection of train by	Frequency of detection current	Insulated rail joints	Passage of certain points of automatic block							Transmission of date between signals	Centralisation of block processing of messages	Possibility for operators to stop a train
							Train protection and cab signalling function				Trackside signal regulations				
							Number of fully free block sections	Continuous train control	Speed limit on the end of section, kmh	Intermittent train control	Speed limit after passage of track element	Related trackside signal aspect	Speed limit for passage the signal			
Czech Republic/Slovakia	LS	3 kV DC, 25 kV AC 50 Hz	1000-1500 m	One track circuit per block section	50 Hz, 75 Hz	yes	2 and more	FM 5.4 Hz	160	no	-	green	160	track circuit	no	no
							1	FM 3.6 Hz	120			yellow	120			
							0	FM 0.9 Hz	stop (up to 30)			red	stop			
East Germany	AB70	15 kV AC 16.7 Hz	about 1000 m	One track circuit per block section; additionally short track circuit or (and) spot wheel detectors at block stations	42 or 100 Hz, 10 or 16 kHz	yes	2 and more	no	-	no	-	green	160 (100 without train protection)	cable	no	no
							1			INDUSI	160	yellow	160 (100 without train protection)			
							0			INDUSI	Emergency braking to stop	red	stop			
France	TVM 300 TVM 430	1.5-3 kV DC, 25 kV AC 50 Hz, 15 kV AC 16.7 Hz	about 2000 m	One track circuit per block section with capacitors every 100 m	1700 and 2300 Hz for one track of a double line; 2000 and 2600 Hz for the other track	no	5 and more	FM 10.3-29 Hz in TVM 300, Data telegram in TVM 430	300	no	-	no	-	cable	yes	no
							4		270							
							3		230							
							2		170							
							1		(up to 35)	Coded loop in front of stop signal	stop, if the next signal has absolute means					
Italy	BACC	3 kV DC, 25 kV AC 50 Hz	1350 m	One track circuit per block section	50 and 178 Hz	yes	4 and more	FM 1.25-4.5 Hz for every track circuit frequency	260	no	no	green	200 without train protection on the frequency 178 Hz	track circuit	no	no
							3		230			green	200 without train protection on the frequency 178 Hz			
							2		200			green	200			
							1		125			yellow	120			
							0		0 (up to 50)			red	stop			

▶

10 Line Block Systems

Country	System	Voltage	Block length	Track circuits config	Frequency	Continuous	Code	Data/signal type	Value	Special	Signal	Speed	Medium		
Japan	ATC	25 kV AC 60 Hz	about 3000 m	Two track circuits per block section	720 and 900 Hz for one track of a double line; 840 and 1020 Hz for the other track	no	4	Data telegram	260	no	no	-	cable	yes	yes
Nether- lands	ATB-FG	1,5 kV DC	1000- 1800 m	One track circuit per block section	75 Hz	yes	3		230	Coded loop in front of stop signal		stop	track circuit	no	no
							2		170						
							1		120						
							½		30						
							0		0						
Germany Austria Spain	LZB	15 kV AC 16,7 Hz 25 kV AC 50 Hz	variable	Axle counter or track circuits	-	no	2 or more	FM 1,6 Hz	140	no	green	140	cable	yes	yes
							1	FM 3,7 Hz	140		yellow	140			
							0	No code	0 (up to 40)		red	stop			
Countries of former USSR	ABK	3 kV DC 25 kV AC 50 Hz	1000- 2600 m	One track cir- cuit per block section	25 or 50 Hz	yes		Data telegram 36 kHz to train 56 kHz from train through cable loop	up to 300 depending on track and train	INDUSI or ASFA (as reserve)	as re- serve	160 or 180	track circuit	no	no
							2 or more	Impulse code green	140		green	140			
							1	Impulse code yellow	60		yellow	60			
							0	Impulse code red-yellow	0 (up to 20)		red	stop			
Russia	ABTC	3 kV DC 25 kV AC 50 Hz	1000- 2600 m	3-6 track circuits per block section according to length	420-780 Hz	no	2 or more	Impulse code green, if the train is coming	140	no	green	140	cable	yes	no
							1	Impulse code yellow, if the train is coming	60		yellow	60			
							0	Impulse code red-yellow, if the train is coming	0 (up to 20)		red	stop			

Table 10.2: Comparison of basic specifications for some block systems

10.3 Decentralised Block Systems

10.3.4.2 Number of Tracks and Traffic Directions

Figure 10.13 presents principles of traffic directions on single and double lines. Most types of automatic block with continuous unblocking can also be applied on lines with more than two tracks which, however, are usually operated like several parallel single and double lines. With certain efforts and compromises, each double line can allow contraflow traffic, at least on the principle 'one train between stations' with telephonic messages. However, there are three possibilities for normal operation on double lines (see also chapter 3.2.4):

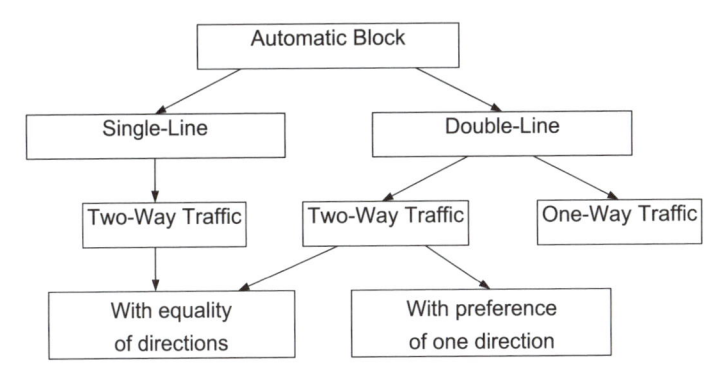

Figure 10.13: Traffic direction in automatic block with continuous unblocking

- Only one-way traffic is supported by the block system. Contraflow movements have to be done by methods of degraded mode operation.
- Bi-directional traffic is enabled, but one direction is preferred. One possibility of preference is that traffic for one direction can be controlled via track and cab signalling, but for the other direction only by the cab signal. Another version is that the length of block sections is shorter for the preferred direction.
- Both directions of traffic are supported equally.

The change of direction between two train sequence stations may only be done if no train is on line. It is initiated through command from one or both station operators. For the case that the open line is clear but detection devices are defective, there is usually a special registered command provided.

Figure 10.14: Automatic block station in Russia

10.3.4.3 Block Sections

Most block systems with continuous unblocking use three-aspect-signalling, which means that the length of a block section is in proximity of the braking distance and each signal works as distant signal for the signal beyond (chapter 7.3.3.2). Block sections can have a constant or different length. Provided that the train moves with constant speed and therefore has constant braking distance, each block section can basically have the same length (in Europe typically between 1,000 and 1,500 m; in spite of some variations due to local specialities). But train acceleration after stations, braking in approach to stations and speed variations on the open line because of terrain gives an advantage to systems in which the length of the block section is adapted to the speed. In the ideal case, time-constant block sections can be achieved, where the occupation time of all block sections is the same, providing an optimal allocation of resources for best line capacity.

On high-speed lines the braking distance is long, therefore it usually stretches over several block sections with different speed indications in each block section in approach to a stop.

10.3.4.4 Communication between Block Stations and Interlockings

The working principle of decentralised block with continuous unblocking is shown in figure 10.15. Each block station continuously sends the aspect of its own signal to the block station in rear. There this information is combined with the track occupancy information and the aspect of this

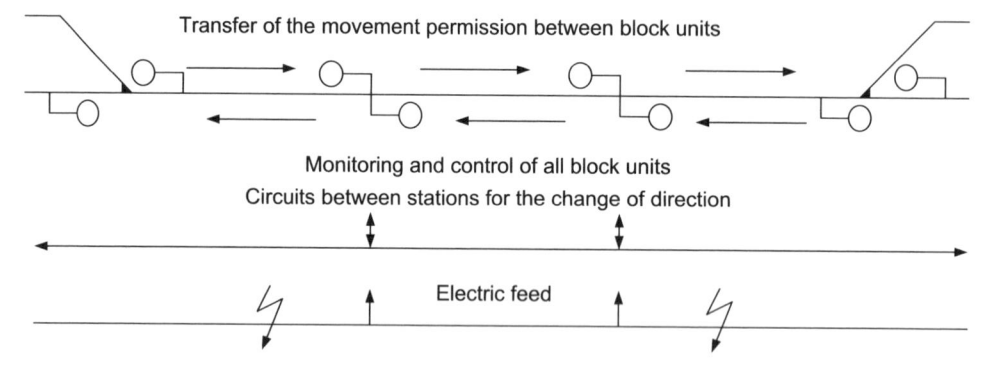

Figure 10.15: Principle of decentral automatic block

signal is created. In the oldest systems, the track circuit occupation and coding is the only information transmission, and information is only transmitted between two neighbouring block stations against the direction of travel. Newer systems often also imply other transmission functions, such as of the blocking information into the direction of travel, and use additional lineside cables. From the train sequence stations some commands simultaneously act via common wires to all block stations (e.g. for change of direction) and some diagnostic information is sent from several block stations to the nearest station by common wires. The main preconditions for a proceed signal are:
- Block section is clear.
- The proper direction is established.
- The following signal is alight (checked in the majority of systems).
- The overlap is clear (if this is required).

Transfer of the information between block units can be via
- cables;
- rails;
- combined via wires and rails.

Data transfer through rails can pass with use of pulse, frequency or phase attribute or modulation. All kinds of modulation as well as pulse and frequency signal attributes allow using the information simultaneously for cab signalling. The phase signal attribute is used seldom because it requires in-phase feed of all block units, which is technically difficult to solve.

As in absence of ground staff at most block stations, no signaller can influence the block signal if it malfunctions; decentralised automatic block is mostly permissive. It means that the driver can pass any red signal on sight, at his own responsibility (chapter 7.3.2). The normal position of the block signals (if line is clear and no train approaching) is proceed.

This decentralised principle requests electric feeds on the line (e. g. 6 or 10 kV) and transformers at each block station. To increase availability of power supply, redundant power supply can be used, either by accumulators or by second external feed. On AC electrified lines, electric feed for the block equipment is often taken from catenary via transformers.

10.4 Centralised Systems for Safety on Open Lines

10.4.1 Overview

In centralised line block systems, the detection and train position data is sent to a central controller, locking functions are performed there and block signals controlled from the centre, too. This block centre can be an interlocking of a neighbouring station or a separate block centre.

System	main disadvantages
token block (chapter 10.3.2)	slow processes; necessity to stop in each station
decentralised manual tokenless block (chapter 10.3.3.3)	slow processes; high efforts for staff
decentralised automatic tokenless block (chapters 10.3.3.3, 10.3.3.4)	high effort for decentralised maintenance
centralised block systems (10.4)	high effort for communication (cables or radio)

Table 10.3: Economic disadvantages of different kinds of block systems

These systems were developed historically from different origins in order to avoid certain economic disadvantages of the decentralised block systems (table 10.3). However, the more complicated communication between the centre and the trains emerged as a new disadvantage of centralised systems. The historical origins of the systems are:
- Extension of the interlocking areas and usage of route locking functions also on the open line (e. g. Germany)
- Concentration of the evaluation units of block stations in a central place (e. g. French high speed lines, Russia)
- Replacement of tokens by electronic information and their central processing (e. g. Britain)
- Improvement of safety overlay systems for unsignalled operation (chapter 10.2) to a full interlocking system (e. g. Germany)
- Development of new high speed systems with particular block system (several European countries)

Based on these different origins, the locking functions are solved logically in different ways (compare chapters 4.1.3 and 10.3.1; figure 10.16):
- Token logic. Virtual tokens are generated and exchanged with the train to permit entry of a particular section.
- Tokenless block logic. Within the central controller, block information is generated and exchanged like in a block system. The clear status of the line section is checked singularly upon clearing or continuously.

10 Line Block Systems

- Route logic. The open line sections are controlled like station areas with route locking. This is possible because all safety functions of block information can also be performed by routes, and because with the presence of continuous track clear detection, the reasons for sharp distinction of interlocking areas and open lines becomes obsolete (chapter 4.1.3). Besides, using the route logic on open lines can imply a gain of safety in certain cases of technical failure.

Decentralised Tokenless Block (for comparison):

N = entrance to the block section
X = exit from the block section

Centralised Tockenless Block:

Open Line Routes:

Radio Electronic Token Block:

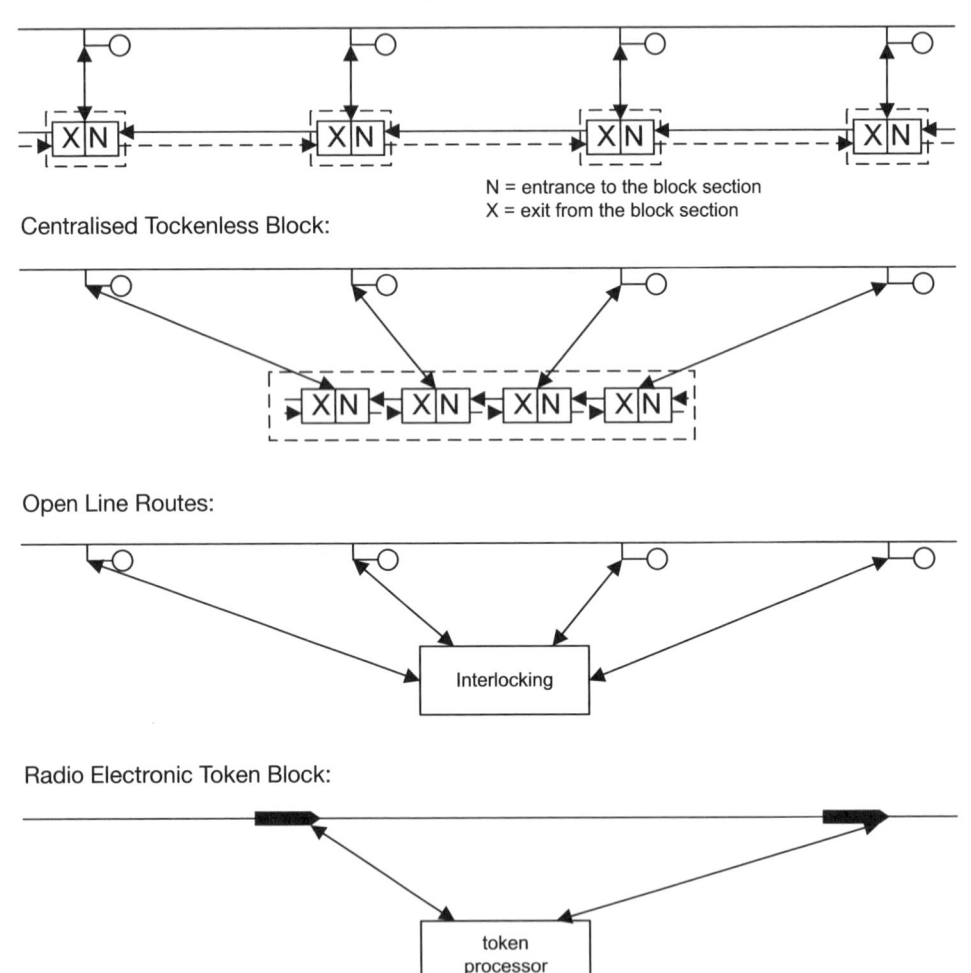

Figure 10.16: Different forms of centralisation of block systems

When using route logic on the open line, the block signals are red in normal position (= if all sections are clear), whereas with block logic with continuous unblocking, they are mostly at green in normal position. As operational decisions are not necessary before route selection on open lines, routes are usually set automatically when a train is approaching or after a exit route from the station has been selected. Due to improved communication possibilities in case of technical failure, absolute instead of permissive stop signals are often used on the open line.

10.4 Centralised Systems for Safety on Open Lines

10.4.2 Centralised Block Systems for Secondary Lines

In newer time, several centralised systems have been developed for secondary lines. The approach was to reduce costs by implementing simplified functions and cheaper technologies. One solution is so-called 'Signalised Track Warrant Control' ('Signalisierter Zugleitbetrieb' SZB) in Europe, which originated as a further development of safety overlay systems (chapter 10.2) and has now become a full electronic interlocking with route logic on the open line (chapter 10.4.4), with non-safe operation control and reduced comfort in degraded mode operation (e.g. driver has to confirm the clear status of a track section on a safe number keyboard on site, see figure 9.36).

A trial performed around 2000 was Funkfahrbetrieb (FFB) in Germany, an electronic interlocking where track clear detection is replaced by position messages from the train, routes are selected by the train itself and cab signals instead of trackside signals. Data transmission between the interlocking and the train was done by radio. However, this system did not gain practical importance.

Another system, the Radio Electronic Token Block, is described in the following.

10.4.3 Radio Electronic Token Block

Radio Electronic Token Block (RETB) originates from secondary lines in Scotland. It has its name because it simulates the handing over of tokens (chapter 10.3.2). Instead of physical tokens, virtual tokens are used in form of electronic messages and are centrally controlled in an electronic token processor. An advantage is relatively low costs for trackside signalling equipment combined with a high level of safety. A disadvantage is that the treatment of non-equipped trains is difficult. Therefore these systems are more suitable for closed networks without or with only little exchange of vehicles. RETB, with differences in detail and partly with different names, is used on several secondary lines distributed over the world.

The central component is a token processor for the whole network. This processor is a redundant safe electronic system, comparable with an electronic interlocking. It can either work automatically or be staffed. The token processor generates the tokens (movement authorities) upon request by a train. It is responsible for the exclusivity of the token in one block section. Communication is done by radio. The token information is signalised to the driver by cab signalling and becomes valid as soon as the driver has checked and acknowledged it. After clearing the section, the token is given back either by an operation action of the driver or, if track clear detection or train position detection is applied, automatically.

Radio transmission between the token processor and the train is safety critical. Therefore each onboard unit has a unique identification number and the data telegrams are redundantly coded to ensure safe transmission. Continuous radio coverage is not necessary, but attention has to be given to good radio coverage at locations where token messages are normally exchanged.

Fixed signal boards are placed along the line to mark locations where movement authorities end and where tokens will be exchanged.

Regarding the safeguarding of points in RETB areas, different solutions can be applied:
- Points are manually switched and locked by train or ground staff.
- Route interlocking functions are included into the processor and points are set by machine power.
- Route functions are included and points are switched and locked manually. This requires the application of key locks (chapter 6.5.4).
- Automatic spring points (chapter 6.1.2.2) are used, and each station track is used in one direction only.

10 Line Block Systems

The treatment of station tracks in the line block can be handled differently in the RETB systems (compare also chapter 4.5.1):

a. The station is controlled by a separate interlocking system (most preferable for bigger, complex stations).
b. Each station track is considered as one block section to enable track selectivity of the tokens (figure 10.17). This solution is particularly favourable for stations where trains regularly cross.
c. Stations are excluded from the block system and trains generally move on sight there.
d. Some station tracks (e.g. the straight track) are included into the block system, whereas others have to be used on sight.

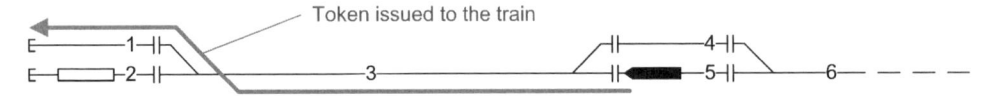

Figure 10.17: Dividing of the line into token sections in RETB (example)

RETB systems can be additionally equipped with track clear detection and/or a train protection system to protect against human error. More information on RETB can be found in (Wennrich 1997), (Jones 2008) and (Hall 2000).

10.4.4 Open Line Controlled from Neighbouring Interlockings

Beginning around 1980 with the appearing of electronic interlocking, several railways are centralising the control of adjacent open line sections into the neighbouring interlockings in new installations (figure 10.18). In most of these systems, on double lines both tracks can be equally used for traffic into both directions. Whereas some railways (e.g. Austria, Russia, France) maintain the concept of exchanging block messages between (virtual) block stations (whose evaluation units are concentrated in the same housing), others (e.g. Germany, Switzerland) abandon this idea and control open line sections based on route interlocking principles like in stations.

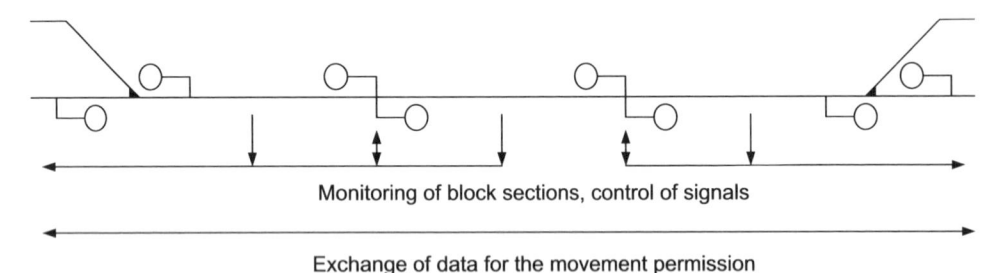

Figure 10.18: Central automatic block for relatively short open line sections

The limited distance (today up to approximately 10 km) for physical control of signals and detection devices is a problem particularly with longer distances between stations. To increase the control length, amplification units (figure 10.19) can be used. Another solution is to locate the element control of each signal on site and transmit only digital information ('logic control') between the interlocking and the block signal control box and feed energy by a ring line (figure 10.20 and figure 10.21). The advantages of the former solution are its usability for all interlocking types and simple equipment on the open line, the advantage of the latter solution is notable reduction of cable expenses.

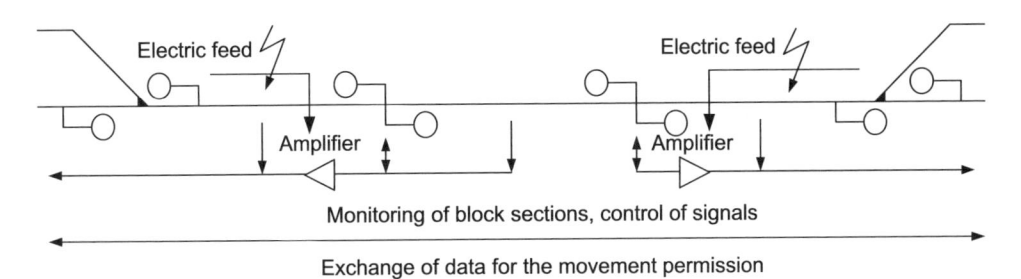

10.4 Centralised Systems for Safety on Open Lines

Figure 10.19: Central automatic block with amplifier

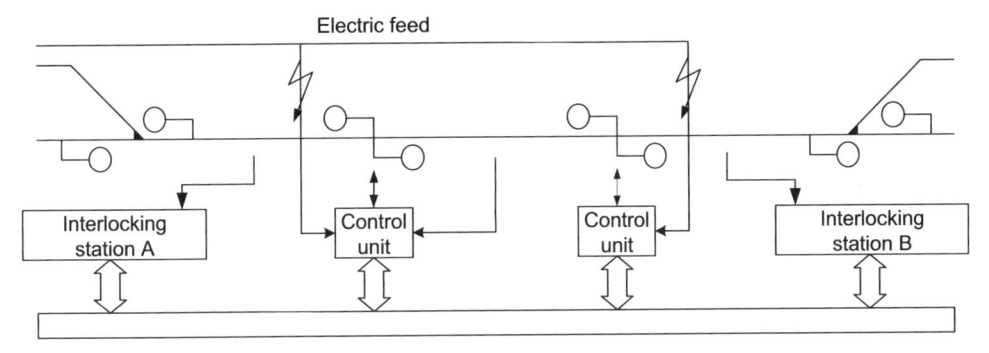

Figure 10.20: Central automatic block with control units of electronic interlocking

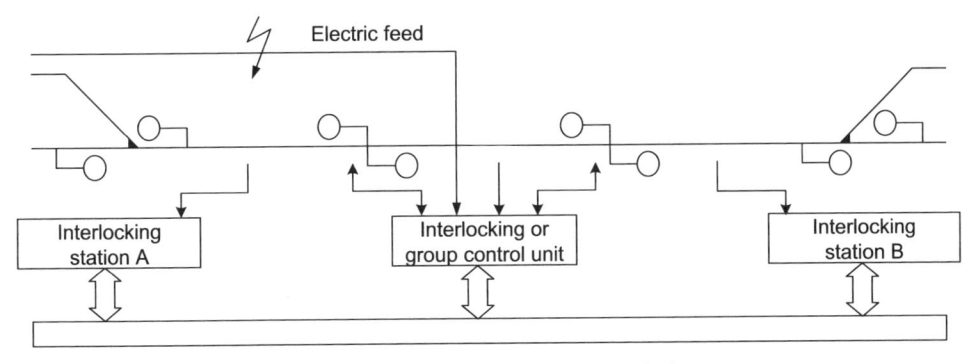

Figure 10.21: Central automatic block with electronic interlocking or group control unit on the open line

10.4.5 Train Control Systems for High Speed Lines

Most train control systems for high speed, such as LZB in Germany, TVM in France, ATC in Japan and ETCS Level 2 (8.3.4.4, 8.3.4.5, 8.3.6, 8.4), include a block system besides the train protection and cab signalling functions. These block systems are mostly controlled centrally. Each block centre is responsible for a longer line section of approximately 100 km or more. Whereas some systems (e.g. TVM 300) exchange block messages within these block centres, most systems (e.g. LZB, ETCS L2, TVM 430) use route logic.

10 Line Block Systems

Block sections on these lines can economically become very short, as no trackside signals are needed. In LZB and ETCS L2, on lines with mixed traffic, both block systems (the one of the high speed train control systems and the one of conventional interlocking) are used parallel on the same tracks, with the block sections of the high speed train control system often being shorter than those of the conventional block systems (figure 10.22, figure 8.37). As mainly high speed trains operate on these lines, high line capacity can be achieved for equipped trains and costs for trackside signals reduced. The disadvantage is that in degraded mode operation without cab signalling or when running unequipped trains, lower capacity has to be accepted.

⌐O Wayside Signal
⌐▣ ETCS L2 block marker

Figure 10.22: Shortening of block sections with ETCS L2

The high achievable line capacity makes these systems also useful for lines with no high speed traffic, but high performance requirements. Examples are some suburban and metropolitan railways.

10.5 Moving Block Systems

High train movement intensity in peak-hour on some lines demands either additional lines or solutions for increase of line throughput. One of the reserves is accuracy of detection: in most system one metre of a block section is detected as occupation of the whole block section. One solution is to reduce the length of fixed block sections to a minimum. Another solution is to detect front and rear end of trains exactly and reduce the distance between trains even down to the length of braking distance (chapter 3.4.2.2). If the distance between trains is more than stopping distance, the second train can move without restrictions. If the trains are nearer, the second train must reduce its speed corresponding to the calculated braking pattern.

Using of 'moving block' started at the end of 20th century on underground and metro lines of Vancouver (1987) and London (1994). The train detection in the system SELTRAC, which is used there, is realised according to principle of LZB (chapter 8.3.6) with self-detection of the trains. The crossings of cable loop as fix points are projected every 25 metres and a precision of detection of about 6 m is achieved. The metro train has a defined length and its completeness is checked by electric wire between railcars. Therewith, the rear end position can be calculated. The train can have reserve equipment for availability redundancy. The moving block realises a time interval between trains of 1–2 min for train speed up to 130 km/h. The use of trackside signals is impossible for moving block systems and these lines usually function without drivers. If any person or large object falls on the track, it will be immediately detected and the train automatically stopped (Friesen/Uebel 1999).

The moving block system designed for some suburban lines of Tokyo has another detection principle. High frequency track circuits 1 kHz with feed and receiver at the same end can detect the distance between track equipment and nearest train axle via the resistance of the track with precision 20 meters (chapter 5.3.7.4). The permitted speed is calculated and transferred to the train via the same track circuit through frequency 3 kHz (figure 10.23) (Watanabe/Takashige 1989). Checking of train completeness is not necessary in this system thanks to track circuits.

Another moving block system, which has not yet been specified and is therefore not yet in use, is the block system of ETCS Level 3 (8.4.2.1). Here the train reports its rear end position and therewith the clearing of a certain portion of track via radio.

10.5 Moving Block Systems

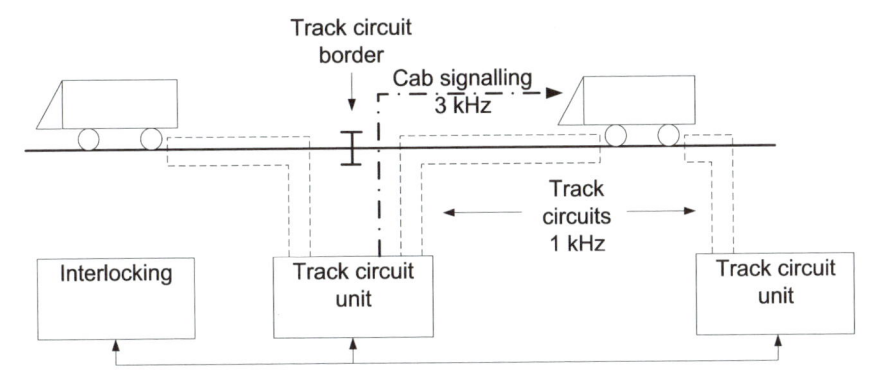

Figure 10.23: Moving block system on base of track circuits

Other new detection systems can also enable train movements according to the principle of 'moving block'. The main requirement is to provide reliable detection and continuity of information exchange.

11 Remote Control and Operation Technology
Carsten Weber, Aleksandr Nikitin, Thomas White

11.1 Remote Control and Monitoring

11.1.1 Types of Dispatcher Control/Monitoring

To keep railway operation under control, an overview of a node or a line is needed. This allows conflicts in the use of the infrastructure to be detected and the appropriate reactions initiated. So it is important in the following explanations that the functions and the people are differentiated. Operation control has to be divided into dispatching and signalling. So a **dispatcher** is responsible for the preview of the operational situation and conflict management. The **signaller** is the person who gives the control commands to the interlocking machine. A person who combines both jobs at his work place is called an **operator**. This term is also used as a collective noun for both signallers and dispatchers.

The information needed to detect the conflicts has to be given to the dispatcher, who is responsible for the railway operation at a line or node. So there are four variants usable to build up an operation control system.

Variant 1. Each of the signallers works as dispatcher for his dedicated control area and has no special equipment for dispatching (figure 11.1). Conflicts in dispatching are solved between neighbouring signallers: For example, if there is only one track usable for the next train in station B and stations A and C are both interested in sending a train to station B, a conflict will arise. So the operator in station B has to detect this problem and to find a solution to keep the railway operation between A and C going. In the time the operator needs to think about the conflict, he might not be able to handle other vehicle movements. So the current trains can be stopped and so new conflicts and delays can rise up. This method is now seldom in use.

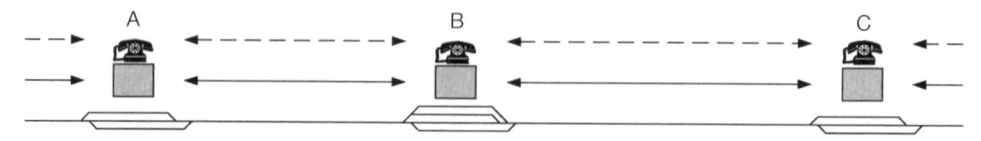

Figure 11.1: Variant 1. Safety information (solid line) and operational information (dashed line) exchanged between stations

Variant 2. If there are a lot of vehicle movements necessary in station B, it might be helpful to relieve the local operator from the dispatching job so that he can concentrate on the safety of the railway operation. He is then called a signaller (figure 11.2) and an additional central dispatcher is employed. For the arrangement of data exchange between those concerned, there are three possibilities:

- The communication between the stations and the dispatcher is achieved through direct telephone calls between two persons. So the dispatcher is an ordinary part of the communication system.
- The dispatcher works as a central node in the communication. So the data is given to the dispatcher, handled by him and issued to the (next) signaller. This way the dispatcher is always up to date and so conflicts can be detected easily, but he also has to do a lot of communications work.
- Only one phone channel is used as a so-called party line. This way all participants can hear every conversation. This decreases the workload of the dispatcher, but loads each signaller with additional work to hear all conversation and evaluate what is addressed to him.

11.1 Remote Control and Monitoring

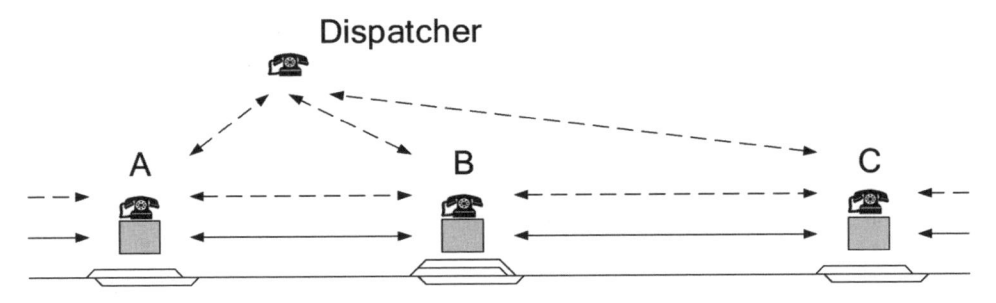

Figure 11.2: Variant 2. Central dispatcher with only telephone communication with the stations

In variant 2, the dispatcher workplace is equipped only with communication facilities. It is used on the low-traffic line equipped with simple automation systems (e.g. with manually operated points, electric token system or semi-automatic block system). The train graph charting as a time/route diagram with real train positions is carried out manually according to the reports of the station dispatchers about the times of the trains passing through.

Variant 3. The dispatcher workplace is equipped with monitoring supervisory systems (without control) on the separate ranges to get necessary information about the train situation (figure 11.3). Field data sets are coordinated with the objects of control and the data are channelled to the dispatcher through special channels (remote supervision). On the symbolic train diagram of the line (shown on monitors) the light indication gives information to the dispatcher about the signals aspects and about track sections occupied by the trains. The dispatcher preserves the right to pass voice instructions and command information to the signallers about the order of the trains passing in the area.

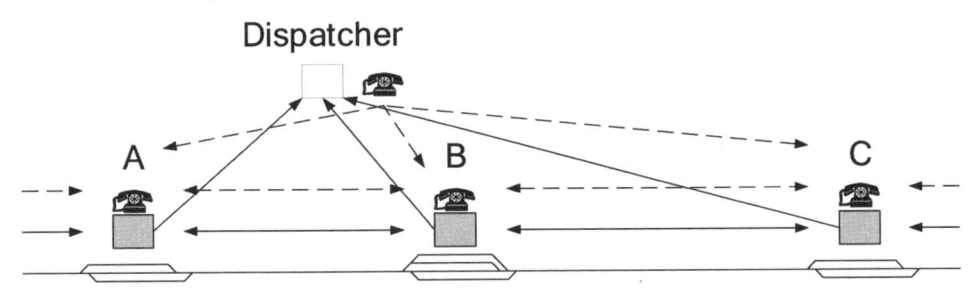

Figure 11.3: Variant 3. Central dispatcher with remote supervision and telephone commanding

Variant 4. The signalling functions, which have been local up to variant 3, are now centralised, too. Either the functions of both signaller and dispatcher are fulfilled by one and the same person for a larger area, or they remain different persons with the dispatcher having a larger area of responsibility than the signaller (figure 11.4). In the latter case, optionally, the dispatcher may

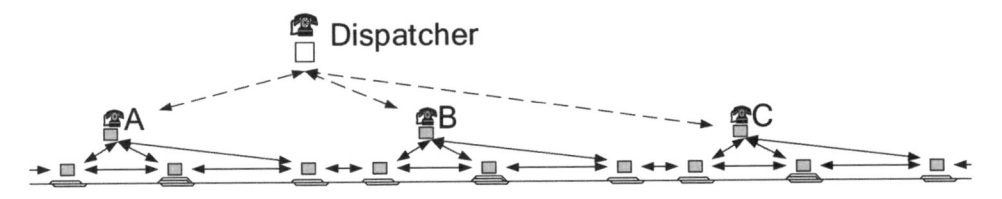

Figure 11.4: Variant 4: Centralised signaller and dispatcher

11 Remote Control and Operation Technology

directly give commands to the interlocking system, evading the signaller. According to operational rules, he can be allowed to do it regularly (e.g. in times of low traffic) or is restricted in doing so to very special (dangerous) cases.

11.1.2 Centralisation of Interlocking Control

Concerning the control of the interlocking system (the work of the signaller), different grades of centralisation are:
- Control and setting of field elements at a centralised interlocking machine
- Control of the whole station by one operator
- Control of the remote zone at the big station
- Control of the neighbouring stations from the basic one
- Centralised control of the small stations along a line
- Centralised control of the stations in a larger area

The last grade of centralisation is used by many railways and is known as CTC (Centralised Traffic Control). Railways worldwide use CTC according to various requirements and operational regulations, but for identical purposes. The principal ones are:
- Increase of train average speed and traffic capacity
- Reduction of operating personnel
- Ensuring safety in the event of disturbances on the line

For instance, on Russian single railway lines, the train's speed increase would be 15–20 %, the traffic capacity increase would be 35–40 % and the number of operating personnel would decrease by up to 60 persons per 100 km through CTC (Sapožnikov/Gavzow/Nikitin 2002).

CTC is the main focus of the subsequent discussion.

11.1.3 Flexible Allocation of Control Areas

Regarding the flexible allocation of control areas to different persons in CTC, different aspects are discussed in the following:

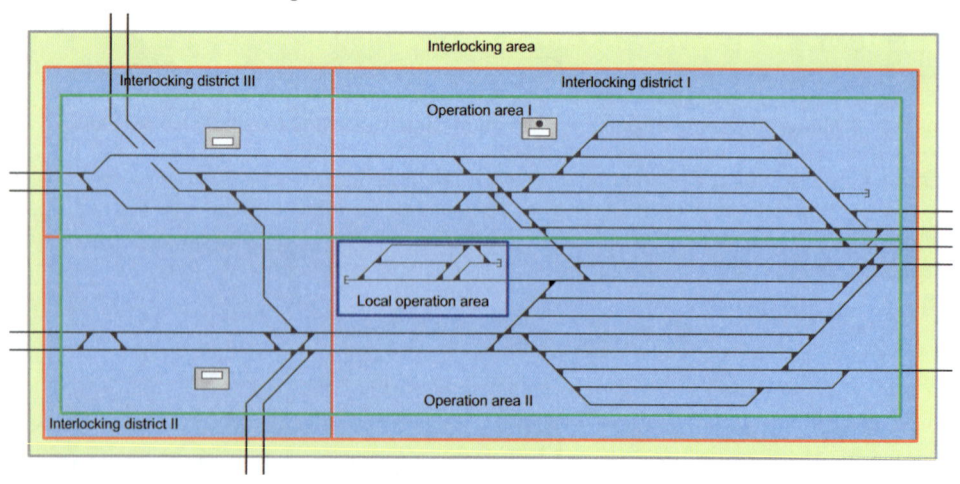

Figure 11.5: Bird view of differences between control and operational borders in centralised railway operation

- The local control of shunting areas
- The size of the area of responsibility of one signaller
- The number of persons to be involved in certain actions.

In areas with a lot of shunting activities it can be useful to separate the areas for trains and for shunting. So the signaller of the station can concentrate on the control of these areas which are mainly dedicated to train movements, whereas separated shunting areas are controlled by a separate shunting signaller or by shunting staff (figure 11.5). This avoids a lot of time consuming co-ordination talks from both kinds of staff. In this sense, local operation areas are installed which can be alternatively controlled from the normal interlocking desk or by local operation devices.

To be able to adjust the number of signallers, the whole interlocking area can be split up into several operation areas. These operation areas can either be controlled separately or as a whole (figure 11.6). Above this there might be a dispatching area, which includes several control areas. This system is very flexible, to manage the operator's workload.

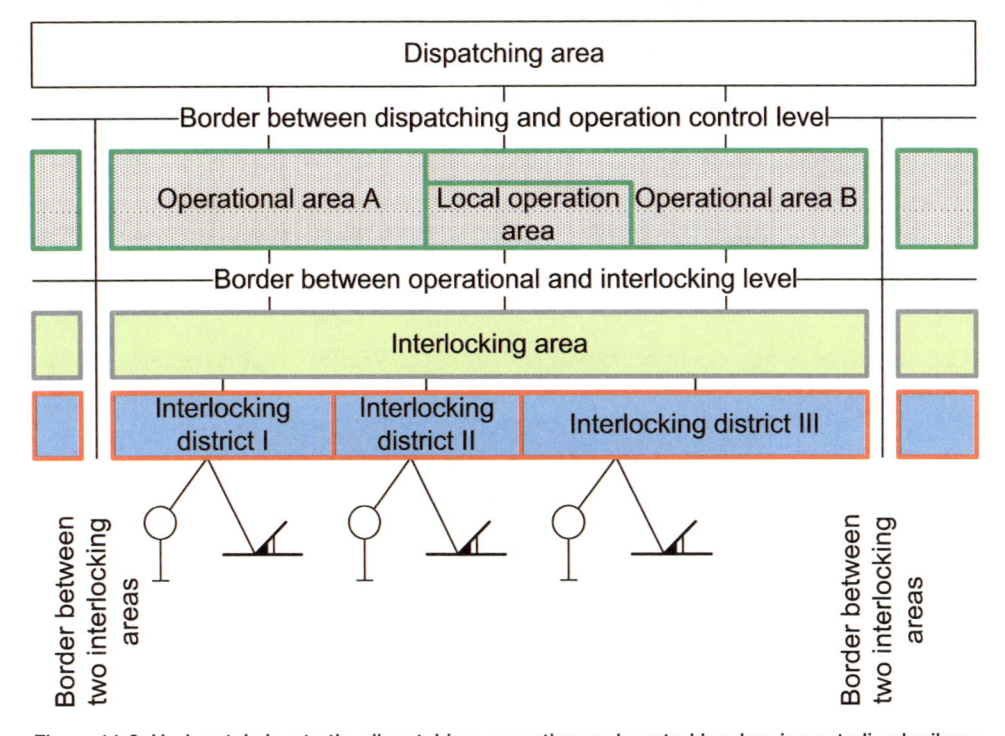

Figure 11.6: Horizontal view to the dispatching, operation and control borders in centralised railway operation

11.2 Processes in Operation Control

11.2.1 Information Input and its Viewing

Depending on the data transfer technology, there are different variants of displaying the operational situation (figure 11.7, figure 11.8):

11 Remote Control and Operation Technology

- At the indicator panel with lamps or LEDs according to track layout (if the data comes through relay equipment)
- Via video projector or plasma-panel display (usually for big rooms)
- Via several individual PC-monitors

Figure 11.7: Operator with an electronic operation desk and an indicator panel in the background (photo: *Siemens*)

The last method of displaying is flexible; the dispatcher can select the data which is important at the moment. The other methods always display the same track layout and usually cannot show all details. So some dispatcher indicator panels in station areas show only main tracks and signals.

If the dispatcher does not control the area or does not give safety critical commands, the display does not need to be safe.

Figure 11.8: Operator with an electronic working place and plasma panels in the background (photo: *Siemens*)

11.2 Processes in Operation Control

As the data transfer through relay equipment and codes from stations to operational centre requires some time and data link capacities, the new information comes cyclically. The cycle time has to be chosen so that the train would not be 'lost' by its passage through the shortest section on the maximal speed. Another solution is to transfer the data immediately, but only from these infrastructure elements whose status has changed.

The modern systems check operational condition of electronic interlocking elements and show any defects to the dispatcher on the track layout (at points, signals, tracks, power feeders etc.) and more detailed in the alarm list. Some railways have centres for the co-ordination of signalling service staff, who are placed near the dispatcher or on the line and use the data of this net.

The modern systems can automatically prepare and demonstrate current traffic diagrams, too.

11.2.2 Evaluation of Operational Situation

The evaluation of the operational situation is based on the data collected and displayed as written in chapter 11.2.1. The dispatcher has to check the situation for conflicts and to give back a reaction, if necessary.

The evaluation of the operational situation is often done in a dispatcher's head based on his expert knowledge. In simple cases, it can be done only in an operator's head, but it is hardly limited. Thus assistant systems, for example a PC, can become necessary to facilitate processes.

One of the simple systems is given by a sheet of paper with or without indicated train lines in the graphic timetable. So the dispatcher fills in the lines for the current trains and checks the differences from the timetable. This way he has a lot of work to keep the train lines effective. The time to think about the operational situation and to find out the best solution is very short, so that the dispatcher can become ineffective because of his work load. Figure 11.9 shows a combined dispatcher's and signaller's workplace.

Figure 11.9: Desktop for a combined job as dispatcher and signaller. The graphic timetable is positioned behind a glass panel, so the dispatcher can paint differing train lines in. At midnight all data of the day will be erased

11 Remote Control and Operation Technology

The next step of the evolution could be achieved by computer-based systems. The computer collects and visualises the current train positions. So the dispatcher could keep his mind on analysing the situation and giving the information to the signallers along the line. These systems were still unable to detect conflicts automatically, so the dispatcher had to be informed on every change at the screen to detect the conflicts himself.

Analysing the current operational situation is quite difficult, so that most railway infrastructure companies are not able to automate this job. There are several problems, which are not easy to handle for technical systems and where therefore human experience is important:

- How much of the reserves, which have been calculated in the timetable, are available in the current situation?
- How much effect does the current imbalance have on connecting trains and for the customers?
- Is it possible to keep other trains away, that the delayed train can run as fast as possible?
- How far can a prediction be done to get usable results?
- How likely is the imbalance of other vehicle movements in the calculated prediction?

Most systems used now try to address the current data as best as possible to the dispatcher. The dispatcher has to think and to decide, what has to be done next to avoid or clear conflicts. The quality of a dispatching system can only be evaluated in situations with many conflicts or emergency cases, where a lot of decisions have to be made in a short time. Most systems get into trouble this way and do not help to reduce the dispatcher's workload in these situations. This effect leads again to the problem of overloading the dispatchers in times of trouble. At the end, the system has to be able to get usable hints or advises if not all infrastructure is available for use.

11.2.3 Command Output

Depending on the adherence to the timetable, different ways of field command controls are possible. The types of command output are:

- **Indirect** – the commands are verbally given from the dispatcher to the signaller, independently of the interlocking machine(s)
- **Direct** – the dispatcher has a panel to give commands directly to the interlocking machine(s)

The indirect system can look very simple, being down to direct dialogues between dispatcher and signaller or up to phone calls or news systems by PC networks. The most important attribute is that there is no connection between the dispatcher's workplace and the interlocking machine. This way all the commands to the interlocking system are given by the signaller, so that it is clear who gave instructions to bypass any safety critical status.

In opposition to these systems, a direct connection to the interlocking machine is helpful to avoid time consuming discussions. To block illegal commands, the dispatcher could get a control interface without safety critical commands.

To send commands to the interlocking machine, there are three different ways possible:

- **Individual** – the dispatcher is able to change e.g. points positions individually
- **Route** – the dispatcher sets routes directly from his work place
- **Program** – the dispatcher changes the program of automated route setting systems

There are different ways of setting program functions. The program can be taken from a table similar to the timetable which combines train numbers and routes, or it can be given by the dispatcher, who programs the whole route of the train. The first option is used in Germany and the other option is typical for US systems.

The signaller, in some cases also the dispatcher, needs to control the whole interlocking machine. In some cases, safety conditions cannot be checked technically due to the failed devices, and the signaller/dispatcher has to take safety responsibility by special actions such as:
- Control of call-on signals
- Emergency release
- Emergency change of traffic direction on the block for single-track line
- Other similar degraded mode operations

For the safe translation of critical commands from CTC to interlocking, the code commands channelling is carried out in some steps with repeat. In some CTC (e.g. in Russia) the channelling of critical code commands is only possible when the co-ordinated and concerted action of two operators occurs.

11.3 Data Transmission in Remote Control Systems

11.3.1 Types of Communication between CTC and Stations

Due to the control zone enlargement (figure 11.6), remote data transmission is required in systems of centralised traffic control. This information interchange in the dispatcher control systems provides information and instructions transfer between the dispatcher office and remote stations through the communication link. Possibility for the monitoring and the control depends on the interlocking type (table 11.1).

Interlocking type	Monitoring	Control
mechanical	through special interface	impossible
electromechanical	through special interface	impossible
relay	through special interface	through special interface
relay-electronic	yes	yes
electronic	yes	yes

Table 11.1: Possibilities of the communication between CTC and interlockings

In the two latter interlocking types, the data transfer between the interlocking and the related local operation desk is realised on the electronic technology, and the dispatcher can use the same data (figure 11.10). Because of large distances between CTC and stations, the line equipment can have amplifiers. Moreover, the communication links usually are redundant. Thanks to net functionality (e.g. Ethernet), the workplaces of dispatchers and signallers are flexible. So signaller and dispatcher can control the area from different access points of the data network.

The communication link between dispatcher and mechanical or electromechanical interlocking stations needs difficult interfaces, but it can only ensure monitoring and is therefore little used. CTC for lines equipped with relay interlockings is often used. The communication between dispatcher office and relay technology is described in the following parts as examples of the special interface.

11 Remote Control and Operation Technology

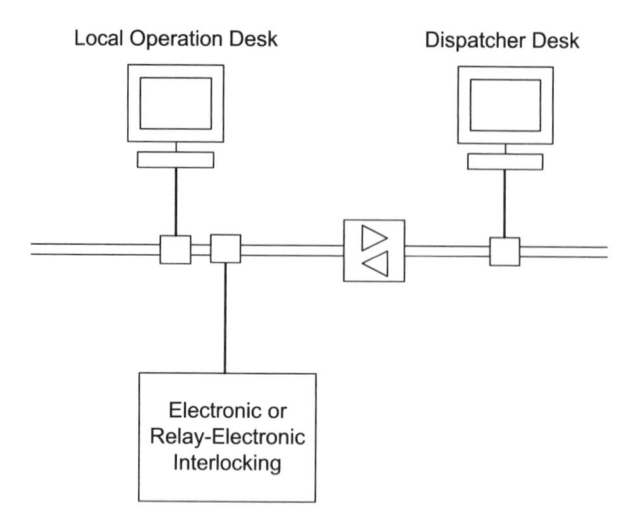

Figure 11.10: Communication between CTC and electronic or relay-electronic interlocking

11.3.2 Interface between CTC and Relay Interlockings on the Relay Technology

The communication link between the signaller's desk and the relay appliances is usually realized through many wires according to the number of control and monitoring elements. If signaller and interlocking equipment are in the same building, it is not difficult. But for the remote transfer of data, this solution is not suitable because of cable cost. Therefore the monitoring and control information is transmitted through two wires.

In many countries the CTC was created for the lines with existing relay interlockings, moreover electronic equipment was that time not in use on the railway. Therefore special relay interlocking coders/decoders (for every station) and electronic CTC connection devices were developed (figure 11.11, figure 11.12). The communication order for this case is described below.

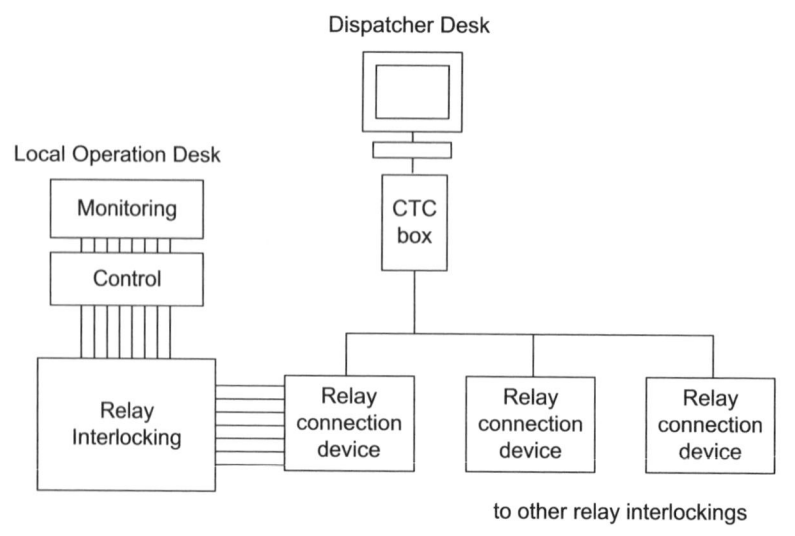

Figure 11.11: Communication between CTC and relay interlocking built on the relay technology

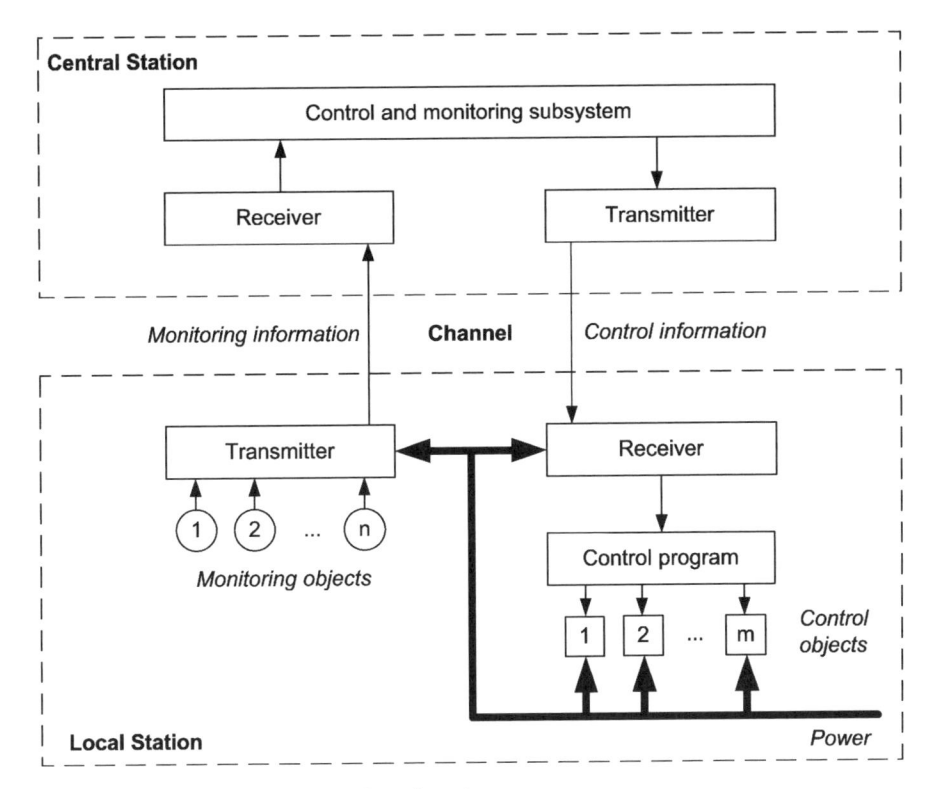

11.3 Data Transmission in Remote Control Systems

Figure 11.12: Equipment of central and local stations

According to the way of data communication through one line, the types of date transmission systems are:
- **Sporadic** – signals are transferred when appearing
- **Cyclical** – signals are transferred during the periods of time (cycles)

The first one aims to achieve the effective use of the communication channel capacity, while the second one aims to correct data errors automatically at the next cycle of messages entry.

The control information is translated sporadically, whereas the monitoring information can use both transmission types (chapter 11.3.1). The information consists of telegrams which contain bits for synchronisation, station code and elements states (or command for one element). The signalling element usually has one of two states (e.g. signal on/of, track circuit clear/occupied), which can be translated as a bit. Impulse-built signals with different attributes are used for the telegram preparation. The process of the signal parameters change is called modulation. The conversion process is called manipulation if only two discrete values corresponding to designations logical to '0' or '1' are used in the system of data transmission.

The most widely used attributes for the telegram are amplitude, phase, pulse width, polarity and frequency:

Amplitude attribute. The attribute is characterized by the current magnitude or by the pulse voltage value (figure 11.13, a). It is used likewise for AC and DC. However, the amplitude attributes have low noise immunity compared with others.

Pulse-width attribute (figure 11.13, b). Different pulse duration is the attribute in this case. The time attributes are not only the pulses, but also the intervals separating the pulses. Type

11 Remote Control and Operation Technology

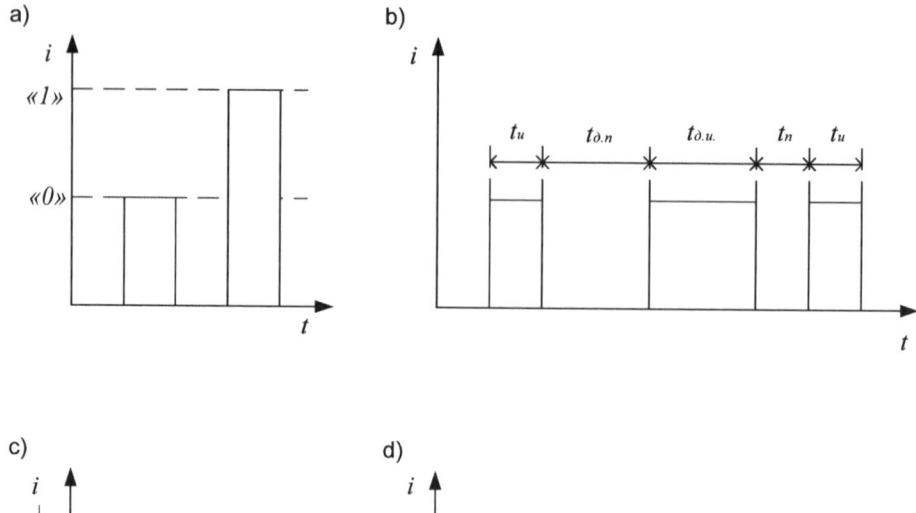

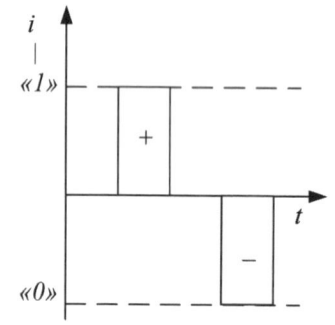

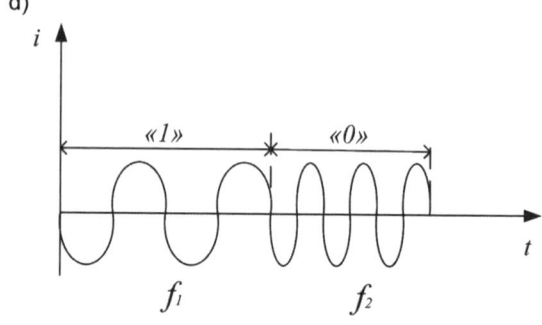

of current is of no importance for the time (pulse-width) attribute and it allows the transfer of pulses characterised by the time (pulse-width) attribute data through any communication channel including the wireless one. Practically, two values are used for binary data transfer: short and long pulses.

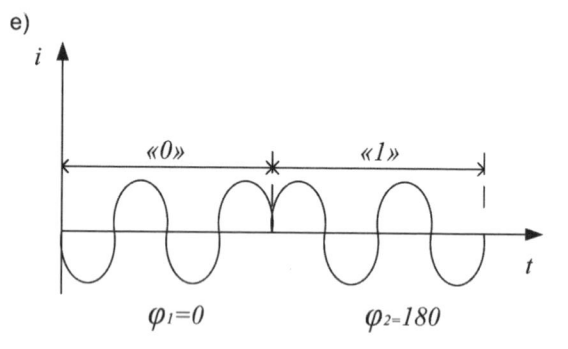

Figure 11.13: Forms of attributes for data coding

Polar attribute (figure 11.13, c). The current polarity is used for the direct current impulses as its attributes. The change of the current direction in the circuit provides high noise immunity of the code sending. However, the use of this attribute is only possible with wires and not radio communication.

Frequency attribute (figure 11.13, d). The frequency of the current oscillation is used as attribute for the AC pulse forming.

Phase attribute (figure 11.13, e). The change of phase shows the state of the next bit.

11.3.3 Interface between CTC and Relay Interlockings on the Electronic Technology

In several countries, CTC was developed at the same time as electronic interlocking. Therefore the date transmission was initially adapted for communication between computers. If there are relay interlockings, they can be connected to the existing net through the special connector (figure 11.14).

If the dispatcher has the possibility of giving safety critical commands and needs safe indication, the relay-electronic connector has to be safe for the data transfer in both directions. Therefore it usually has two channels (figure 11.15). The vital comparator controls the relay current; the position of the relay is read from inverse contacts. With electronic equipment the data exchange usually is faster than with relay equipment: Typical cycles of data exchange are 4-5 secs for relay technology and 1-2 secs or below for electronics. The maximum value of ca. 5 secs results from the fact that the train shall not get lost when passing through the shortest section of track clear detection. Moreover, electronic technology is more flexible regarding the policy on how to deal with safety and availability. For example, if a command or a message is safety related, it can be foreseen that it will only act after correct repetition.

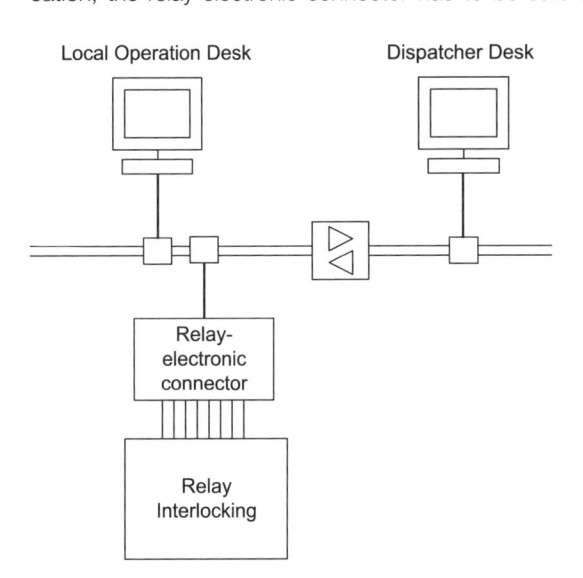

Figure 11.14: Communication between CTC and relay interlocking built on the electronic technology

11.4 Operator's Workload

Too high or too low workload of persons involved in the process might cause human errors by forgetting safety critical facts, and therefore lead to accidents. Therefore, an optimum of workload needs to be achieved.

The operator's workload depends on the existing infrastructure, the number of vehicle movements (train or shunting movements) and the operation programme in the area the operator has to control. In case of older interlocking machines, one of the limitations of the controlled area was set by the visibility range. If the positions of the controlled elements have been concentrated close to the interlocking

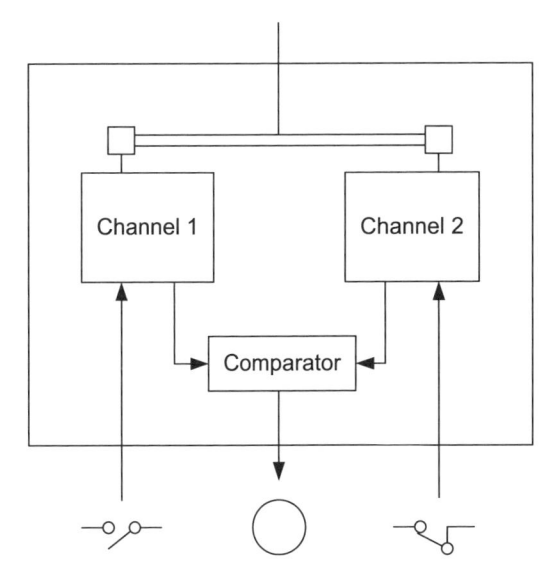

Figure 11.15: Relay-electronic connector

11 Remote Control and Operation Technology

and a lot of vehicles have been moving around, it could be possible that more than one signaller was needed at this interlocking. This raises the question: How can I find out how many signallers are necessary to organise and operate safely within a given area?

As written above, the operator's workload is in relation to:
- the usable technologies to achieve the tasks,
- the dedicated infrastructure,
- the operation programme and
- the number of vehicle movements.

11.4.1 Influence of Technology

The usable technology reflects the state-of-the-art at the time of construction of the interlocking machine. This can only be changed over long periods by replacing the technology. The operator has to know the way to operate the interlocking machine and to know at which position which element is being controlled. Depending on the technology, he might have to move inside his signal box to operate different points and signals or he works at an operation desk or at a computer. These different characteristics of operating an interlocking machine have an influence to the operation time to set and to cancel a route, as seen in figure 11.16.

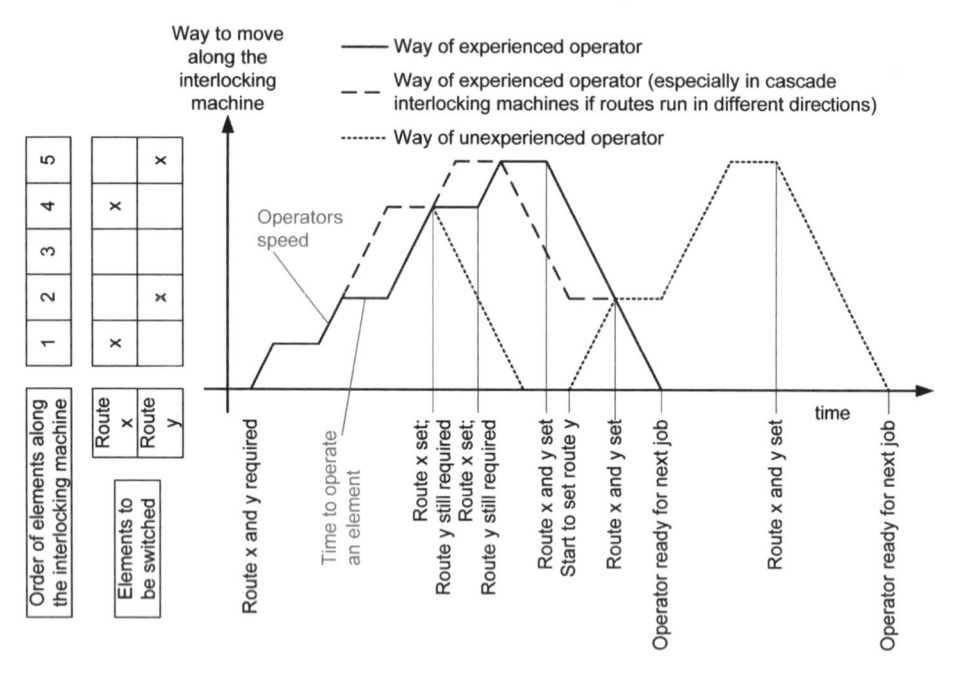

Figure 11.16: Required times to set two routes by different operators and operation technologies at the same kind of older interlocking machines

11.4.2 Influence of Size of Infrastructure

Quite close to the interlocking machine is the dedicated infrastructure, which is connected to the interlocking machine and so controlled and operated by the operator. Changes in infrastructure are expensive because of their linkage to the interlocking machine and so they are seldom done. The link between the work load and the infrastructure is quite simple: less infrastructure means less operator workload. An operator at a simple block station has to handle only his block instruments and an operator at a level crossing only his crossing. In contrast, where the signaller has several stations under his control, his workload is much higher.

11.4.3 Influence of Operation Programme

The operation programme changes at least once a year. If freight traffic dominates the railway, it can even change more frequently. Depending on the position inside the railway network, the changes can be marginal or the complete schedule can be changed. So if there are a lot of train and shunting movements operating in less time, the operator might have a lot of work and get into trouble this way.

To handle the problem of changing operation programmes for the dimensioning of new operation areas by replacing interlocking machines, a capability analysis should be done. The problem will be to fill in the right number and ways of shunting inside the capability analysis, because they are often unknown at the time of setting the operation areas.

If most of the trains are passenger trains, there will be fewer train runs in the night hours than during the daytime. This break in the operator's workload can have the following results for the railway system in these hours:

a) closing the line for any train run (no operators required but no runs possible),
b) keep all the stations manned (the operators have nearly nothing to do but ad hoc trains could run on the line),
c) deactivating the stations which are not required (trains can pass the stations, but there is no crossing or overtaking possible; (see also chapter 4.3.11: fleeting and automatic route calling),
d) remote control of the stations (another operator gets this operation area in addition to his own area) or
e) adding neighbouring operation areas to a large operation area if the areas are regularly remote controlled.

The options c) to e) require special equipment inside the interlocking machine. The options c) and d) are often retrofit equipment which has to be justified by the economics, given the hours during which no operator is required.

The options d) and mainly e) are possible only in relay and electronic interlocking machines. In case e) the **interlocking area** has to be designed to be as large as possible, but not larger than what one operator can handles in times of low traffic (figure 11.5). It is possible to split the interlocking area into **interlocking districts**, if the distance between the field elements and the interlocking machine is too large for one interlocking machine. During hours with high traffic levels, the operator is able to hand over **operation areas** to colleagues. Some interlocking machines are also able to be operated in their whole interlocking area from different operation desks at the same time. That means in these cases that the operation area can be operated from more than one operation desk.

11.4.4 Influence of Moving Vehicles

The current number of vehicle movements depends on the operation programme too, but it is affected by the imbalances at the moment. Thereby a situation can emerge in which timetables cannot be used any longer for route setting, except for the train's running direction. In some cases the written times, track usage and train order can not be kept alive, so that the signaller has to ask the dispatcher or has to do dispatching by himself for example for shunting. Depending on the experiences of imbalances or an estimation of them, these have to be considered, too.

11.4.5 Influence of Disturbances

It is helpful to keep the disturbances of the interlocking system seldom and to ensure that the information the operator can use at his work place is really correct. This way he can analyse which elements are in trouble and what possibilities exist to interact the elements.

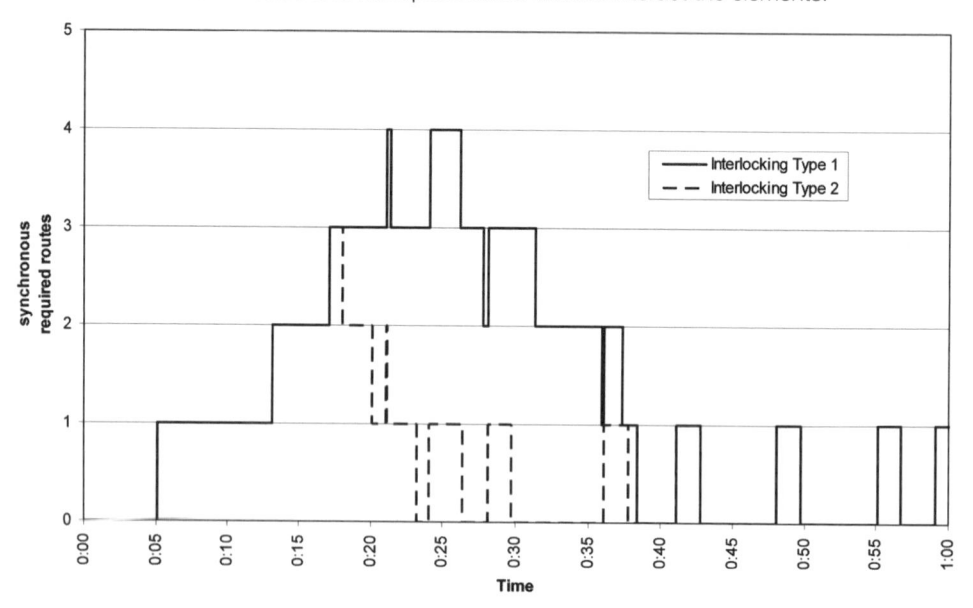

Figure 11.17: Comparison of synchronously required routes and their influence to the workload by using different interlocking types

Figure 11.17 shows the result of comparison of two fictive interlocking types used in the same operational rules independent from what particular types and technology they are. At 0:05 the signaller starts calling the route for the first train. The route does not work, so that the signaller has to build up the route without any automatic support of the interlocking machine. While the signaller is busy with this route, the second (0:12) and the third (0:16) routes are requested. Shortly after that in interlocking type 2 the first train can start running (0:18), as the signaller has then done the job for the first route, and can concentrate on the other routes now. In the interlocking type 1, in contrast, the first route keeps the signaller busy for a longer time, until 0:22, due to more difficult operation procedures in this interlocking type. Therefore, more delays are the result in interlocking type 1 than in type 2. As seen in figure 11.17, the interlocking type used can influence the imbalances of every train and at the end the sum of the trains. Different types of interlocking mean also different workload for the operator.

The operational rules have their influence on operator's strain, too. Some railway infrastructure companies have many fallback modes to keep on the traffic so that it is really difficult for the operators to find out the right rule in the right time. This way assistants inside the interlocking can help by giving detailed hints to the signaller. A good structure diagram with an easy view will also make sure that the operator checks everything until he allows the train to start his run. For analysing the operators' workload it is important to know all the influences of the railway traffic levels.

11.4.6 Results and Conclusions of Calculating Operator's Workload

The whole analysing process ends if the sum of the imbalances is acceptable or there is only one job to do at one time. In some combinations it might be possible to do two or more jobs at the same time, but some combinations are impossible for example if the operator should phone two persons at the same time. To avoid this problem, there are two solutions possible: At first doing two jobs at the same time can be prohibited, or as a second a matrix of jobs which can be done at the same time and their number can be defined.

If it is not possible to keep the workload of the operator below a certain limit, there are some options:
- More automated functions to support the operator
- Changing the borders of the operation area
- Operator assistants for taking special jobs (passenger information etc.)
- More operators

The operation programme can be changed too, but this will not be helpful because of less use of infrastructure and therefore less benefit.

11.5 Examples for Operation Control Systems

11.5.1 Centralised Traffic Control in the USA

11.5.1.1 Background

A central controlling authority has been the basis of North American railway operation since 1851. Original traffic control methods involved local station staff operating interlockings and/or writing the dispatcher's instructions to trains and delivering them when the train passed or stopped. 76 years later, Centralised Traffic Control changed the method of communicating authority, but little else. Originally only new interlockings, with points formerly hand operated by train crews, were integrated into the CTC. The organisation and methods of transportation management remain much the same as they have for over 150 years, with some changes in title for some functions and, of course, fewer personnel.

11.5.1.2 Organisation

A typical control centre in North America employs several train dispatchers; in some there are more than one hundred. Each train dispatcher is assigned a territory that can range from approximately 30 line km of heavy traffic to 1,600 line km of light traffic. A manager, generally called the chief dispatcher, is assigned a territory that includes two or more train dispatchers. Train dispatchers generally work autonomously, applying to the chief dispatcher for questions of policy regarding train movement. The chief dispatcher also typically co-ordinates main line and terminal operation, arranges planned maintenance of way possessions, and may manage work assignments of train crews. The territory of several chief dispatchers is generally supervised by a manager, often called corridor manager or some similar title (figure 11.18). The corridor manager generally co-ordinates the operation of the chief dispatchers' territories, and issues instructions regarding traffic management policy or specific handling instructions for individual trains when needed. A typical control centre also includes one or more locomotive distribution managers, one or more maintenance of way possession planners, and possibly train and engine crew personnel management positions. These positions are often parallel to, and not under the control of, network traffic management managers. A control centre may also include several commodity group managers. From the transportation management point of view, traffic on a large North American railroad is similar to traffic of several train operating companies on a privatised national network. The handling of specific types of traffic, such as coal, grain, intermodal, and automotive, is assigned to a management group that co-ordinates the activity of its own trains. A commodity group manager will generally provide handling information and instructions for its trains to the corridor manager, who issues instructions to the chief dispatchers after integrating the commodity group's handling instructions into the overall traffic management plan.

TYPICAL NORTH AMERICA CONTROL CENTRE STRUCTURE

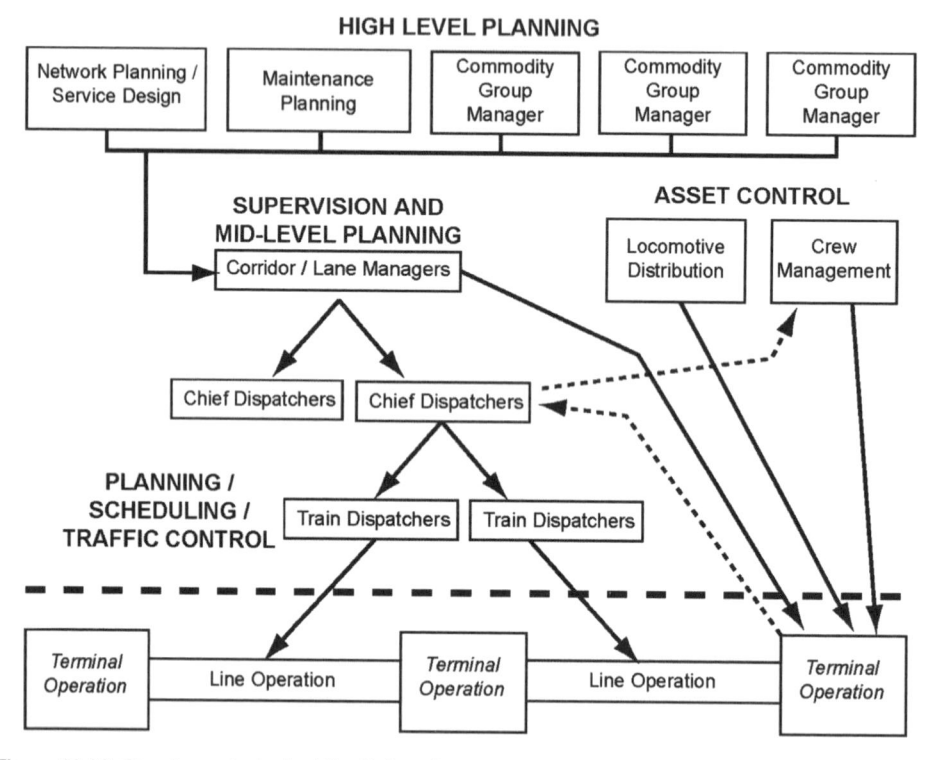

Figure 11.18: Structure of a typical North American control centre

11.5.1.3 Information Systems

Control centre management depends upon several, generally independent, data systems. Train and car consist and location information is often maintained in a separate information system. Train location information is generally updated by manual entry at terminals and by passing reports at Automatic Equipment Identification (AEI) readers (chapter 12.3.5). These systems may also receive train location data from the traffic control system. Locomotive management and crew management systems are often separate from the train and car information system, receiving train location information from that system. Maintenance of way planning generally involves only extended possessions in the future and not day to day maintenance operations, so the system generally uses only schedule information from the train and car data system. There may also be a separate system for speed restriction and track condition information management.

11.5.1.4 Control Systems

There are almost no locally controlled interlockings left in North America. A very small number of large interlocking plants remain, with plans for centralised operation in the future. Many of these share the added centralisation complexity of including the tracks of two or more railway companies crossing at grade, increasing the cost, involving potentially complicated cost-sharing agreements, and present the complexity of integrating the different control systems of the railway companies involved. Most of the remaining local interlockings generally control movable bridges. In virtually all cases, the bridge interlocking has little or no association with traffic management. Marine navigation has right of way over rail traffic, by law. Trains are stopped for marine navigation or proceed without delay.

North American train dispatchers operate the CTC controls personally, and do not convey instructions to a CTC operator. CTC installations virtually all use microprocessor control stations. Traffic on a significant amount of the North American network is controlled by written instructions issued by the dispatcher by radio and copied by the crew of the train. These segments may have block signals, but not all do so. Generally, authority management software ensures that no overlapping authority is issued. The train dispatcher enters the proposed movement limits and the system generates an authority to read to the crew. These systems are generally separate from the centralised traffic control system; however, they may be presented to the train dispatcher in the same windowed display as the CTC system and the train and car information system.

With very few exceptions (the local interlockings have not yet been centralised) there are no station staff. At large freight terminals, the staff are engaged solely in yard management. For traffic control purposes, most stations, other than major terminals, consist only of a main track and a passing loop on a single track line, or crossovers between main tracks on a multiple track line. On lines without CTC, points are manually operated by the train crew. These points are not interlocked and generally not equipped with electric locking, but point position is included in automatic block signal detection. In CTC territory, virtually all points that are not used for traffic control purposes (e.g. industrial tracks and local industrial support yards) are manually operated by train crews and are generally equipped with electric locking or an automatic signal for entry into the main track. Trains typically remain on the main track when stopped for work at these locations.

11.5.1.5 Communication

Virtually all train movement communication is conducted directly between the train dispatcher and the train crew by radio. The exception is the written information the crew must receive when they report for duty. Train dispatchers maintain the records of temporary speed restrictions, tracks out of service, maintenance of way protection, and changes to the rules or timetable that have not yet been published in permanent form. The train dispatcher transmits the information to each train before the crew comes on duty. The information may be sent directly to a remote computer network printer, to a fax machine, or to a data storage and retrieval system. For the latter, the crew enters an identifying code and the documents are printed. The crew will call the train dispatcher on the radio when ready to leave. If conditions require

11 Remote Control and Operation Technology

changes to the restriction information received by the crew at the beginning of the trip, the train dispatcher will issue the new information to each train by radio.

Train location information may be made available to terminal managers and the dispatchers of adjoining districts by passive CTC displays. These displays are identical to the train dispatcher's display but have no input capability.

11.5.1.6 Dispatching Procedures

North American railways are generally unscheduled, at least in the pure sense of the word. North American trains generally have transportation service plan schedules that provide the expected times at terminals, or sometimes only the expected running time between terminals. The train dispatching criterion is generally priority. Trains must generally not be delayed by trains of lesser priority. Priority is established by the transportation service plan schedules or by instructions of the network managers. The train dispatcher performs all scheduling as trains are introduced, and rescheduling when conditions change. Most of the scheduling process is achieved mentally. Traffic planning software is just beginning to be introduced in North America.

North American train dispatchers also arrange daily track maintenance. When a maintenance crew arrives at their jobsite, the foreman calls the dispatcher on the telephone or radio and explains the required limits and time. If traffic allows, the dispatcher will transmit written authority verbally to the foreman of the maintenance crew, describing exact track and time limits. The train dispatcher is responsible for protecting the maintenance of way work. The CTC control workstation has provisions for blocking signals to prevent train movement. This feature may be connected to or a part of the authority issuing system to ensure that the work limits are protected before authority can be issued.

Alternatively, for maintenance of way work that will occupy a single location or small area for an entire workday, the dispatcher may issue a written instruction to be delivered to each train at its initial station on that segment of the line. The written authority requires trains to stop short of the maintenance of way activity and not proceed without specific authority of the foreman in charge of the work. Under these conditions, trains are allowed through the work limits at the convenience of the foreman in charge of the work. The train dispatcher performs rescheduling as needed after trains are delayed by the maintenance work.

11.5.1.7 Decentralization

The capabilities of current signal, communication, and data processing systems may exceed the limits of practical application. It is technologically possible to control 37,000 line km of railway from one room. It is also possible to develop system redundancy and backup to allow continued operation if a segment of the technology becomes inoperative. Regardless of the redundancy and backup, all of the skilled personnel are located in the same place, all subject to the same risk of natural or man made disaster. The logistics of covering all positions with a qualified person every shift can be quite difficult. Providing the dispatchers with the ability to see their territory personally and develop essential route knowledge is virtually impossible. Railways in the US that developed large control centres for the entire system have split at least parts of the territory into smaller regional dispatching offices. One North American railroad never consolidated control functions into a system control centre, determining that smaller, regional control centres were preferable.

11.5.2 Operation Control Centres in Germany

The development of operation control in Germany is quite different to the systems in the USA or in Russia. One of the reasons is the base of the communication between stations. The Ger-

11.5 Examples for Operation Control Systems

man system is traditionally based on blocking systems using electro-mechanical block instruments (chapter 10.3.3.2) in contrast to track circuits as used in the USA or Russia. The traditional German manual block without technical track clear detection on open lines requires staff along the line especially to check the completeness of the train. This caused a late start of centralisation in Germany, compared with the US and Russia. This was also supported by the fact that in Germany the density of population and therefore the availability of local staff is much higher. The most important step of centralisation in Germany is the operation control centres, which were being introduced from the 1990s.

The operation control centres for centralised traffic control of German Rail (*Deutsche Bahn*; *DB*) show a modern way of centralisation of railway infrastructure operation. The whole system is based on electronic interlockings (chapter 9.4.6). It is possible to control relay interlockings from the operation control centres too, but it is not used frequently. Stations with older interlocking machines are locally operated but centrally dispatched, so that the whole supervision of the railway operation could be centralised.

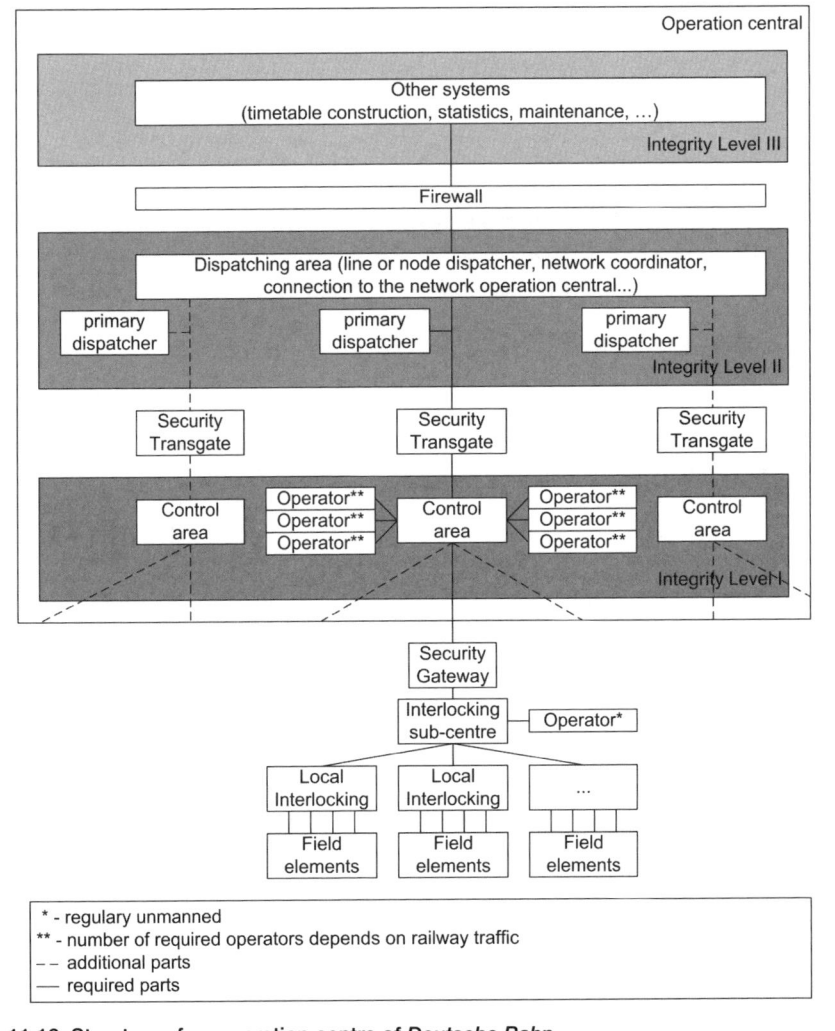

Figure 11.19: Structure of an operation centre of *Deutsche Bahn*

11 Remote Control and Operation Technology

There are seven operation control centres used for controlling main lines and dispatching branch lines. The electronic interlocking machine has to be supplemented by several computers to be able to do remote control. Inside the operation control centre the connected interlocking machines are combined as a control area (figure 11.19). These areas are strictly separated, so that they cannot be changed after installing the interlocking machine. The areas can be characterised as nodes or lines, depending on the best way of dispatching them. Each control area has an own primary dispatcher as the lowest position of a dispatcher and some operators (locally responsible signallers) to control the interlockings from the operation centre. The control area can be divided into several operation areas to level the personal to the current operational situation.

The interlocking sub-centres (chapter 9.4.6) are full interlockings, which can be operated alone. To be able to undertake remote control, several computers have to be added to the interlocking sub-centre. So it is possible to build a connection to the so-called security gateway into the operation control centre. This way all orders and feedbacks are sent between the two systems. Several interlockings are combined by the control area and there is no difference in safety and in handling between the operation from the operation control centre and local operation on site. Every control area includes an integration component, which handles the data inside the control area. Each operator's desk requires an operator's computer too. Another computer is only used for checking the integrity of the data inside this network level. All signallers desks which are in the operation control centre are connected to this level called integrity level 1.

Depending on the planned and the current situation of railway traffic inside of the control areas, it is possible to adjust the number of required operators. This is the most important advantage of centralising the whole interlocking system into the operation control centres. As a minimum, one operator is needed for each control area.

Dispatchers in the lowest level are situated in the integrity level 2, which is linked to the integrity level 1 by a so called security transgate. This gate sends all relevant data from the control level up to the operation level, but blocks all commands which are not allowed to be sent to the control level. So the dispatcher can see the current state of a track or a point position and so on but he is not able to switch the position of the elements. Allowed commands are only these for calling routes from the interlocking machine. These commands can also be sent by a train routeing system. Every control area has its own primary dispatcher. If the whole system works, the signaller himself is only needed to initiate actions in degraded mode operation. The regular actions can be started by the dispatcher or the train routeing system too. The train routeing system can also be used for regular shunting runs.

Area dispatchers work above line dispatchers to supervise the traffic in larger areas and are positioned in the same integrity level. Above him, there is a so called network co-ordinator, who is responsible for the traffic in the whole area of the operation control centre. For example in the case of route conflicts this person is allowed to take a decision on how to keep the trains rolling. This person is also called if trains are in conflict, which leave the area of the operation control centre to arrange the new slots inside the next operation control centre, or foreign railway infrastructure company.

Outside these systems in the integrity level three, there are all other systems which have no safety specification. They are separated from the integrity level two by a firewall. In this level the timetable creation systems or statistic functions are situated.

A step above the seven operation control centres in Germany there exists a network control centre in Frankfurt (M). Its only job is to control the traffic in the network as such especially for long distance trains. It is not possible to influence an interlocking machine from there, but they get a view of the current traffic situation and so they keep the overview on the railway traffic of *Deutsche Bahn* in Germany.

11.5.3 Operation Control Centres in Russia

On the Russian railways there are two trends to increase efficiency of the railway transport system control.

The first one is the creation of automatic traffic management systems of long-term planning of the train, locomotive and car models to control and forecast technological process of the railway and marshalling yards. The developments of such systems on the level of local railway divisions started in the fifties of the last century and it resulted in the development of data-computing centres (DCC) of regional railway divisions. The information exchange between the data-computing centres of the neighbouring railroads includes data about trains. The hierarchial coordination is provided by the Main Computer Centre (MCC) in Moscow.

These information systems were created as separate applications interconnected only with the data model. The new principle of the automatic control systems tasks uniting is realised by means of the creation of the Corporate data warehouse of Russia within the structure of MCC. The functioning of the main CTC is realized on its basis. The beforehand processed information from operation control centres of all 17 regional railway divisions of Russia is collected and stored in this centre.

The second trend includes automation of dispatcher operation control and its application to control train traffic at the stations, rail junctions and areas. The first centralised traffic control devices (on the relay equipment) were used in Russia in 1936 on the area of 65 km Lubertsy-Kurovskaya. The first Russian CTC computer-based system 'DC-MPK' was put into operation on the dispatcher range Saint-Petersburg–Sestroretsk in 1995. Nowadays the new CTC systems DC-MPK, 'Trakt', 'Setun', 'Dialog', 'Jug' are used on the Russian railways.

Practically, the three-tier model of operative control is already used on the Russian railways (figure 11.20).

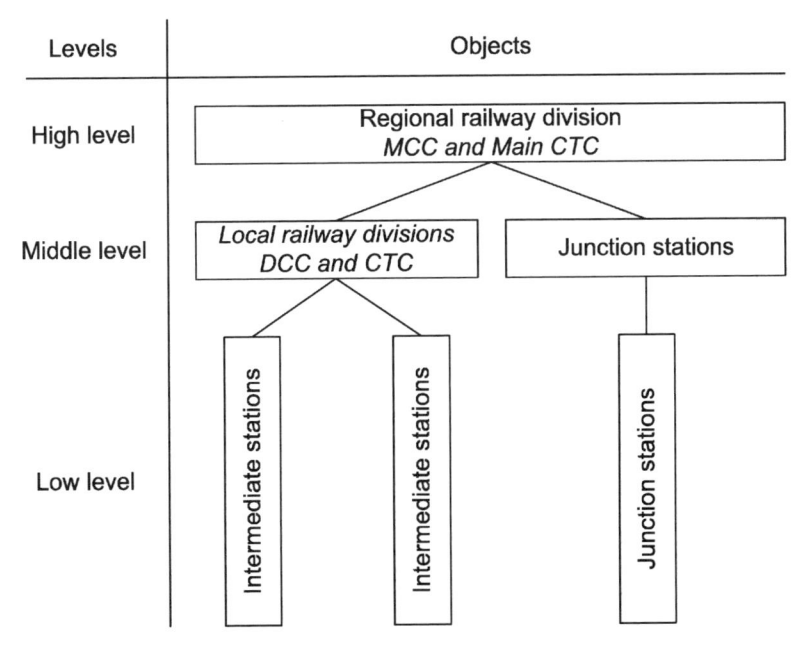

Figure 11.20: Three-tier model of Russian CTC

11 Remote Control and Operation Technology

The collection of information about train traffic in the actual time scale is realised by the centralised traffic control and its further transmission is realized through the gateway among servers of railway division. Further the information and data are transmitted to the main railway centre (MCC). The strategic control is realised on the high level in MCC and operative control of traffic is realised on the middle level in CTC centers of the regional railway division.

The workplace of the train dispatcher includes two computers with graphs similar to the information models of data-computing centre and two or more LCU control monitors. One monitor displays an area overview picture, whereas the second displays a zoom picture of the station or several stations where the signaller is currently working.

The software of the dispatcher's workplace allows the displaying of the operative technological situation (train location) of the neighbouring areas and at the borders of areas of responsibility of different regional railway divisions.

Moreover, the maintenance workplace can be connected with any channels of remote monitoring and control by means of multiplexer and the signals from code line can be displayed. The data of oscillography can be logged on winchester of automated workplace and then viewed by the maintenance staff when necessary.

The feedback from the operation control centre to DCC includes the transmission of data about the time of trains passing at stations based on the sequence of track occupation and clearing. The station CTC equipments are connected with the systems 'Palma' and 'Lotos', which perform similar functions as Automatic Equipment Identification in North America (compare chapter 12.3.5).

50-70 % of Russia railway areas are considered to have low traffic. The systems of dispatcher control should be used all over the railway network. Concerning the fact that the maximal economic efficiency is achieved by the dispatcher remote control, all intermediate stations equipped with relay or electronic interlocking should be included into the coded data transmission. The engineering problem is the areas equipped with semi-automatic blocking where not all lines are equipped with technical track clear detection. This lack requires additional control of train completeness (for example by axle counters).

12 Safety and Control of Marshalling Yards

Thomas Berndt, Peter Márton, Vladimir Ivančenko, Igor Dolgij, Dmitrij Švalov, Thomas White

12.1 Principles of Marshalling of Trains

In every railway system the marshalling of trains, exchange of wagons between trains and splitting up of trains must be organised. To do this different shunting methods are in use. Most important shunting methods (based on the physical process) are:
- switching over (North America: switching),
- pushing off (also fly shunting, throwing, kicking),
- gravity shunting.

In the national rules more detailed shunting methods can exist.

To **switch over**, the wagon or group of wagons will be coupled to a locomotive. After that the locomotive moves the wagons to the target track. This shunting method is (relatively) safe and can be used to move every kind of wagon (or other locomotive), but it is not efficient.

To **push off**, an uncoupled wagon or wagon group will be accelerated by a locomotive. Then the locomotive stops and the wagons move freely into the target track. This shunting method is more efficient. The safety is lower because the wagons are free moving. The wagons can be stopped by brake shoe (also called slipper or in North America skate) or by other wagons, standing in the target track. Finding the proper speed for pushing off is not easy. If the speed is too low, the wagon does not reach the target location. If the speed is too high, damage to standing wagons is possible. Therefore this shunting method cannot be used for every kind of wagon (forbidden for wagons loaded with dangerous goods).

For pushing off or switching over no special equipment is needed. Minimum is three tracks connected with a set of points.

To use **gravity shunting**, a hump in a gravity yard is necessary. Additional technical facilities (see chapter 12.3) help to get high shunting capacities. In general this shunting method is more efficient than the others, but cannot be used in every case. It is forbidden if dangerous or sensitive goods are loaded in the wagons. In other cases technical parameters of wagons do not allow it (e. g. too long a wheelbase). Most modern passenger carriages cannot to be hump shunted.

Shunting methods are used at different places in railway systems. The most important places are:
- collecting and distributing of wagons in industrial sidings, port railways, container terminals and
- exchange of wagons between trains in marshalling yards.

The following discussions concentrate on gravity shunting as the most complex form of shunting.

12.2 Parts of Marshalling Yards and their Function

12.2.1 General Structure and Functioning

The technology of splitting up trains and reforming them by gravity shunting at marshalling yards is based on specially designed yards. The main components are the receiving yard (also called arrival or reception yard), the hump, the classification yard (North America: bowl yard)

12 Safety and Control of Marshalling Yards

and departure yard. Inbound trains will arrive on one or more receiving tracks. Then the trains will be prepared to splitting up (including inbound inspection and preparation of a hump list). After that trains will be humped and wagons run to the classification tracks. Outbound trains are assembled by moving classified blocks of wagons from the classification tracks, placing them on one of the departure tracks and coupling them to trains there. Assembled trains receive an outbound inspection and brake test prior to departure. Ideally, the hump is located sequentially between the receiving yard and the classification yard. In this case the engine will proceed to the far end of the track and push the wagons (the cut) over the hump for classification. An example of this kind of marshalling yard including work routines is shown at figure 12.1.

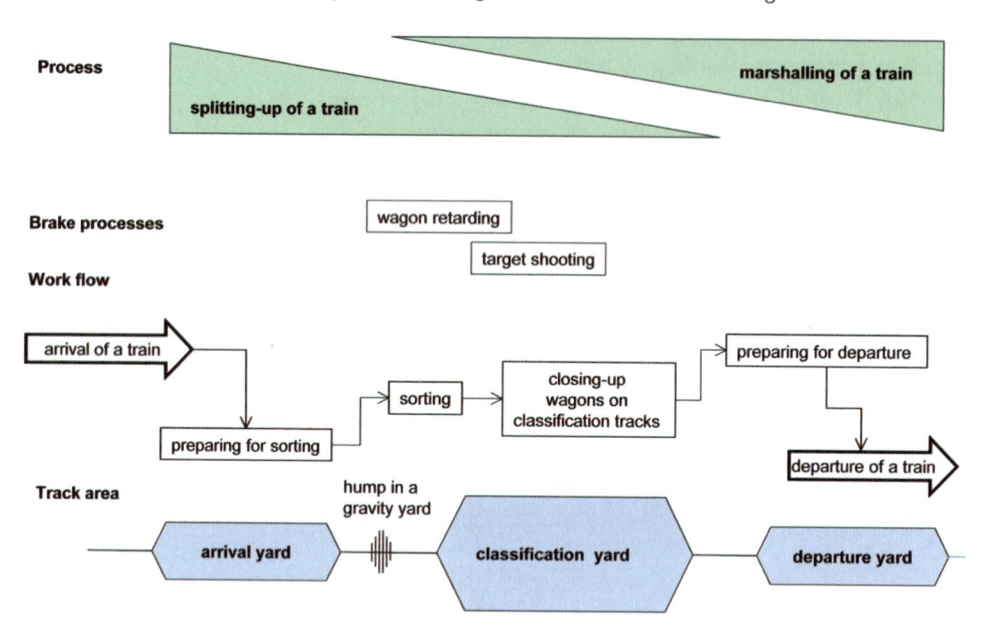

Figure 12.1: Work routines in marshalling yards

Marshalling yards have to meet a number of special requirements. These requirements are for example: efficiency of use, quiet operation of retarders and point drives, special sorting regulations and limited access to the area. For this reason in detail a lot of differences exist in
- construction form,
- level of automation and
- technical equipment.

12.2.2 Layout Variants

Some criteria to distinguish construction forms of marshalling yard are shown at figure 12.2.

The differentiation according to **height profile** refers to the inclination of the various components. **Flat marshalling yards** are established in the level area.

By contrast, almost the whole rail area of the **marshalling yard on a continuous slope** has a declining gradient in the direction away from the hump. The gradient relations in the yards are different therefore from flat marshalling yards, but the kinds and arrangements of yards do not differ basically. In both cases the potential energy of the wagons is used for the sorting of the trains or wagon groups.

12.2 Parts of Marshalling Yards and their Function

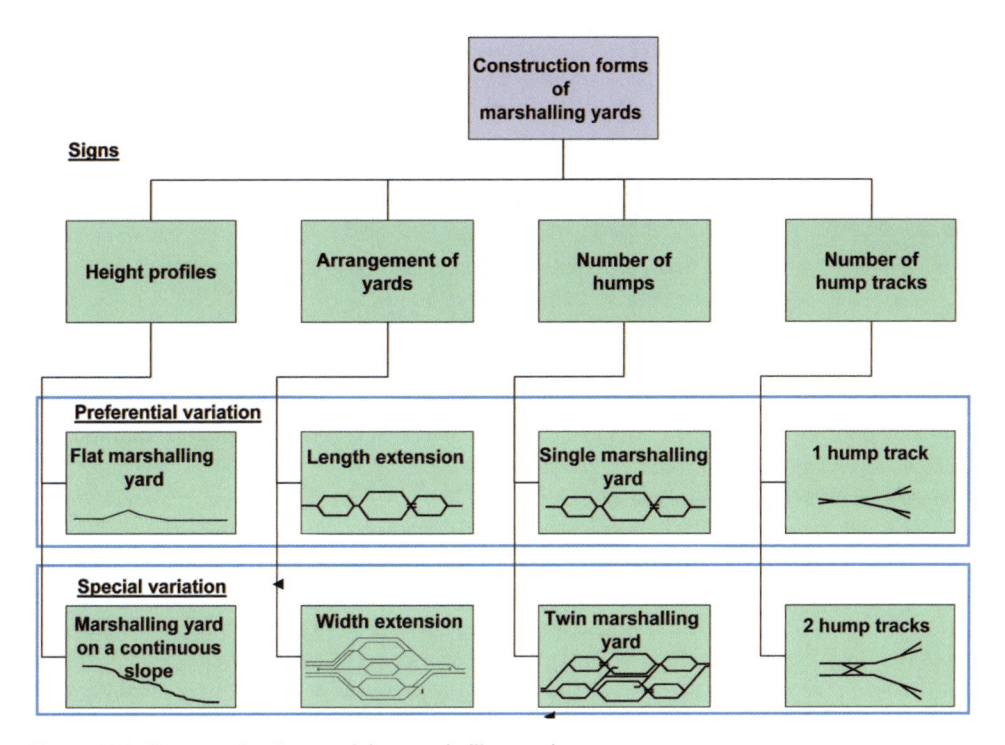

Figure 12.2: Construction forms of the marshalling yards

In marshalling yards on a continuous slope, the wagons roll under gravity. Thus shunting locomotives to push the wagons over the hump and close up the wagons in the classification yard are not necessary. These advantages are compensated by the disadvantages of special costs of construction and operation.

The most favourable **arrangement** is to place the yards in a line (**extension in length**, figure 12.1). Due to restrictions in the availability of land or due to very long trains, this is not always possible. In these cases the track groups can also be arranged side by side (**extension in width**, figure 12.3). A typical North American freight train is two to three km long, therefore extension in width is often used here. In American marshalling yards often only arrival and departure yards are long enough to accommodate whole trains. The classification yard is shorter and contains only parts of the new train.

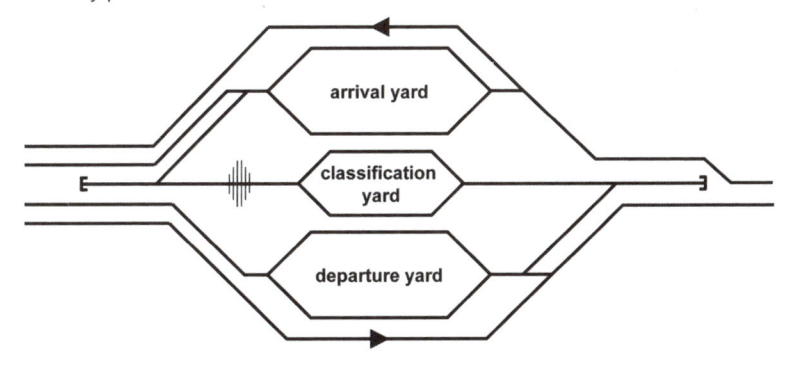

Figure 12.3: Marshalling yard – extension in width

12 Safety and Control of Marshalling Yards

In case of extension in width the locomotive must proceed to the far end of the yard, pull the wagons out (generally onto a dedicated track called a 'pullback track', then push them to the hump. The track approaching the hump from the receiving yard is on an ascending grade to the apex, or top. At the apex, the gradient quickly changes to a steep descending grade. As the uncoupled wagons reach the apex, they begin to roll freely into the classification tracks. Wagon retarders slow down the free rolling wagons to an appropriate speed before entering the classification track. (More details on retarders see chapter 12.3.2).

In **marshalling yards with two yard systems** all kinds of yards exist twice. Usually the yards are arranged in opposite directions in both systems. It is possible to increase the efficiency, if high capacities are needed.

The use of **two hump tracks** can achieve higher hump capacities.

The principles of regulating speeds of descent and track occupancy control can differ greatly at marshalling yards in different countries. These differences can be explained by a number of reasons, including:
- traditions of developing automation systems;
- differences in methodology, theoretical approaches and control algorithms;
- application of different computer control complexes, field equipment and technical solutions.

In general, the physical process of gravity shunting is very difficult because the wagons can have very different running qualities and meteorological influences can change. Various wagon types and their parameters (kind of wagon, whether it is full or empty, wagon length, wagon construction, mass of wagon, number of axles, quality of axle bearing) is the cause of the wide ranges of riding quality.

12.2.3 Automation

To make the complex process of gravity shunting more efficient, the aims are ease of control and safety. Automation is the way to achieve this.

The **lowest level** (points operated by hand, use of brake shoes) is the technology of former times. Control of the splitting process was very difficult at this level. Communication between hump foreman, locomotive driver and other workers was based on a humping list and special humping signals. Today this level can be accepted only where volumes are low. It is not only a question of economy. The use of brake shoes is dangerous, slow and not easy to handle.

A lot of new solutions of technical equipment (voice radio communication, retarders, sensors, points) and control systems are now available. The **first level of automation** was the replacement of the use of brake shoes by retarders. Modern retarders are quick working, that means they are more efficient and can have a lot of special features like
- active and inactive position,
- remote control and
- brake force adaptation.

The principles of wagon braking are also very different. The choice of brake facilities is influenced by several parameters. On this level the retarder is controlled by man or a feature of the retarder itself. Point and retarder control is now concentrated on the hump signal box.

The **second level of automation** is that of retarder and point control by hump process control systems. To do this a lot of information has to be processed. Today the most high performance marshalling yards work on this level. The last steps on the way to fully automation **(third level of automation)** are:

- the integration of automatically controlled hump locomotives,
- closing the gaps in the information flow,
- automatic detection of wagons and locomotives in the complete process, weigh-in-motion scales at the crest of hump (in North America and industrial sidings) and
- last not least optimising the algorithms of control and the technical details of marshalling yard.

Examples and more details are shown in chapter 12.3.

12.3 Control of Marshalling Yards

12.3.1 Introduction

Control of gravity marshalling yards with a high level of automation is very complex and must be based on hump process control systems. These systems are interacting with:
- shunting-technical facilities,
- points and
- sensors.

Shunting-technical facilities in marshalling yards are used
- to brake (retarder / rail brakes),
- to promote (handling systems) or
- to detain (e.g. holding brakes, concealable/movable buffer stops).

Points must work very quickly and safely.

To get the **necessary information**, a lot of data must be generated, evaluated and translated in real time using the following objects or influences (see figure 12.4):

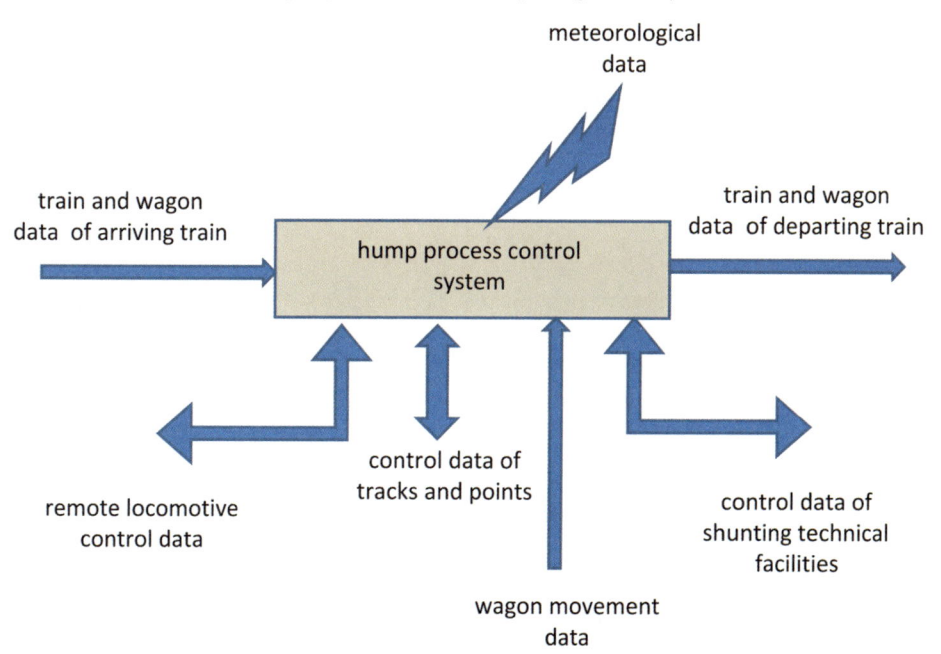

Figure 12.4: Data exchange

- data of arriving trains (number, kind and position of wagons, arrival time etc.),
- data of wagons (technical parameters like height, width, number of axles),
- data of environmental parameters (wind, rain, snow, temperature etc.),
- wagon movement data (speed, position),
- data of points (point control, point occupation),
- data of retarders and other shunting technical facilities (working position and occupation),
- data of tracks (track occupation)

and a lot more.

The data can be generated in very different ways. There are
- messages from neighbouring systems (like data of inbound trains),
- data generated by sensors (meteorological data, track occupation etc.) and
- data from data bases (technical parameters of wagons).

Today a lot of very different sensors are necessary and in use to get information out of the process of gravity shunting. These sensors must have important parameters like quick response, high reliability, resistance against hard environmental conditions and others.
All functions of operation and control are concentrated in hump information systems.

12.3.2 Retarders

12.3.2.1 Kinds of Retarders

Retarders generate braking force which works on the wagon wheels. The braking force can be generated in diverse ways. This allows the adaptation of brakes to individual demands of operation and control (see figure 12.5). That is why there are several forms of retarder in use.

In places with extreme climates such as Russia or the North of China, retarders powered by compressed air are used to be able to guarantee their effectiveness in temperatures from −50 to +50 degrees Celsius. The use of compressed air requires suitable facilities for production

```
                    Brake force generated by ...
     ┌──────────────────┬──────────────────┬──────────────────┐
pressing beams on   passing pneumatic   passing elastic   passing magnetic
 flanks of wheels       dampers             rails              fields

• Clasp retarder    • Piston retarder   • Elastic rail brakes  • Electric dynamic
                                                                  rail brakes

                    passing hydraulic
                        dampers

                    • Piston retarder
                    • Hydraulic spiral retarder
```

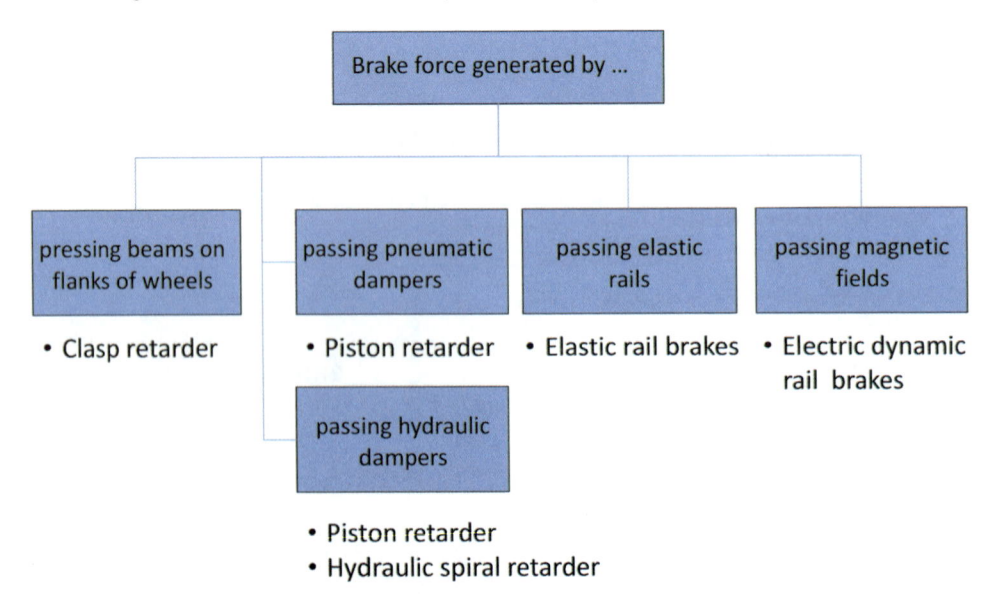

Figure 12.5: Physical principles of braking and kinds of retarders

12.3 Control of Marshalling Yards

and distribution of the air pressure. The control and supervision of the air pressure arrangements is also performed by the control system. In other regions with more favourable climatic conditions, hydraulic systems are preferred.

Beam or clasp retarders consist of two moveable beams on one (figure 12.6) or both rails (figure 12.7). To generate brake force, the beams will be pressed on the flanks of wheel tyres of running wagons (pressing at two points = two power retarder). Other clasp retarders use an additional damper at outside or inside beams (pressing at three points = three power retarder).

The brake force is controllable. The retarders have a working and a non working position. In non working position the retarder can be passed by vehicles including locomotives. The retarder drive can be electrical, hydraulic or pneumatic. The variation of important

Figure 12.6: Single-rail pneumatic beam retarder in Zvolen (Slovakia) (photo: *Peter Šoltys*)

Figure 12.7: Two-rail hydraulic beam retarder in Mannheim (Germany)

parameters like length, time of reaction, brake force and a possible maximum entry speed of wagons allows the adaption to special needs.

Electric-dynamic rail brakes are based on eddy current principles. The brake force is controllable via electric power regulation. In the non-working position (electric power off) the retarder can be passed by most types of locomotives.

In the area of **elastic rail brakes** the steel rails are replaced by rubber brake elements (special rubber vulcanised on metal sheets – see figure 12.8). Rubber parts are elastically deformed by the wheel running over them. This transforms kinetic energy into other forms of energy, and therefore brakes the wagon. This kind of retarder is not controllable directly. If the working position is not changeable, locomotives cannot pass over them.

Hydraulic spiral retarders brake the wagon(s) to an exactly defined maximum. When a wagon passes over such a retarder

Figure 12.8: Elastic rail brake in Munich (Germany)

12 Safety and Control of Marshalling Yards

the wheel flange interacts with the cylinder spiral ledge (figure 12.9), the latter performing one turn-over. If the wagon speed is lower than the one for which the retarder is adjusted, the retarder's valve doesn't prevent liquid from flowing from one cavity into another. In this case braking doesn't take place. When the speed exceeds the adjusted one, the retarder produces maximum braking effort. Some kinds of hydraulic spiral retarder allow alternative working positions (active or inactive). In this case locomotives can pass this kind of retarders in the inactive position.

Figure 12.9: Hydraulic spiral retarder in marshalling yard Košice (Slovakia) (photo: *Peter Bado*)

Piston retarders are based on oil-hydraulic or pneumatic principles. Their brake effect takes place on wagon wheel flange running over the retarder's piston fixed on the rail (figure 12.10). The excess kinetic energy is reduced due to the piston travelling down when the wagon is rolling over. Oil-hydraulic piston retarders (in Europe also called Dowty elements) do not work if the speed is lower than 1 m/s. Some kinds of piston retarders are controllable and allow working or non working position.

Figure 12.10: Piston retarder in Harbin (China)

12.3 Control of Marshalling Yards

12.3.2.2 Use of Retarders

Criteria for the choice of the rail brakes are
- brake force to be performed,
- fulfilment of the requirements from the shunting-technical procedure,
- conditions of the application place as well as
- economic efficiency.

The arrangement of retarders in marshalling yards is a result of special **retarding concepts**. The retarding concepts must be developed in connection with construction details of the marshalling yard (especially gradient), retarder characteristics, quality and quantity of wagons in operation and environmental conditions. Modern hump process control systems can be adapted to control a special retarding concept. Basically two retarding concepts are known: **Continuous speed control methods** based on piston retarders (see examples from China in chapter 12.3.7.4) and **target shooting methods** in a wide range of variations based on other kinds of retarders (see following examples). Target shooting method means that the wagon, leaving the last retarder, have exactly that speed which is needed to stop at the right position in the classification track. That means different speeds depending on the changing distances between the last retarder and the target position. In general retarders can be used:
- to hold wagons or groups of wagons (cuts) in a defined track (e.g. holding brakes in marshalling yards on a continuous slope),
- to guarantee speed limits in the point zone,
- to keep distances between rolling down shunting units in the gravity incline and
- to control the speed to achieve the right position of wagons in classification tracks.

Combination of kinds of retarders (and handling systems) allows a lot of modifications of target shooting method. In detail there are also differences in the braking positions and gradient ratio. In Europe, retarders are normally at the positions shown at figure 12.11. Retarders can be called by working principles (see chapter 12.3.2.1) or by function in use. The retarders at braking position 1 are called 'ramp retarders'. Often two-rail hydraulic beam retarders are preferred

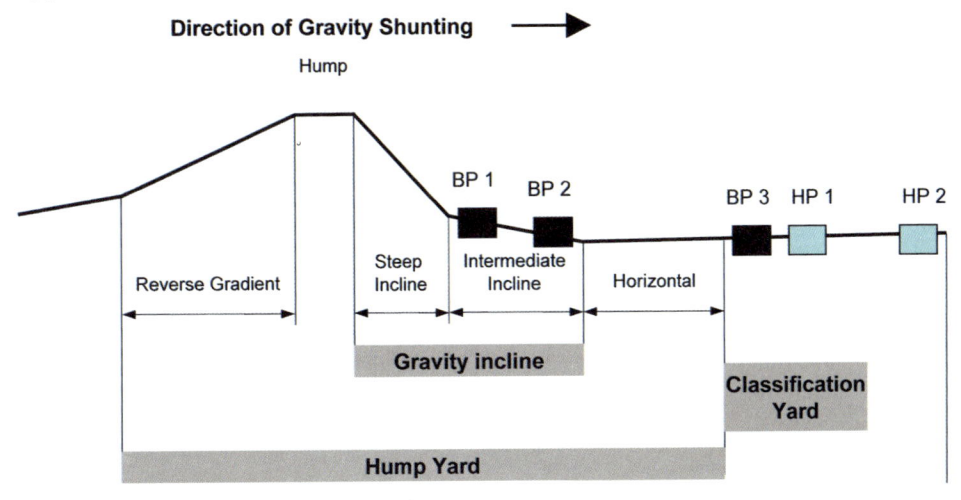

Legend:
BP - position of brakes
HP – position of handling system

Figure 12.11: Braking positions and positions of handling systems in Western Europe

12 Safety and Control of Marshalling Yards

there. The retarders at position 2 are the hump or main retarder (in UK called king retarder and in North America master retarder). The real target braking takes place in the classification yard (position 3). These retarders are called secondary retarder or in UK queen retarder (North America: group retarder). At this position single-rail hydraulic beam retarder, electric dynamic retarder or elastic rail brakes are preferred. Piston retarders are also possible.

Figure 12.12 shows an example according the national rules in Russia. In this case the retarder at the first braking position ensures the required intervals between shunting units in the area from the first braking position up to the second one. The retarder at second braking position can realise the intervals between shunting units to avoid collisions and enable setting of points during passing by vehicles. The purpose of the retarder at the third braking position is to reduce the speed of shunting units if necessary. In the classification tracks the wagons must achieve their position, considering the approach to standing wagons with the safe collision speed. This speed on hump yards is low (e. g. Russia 5 km/h).

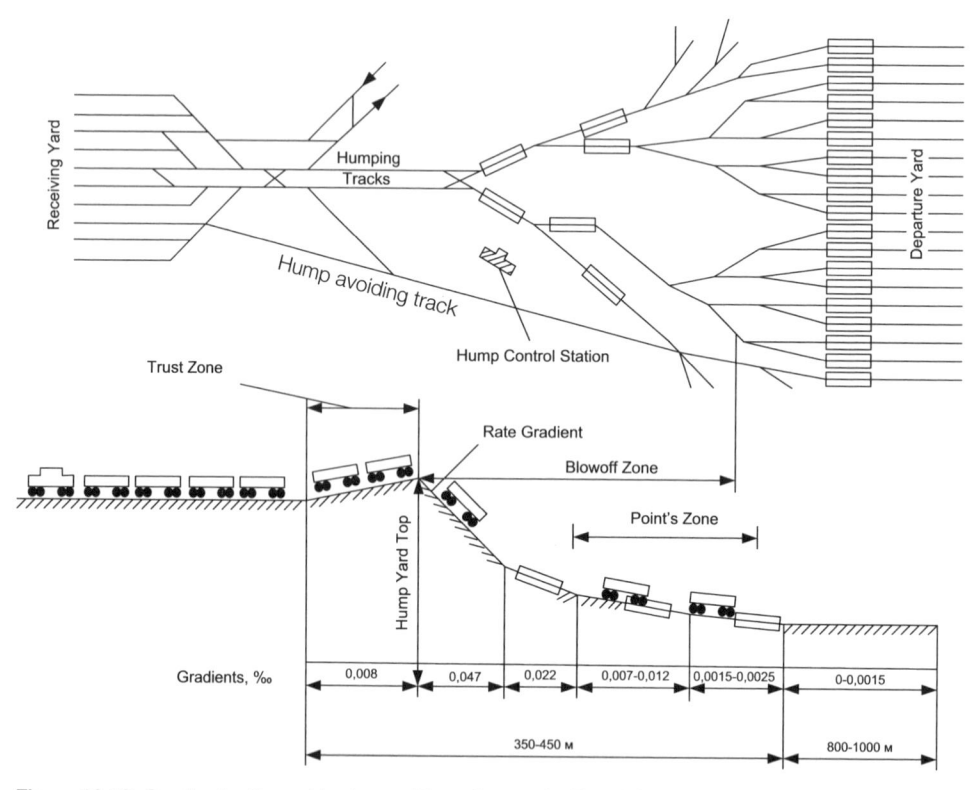

Figure 12.12: Gradient ratio and brake positions (example: Russia)

12.3.3 Handling Systems for Freight Wagons

The standard European screw coupling is not self coupling. After gravity shunting the wagons are standing in the classification yard and must be coupled manually. But often coupling is not possible because there are small gaps between the wagons. To close the gaps, shunting locomotives or special handling systems can be used. Modern European marshalling yards with enormous marshalling operations are equipped with these systems. If there are automatic

Figure 12.13: Handling systems in Munich (Germany)

coupling systems in use (e.g. America, Asia, Russia), this kind of equipment is not necessary. Handling systems are located inside the track and move by means of automatically controlled ropes (see figure 12.13). The systems can be installed at the end of classification tracks (called clearing sweeper) and additionally at the beginning of classification tracks (called rope haulage sweeper). **Clearing sweeper** helps to close the gaps between wagons or wagons groups on the end of sorting tracks to enable coupling. **Rope haulage sweeper** can be necessary additionally to clear up sorting tracks at the beginning, behind the points and last retarder. Both systems are different in technical details. In general the use of handling systems is part of the retarding concept. So it is possible to use the classification tracks more efficiently and locomotives are not needed in this yard.

12.3.4 Points

Points influence the quality of the shunting process. Important parameters are throwing time, point position control and kind of operation. There are point machines available for points in shunting areas for normal and slow points up to very fast points. Short throwing times of very fast points (0.5 secs throwing time) helps to get higher shunting quality. That's why normally very fast points are in use at the hump area. Points with lower throwing times are acceptable in other areas of marshalling yards (arrival and departure yard) and in case of low performance demands. Modern hump process control systems are able to control various point machines. Point position control is also important for safety.

In the past, mostly interlocking controlled mechanical points were used. The application of power switches in yards was generally limited to the entrance of arrival yards and the exit of departure yards due to the too long distance for mechanical switching. The availability of electrical points is the basis of new concepts of operation in shunting areas and the distance is no longer a problem. Electric points can be operated by a control panel or by a pushbutton mounted on a post adjacent to the points.

Control panels can be integrated into interlocking cabins or mounted in the field like the push-buttons. If the operating facilities are placed in the field, a member of the train crew or a shunter would walk along, pushing the required buttons for the route.

This application, in comparison with centralised control from an interlocking, requires less infrastructure (e. g., cabling, control panel). This technology is available and in use in Europe and North America. It allows a high level of flexibility by scalable solutions from low level in industrial sidings to high level in marshalling yards. More details see chapter 9.4.10.

12.3.5 Sensors

Sensors are used to measure
- static data of wagon (e. g. height, length, mass) and
- dynamic data of wagon (e. g. speed, position), track elements (occupation, work position) and weather (e. g. wind, temperature, humidity).

The measurement takes place before or during the shunting process.

The **area of the wagon where wind can act** to retard the wagon speed is a result of length, height and width of the wagon. The measurement of **length** is possible with Doppler radar devices. To get **height** and **width data**, light grids (light curtains) can be installed. Additional light grids are used for detecting the **space between cuts. Axle loads** can be measured directly at the rail by bending torque evaluation.

For **speed detection** Doppler radar devices can be used also.

Weather, especially wind, can influence the wagon speed dramatically. Real time **weather data capture** (wind measurement, temperature, humidity) is an important factor to find out the riding quality of wagon and to optimise control of gravity shunting.

The reduction of the necessity of measurements increases efficiency and helps to reduce the complexity of systems. One way to do it is the implementation of automatic equipment identification systems.

In general the **identification of equipment (wagons and locomotives)**, often in combination with databases, is very important to organise railway operation in different processes like
- organisation of efficient use, maintenance and repair of equipment,
- preparing splitting up of trains in marshalling yards and
- customer information.

Technical solutions are
- manual identification by human (reading wagon numbers),
- automatic local identification (infrared, video and others),
- Automatic Equipment Identification (AEI) by tags,
- Radio control by satellites.

Manual solutions are expensive. **Local identification** by infrared, video or other is technically difficult. Technical solutions based on AEI and radio control by satellites are in use in Europe. Wagons with dangerous goods or valuable freight are routinely equipped. Today most of the railways use a mixed system of data messages and inspection. That means if trains are leaving a station, the data messages are sent to the information system. From this information system, the terminal marshalling yard of the train can download this data. Thanks to that, cut lists can be generated before the train arrives. When the train arrives, inbound inspection based on computer generated lists begins. This is necessary to check for changes (wagons on other positions, wagons added or switched). Every kind of data change (status and location of wagons) must be fed into computer systems or is a result of a computer program.

North American railways have developed a system, **Automatic Equipment Identification (AEI)** that provides a direct connection between each wagon and the information systems. Each freight wagon and locomotive is equipped with two AEI tags, affixed to the sides of the vehicle. The tag is a specially created transponder and memory to store basic information about the wagon. In North America, each wagon is identified by reporting marks (a combination of one to four alpha characters representing the owner, also known as initials) and a number of one to six digits. Each locomotive and freight wagon (including containers and highway truck trailers) in North America has a distinct combination of reporting mark and number. Every piece of equipment is listed in a database called **UMLER (Universal Machine Language Equipment Register)**.

AEI readers are stationed along the tracks, generally on main lines near terminals and at other places where train consists may change significantly. The tags are passive and the reader is active. The reader can work properly at any normal freight train speed, so there is no speed restriction associated with the process. As each vehicle passes the reader, the reporting marks and number of each are accumulated into a list that is sent to the information system that handles wagon movements. That system can associate its own internal destination and blocking information with UMLER basic wagon data and produce accurate train lists that may be used by yards and terminals for planning, train information queries, and electronic transfer to computers handling hump yard automation, and to connecting railroads through a process called **Electronic Data Interchange (EDI)**.

AEI readers are generally located along the main tracks just after a train leaves a terminal, where the accuracy of the train list generated by the yard computers may be checked, and on the approach to terminals, to detect changes in consist since the train left the last terminal (e. g. industry work, setting out defective wagons, picking up repaired wagons).

Not all systems use all kinds of sensors. The use of sensors is part of a construction concept of every marshalling yard control system.

12.3.6 Track Clear Detection

The purpose of track clear detection in marshalling yards is to prove that a moveable track element is clear of rail vehicles before being switched. Technical solutions to do it can be:
- track circuits,
- infrared scanner,
- radar scanner,
- axle counter (induction loops).

Technical details of track clear detection are described in chapter 5.

12.3.7 Yard Management Systems

12.3.7.1 Framework Conditions

Yard management systems control marshalling yards at high levels of automation. The heart of these systems is computers and their specially developed software solutions. These systems must be able to be adapted to various user demands like

- construction forms of marshalling yards,
- shunting qualities,
- retarding concepts and
- kinds of sensors and detectors.

Most of the known systems use the same principles. Differences exist in technical details, flexibility in use and level of automation.

To control the work flow in marshalling yards, a huge volume of necessary information must be processed in very short time precisely and reliably. The basis for control is the measurement values of the sensors as well as other relevant data received via data links (e.g. preregistration data of running trains). Such control systems are integral components of modern hump interlocking systems for gravity shunting. The essential task is the purposeful influencing of the wagon cuts run from the hump into the sorting tracks. Besides, the following tasks are to be solved:

- the journey control for running off wagons,
- the tracking and tracing of the journey of the wagons across the whole yard,
- the control of the hump locomotives and
- the speed regulation of the wagon cuts.

A description and demarcation of the yard management systems develops difficultly, because internationally very different products are in use. The variability results from:

- technological,
- economic and
- operational basic conditions.

Technological basic conditions are
- operational programs to be used (and with it brake programs),
- specific technical requirements (e.g. climate suitability),
- adaptation to available technical facilities and vehicles (yards, rail brake technology, vehicle couplings etc.),
- national and international standards, legal regulations etc.

Economic basic conditions can be
- economic characteristics of the respective railway company (e.g. available means for investments, demanded return on investment of used means),
- calculated useful life duration and efficiency of the whole yard.

Operational basic conditions are
- operational standards/rules,
- operational concepts (e.g. production procedure).

12.3.7.2 Structure

In general yard management systems can be arranged of the following components
- the control system itself (hump process control system),
- diagnosis systems and
- management information systems.

Not all subsystems (beside the control system) always are part of a yard management system. Besides, the names of the manufacturers differ for the systems and their subsystems.

The **control system**, which is actually an electronic interlocking with extensive automation functions, serves for the optimum control of the hump process. Problems are
- the protection of every wagon cut against the following and advance-running wagons,

12.3 Control of Marshalling Yards

- the exclusion of points moving under vehicles,
- the possibility to bring the wagons to an easy to couple position in the planned sorting track and
- to enable a very high performance.

Additional functions are often control of automatic hump locomotives, handling systems and retarders.

The control systems often are complemented by **Diagnostic Systems**. With their help, the following can be supervised:

- infrastructure elements (points, signals),
- shunting-technical facilities (retarders, handling systems, brake test facilities),
- hump locomotives and
- information and communication systems.

This supervision serves to guarantee operational safety. However, it can also be used for a purposeful servicing and care of the concerning components to avoid technically caused disturbances or possibly even failures.

Management Information Systems utilise information of the control and diagnosis systems. They do not intervene actively into the cut process. Nevertheless, they also support the work flow by the supply of information. The following belong to the possible tasks:

- long-term planning and forecasting of future situations,
- handling of inbound trains (takeover of train data of the inbound trains from offshore operating locations, data capture of inbound trains, generation of cut lists and their transference to the control system),
- handling of outbound trains (supply of documents for the train formation and the departure operation of trains e.g. brake test, train accompanying documents, handed over by train data of the outbound trains to the following operating locations and if necessary to the customers),
- optimising and monitoring of all yard operation,
- operations management,
- personal deployment and
- production of compressed management information about the achievement assessment (statistical evaluations).

12.3.7.3 Systems in Europe

One of the systems with a high number of installations especially in Western Europe is the **MSR 32 Microcomputer System** of *Siemens* (Siemens 2008). MSR 32 allows the integration of a complete range of applications for the automation of marshalling yard. The system is scalable and modular.

The technical system consists of (see figure 12.14):

- hump microprocessor,
- sensors like:
  - low-cost wheel detectors for track clear detection and approach control,
  - Doppler radar devices for speed and length measurement,
  - Light grids (light curtains) for detection of wagon cuts and the space between,
  - bending torque evaluation directly at the rail for weight measurement,
  - weather data capture (wind, temperature, humidity),
- point machines for different throwing times,
- control of various retarders and of handling systems,

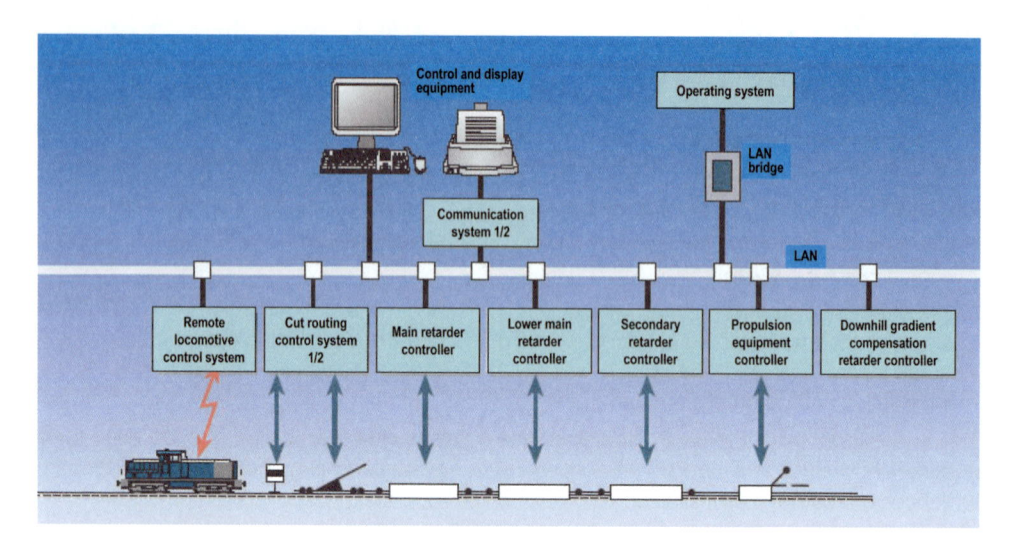

Figure 12.14: MSR 32 – System architecture (graphic: *Siemens*)

- and a lot of other features (e.g. Interface to planning systems via LAN bridge, other interface options, lightning and overvoltage protection, connection to radio clock).

For low performance demands the MSR 32 EOW Control System is available (compare chapter 9.4.10). This system can be used to control decentralised and centralised electric points. MSR 32 EOW also can be used in connection with MSR 32. In this case control of arrival yard and departure yard is done by MSR 32 EOW and classification yard by MSR 32.

In Europe a complete Automatic Equipment Identification (chapter 12.3.5; like the North American System) is not in use. To get train and wagon information of arriving trains, data messages and inbound inspections can be used. Additional information systems can support the process of inbound inspections and generating of wagon data, e.g. the Automatic Composition Checking System for Freight Transport (ARKOS) can be integrated in MSR 32 systems.

Another system for marshalling yard automation is installed in Scandinavia. It's called **Alister-Cargo**, a product of *Funkwerk* (Funkwerk 2008). This electronic control system for marshalling yards is completely based on standard industrial control components as PLC's (Programmable Logic Controllers).

In former **Czechoslovakia**, a system for marshalling yard automation called KOMPAS was developed. It is a product of *AŽD Praha*. It is a modular system. It is possible to apply it in five different modes, according to capacity and size of marshalling yard. Typical difference from other systems is the usage only of clasp retarders. In classification tracks three stages of target brakes and one stage of holding brakes are used (for more details see Hájek 2006). Further development based on KOMPAS is realized by *První signální*. Name of the system is MODEST MARSHAL (for more details see Zářecký 2008).

12.3.7.4 Systems in Russia

In Russia the first version of the yard management system called **KGM** was worked out in the *Rostov Institute of Railway Transport Engineers* in 1983 and put into operation at Krasnyj Liman marshalling yard. During the next 25 years, the complex was constantly being perfected. It was put into operation at 28 marshalling yards of Russia and CIS countries. The structure of the latest version of the system is given in figure 12.15.

12.3 Control of Marshalling Yards

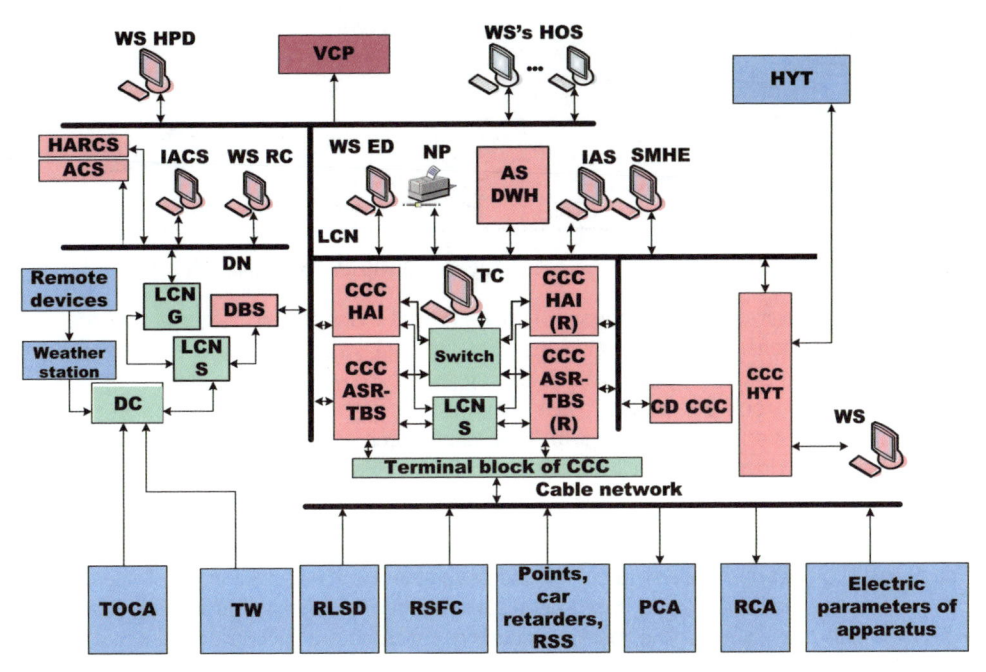

List of abbreviations

ACS	– automated control system
AS DWH	– application server and data ware house
ASR-TBS	– automatic speed regulating with target braking subsystem
CCC	– control computer complex
CCC HYT	– control computer complex of hump yard top
CD CCC	– check-diagnostic control computer complex
DBS	– database server
DC	– data converter
DN	– data network of Open Joint-stock Company 'Russian Railways'
HARCS	– hump automatic radio cab signalling
HAI	– hump automatic interlocking
HYT	– hump yard top apparatus
IACS	– integrated automated control system
IAS	– information-analytical subsystem
LCN	– local computer networks
LCN G	– LCN's gateway
LCN S	– LCN's switch
NP	– network printer
PCA	– points control apparatus
(R)	– (reserve)
RLSD	– radio-locating speed detectors
RCA	– retarders control apparatus
RSFC	– radiometric sensors for freeness checking of measuring sections
RSS	– radiotechnical speed sensors
SMHE	– subsystem for maintenance of hump equipment
TC	– technological computer
TOCA	– track occupancy control apparatus
TW	– tensometric weighers
VCP	– multiple-access visual control panel
WS	– workstation
WS ED	– WS of electrician on duty
WS's HOS	– WS's of hump operation staff
WS HPD	– WS of hump person on duty
WS RC	– WS for remote checking

Figure 12.15: Control System for marshalling yard in Russia (RFNIIAS 2008)

12 Safety and Control of Marshalling Yards

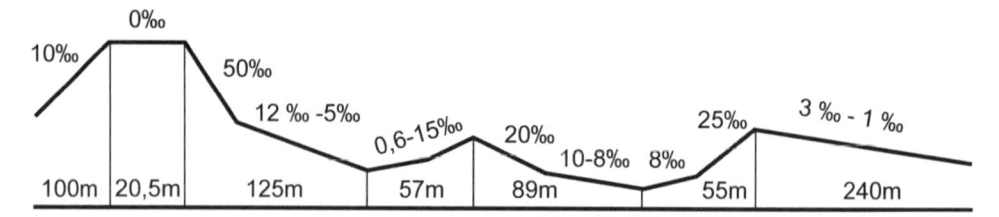

Figure 12.16: Profile of the upgrade speed control system Shenzhen (Xu Zhengli 2003)

The systems consist of all components of modern yard management systems and are designed for the national rules and conditions (e.g. gradient, retarders, retarder control procedures and climatic situation) of Russia. The system also provides for the data transmission to the Open Joint-stock Company *'Russian Railways'* through the data transmission network channels and the dispatchers' access to the branch automated control system of signalling division. For more information see (Ivančenko et al. 2002, RFNIIAS 2008).

12.3.7.5 Systems in China

The *China TDJ System Research Centre* in Harbin created various applications of speed control systems in marshalling yards based on piston retarders. **Chinese marshalling yards** are equipped with this solution. The realised retarding concepts are qualified continuous speed control methods. A special profile (see figure 12.16) and the integration of boosters are interesting parts of this application. The 'TDJ' Boosters can accelerate wagons, if necessary. They are powered by compressed air, separately controllable and may be combined with piston retarders (12.3.2.1). The construction of boosters is similar to the piston retarders. A piston can be pressed out of a cylinder by compressed air to give an impulse to a wheel after passing the booster (details on piston retarders and boosters see TDJ 2008).

The design for this system (called upgrade speed system) allows humping speed of $v_0 = 6$ km/h. The average coupling speed is 4.27 km/h and the coupling rate reaches 100%.

In Chinese middle and small sized marshalling yards controllable retarders are the major speed control equipment. In these applications boosters are not integrated. In Nancha marshalling yard, an example of this kind of applications, 4610 TDJ control retarders are applied. The piston retarders are installed in switching and tangent area of classification yard consisting of 16 tracks (Xu Zhengli 2003).

12.3.7.6 Systems in North America

In North America several systems are in use. Systems like Trainyard Tech, PROYARD III and STAR NX support a wide range of demands. Differences to other systems are
- the automatic identification of wagons by tags (no more by wagon numbers) and
- wireless yard control (e.g. STAR NX- System).

For more information on North American yard management systems see (Judge 2007).

13 Level Crossings
Gregor Theeg, Dmitrij Švalov, Eric Schöne

13.1 Requirements and Basic Classification

On level crossings between rail and road, the partly contradictory characteristics of both transport systems meet (table 13.1). As a result, safety problems arise. The high kinetic energy of a train and the impossibility of stopping on sight when seeing an obstacle meets the relatively low safety discipline of road traffic. Accidents at level crossings typically cause around one third of all fatalities in railway operation accidents. The vast majority are caused by the inappropriate behaviour of road users. However, in relation to the total number of road accidents, accidents on level crossings represent only a low percentage of the total (less than 1 %). Due to the high kinetic energy on railways, the average severity of an accident on a level crossing, measured in killed and injured persons and damage to equipment, is much higher than in other types of road accidents. It is mainly road users, but railway passengers and staff are also endangered. Further dangers arise when dangerous goods are involved.

Criterion	Rail	Road
mass of vehicles	high	relatively low
acceleration and deceleration rates	low	relatively high
stopping distance	long	short
spacing method	signalled (fixed block)	on sight
driving style	controlled	individual

Table 13.1: Characteristics of rail and road traffic

To increase safety, many countries follow a strategy of abandoning level crossings and replacing them with costly grade separation solutions on two levels (chapter 13.5). Nevertheless, a large number of level crossings remain and that is likely to be the situation for the foreseeable future.

A result of the comparison of stopping distances of road and rail traffic is to give priority to rail traffic, which shall not be obstructed by the level crossing in normal operation. Therefore, the road user must be warned about an approaching train and be able to stop at the level crossing. This warning can be done either by direct optical or acoustical perception of the train, or by special signals installed at the level crossing. If stopping is impossible because the road user is already closer to the level crossing than his stopping distance, or he is already on the level crossing, he must be able to pass over the level crossing completely without conflict.

Road users should have the permanent possibility of escaping from the conflict area, which is that area used in common by both rail and road traffic. If this is not possible, supervision to ensure that the conflict area is clear is obligatory in many, but not in all countries (chapter 13.4.4.4).

The installations must work safely, which means that in case of failure they have to be fail-safe. To maintain operation in failure cases, solutions for degraded mode operation need to be provided (chapter 13.4.5).

Last, but not least, neither rail nor road traffic should be obstructed more than necessary. Long obstruction times for road traffic can even become safety critical, as the discipline of road users decreases. On the other hand, the obstruction of rail traffic impairs the capacity of the line and makes timetable operation more difficult.

A basic classification of level crossings by the *ERA (European Rail Agency)* is as follows:
- **Passive level crossings:** These always appear to the road user in the same way, irrespective of whether or not there is a train approaching. Therefore, the road user has to look for trains himself.
- **Active level crossings:** These indicate to the road user whether a train is approaching or not.

13.2   Static Roadside Signs

The purpose of static roadside signs is to make the presence of a level crossing clearly visibly to road users and to attract their attention. The roadside signalling is mainly regulated by the highway authorities of the country concerned. Although general road signs are widely harmonised internationally, in the particular case of level crossings the situation is different. In spite of some similarities, in each country the level crossing appears differently to the road user and the rules for using the level crossing differ even more.

In some countries there is a general speed restriction on level crossings or even the general obligation to stop on reaching them. In others there are no such restrictions, or the road user is just obliged to slow down appropriately.

The most widely used sign to indicate the presence of a level crossing to road users is the distinctive St. Andrew's cross. This strengthens its importance to give absolute priority to rail traffic. In some countries, all level crossings are equipped with the St. Andrew's cross; in others only those without barriers (Hahn 2006).

In all cases, the St. Andrew's cross obliges road users to give absolute priority to rail traffic and forbids stopping on the level crossing. Besides, the St. Andrew's cross marks the proper (safe) stopping place in case the exit from the level crossing is obstructed. Often private roads, footpaths, field paths and forest tracks do not have St. Andrew's crosses, but the regulations give absolute priority to rail traffic at these level crossings as well.

A basis for international standardisation of the St. Andrew's cross was given by the *Vienna Convention on Road Signs and Signals* in 1968. However, the design varies between countries regarding its exact form, colour, use of text and degree of duplication (Hahn 2006) (figure 13.1):
- In most countries, a profiled St. Andrew's cross is obligatory to enable perception even in unfavourable weather conditions (snow etc.). Only a minority of countries use the St. Andrew's cross painted on a rectangular board.
- In many European countries, a distinction is made by the number of tracks to be crossed by the road user. In these countries, the St. Andrew's cross with two or more tracks to be crossed is doubled. In other countries (e.g. USA), the number of tracks to be crossed is stated by an additional number below the St. Andrew's cross.
- In all European countries, but also others such as Canada, the St. Andrew's cross is painted in colour, with red as one and white or yellow as the other colour, but without text. In other countries such as the USA, Mexico, Australia, China and Saudi Arabia, the words 'RAILROAD CROSSING' (in the respective language) appear in black text on a single colour (usually white) background.
- The geometrical shape of the St. Andrew's cross is rectangular (most countries outside Europe), with right and left acute angles (most European countries) or with upper and lower acute angles (e.g. Germany).

13.3 Passive Level Crossings

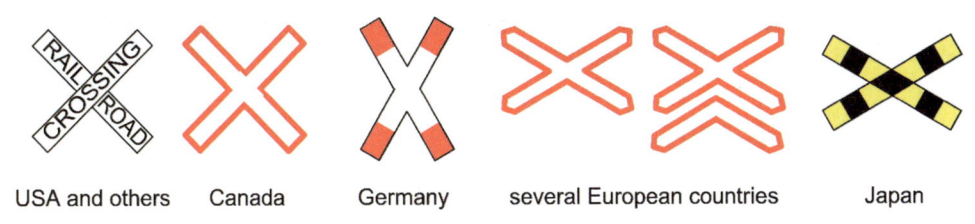

| USA and others | Canada | Germany | several European countries | Japan |

Figure 13.1: Forms of St. Andrew's cross on level crossings (not in scale)

In addition to the St. Andrew's cross, in many cases other signals such as text boards are installed immediately in front of the level crossing to warn the road users and give instructions.

In many countries, warnings of the approach to level crossings are given by road signs few hundred metres (usually 50 to 250 m, depending on local situation) in advance to give the road user the ability to prepare. In some cases, these signs distinguish between active and passive level crossings or between level crossings with and without barriers. A frequently used form is a triangular road sign with red rim and a steam locomotive or a modern train inside. In the USA, it is a circular sign with a black rim and a black X inside. Frequently the distance between the warning sign and the level crossing is measured by countdown markers with three stripes, two and then one. An example is given in figure 13.2.

Besides road signs, the warning of level crossings is often supported by pavement design.

Figure 13.2: Road sided warning signs and countdown markers (Sweden as example)

13.3 Passive Level Crossings

In passive level crossings, the road user is responsible for observing the railway line and recognising an approaching train directly. The most important measure to ensure the perception of the train is to keep the approach sight triangle clear of obstacles. The approach sight triangle is formed as follows, primarily described for the case that the road user is allowed to pass a clear level crossing without stopping or slowing down:

As described in chapter 13.1, the road user, when arriving at the permitted speed, must be able to stop at the level crossing when recognising an approaching train. Or, if stopping is not possible because he is already within the stopping distance from the level crossing, he must be able to pass the level crossing safely. The necessary sighting point A (figure 13.3) is the latest point where the road user must decide whether to stop in front of or to pass over the level crossing. It is determined by the stopping distance of the road user, which varies with the initial speed, the braking deceleration and the reaction time of the driver and the vehicle. The neces-

13 Level Crossings

sary sighting distance from the sighting point A to the stopping point, which is usually at the St. Andrew's cross, can be calculated as follows:

$$l_A = t_r \cdot v_v + l_b$$

and the complete clearing length as follows:

$$l_C = t_r \cdot v_v + l_b + l_{lc} + l_v$$

with:
l_b: braking distance of the road vehicle (speed-dependent)
l_{lc}: length from the stop position to the end of the conflicting area of the level crossing
l_v: length of the road vehicle
t_r: reaction time of driver and vehicle
v_v: speed of the road vehicle

Accordingly, the clearing time can be calculated as follows:

$$t_C = t_r + \frac{l_b + l_{lc} + l_v}{v_v}$$

The minimum approach time to avoid conflict is:

$$t_a = t_C + S$$

with
S: safety margin [s]

The approach length l_B of the train is therefore as follows:

$$l_B = v_t \cdot t_a = v_t \cdot \left(t_r + \frac{l_b + l_{lc} + l_v}{v_v} + S \right)$$

with:
v_t: speed of the train

Typical value ranges of variables are shown in table 13.2.

Variable	Typical value range
l_b	5 to 100 m (depending on vehicle speed and brake deceleration)
l_{lc}	5 to 20 m (depending on number of tracks and crossing angle)
l_v	up to 25 m (depending on national upper limit for vehicle length)
t_r	1 to 3 s
v_v	1 to 30 m/s (depending on general or local speed restriction)
S	2 to 5 s

Table 13.2: Value ranges of variables

In some national cases or where special regulations apply, stopping is obligatory even if no train is approaching. In this case, t_r and l_b can be set zero, which means that the sighting point A is the stopping point (the position of the St. Andrew's cross). In this case the clearing time t_c must be higher because the acceleration of the road vehicle (starting up) must be considered. Therefore the approach length l_B can also be longer.

For a real level crossing, particularly where stopping in front of the level crossing is not mandatory if no train is approaching, the differing speeds of road users have to be considered: Whereas the sighting distance l_A increases with increasing speed of the road user, for the ap-

13.3 Passive Level Crossings

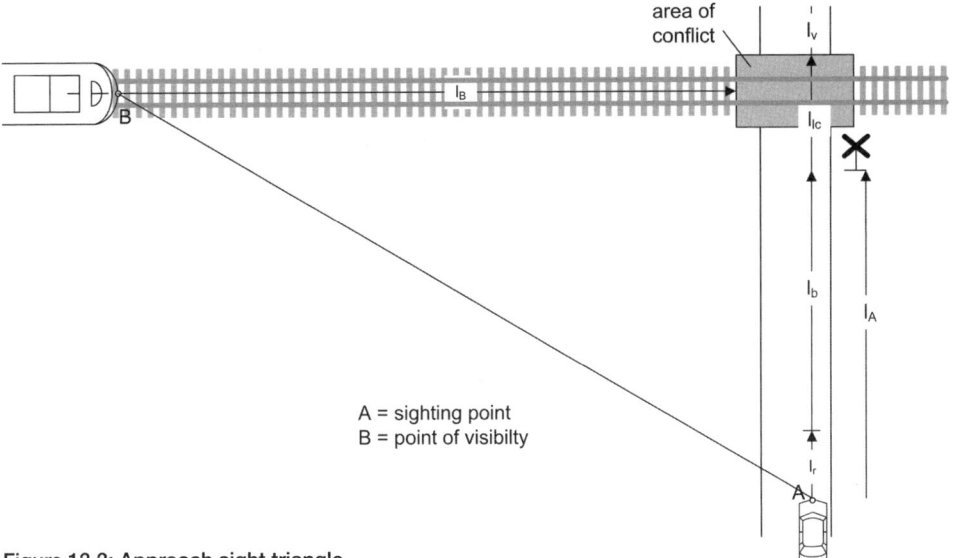

Figure 13.3: Approach sight triangle

A = sighting point
B = point of visibilty

proach distance l_B the situation is more complex. Faster road users need longer stopping distances and therefore longer time for the braking process, but pass over the level crossing itself faster. In practice, depending on the choice of the parameters, a function similar to that in figure 13.4 appears. From this diagram and from testing with different parameters, the following can be concluded:

- Among different kinds of road vehicles, the longest permitted vehicles are relevant.
- For speeds higher than approximately 30 or 40 km/h, the required approach distance changes only slightly, whereas it changes sharply for low speeds. Therefore, among road vehicles, the slowest are usually those with the longest required approach distance on the rail side.
- Pedestrians, although usually being the slowest road users, are not relevant for this calculation in most cases thanks to their low 'vehicle length'. However, they can become relevant in special situations such as extreme lengths of the conflicting area, which can be either much more than a double line to be crossed or a crossing at a very acute angle. They can also be an important factor if volumes are unusually high, as can happen in town centres.

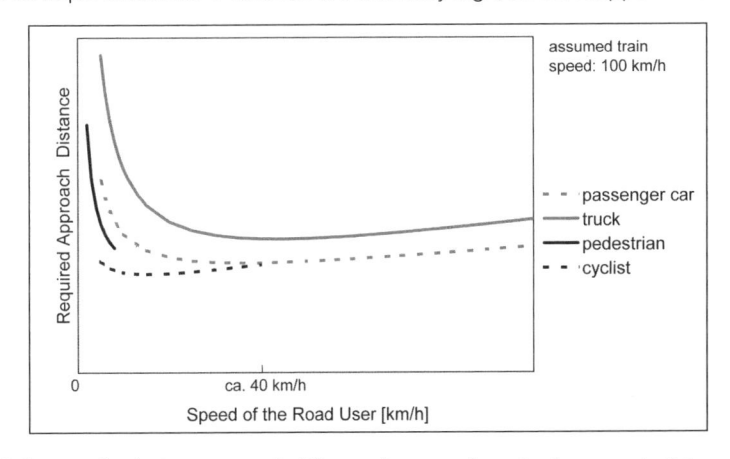

Figure 13.4: Connection between speed of the road user and required approach distance of the train

13 Level Crossings

Therefore, for planning of the approach sight triangles of a level crossing, the longest permitted road vehicle has to be assumed, with speed from the minimum assumable (usually around 10 km/h) up to the permitted speed of the road, and the required approach sight triangles to be added (figure 13.5). With very long conflicting areas, pedestrians also have to be taken into consideration. They can be assumed without braking distance, therefore observing the level crossing short distance from the St. Andrew's cross.

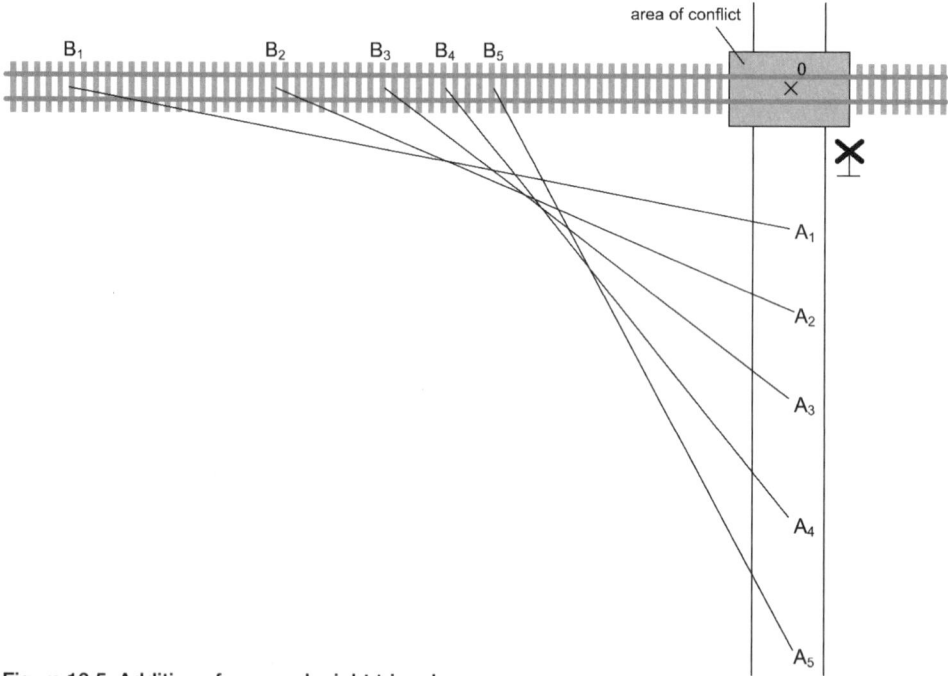

Figure 13.5: Addition of approach sight triangles

By reducing the permitted speed on the road or on the railway (figure 13.6) or even obliging the road user to stop in front of the level crossing, the requirements according to the size of the approach sight triangle can be reduced. This has the disadvantage of the extension of travel times and the reduction of capacity of the affected transport system. Generally, the possibilities of speed restriction are used as follows example of figure 13.5 with deciding points $A1/B1$ for the slowest road vehicle and $A5/B5$ for the fastest road vehicle):
– When sight obstructions occur in triangle $0 - A5 - B5$, it is necessary to limit the speed of road traffic.
– When sight obstructions occur in triangle $0 - A1 - B1$, it is necessary to limit the speed of rail traffic.
– When sight obstructions occur in intersection between triangles $0 - A5 - B5$ and $0 - A1 - B1$, both speeds have to be limited.

Additionally, in most countries road users are warned by audible signals given by the train at defined locations in the approach to the level crossing. In some countries it is permitted to warn road users only by audible signals without any visibility of the train, but only at level crossings with low road traffic levels.

To increase safety particularly for pedestrians and cyclists, special fences are often applied to force persons to look in both directions and to reduce speed (figure 13.7).

Figure 13.6: Speed restrictions for rail and road traffic on level crossings

Figure 13.7: Pedestrian's fences at passive level crossings (photo: *DB AG/Stefan Klink*)

13.4 Active Level Crossings

13.4.1 Overview

Active level crossings include all those which give different indications to the road user depending on the approach of a train. This includes technical safeguarding with light signals, barriers and others as well as manual safeguarding by hand signals of a level crossing post. The following explanations concentrate mainly on technically safeguarded level crossings, but many are also applicable to level crossings safeguarded by a person. Technically safeguarded level crossings can be distinguished by multiple criteria. Some of them are:

– the form of roadside safeguarding (e.g. light signal only, half barriers, full closure) (chapter 13.4.2)
– the procedures of opening and closing (chapter 13.4.3)
– the form of supervision (chapter 13.4.4).

13.4.2 Dynamic Roadside Safeguarding

13.4.2.1 Overview

In contrast to the static roadside signing described in 13.2, the dynamic roadside safeguarding of active level crossings gives signals to the road users or blocks the level crossing mechanically depending on the approach of trains and the status of the level crossing. Besides closing a crossing by the hand signals of a person, different technical solutions (in different combinations) can be applied additionally or alternatively. There is a large variety among the countries in the details of roadside signalling, what makes the orientation of car drivers in another country often difficult. The most common devices are (figure 13.8):

- Light signals (13.4.2.2) differ between the countries. They can have the form of a steady red light, a red flashing light, two alternately flashing red lights or others, in some countries also including yellow lights, for signalling Stop to the road user. The lamps are usually placed inside, above or below the St. Andrew's cross.
- Mechanical closure of the road. The most common solutions are half barriers and full closure (either by full barriers or by two pairs of half barriers) (13.4.2.3). In some countries also other mechanical obstacles such as road blockers which can be sunk in the road are applied (13.4.2.4). Historically in Britain, when train speed was low, swinging gates were applied to block alternately the road or the railway.
- Additional audible signals, either given wayside or by the train. Wayside audible signals can be giving either during the closing of the barriers or continuously until the train arrives.
- A person who blocks the level crossing by defined hand signals, which is mainly applied as a temporary solution or a supplement to technical warning devices.

An important issue for roadside signalling is the warning time. The warning time is defined as the time from the appearance of the first signal which obliges the road user to stop until the train reaches the level crossing. The warning time depends on the technical solution for closing, proving and reopening the level crossing. It must not be too long, as, besides obstructing the road traffic, increased warning time reduces the discipline of the road users and therefore the safety. Approaches for calculation are stated in chapter 13.4.4.5.

In most modern systems the optical and audible signals are switched on a defined time before starting to lower the barrier arms to give road users who are closer than stopping distance to the level crossing the ability to pass without being obstructed by the barriers, see chapter 13.4.3.3.

The main advantage of full closure over half barriers is that the discipline of the road users is higher, as the entry of vehicles by driving around the barriers is completely prevented. On the other hand, in installations with full closure the road user can be trapped. To avoid this, there is usually a time gap between lowering the barrier arms on the entrance and on the exit side if the full

Figure 13.8: Example of a level crossing (Germany) (photo: *DB AG/Christian Bedeschinski*)

13.4 Active Level Crossings

closure is realised by two barriers in both directions. Full closure also increases the necessity of a kind of supervision in the area of conflict (chapter 13.4.4.4).

In some countries, level crossings can also indicate other signals than simple 'Stop' to the road user: Some possibilities are:

- A pre-indication comparable with the yellow signal at road traffic lights to indicate that red will soon appear.
- An active indication for the passable status and the proper working of the level crossing (e. g. white light). If this active free signal is not alight, the road user has to behave as if on a passive level crossing. This can cause problems if only some level crossings are thus equipped in any given country.
- An additional signal to specifically indicate a second train on a double line, after the first train has passed (often used at active level crossings without barriers). The purpose is to solve the additional problem of discipline with road users passing the level crossing after the first train has left it, not regarding the possibility of a second train approaching immediately after.

The choice of one or other kind of roadside safeguarding depends on different factors, which are variable from one country to the other. Such factors are:

- volume of rail and road traffic
- speed of rail and road traffic
- local requirements
- philosophy in the respective country or railway company

Figure 13.9 presents a Russian example of the allocation of equipment for level crossing. In figure 13.9 the following equipment is shown:

- level crossing light signals with automatic barriers (A, B)
- devices for blocking of level crossing (automatic road blockers) (AB1-AB4)
- control units for level crossings and automatic road blockers placed into relay boxes or transportable modules
- battery box with a number of accumulators placed inside (BB)
- devices for train detection, in this case track circuit equipment (TC)
- supervision light signals for train driver (C1, C2)
- barrier keeper's lodge (KL)

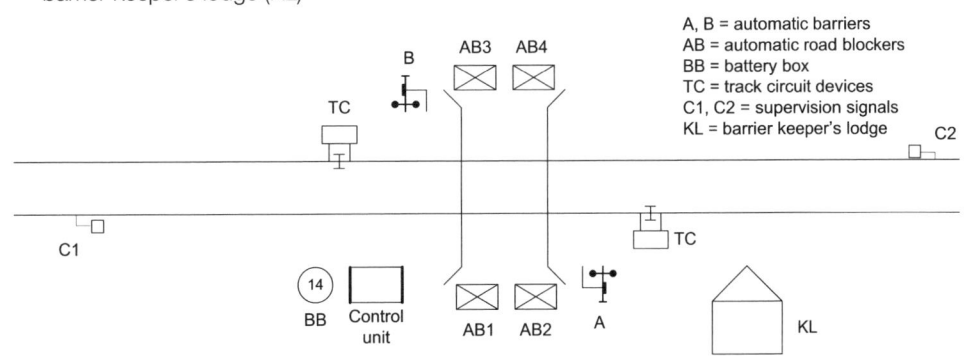

Figure 13.9: Equipment plan of a level crossing (example)

13 Level Crossings

13.4.2.2 Road Light Signals

Level crossing light signals are mounted on a signal post. The following equipment is placed on the signal post, if the level crossing is to be equipped with these items:
- light signal heads with the required colours, usually red, in some cases also yellow and/or white
- signal background(s)
- audible signal giving devices, e.g. a bell
- static road signs such as the St. Andrew's cross
- Cable sleeves used for cable termination

Flashing road light signals must have defined lit and unlit times with a defined tolerance which can differ between countries. For example, in Russia, both times are 0.75 seconds ±20%. The signal optics must provide a defined sighting distance (optical range). In Russia, for example, this is not less than 100 m on a straight road and not less than 50 m at curved sections of roads.

Light signal heads of two types are used, those with incandescent lamps and those with light-emitting diodes (LED). With use of incandescent lamps as a signal source, many railways require safety redundancy, particularly for those lamps which are used to indicate Stop. This is usually solved by lamps with double filaments. In the case of primary filament failure, the feed circuit switches to the reserve filament automatically.

13.4.2.3 Barriers

Barriers can be placed in the same location as level crossing light signals. When placed in different locations, the barrier is placed not further from the railway than the light signal. Several railways require a clear area between the profile of the train and the barrier arms as safe area for a possibly trapped person. Another possibility is a special escape route beside the level crossing (figure 13.10).

Figure 13.10: Escape way beside level crossing (photo: *Carsten Weber*)

Figure 13.11 presents an example for the structure of an automatic barrier with the following equipment:
- electric drive placed on foundation and pedestal
- barrier arm with counterbalance

Barrier arms can contain warning lights or retro-reflectors to improve visibility to road users.

Some systems require the ability of the barrier arms to be overrun by road vehicles. This enables the escape of vehi-

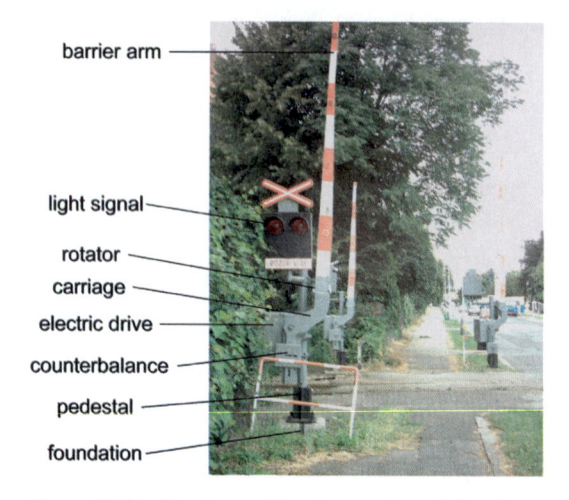

Figure 13.11: Structure of barriers and road signals (example from Czech Republic)

cles from the space between the barriers, even if the barrier was lowered e. g. on the roof of a road vehicle or between a tractor unit and its trailer. Besides, it reduces the material damage in the case of a road vehicle colliding with the barrier arm. For this purpose, the barrier arm is fixed rotating with high friction in carriage with the possibility to turn the barrier arm at right angle (90°) either upwards or to each side in the case of automobile collision. Providing a predetermined breaking point on the barrier arm is another possibility to avoid greater damage and dangerous situations.

Some systems also provide a detection of breakage of the barrier arm. For this purpose, a breakable electrical loop is installed inside each barrier arm.

An electric motor or a hydraulic system controls the lowering and rising of the barrier arm. For electric motors, AC mono-phase, AC three-phase or DC motors can be used. Usually these machines are designed that way that they can be powered by the public electric supply network.

During barrier arm lowering, its potential energy is transformed into kinetic energy. This has to be compensated in order that the barrier arm is not lowered too fast or even striking the pavement. A hydraulic shock absorber as constituent of the electric drive is used to guarantee a uniform rate of lowering and to provide a smooth stop at the end.

Modern types of barriers are equipped with systems which fix and/or supervise the barrier arm in its extreme position. Special requirements on the bearing of barriers exist if they are supposed to lower automatically in case of failure or breakdown of power supply.

The barrier arm must usually cover at least half the width of the road on the side of approaching vehicular traffic. However, when half barriers are used, a certain minimum width (e. g. 3 metres) of the road on the other side must be clear.

13.4.2.4 Automatic Road Blockers

In addition to light signals and barriers, the countries of the former Soviet Union use road blockers to prevent the passage of a level crossing by road vehicles. These devices for blocking of the level crossing consist of four covering lids which are placed in a bed (figure 13.12).

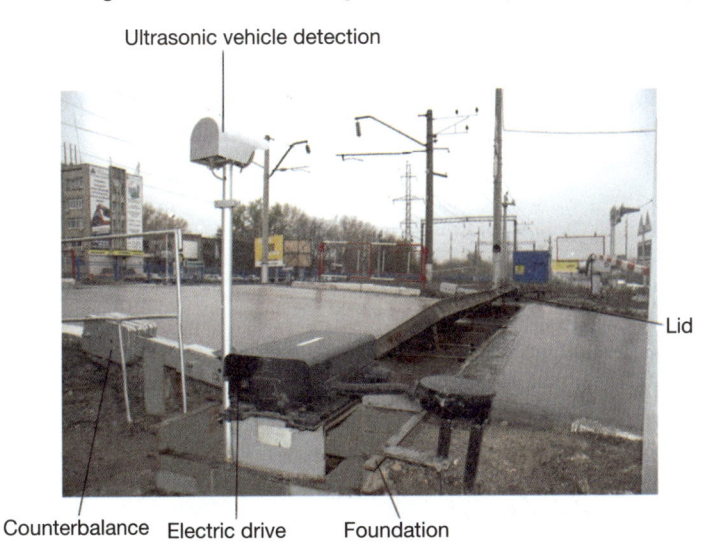

Figure 13.12: Mechanical road blockers

With a sunken lid, the road blocker does not obstruct road traffic. Upon the approach of a train, the lids are raised, thereby preventing road vehicles from entering the level crossing. The raising and sinking of lids is by electric drives.

Vehicles on the level crossing after the raising of lids are able to leave the level crossing, since the lid sinks when a vehicle runs over it coming from the railway side and rises again under the action of counterbalance when the vehicle has left it.

Ultrasonic vehicle detection sensors are used for a presence check of vehicles in lid zones. In the case of appearance of a vehicle in the check zone of the sensor during the raising of the lid, the electric drive is turned off and the raise is stopped until the vehicle has cleared the check zone.

The road blockers, likewise barriers in Russia, are provided with the opportunity of dual control: automatic control under approach section occupation by train and manual control by buttons.

13.4.2.5 Safe Usage by the Road User

The regulations for road users whether or not they can rely on the proper working of the safeguarding differ, and depend on the kind and quality of applied technical solutions. The following principles can be found:

- The road user must always check visually if no train is approaching before passing over the level crossing. This means that signals and barriers are only an overlay system, but do not ensure safety in itself. This solution was historically widely used, when the reliability of the technical systems was low, speeds of rail and road traffic as well as the volume of road traffic were relatively low and it was acceptable to let each road vehicle brake at the level crossing. A problematic issue in these systems, as generally in safety overlay systems which work with the view of users, is that the unlit signals and open barriers tempt road users to rely on them erroneously.
- The road user can rely on the technical safeguarding systems. This principle is widely used today as on the one hand speed and traffic volume have increased and on the other hand highly reliable systems for technical safeguarding are available today. This principle requires that in cases of technical failure either the road has to be closed safely for road traffic or rail traffic has to be stopped (chapter 13.4.5).
- Distinction is made in roadside light signals between a status which guarantees the road user that the level crossing can be safely crossed and a failure status where the road user has to check that the railway is clear visually before crossing.

13.4.3 Opening and Closing of Level Crossings

13.4.3.1 Definition of Normal Position

Each level crossing has a defined normal position which can be either the open or the closed status for road traffic. The choice of the one or the other is roughly made upon the criterion if the volume of road traffic (measured in number of vehicles/pedestrians per time) is higher or lower than rail traffic (measured in number of moving units in the same time):

- Level crossings which are open to road traffic in normal position and closed upon request by an approaching train are the solution applied most frequently and the following descriptions concentrate mainly on them.
- Level crossings which are closed to road traffic in normal position and opened upon request by a road user if no train is approaching are used in situations with very low road traf-

fic. A typical example is a private road crossing a railway with relatively high traffic volume. Such level crossings are avoided, but in some situations they are applied due to lack of alternatives. This solution can only be used with barriers (full closure).

In both cases, priority is to be given to rail traffic, which means that trains will not be obstructed at the level crossing unless an error occurs.

13.4.3.2 Basic Methods and Principles of Opening and Closing

Regarding the technology of opening and closing of the level crossing, the following types can be distinguished:
- Staff operated level crossings, which can be:
  - manual by muscle power or
  - machine operated under control of a person.
- Automatically operated level crossings, where the operation is initiated automatically by the approaching/clearing train or by route calling in an interlocking system.

For staff operated level crossings, the operator has to be informed about the train to be expected. Depending on the situation, he either has to close the level crossing immediately upon receiving this information or after a certain time.

For automatic installations, the closing and opening is usually initiated by the train passing a detector in a defined location. Suitable detectors are different kinds of spot wheel detectors (chapter 5.2.2), track circuits (chapter 5.2.3.2) and magnetic inductive loops (chapter 5.2.4.3). To increase the reliability, the detectors are often doubled for redundancy. In autonomous installations without feedback information to the train driver (chapter 13.4.4.1), this redundancy is even an issue of safety. In these cases, the evaluation of the detection has to be done as follows:
- The level crossing must be closed whenever one of both redundant detectors has detected an approaching vehicle.
- The level crossing must only be opened if both redundant detectors have detected a clearing vehicle.

For route dependent level crossings, closing of the level crossing is often not initiated immediately when requesting the route, but later in the process in order to reduce closing time. After requesting the route, a proper solution is first to bring all points to the correct position, prove all other conditions and bring the signal into readiness to clear, but to close the level crossing and clear the signal later when the train reaches a special approach point.

13.4.3.3 Algorithm of Opening and Closing

Figure 13.13 presents an algorithm of opening and closing of a level crossing equipped with light signals, with white lights for the open position, automatic barriers and automatic road blockers (13.4.2).

In normal state, when the approach section is free from rail vehicles (figure 13.13a), the signalling equipment is in the following state: light signals unlit (white lights flash), barrier arms in the up position.

The level crossing begins to close at that time when the train is detected at a defined distance from the level crossing. That distance named 'effective length of approach section' or 'approach length' must suffice so that
1. a road user who is too close to the level crossing for stopping will have enough time to vacate the area of conflict of the level crossing before the train arrives, and

13 Level Crossings

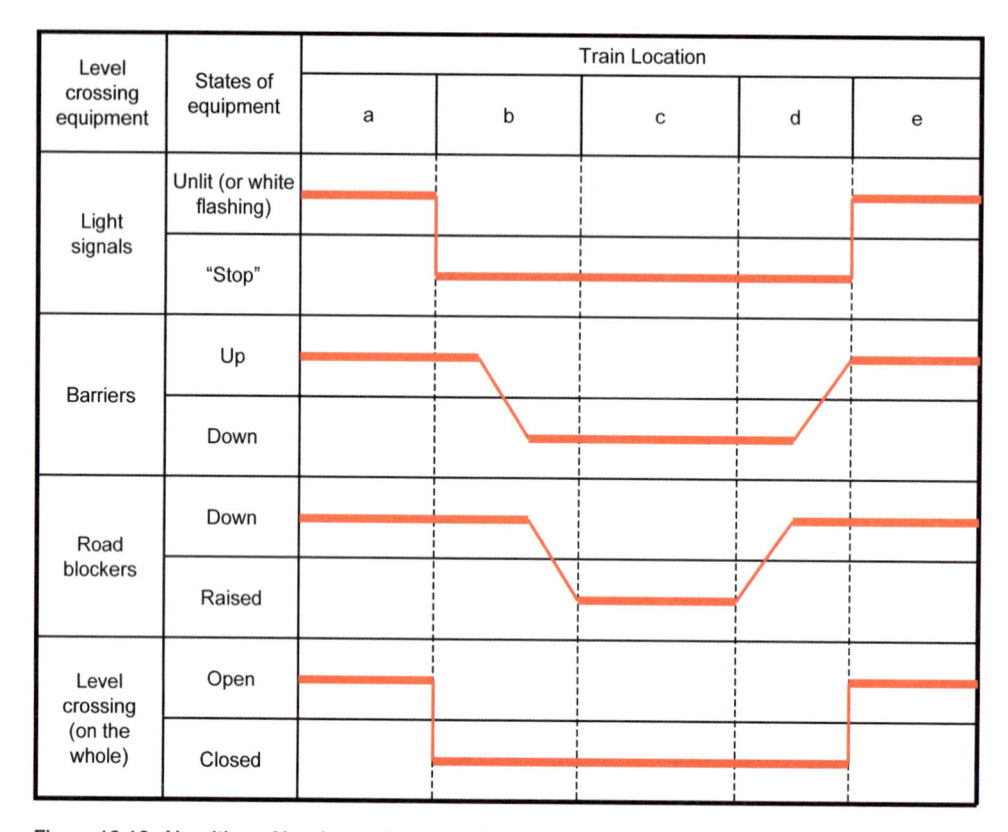

Level crossing equipment	States of equipment	Train Location				
		a	b	c	d	e
Light signals	Unlit (or white flashing)					
	"Stop"					
Barriers	Up					
	Down					
Road blockers	Down					
	Raised					
Level crossing (on the whole)	Open					
	Closed					

Figure 13.13: Algorithm of level crossing operation (Russian example)

2. depending on the kind of supervision, the train can be stopped in case the level crossing fails to close.

The calculation of the approach length is described in chapter 13.4.4.5.

Upon approach section occupation by the train (figure 13.13b), light signals indicate 'Stop!' towards the road. At a defined time after that, the barrier arms begin to lower into the horizontal position. This time is different in each country and depends on the properties of road traffic.

After the lowering of the barrier arms is complete, the automatic road blockers (if existing) rise into the up position. The level crossing is now closed and remains so during the train movements which occupy its area (figure 13.13c).

After clearing by the train (figure 13.13d), first the automatic road blockers sink to their normal position. Then barrier arms rise to up position and light signals switch off. The order of these two actions differs: In most countries the lights switch off upon the barrier arms reaching the up position, whereas in some countries they switch off together with the barrier arms starting to rise. If white lights are used to indicate the open position, they are switched on in the same moment or after the train has moved away from the level crossing for a specified distance (e. g. 150 metres in Russia on open lines) to prevent dangerous situations caused by rolling back of rail vehicles onto the level crossing. In some cases with more than one track to be crossed, the opening of the level crossing is suppressed if a train is expected soon on another track. This prevents extremely short open times, but it can also result in very long road closures due to a third train approaching on the original track, for instance.

13.4 Active Level Crossings

13.4.3.4 Treatment of Scheduled Stops and Different Train Speeds

Where trains have a scheduled platform stop in the approach to the level crossing, which often occurs on secondary lines, a suitable solution is that train staff, when ready to depart, initiate the closing of the level crossing. This can be done by buttons to be pushed on the platform or by special radio transmitting devices to be operated by the train driver (figure 13.14). In the USA, motion sensors are used to detect when a train starts moving from a stop, and to close the level crossing then. On many secondary lines which are used by one category of passenger trains only, this can be solved rather easily. Either there is a platform stopping point in the approach to the level crossing or not.

Difficulties occur where some trains have a scheduled stop at a platform in the approach to the level crossing and others do not. In unfavourable cases, automatically closing the barriers before the train reaches the platform would mean that passengers cannot get on their train which stands at the platform. Also, different speeds of trains cause difficulties on lines with mixed traffic. The initiation point for closing the level crossing would have to be adjusted to the fastest train, which means unnecessarily long waiting time for road users in the case of slower trains.

Figure 13.14: Wayside infrared detector for closing the level crossing by the train driver (on top of left post) with button for auxiliary operation below

A solution is to set the initiation point by train categories or to calculate it for each individual train. This requires reliable information about the category or speed of an approaching train. This distinction can either be made by local staff, train staff, or automatically. An example of an automatic solution can be found in Sweden, where normal closing positions are adjusted to a speed of 160 km/h, whereas for high speed trains on the same line advanced closing applies, initiated by radio signals sent by the train (Hagelin/Strindh 1997). With automatic train routeing systems and if these an sufficiently safe, the train category can be used to distinguish the initiation points for level crossings by train category. Even stopping and non stopping trains can be distinguished automatically (Akuzawa 1982).

13.4.3.5 Power Supply

Although level crossings inside or near interlocking areas are usually power supplied together with the interlocking, most level crossings on the open line need autonomous power supply. The high availability of power supply is very important, so most railways require redundancy with two independent power sources – basic and reserve. Commonly as basic sources high-voltage lines are used. As reserve power supply accumulator batteries are often used. The battery system must be able to sustain the level crossing in operation for some hours and to reload automatically when basic power supply is again available.

13 Level Crossings

13.4.4 Supervision of Level Crossings

13.4.4.1 Basic Supervision Types

For many years, no feedback information about the closure of the barriers was usually given before permitting the train movement. Thus safety depended on the correct and punctual actions of the barrier keeper after being informed about the approaching train. This was a widely applied solution until after World War II even in highly developed countries. With the increasing speed of trains and road vehicles, this low level of safety resulted in a high number of accidents.

Therefore, the development of safer methods was urgently necessary. The first form of feedback information was given in regulations saying that a signal for a train must not be cleared before all level crossings until the next signal have been proved to be protected by the signaller or reported from barrier keepers to the signaller. First, this method was applied for level crossings in station areas, later also on the open line. Especially in long block sections, this solution resulted in extremely long waiting times for road users.

Much later than moveable track elements and the clear status of tracks, level crossings were brought into technical dependence with signals, which means that the signal can technically only be cleared if the level crossing is protected.

For technically supervised level crossings, two basic categories, with certain variations, can be found in almost all countries (figure 13.15):

- **Route-dependent level crossings** are included in a route in an interlocking area. The main signal can only be cleared if the level crossing is protected, after being closed by route request or later by the approaching train (chapter 13.4.3.2).
- **Autonomous level crossings** are supervised independently from the routes (but can also be geographically located in interlocking areas).

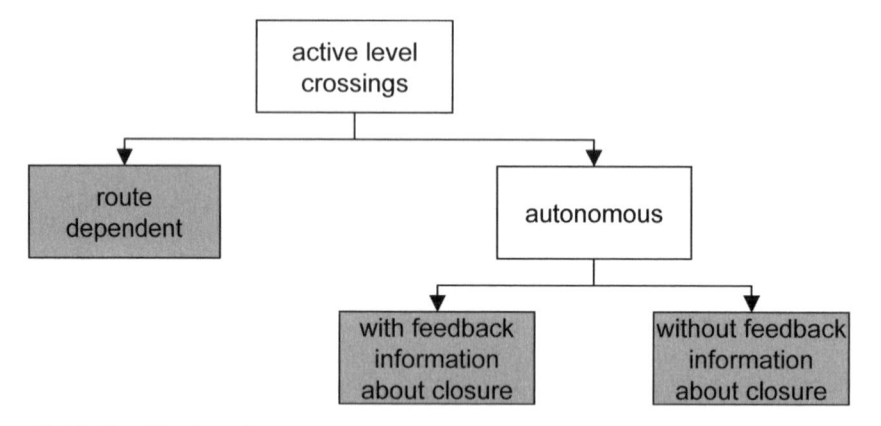

Figure 13.15: Classification of active level crossings by the kind of supervision

Route-dependent level crossings are usually applied in situations where a main signal is immediately in rear of the level crossing. This refers to level crossings which are situated in or near station and junction areas, but also level crossings on the open line in proximity to block signals. Autonomous level crossings are used in all other cases. As lines with high density of rail traffic and with short block sections usually do not have many level crossings, autonomous level crossings are the most frequent case on the open line. Depending on the local situation, also level crossings are applied which are supervised autonomously from one direction, but route-dependently from the other direction.

Most autonomous level crossings indicate the protected status to the train driver by level crossing supervision signals. Recently, some railways have introduced systems which, by redundancy, are so highly reliable in closing the road that no supervision signals are required. As technical problems are not revealed directly to the driver or a local person, these level crossings have to be supervised remotely, usually from a signal box. If the diagnostic system detects an error which reduces safety (e.g. failure of 1 out of 2 channels), trains have to be stopped at the next main signal in rear. Nevertheless a slight risk remains in that a train passes an unsafe level crossing in the (very improbable) case of system failure. The main advantage of these systems is that the closing time of the road is minimised, as the level crossing has to be protected when the train reaches the level crossing, not in stopping distance before.

Combining this type of level crossing with different initiation points determined by train category (13.4.4) causes additional difficulties, as the train sorting becomes safety critical.

13.4.4.2 Safety Criteria

Generally, signal lamps, barriers and other technical means can be supervised in their functions. The railways differ in to what extent partial failure of the safety equipment, e.g. the red lights lit, but the barriers open, can be tolerated and the level crossing can still be considered as passable safely by the train. In the same sense, as closing a level crossing especially with barriers is a process of some seconds in duration, the arrangements when a level crossing is regarded as protected and the railway signals can be cleared differ between countries. This is related to the closure time for road traffic, so the solutions found between safety requirements and a desirable short closure time differ. Possible criteria for a level crossing to be considered as 'protected' are:

- The level crossing is considered as protected when all existing signals which indicate the closed status are lit and all existing barriers are completely closed.
- The level crossing is considered as protected when at least parts of the signals and barriers are closed. This means that, for example, the level crossing can be regarded as protected when the light signals are lit, but the barriers have not yet started to close, or when the barrier arms have started but not completed the closing process. In such cases, the level crossing can even be regarded as protected in cases of partial technical failure, e.g. of the barriers.
- The level crossing is considered as protected if it has the ability to close. Using this criterion minimises the closing time, but the system must be highly reliable in closing to ensure safety, which is usually solved by redundancy. Here, in case of detection of a road user on the level crossing, stopping the train on time is not possible. This principle is used in modern systems without supervision signals (chapter 13.4.4.1).

In some countries, a level crossing is no longer considered safe if it has been closed longer than a certain maximum time without being accessed by a train, as the discipline of road users decreases. This feature is provided by modern automatic level crossings and leads to the quick revelation of malfunction with lowered barriers. The time can depend on the type of level crossing, e.g. in Germany the maximum warning time for crossings without barriers is 90 s, for crossings with half barriers 240 s.

Modern control and supervision units are either solved by relay or microelectronic technology. If relay technology, safe signalling relays must be used at least for the supervision functions. Electronic systems achieve safety (and often also availability) by redundancy and checking functions.

13 Level Crossings

13.4.4.3 Level Crossing Supervision Signals

For autonomous automatic level crossings, special supervision signals are used to indicate the status to the train driver, who is obliged to stop at the level crossing or pass over it on sight if no safe status is indicated. In most modern systems, the supervision signals are controlled automatically by the closure of the level crossing. Any technical error is here revealed to the train driver, who then informs the responsible staff. In other (mostly older) systems, a person is involved who checks the status of the level crossing and then controls the supervision signals, or has just the possibility to release it to Stop position in case of danger.

The level crossing supervision signals are, similarly to signals for train spacing, applied in form of main and distant supervision signals. The main supervision signal is situated at the level crossing (often without providing an overlap or with short overlap), whereas the distant supervision signal is situated the stopping distance in rear of the main (figure 13.16). In many railways generally or in some cases only one of these is used, which is usually the distant supervision signal, or in cases of very low speed the main. Some signal instructions additionally provide repeater signals for topographical cases of bad visibility.

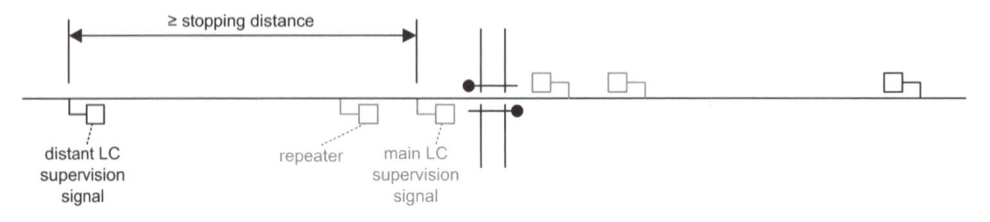

Figure 13.16: Position of level crossing supervision signals

The signal aspects of the level crossing supervision signals are coded very differently between the countries, using red, yellow, green and white colour and colour position light signals (figure 13.17).

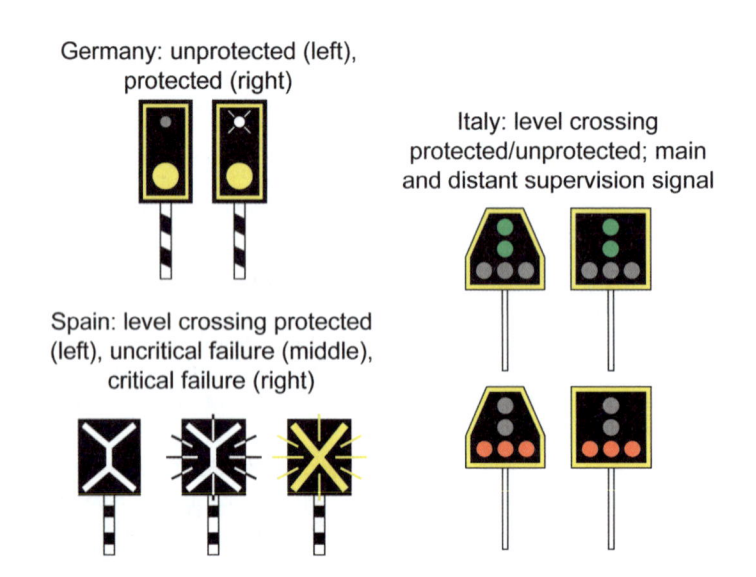

Figure 13.17: Examples of supervision signals

13.4.4.4 Detection of the Conflicting Area

In different cases, the area of conflict has to be proved clear. This depends on the local situation, the possibility of trapping road users (full closure or not) and the regulations of the respective country. The following basic solutions are applied:
- Visual proving by an operator. The person (often a signaller) proves the clear status of the level crossing either directly or remotely via camera and screen (chapter 5.2.5.1).
- Automatic proving by different technical solutions. These solutions can be divided into intrusive systems (below the road surface) and non-intrusive systems (over the road surface) (Lazarevic et al. 2007). An example of intrusive technologies is inductive loops, which react to the iron mass of vehicles (but not pedestrians). Other examples for intrusive technologies are pneumatic road tubes and piezoelectric cables, reacting to the weight of an object of not too low mass. Examples for non-intrusive technologies are radar (figure 13.18), infrared and video image processing.

In most cases where there is supervision of the conflicting area, there are the following possibilities of evaluation of this information:
- The clear status is a precondition for clearing the respective railway signal. In systems with visual supervision, the clear status has to be acknowledged by a (safety critical) operation action for this purpose. In cases with automatic supervision, a technical dependence similar to interlocking functions applies.
- The clear status is continuously supervised as long as the train movement is permitted, and the train stopped in case of emergency. In cases with visual supervision, this can be done by an emergency call to the train driver or by manual emergency closing of the signal. With automatic supervision, the train is stopped by an alarm in the cab signal or by automatic closing of the signal.
- Both above criteria combined.

Figure 13.18: Radar scanner for technical monitoring of the conflicting area (manufacturer and graphic: *Honeywell*)

If there is no technical or manual supervision of conflicting area, in most countries the driver of a road vehicle must be able to leave the conflict area at any time. This can be ensured by using half barriers instead of full closure.

13.4.4.5 Warning Time and Approach Distance at Automatic Level Crossings

The warning time t_W of an automatic level crossing is the time from the first warning of road users (usually switching on of the warning light or, without light signals, the starting of the barriers to close) until the appearance of the train at the level crossing. The approach distance l_A is the distance on the railway line which the train passes from initiation of the level crossing until reaching the level crossing. The time delay from initiation until the occurrence of the first warning signal is very low in automatic systems and can therefore be ignored. The relation between warning time t_W and approach distance l_A is:

$$l_A = v_t \cdot t_W$$

with v_t: speed of the train.

13 Level Crossings

If train speeds differ in approach to a certain level crossing, either the fastest train has to be considered or train categories be distinguished (chapter 13.4.3.4).

For calculation of the warning time and approach distance, two criteria must be fulfilled, the maximum of both being the decisive one:

1. Road traffic must be warned early enough that a road user has the possibility of either stopping safely in front of the level crossing or, if stopping is impossible, passing over it completely. The longest and slowest road users and a certain safety margin have to be considered. The calculation is done similarly as in 13.3 for passive level crossings and is therefore not repeated here.
2. Depending on the type of supervision (chapter 13.4.4.1), the train will not be obstructed if the level crossing works properly. This criterion usually results in longer warning times. It is calculated as follows:

Based on the three supervision cases described in chapter 13.4.4.1 and figure 13.15, the approach length and warning time if no stop is scheduled in the approach to the level crossing can be determined as follows (figure 13.19):

- In autonomous level crossings without feedback information about the protected status, the warning time equals the closing time. This means the time from initiation until full closure of the level crossing (chapter 13.4.3.3), if it is accepted that the train appears on the level crossing as soon as the barriers have completely closed.
- In autonomous level crossings with feedback information about the protected status to the train, the approach distance is the sum of the following: the distance the train passes during safeguarding of the level crossing, the sighting distance of the supervision signal and the distance from the supervision signal to the level crossing (at least the braking distance). If full closure of all barriers is required before switching on the supervision signal (chapter 13.4.4.2), the safeguarding time equals the time needed for the closing process. Otherwise, e.g. if the criterion is that road signals have to be on and barriers begun to lower, this time component is shorter.

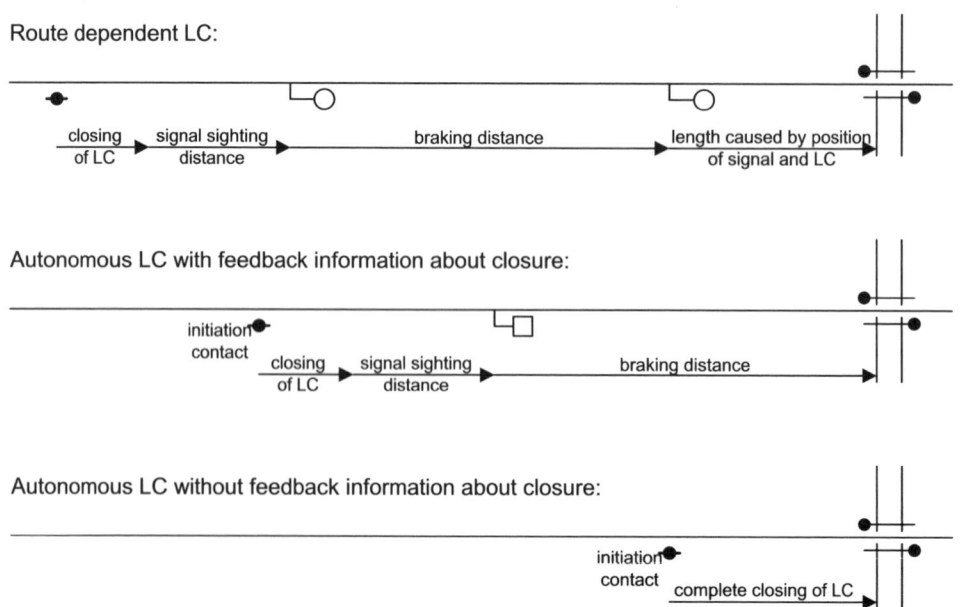

Figure 13.19: Approach length of level crossings with different kinds of supervision without scheduled stop in approach to the level crossing

- In route-dependent level crossings, the distance from the main signal to the level crossing has to be added to the above case with level crossing supervision signal.

Therefore, as can be seen in figure 13.19, route dependent level crossings require the longest warning time and level crossings without feedback information the shortest. However, this can change towards the route-dependent level crossing as an advantageous solution if there is a scheduled stop in approach to the level crossing.

13.4.5 Possibilities of Degraded Mode Operation

Failure of the safeguarding of the level crossing in safe systems can be revealed primarily to the following persons:
- Revealing to the train driver by main signals, level crossing supervision signals or others. This is the most frequently used method.
- Revealing to ground staff, who are then responsible for stopping either rail or road traffic.
- Revealing to the road users by particular roadside signalling or by permanent closure. For this purpose, in some countries a special positive light signal indication (usually white flashing) is used to indicate the proper working of the level crossing and the non-approach of a train. The absence of any light indicates the disturbed status to the road user.

Procedures to maintain railway and road operation in these failure cases can allocate the responsibilities for auxiliary safeguarding to following persons:

- To local staff: Barrier keepers, signallers or other local staff are responsible for safeguarding the level crossing by stopping the road traffic before permitting the train driver to pass by written instruction, auxiliary signal or hand signals (figure 13.20), usually at reduced speed. This solution is usually applied for staffed level crossings, often also in those route-dependent level crossings in proximity to staffed signal boxes.
- To train staff: The train driver has to move very slowly, warning the road users by audible signal, ready to stop if conflicts occur. In other solutions, train staff members leave the train and stop road traffic by hand signals or kinds of auxiliary barriers.

Figure 13.20: Protecting level crossing by hand signals (photo: *Martin Hahn*)

- To the road user: In systems where the failure is revealed to the road user by roadside signals, the road user can be obliged by highway regulations to use the level crossing like a passive one, often combined with a speed restriction for the trains. This requires a certain area of sight for road traffic, similar to passive level crossings (chapter 13.3).
- An alternative, which is only acceptable for roads with minor importance and with possibility of diversion, is to maintain permanent closure of the road until repair.

On level crossings with very simple circumstances (low road and rail traffic, slow speed of rail vehicles) the aforementioned measures can be used permanently as regular protection. This can be found often on feeder lines or on industrial railways.

13 Level Crossings

13.4.6 Combination with Road Junctions

13.4.6.1 Identification of the Problem

Special safety issues occur when a level crossing is situated in the neighbourhood of a road junction, especially if the distance between the conflicting areas of the level crossing and the road junction is shorter than the longest vehicle permitted on the respective road. In this situation, it is possible that a road vehicle which has to wait at the road junction cannot leave the level crossing immediately, causing the danger of collision with a train. Likewise, the road junction could be overcongested by road vehicles waiting at the level crossing.

In other situations, the distance between the conflicting areas is long enough to accommodate the longest road vehicle, but the probability of the level crossing being overcongested by road vehicles waiting at the road junction is high. Although road users are obliged not to enter the level crossing if they cannot quickly pass completely over it, due to low discipline this often occurs and endangers traffic.

13.4.6.2 Solutions for Road Junctions not Controlled by Traffic Light Signals

For uncontrolled road junctions, the simplest solution is to make the road with the level crossing the priority road at the junction and, if necessary, forbid left turns (in Britain and some other countries right turns). But this is not always desirable from the point of road traffic. Also the problem of overcongesting the road junction with vehicles waiting at the level crossing cannot be solved by this.

Another solution is to stop the road traffic on the priority road by advanced light signals when the level crossing is being closed to enable traffic of the minor road to clear the level crossing safely (figure 13.21).

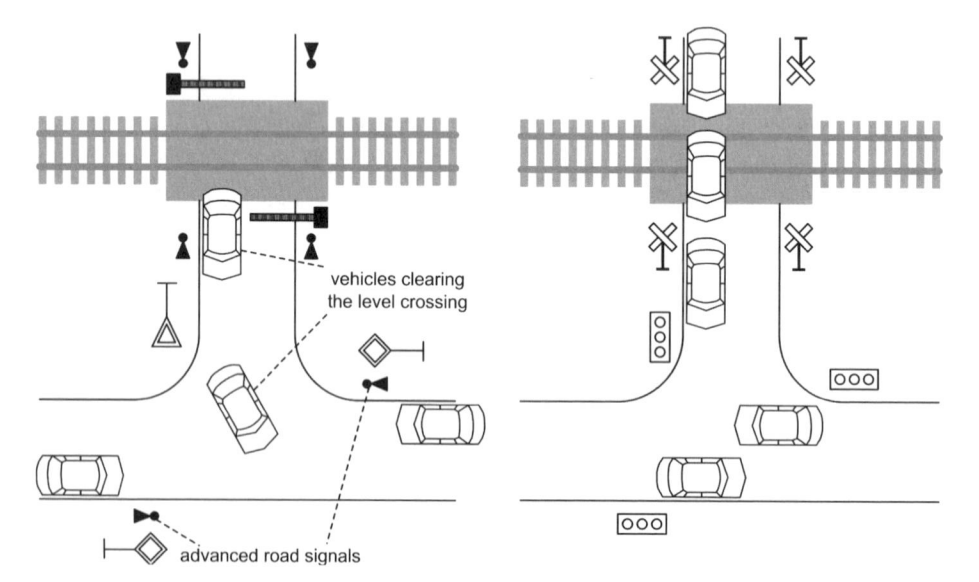

Figure 13.21: Advanced road signals for level crossings in proximity to a road junction

Figure 13.22: Problem of signalized road junction in the neighbourhood of a level crossing

13.4.6.3 Solutions for Road Junctions Controlled by Traffic Light Signals

The solution for signal-controlled road junctions (figure 13.22) is coordination between the level crossing and the road junction to prevent the described dangerous situations. To enable vehicles to escape from the section between the road junction and the level crossing, the entry has to be closed a defined time before the exit and remain closed until the exit is free again. This, however, can result in longer closing times of the level crossing (figure 13.23).

As road traffic light signals usually do not meet the safety requirements of railway installations, the case of failure of the road traffic signals has to be considered. If the level crossing is situated in the minor road of the road junction, an additional emergency red analogous to the advanced road signals in figure 13.21 will need to be provided.

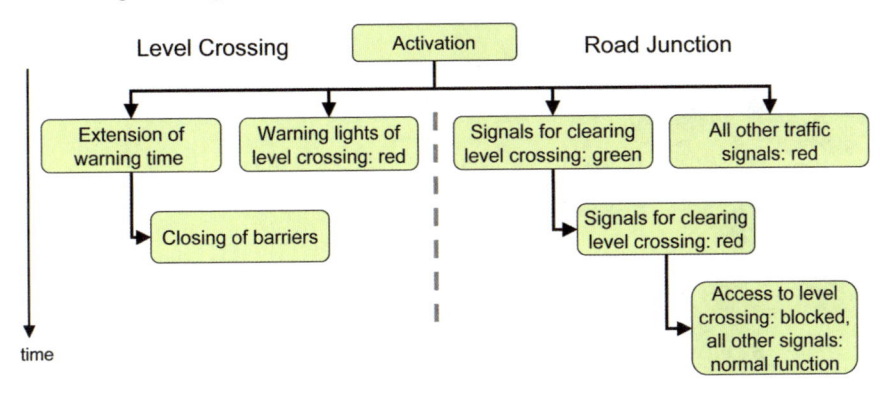

Figure 13.23: Function of level crossing combined with road traffic light signals

13.5 Removal of Level Crossings

Many countries have started initiatives and programs to remove level crossings. In spite of modern technologies to protect level crossings, they remain a dangerous point within the railway system. Furthermore the road capacity for road traffic is considerably affected by frequent and/or long closing of level crossings with high railway traffic.

In many countries, new places where road and rail are to cross have to be planned and realised as grade-separated crossings (under- or overbridges). Exceptions can be made for low use crossings. To reduce the number of existing level crossings as well, there are different possibilities:

- removing level crossing without substitution, if other ways can be used (only for very low use level crossings, e.g. agriculturally used field paths);
- concentrating the road traffic of several level crossings at one level crossing (only for low use level crossings, often a means of reducing costs when automating manually operated level crossings);
- substitution of level crossing by grade separation;
- combination of aforementioned possibilities, e.g. concentration of road traffic at one remaining level crossing and substitution of level crossings by pedestrian bridges because of the higher sensitivity of pedestrians to detours.

As a result of high expenses for constructing grade separated crossings and other extensive building measures, many existing level crossings will remain for the foreseeable future. The target for these level crossings must be an improvement of safety by technical, organisational and educational measures.

www.eurailpress.de/sd

7+8/2009
Juli+August | € 15 | C 11180
www.eurailpress.de/sd

SIGNAL+DRAHT
Rail Signalling and Telecommunication

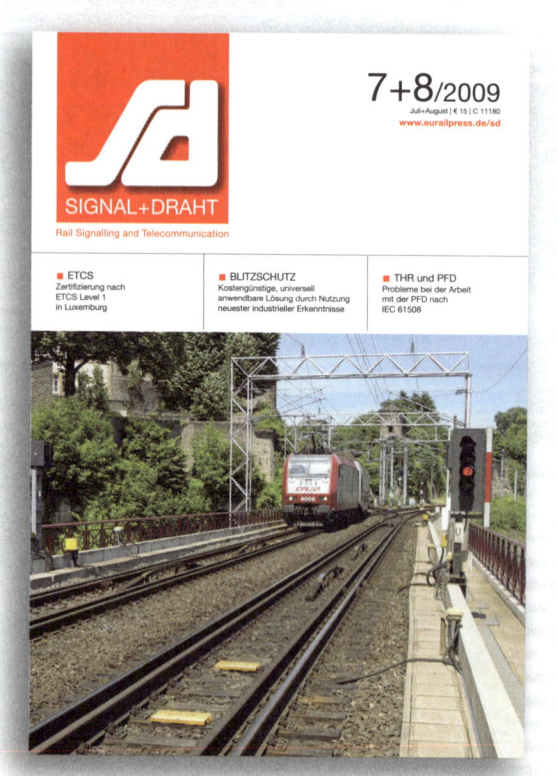

- ETCS
  Zertifizierung nach
  ETCS Level 1
  in Luxemburg
- BLITZSCHUTZ
  Kostengünstig, universell
  anwendbare Lösung durch Nutzung
  neuester industrieller Erkenntnisse
- THR und PFD
  Probleme bei der Arbeit
  mit der PFD nach
  IEC 61508

Weltweit führendes Fachwissen in Signaltechnik, Telekommunikation und Fahrgastinformation

The world's leading expert source of signalling, telecommunications and passenger information

SIGNAL+DRAHT ist seit 1906 das international führende Fachmedium, aus dem sich Ingenieure, Fach- und Führungskräfte in aller Welt über Signal- und Sicherungstechnik, Telekommunikation, Betriebsleittechnik sowie Verkaufs- und Informationstechnik im gesamten Schienenfern- und -nahverkehr informieren.
Die Zeitschrift erscheint zweisprachig in deutsch und englisch.

Bestellen Sie jetzt Ihr kostenloses Probeexemplar!

Since 1906 SIGNAL+DRAHT is the acknowledged leading international trade medium which is the source of information for engineers, experts and executives from railway, industries and the sciences worldwide.
The magazine is published bilingual in German and English.

Please order your free copy !

DVV Media Group GmbH I Eurailpress • Nordkanalstraße 36 • 20097 Hamburg • Germany
Tel.: +49 40/2 37 14-292 • E-Mail: service@eurailpress.de • www.eurailpress.de/sd

DVV Media Group

14 Hazard Alert Systems
Andreas Schöbel, Dmitrij Švalov

Although railways are safe in comparison to other carriers, hazards exist due to the operation process. Safety is guaranteed during the life cycle of each component by taking measures in design and ongoing maintenance. Nevertheless, dangerous situations will occur, since imperfections in the railway machinery and the human factor are included as well. So safety is often defined as free from unacceptable risk (CENELEC 1999).

It is however difficult to define the term risk formally, because it is the product of probability and severity combined. Even if a suitable definition can be devised, the bigger problem of getting significant statistical data of past happenings occurs. So for practical work, a definition of risk is essential. For the sake of completeness, a distinction between terms has to be made – in contrast to the term safety, security is used exclusively for man-made, deliberate hazards like terrorism, vandalism and other crimes.

Not until the hazards are known can a purposeful and effective design of measures against this hazard start. There is a difference between the development of new components for monitoring hazards and long term used inspection systems for already well-known hazards. New components have to be designed by the state-of-the-arte rules, but existing technologies which are working well were sometimes developed without following modern principles.

Due to the diverse responsibilities within a national railway company, safety is related to a large number of disciplines. This issue becomes relevant if the business organisation on a national scale is changed due to international harmonisation. Another important aspect of spreading of measures used is the quota of international traffic in a national railway network because this influences the strategy of national railway companies. In general, a high proportion of international traffic leads to increased investment in infrastructure measures. Fundamental risk reducing measures can be divided by functionality (Brux 2002):

– Event-avoiding
– Damage-reducing
– Rescue-supporting

Event-avoiding measures aim of preventing hazardous events. Important here is the coordination of the responsibilities of railway undertakings and infrastructure managers. But it is not possible to reach sufficient protection by preventative measures only, because it will not be financially feasible. Therefore damage-reducing measures also have to be planned, to minimise loss in case of an accident.

After one accident has happened, rescue measures become important for saving people. Hazard alert systems are related to the event-avoiding and damage-reducing sector of risk reducing measures. From the early days, technical measures were always designed to improve safety and to reduce costs. The disappearance of mechanical signalling meant a reduction of posts for the supervision of trains with technological progress and managerial initiatives further accelerating this trend. Thus the importance of an effective and comprehensive use of hazard alarm systems has increased permanently.

14 Hazard Alert Systems

14.1 Hazards in Railway Systems

14.1.1 Safety Related Hazards

The following hazards need be monitored:
- Derailed cars or wheelsets
  Even one derailed axle can lead to derailment of a whole car. If the derailment is stable, the length of damaged or destroyed infrastructure can be high. Only if one derailed axle hits an obstacle (e.g. points), which causes the train to divide, will the derailed train be stopped by a rapid decrease in pressure of the brake pipe.
- Hot (damaged) boxes
  The dangers caused by hot axleboxes are well known and in conjunction with irregular distributions of loads within vehicles. Hot boxes can lead to a derailment. The best indicator for damaged boxes is the temperature of the box itself.
- Blocked brakes
  Because of the generated, massive frictional heat, blocked brakes can produce the similar damage. Moreover, they can cause fires in the bogie construction due to sparks. These sparks can also enkindle vegetation beside railway lines.
- Flat spots
  Flat spots on wheels can occur during the braking process and lead to rail damage. Moreover, long flat spots can increase the probability of derailment on points.
- Broken axles
  If a hot box is not detected in time, the axle will also get hot. This heating can lead to a break and then to a derailment. There are also cases of cold axle breakage influenced by metallurgical reasons which can only be detected by ultrasonic measurement.
- Broken bearing springs
  Bearing-springs can break due to overloading, which leads normally to a derailment.
- Displaced cargo
  If cargo is fixed inadequately or incorrectly, it may obstruct the clearance gauge or be completely displaced from the wagon.
- Exceeding the clearance profile
  The risky situation caused by exceeding the clearance profile can have a number of consequences:
    - collision with masts for overhead electrification systems, signal posts or other railway infrastructure
    - collision with other trains
    - collision with passengers on platforms, especially from swinging doors.
  If overhead line equipment is involved, flashovers from the contact wire may take place.
  A question for the design of the inspection system is what kind of elements will exceed the clearance profile. Besides displaced cargo, loose fastener of cargo or even whole wagons due to derailment can exceed the clearance gauge. Another specific problem of clearance gauge violation concerns truck antennas on low-loader wagons. In detail, low-loader wagons offer the transportation of trucks on railways. Here, to gain good reception radio antennas are extending automatically during their passage on the low-loader wagons. Given that overhead contact wires are lowered in tunnels, the possibility of flashovers rises significantly.
- Open doors
  Doors on passenger cars are sometimes not closed even if an automatic door closing system is installed onboard. So passengers can be sucked out while standing next to an opened door. In cargo trains open doors can lead to a loss of goods. There are also some special cargo cars where open doors can lead to an exceeding of the clearance profile.

- Fires
  As mentioned, fires can be the consequence of hot boxes or blocked brakes. Inside the car, interior decoration or goods may burn. Although fires are rare in railway operation, fires in tunnels are an important issue with new lines. These can quite probably result in fatalities, loss of vehicles and loss of integrity of the tunnel structure itself. The rapid temperature rise and smoke generation lead to conditions which are not controllable by conventional fire fighting measures. Upon arrival of the fire-brigade at the tunnel portal, these conditions are likely to deter them from accessing the scene of the accident within the tunnel.

Additionally there are some other characteristics which seem to be more difficult to monitor but also necessary for an overall train inspection:

- Guiding force (Y), wheel load (Q) per axle and bogie conditions
  The percentage of Y and Q forces per wheel allows calculating the probability of derailment by the criteria of *Nadal* ($Y/Q < 1,2$).
- Profile of running surface of wheels (eccentricity, rewelding, spalling)
  These properties are interesting for the maintenance of the infrastructure.
- Uneven running of wheels
  Uneven running wheels can damage the rail and should be therefore rejected.
- Emissions from dangerous goods
  If cars are not leak proof enough, the environment can be polluted.
- Contact pressure (between pantograph and contact wire)
  Incorrectly adjusted contact pressure leads to unusually high abrasion of pantograph and contact wire. For optimising the lifecycle of both components, monitoring the contact pressure is necessary.
- Broken pantograph contact strip
  Broken pantographs have to be recognised as soon as possible, because otherwise the contact wire will be destroyed.
- Natural hazards (flooding, mudslide, rock fall, avalanche and earthquake)
  These have to be included in the term of safety related hazards. Railway lines often follow rivers, because this is a comparatively cheap option under difficult topographic conditions. Thereby a hazard situation for railway lines results from extreme precipitation (flooding, mudslide, rock fall and avalanche). For populated areas there are often means of warning the inhabitants, but railways are mostly not informed directly. Thus it is necessary to specify the critical levels for railway lines.

14.1.2 Security Related Hazards

Security related hazards can be described as wilful damage of infrastructure, rail cars, goods or persons. So terrorism and vandalism is subsumed under this topic but also crime in passenger trains (theft, aggression, personal injury) which is often prevented by video surveillance in rail cars or on platforms. Another related subject is the illegal transport of dangerous (radioactive, explosive, polluting, harmful) goods on rail cars which are often not declared in the right way.

14.2 Solutions for Hazard Detection

The implementation of solutions is influenced by several interests. Beside monetary aspects, the judicial situation has to be considered. A national regulation of one infrastructure manager can be easily cancelled by the national rail regulator on the basis of free access to the railway market. So there is a need for an international harmonisation and standardisation which must be done by an international consortium like UIC. This approach is suitable only for longer periods.

14 Hazard Alert Systems

Originally, station inspectors were not the only employees among those responsible for the operation of trains who had to deal with train supervision. Interlocking, block and/or level-crossing attendants had to monitor the condition of the rolling stock, too. The locations for this task had been defined so as to enable train supervision to be carried out on both sides of the track. As a result of the ongoing reduction of station inspectors and other staff in terms of automation of railway operation, the railway system has lost a decisive link of well-established organisational and technical processes.

Due to the gradual implementation of this strategy, aiming at a higher productivity, visual train supervision will not exist in the near future in many European countries. The consequences of this development can be mitigated by the introduction of technical solutions. This tends to result in less labour-intensive and more efficient processes. The role of the traditional train supervision can be split up to diverse challenges of train inspection. However, this is a good example to illustrate that technical solutions, currently available on the market, were designed in a similar way. They do not provide the whole solution for train inspection but rather fit to special tasks. Thus it is not possible to replace one station inspector by one special hazard alert system. Therefore sophisticated strategies for a demand-oriented applying of these systems have to be adopted.

14.2.1 Ways of Inspection

For monitoring hazards there are three possible ways for realisation:
- Vehicle-based
- Track-based
- Combined inspection (vehicle-based and track-based components)

Theoretically the vehicle-based way of monitoring has the advantage of the possibility to retrieve sensor data from almost everywhere on the wagon for optimal indication of hazards. Practically, this is not possible. Only on special rail cars it is economically justifiable to install specific onboard facilities (e.g. rail cars which are used for hazardous materials transportation are often equipped with derailment detectors). As a rule, onboard systems are only used if there is a power supply onboard because otherwise power supply must be installed before adding any other component. Equipped cars are common in passenger traffic but not in freight traffic. Only special freight cars will offer the availability of power supply.

Vehicle-based measures always have to deal with the implementation either on the vehicle itself or at an operation control centre. An easy way of such integration on the vehicle side is the combination of detectors with the braking system. In the case of a recognised alarm, the car can be stopped immediately. Because of the direct manipulation of the braking system, additional wiring over further rail cars is omitted. Another approach consists in the combination with telematics applications, which are suitable for reasons of practical functionality. Onboard generated measuring data will be sent in the course of an alarm to an operation control centre.

Mostly wayside monitoring applications make sense if the measurement categories do not change too rapidly. Also, if a lot of cars in different ownerships have to be checked by an infrastructure company, this approach is preferable. Such a situation is typical for railway traffic in Central European countries. Technical systems for automatic train supervision are able to check both sides of the train at the same time, unlike humans and their locations can be varied. These advantages will result in fewer locations needed in total.

In addition, the technical solutions are able to detect faulty conditions of the rolling-stock, which can not be discovered even by well-trained station inspectors. Therefore, these systems are important for prevention or at least for early accident identification and are expected to bring about higher productivities, too. Many wayside measuring systems use an axle-oriented

14.2 Solutions for Hazard Detection

data structure for storing information. So after passing wayside facilities the outcome is axle-related, but for optimal operational handling the rail car number is needed. Therefore the automatic rail car identification has also to be taken into account to guarantee that the results of inspections are related to single rail cars.

Another classification for the monitored parameters is given by the differentiation of the time of determination of rail car properties. Some (mostly with static character) must be collected before a train starts (or soon after) and other properties are only ascertainable during the progress of a train (e.g. bearing temperature). This is analogous to the task of train observation being shared between the examiner and movement inspector.

For high speed traffic, vehicle side and wayside measures are often combined to achieve the safety targets and to optimise the costs for maintenance in consideration of the availability of rail cars (Maly et al. 2001).

14.2.2 Fault States to Monitor

By using sensor systems – available on the market – the following fault conditions with static character can be checked:
- One-sided loading
- Displacement of the load
- Flat spots
- Open doors (doors are not closed before departure)
- Axle load, load per metre (overload)

For prevention, a reasonable approach is to check trains at a location where many trains start or end their journeys. This kind of inspection can also be required at national borders or where the infrastructure operator changes. Furthermore, there are some other hazards resulting from the following rail car characteristics, which can only be monitored with high technical and economic effort – but they are important for estimating the risk of derailment of one vehicle:
- Buckling of vehicles
- Rolling of vehicles
- Torsional moment
- Instability
- Maximum load before track displaces sideways
- Wheel geometry
- Kinematic gauge

Additionally, during the run of one train the following rail car properties have to be monitored for safety reasons:
- Defective axle-bearings
- Displaced cargos
- Already derailed wheelsets
- Temperature of axles and brakes
- Flat spots
- Broken bearing surface
- Axle breakages
- Blocked and defective wheels
- Open doors (doors open during the run)
- Fires

Depending on the possible consequences of a fault state, different types of measures must be taken. The lowest level of action is sending information to a maintenance centre. Next level can be defined as 'warning' when a monitored fault state leads to a train stop at a predefined point. In case of an emergency the level of 'alarm' is reached, so a train has to be stopped immediately to prevent further damage. These three levels of fault state and consequential actions can be defined differently by every infrastructure manager according to the specific operation processes.

14.2.3 New Approach of Inspection – The Checkpoint Concept

Checkpoints (Sünder et al. 2006) can be defined as trackside locations containing an accumulation of technical systems, which are required to enable the substitution of the traditional train supervision. Prototypes of this kind of an overall train inspection are developed in some European countries (e.g. Austria, Italy). Each sensor system is used for the supervision of one or several train conditions.

The significance of a single measurement can be increased by the conjunction of collected data. The average temperature of an intact wheel bearing is, for instance, related to the wheel load. In case the wheel load is high, a higher average temperature of the bearing and/or a faster increase of it have to be assumed. To avoid misjudgements of the actual bearing's condition, this physical correlation can be shown by calculating and evaluating a weight-compensated bearing's temperature.

This means that stricter limits will be applied to bearings with low wheel loads. Consequently, the contrary will be carried out if the wheel load is high.

The core of each Checkpoint is the so-called data concentrator. A dedicated software package is available for the above-mentioned functionality. Data, which are transmitted from sensor components to the data concentrator after passage of a train, are converted into a uniform format. If necessary, additional sensor and/or environmental information are added.

In doing so, abstract data objects are generated. Together with positioning data, they are displayed on the train model, which serves as the basis for data conjunction.

Then the results are compared with the rules. In case a threshold is exceeded, actions defined by the infrastructure operator have to be set. The range of alternatives comprises a short message to the owner of the rolling stock, but also a signal showing a stop indication. This step can only be executed by means of the operations control system and/or the interlocking, almost always an electronic one.

Assuming that a single sensor registers the exceeding of a threshold, the whole procedure will then be shortened. In this case, immediate measures are initiated.

In contrast to interlocking, not all sensor components achieve the necessary safety values. This shortcoming can be compensated by redundant or diverse measurements. An increase in data quality and an early detection of faulty conditions can be achieved by networking the various Checkpoints, too.

In this case, all data concentrators are directly linked to the Checkpoint centre. There, important data of trains are analysed and stored whereas current and already finalised train routes are distinguished. The advantages of networked Checkpoints are obvious but the costs for a Checkpoint centre and all the required sensors can be high.

There are several features of a checkpoint system, which may bring some added value for an overall train safety:
- The technical Checkpoint concept is designed on a modular basis. Using cost-benefit considerations, the range of sensor components can vary from one Checkpoint location to another. Therefore, new or additional sensors can be integrated into an already existing structure. The same applies to data of virtual sensors. Assuming that Checkpoint B has no

dynamic scale, data which was collected at Checkpoint A can be transmitted to the data concentrator of Checkpoint B via the Checkpoint centre. At Checkpoint B, this data can be used for conjunctions, too. This feature is limited to values, which are subject to no changes or not more than small changes, only (e.g. the weight of the rolling-stock).
- Trend analysis is another feature of networked Checkpoints. On the one hand, critical situations can be detected at an early stage. This target can be achieved by supervising the evolution of important measured characteristics of a train while passing over the infrastructure's network. On the other hand, these analyses enable a check to be made of the plausibility of data by comparing single measurements with the appropriate series.
- In contrast to decentralised single systems, networked Checkpoints have a further advantage. The status of all data concentrators and their connected sensor components is permanently supervised. In case of troubles and/or failures, short messages can be generated and automatically sent from the Checkpoint centre to the central trouble management system. The latter is responsible for arranging and managing all measures, which are required for the removal of the problems.

14.2.4 Extract of Available Technologies and Products

For a better understanding of hazard alert systems, some systems in use are explained in this Chapter. If different products are currently available against one hazard, the description is only a summary of the basic principles. Technical details may also differ. It is not possible to describe technologies or products which are currently under development.

14.2.4.1 Detection of Illegal Goods

Illegal transport of dangerous (radioactive, explosive, polluting, harmful) goods on rail cars can be discovered by x-ray scanners which are able to check the interior of rail cars. Using a high-energy electron linear accelerator as its radiation source and with specially designed fast-sampling and speed self-adaption technology, this kind of system is an ideal solution for non-intrusive and rapid inspection of railway vehicles at fixed locations in the railway systems e.g. railway stations, border crossings (Nuctech 2007).

14.2.4.2 Hot Box and Blocked Brake Detection

For the past twenty years, hot box detectors and fixed brake detectors are used by most European railway infrastructure companies (Eisenbrand 2001, Mironov et al. 2002, Mironov 2006, Rottensteiner 2003, Schöbel et al. 2006, Švalov et al. 2005). As a proxy for the risky situation, the temperature of the axle bearing and the brake disc are measured. As infra-red sensors are easily available, there are many different manufacturers. The varieties can be found in the sensor geometry used for measuring and the consequential ascertainable types of bogie constructions.

A wayside hot box detection system generally consists of the functional elements:

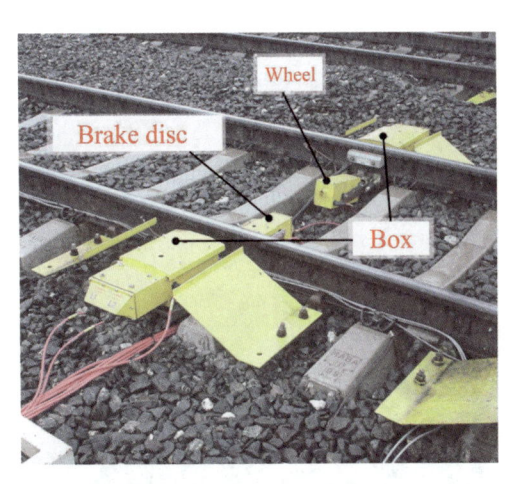

Figure 14.1: Hot box detection system with brake disc and wheel detection (photo: ÖBB Infrastruktur Betrieb)

14 Hazard Alert Systems

- Data acquisition by sensors (inc. axle counters)
- Evaluation, control and alarm generating
- Data transmission
- Visualisation of results and alarms

Typically, two hot box detection sensors, one provided on each side of the track, measure the axle boxes (figure 14.1). Simultaneously, a further infra-red sensor measures the temperature of the brake disc for fixed brake detection and a hot wheel detection sensor measures the temperature of the wheel flange to detect critical temperatures of blocked brakes. In most cases axle counters are used to allocate measurements to individual axles. Visualisation of the results from measurements is often done by a customary PC with WINDOWS operating system. Moreover, all data transmitted from track-side equipment, can be stored in a centre and if necessary exchanged with other systems.

So, in case of an alarm one infrastructure manager has to inform the driver of a train that a wayside hot box detection system has recognised a temperature exceeding a warning limit. It is also possible to declare two limits of temperature for warning and for triggering an alarm. In both cases an inspection of the axle is needed. This will be done by technical inspectors where available or by the driver of the train. Important for the braking process is the normal use of brake power and to prevent an emergency brake because the forces involved could cause a derailment.

A driver can only ascertain visually if an axle journal is broken, an axle-bearing is glowing, or an axle-box case is deformed. Even if none of these indicators can be found, the train will continue its journey with reduced maximum speed to the next place where a technical inspector is available. Otherwise – if the driver verifies the defect – the wagon has to be removed even when the alarm has been generated by the locomotive itself.

14.2.4.3 Derailment Detection

Independent from the effective cause of a derailment, an already derailed axle can be reliably recognised by the exceeding of the clearance gauge located between the rails. An over-riding of buffers is an exception for this definition, although this is classified as a derailment, too. Assuming a derailed axle, it is very improbable that there is no derailed wheel between the rails. If one wheel of a derailed axle is lost, the remaining wheel will always run between the rails which can be explained in a technical way by the forces that act on the axle. Accordingly, monitoring this area is appropriate for reliable wayside derailment detection (figure 14.2). After a detected derailment, the mechanical cage has to be replaced because of plastic deformation by the derailed axle. In case of a positive detection by closing or opening an electric circuit at a mechanical cage, a notification has to be sent to an operations control centre to be able to stop that train immediately (Stadlbauer et al. 2007, Inteletrack 2007).

Multi-use wayside derailment detections (figure 14.3) are suitable for use on railways. After a detected derailment, the mechanical cage of this detection is not destroyed. De-

Figure 14.2: Wayside derailment detection (photo: ÖBB *Infrastruktur Betrieb*)

14.2 Solutions for Hazard Detection

Figure 14.3: Wayside derailment and dragging detection

Figure 14.4: Onboard derailment detection (photo: *Knorr-Bremse*)

railment detection can also be combined with dragging detection. Equipment which may be brake system components, chains, cables, or steel bands used to secure lading is checked if it is exceeding the clearance profile. In contrast, for vehicle side detection the standard deviation of acceleration onboard is a common criterion (Hecht/Schirmer 2001) which must be integrated in the braking system of one car (figure 14.4). So in case of a derailment, one brake valve has to be opened to stop the train immediately. Usually this dependency is realised in a simple mechanical way.

14.2.4.4 Dynamic Weighing and Flat Spot Detection

The weight of the cargo is one of the most important values to be observed. Axle overload damages both the rolling stock and the rail. Load asymmetries (caused by skidding cargo) can lead to tilting of the whole wagon. Such effects must be recognised in time to carry out the necessary cargo rearrangements. Manual inspection of these effects is mostly impractical due to the complexity of the examination (closed freight cars, etc.). Today's fully automatic scales perform these kinds of measurements effectively. But it has to be mentioned that there are considerable qualitative and functional differences among them which are also influencing the price. For instance, most of dynamic weighing systems are also in a position to detect flat spots by interpreting force vertex as an indicator for flat spots. Moreover some systems provide additional information to estimate further wheel characteristics. Generally the length of the measuring section depends onto the demands of the functionality and the achievable quality of the output. In view of free network access it is becoming more and more important for an infrastructure manager to check the wheel-rail contact (loading, driving stability, out-of-roundness of wheels) of the trains running on a network. Using wayside equipment installed on transit

Figure 14.5: Dynamic weighing and flat spot detection *ARGOS*® (photo: *HBM*)

14 Hazard Alert Systems

tracks, data which is needed for the evaluation of every single vehicle can be collected with respect to the mandatory regulations. The installation effort of weighing systems varies. Some need no special adaptation of the track and for others even part of the infrastructure has to be changed during the installation (sleepers, ballast bed). Most systems use strain gauges located on the web of the rail and/or at the base of rail (Mittermayer et al. 2005). Additionally or alternatively, the sleepers may be equipped with force measuring strain gauges (figure 14.5).

Beside evaluation of wheel loads of the trains travelling at operating speed, the systems often acquire the whole dynamic condition of the wheel. The flawless state of wheels is an especially important factor in rail traffic because rail surface is destroyed by unbalanced Y and Q forces. Equally with overloaded wheels, irregularities on the running surface lead also to damages at vehicles and stress the track and the rails excessively. Another aspect of flat spot detection is the acoustic emission, because the noise factor is gaining more and more importance in European railways. For train observation by station inspectors, the sound of running rail cars is also a criterion for recognition of flat spots. This mode of measurement by acoustics can also be realised in a technical way (Witt 2007).

14.2.4.5 Silhouette Measuring Systems

Silhouette measuring systems use optical measuring principles for detection of exceeding the clearance gauge. One principle uses for evaluation and validation of the exceeding pictures, which have to be taken with sufficient quality. A subsequent analysis of the 2D-pictures allows an estimate, if there is any excess. Another measurement consists of light barriers made

Figure 14.6: **Wayside silhouette measuring system (photo: *ASE*)**

by lasers, which are arranged around the borders of the clearance gauge. Clearance excess causes a disruption of one or of several laser barriers, which can easily be measured. The probability of false alarm is high due to dispersed material of passing trains. For the reliability of the system a combination of optical camera and laser is recommendable (ASE 2007).

14.2.4.6 Fire Recognition Systems

A possibility for fire recognition in tunnels is a linear heat detection system based on optical fibre sensors (figure 14.7). Due to its measuring characteristics, the fibre sensor becomes especially suitable for protection of objects (installations, buildings, constructions) against fire or massive heat influence. Physical quantities such as temperature or pressure and tensile force have influences on the silica fibre. In detail, they locally change the characteristics of the light waveguide. As a result of the influence, the light in the silica fibres is scattered. This effect allows determining the location and the quantity of the external physical influence, which enables the waveguide to be used as a simple linear sensor (Siemens 2007).

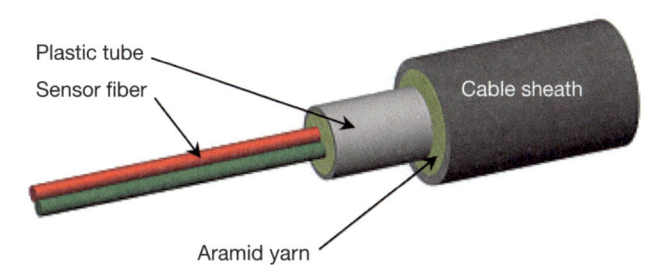

Figure 14.7: Optical fibre sensor cable (graphic: *Siemens*)

Natural convection in long railway tunnels may hinder the system to recognise small or medium fires. Also, airflow of moving trains is critical for measuring. So only the case of a single train standing in a tunnel without any heavy natural convection offers good boundary conditions for reliable fire recognition at an early stage of a fire. Outside a tunnel it is much more difficult to recognise a fire due to weather (sun, wind, rain or snow). This kind of system is often used in combination with a fire fighting system by activating the fire fighting system when passing predefined temperature levels.

14.2.4.7 Fire Fighting Systems

With a stationary fire fighting system (figure 14.8) within the tunnel a fire cannot be prevented at all but the effects of the fire can be reduced significantly.

The principle of water mist technology is based upon many tiny water droplets, equally distributed at the scene of the fire (Aquasys 2007). In contrary to conventional sprinkler systems, water mist utilises the physical phenomena of evaporation of water for the benefit of fire suppression. In detail, the evaporation is responsible for two of the three main advantages of water mist for fire suppression purposes:
– Vast cooling effect: water, when evaporated, requires external energy, which has to be taken from thermal energy of the fire.
– Oxygen depletion: evaporation of water particles leads to a considerable extension of the volume and thus to a reduced ratio of oxygen.

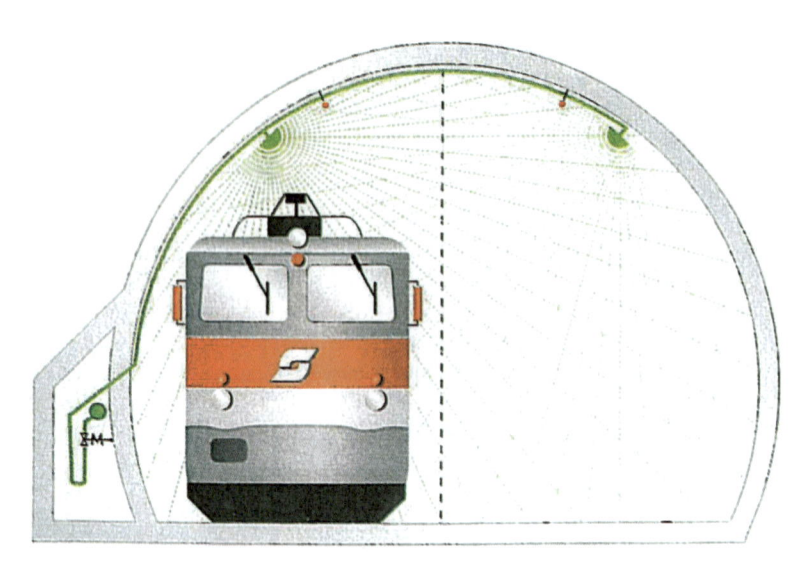

Figure 14.8: Functionality of a Stationary Fire Fighting System (graphic: *Bioversal*)

The third decisive advantage of water mist technique is achieved by all the lots of tiny droplets distributed in the ambient of a fire and acting as many reflectors. The result is to confine the heat to the area of the flames. Remarkably, water mist systems are using pure water as a fire suppression agent, which is harmless to persons and environmentally friendly. Thus, water mist systems can be activated immediately upon detection and do not require any delay for evacuating the affected area or facility. This advantage contributes considerably to the overall loss prevention philosophy in case of a fire. Another type of fire fighting systems is working with water under high pressure (Marioff 2007). This offers the possibility to extinguish a fire in a short time and to save water.

14.2.4.8 Wind Measuring Systems

For reasons of tunnel safety or for reducing the risk of derailment at exposed positions (e.g. bridges crossing deep valleys) a wind measuring system (SST 2007) can be installed. In case of a fire the direction of escape is essential for survival. Also for fire brigade and other rescue workers the information about wind conditions (direction, strength) influences the rescue strategy. At exposed positions the risk for derailment can be increased if back pressure caused by strong wind influences the ratio of Y and Q forces. So the track speed limit must be related to the wind speed for the critical section.

14.2.4.9 Earthquake Detection

For railway lines in areas which are tectonically active a detection of earthquakes can be useful which need to be integrated into operations control. So in the case of an earthquake the speed of all affected trains must be reduced or even brought to a halt.

14.2.4.10 Flood Warning

There are two possibilities for flood warning of railway lines. Either there is an existing hydrological and/or hydraulic forecasting system, or there is no prognosis system. As a result of parallel running of railway lines and rivers, it is recommended to link the forecast organisation to the railway infrastructure manager. Of course, it is helpful to check existing hydrological forecasting instruments for their applicability to warn the railway (Nester 2007). On the one hand local accuracy will not be high enough to mark safe areas but on the other it will be possible to inform responsible employees to check their lines. If there is no existing forecasting system along a river, there are other alternatives. Different events are categorised by discharge and outflow. Based on the analysis some kind of scenario catalogue is created which is provided to the infrastructure manager to help him in his decision process as to whether or not to close a line.

14.3 Choice of Location

Wayside measures cannot be located at every place where once an accident occurred or will possibly happen. After an event has occurred, it is quite simple to design the optimal position for detection and minimising loss in this specific case. But this empirical method will not fulfil economic boundary conditions.

Generally there are two different points of view, the line-oriented and the network-oriented. The line-oriented view allows the calculation of the nearest position to have enough time for stopping a train at a predefined position for further investigation. For the specification of these points where the train has to stop the network-oriented view is helpful. So there is the requirement to define all risky elements in a railway network which should not be passed by a train with irregularities. Furthermore the combination of measures depends on the strategy of an infrastructure manager, which can be described as a mix of event-avoiding systems and damage-reducing components.

With regard to their future locations, there are two fundamental concepts:
1. Whenever traditional train supervision is to be replaced, a technical equivalent has to be installed.
2. The number of locations and/or systems necessary for conducting train supervision can be optimised provided that they are based on cost-benefit considerations. In this case, the number of locations should be lower.

Due to economic reasons only the second approach (Schöbel 2005) is practically relevant and will be discussed in the following section.

14.3.1 Operational Handling

The process in case of a detected fault on a train has to be carefully planned before going into operation. For vehicle-side detection the data transmission to an operation control centre of an infrastructure manager has to be specified. Wayside systems are exclusively the responsibility of an infrastructure manager for the planning of locations until the operational handover. Modern signal box technology enables the integration of the monitoring system. This allows an automatic stop of trains with strong irregularities, whereby the classification of these fault conditions must be done by the infrastructure manager (e.g. an already derailed axle on a train leads to a stop at the next mandatory signal). Basically it can be distinguished between warning and alarm: A warning indicates only an overstepping of threshold value that can lead to a dangerous situation (e.g. overweight of one vehicle, warm box). In case of an alarm there has to be an immediate reaction because of an already existing hazard (e.g. hot box, derailed axle, exceeding of clearance profile). Thus, only highly reliable, available and accurate technical solution will be integrated into operations control because otherwise the reliability of operation would be reduced dramatically. The local position in the railway system is defined according to the last stopping position ahead of a risky element of infrastructure.

14 Hazard Alert Systems

Taking the time behaviour of an integrated sensor system into account, the local position can be calculated in the following way: Starting from the mandatory signal (or stopping point) the nearest location of each sensor component can be found in consideration of the allowed speed limit, the sight on the distant signal and the response time of the sensor component (figure 14.9).

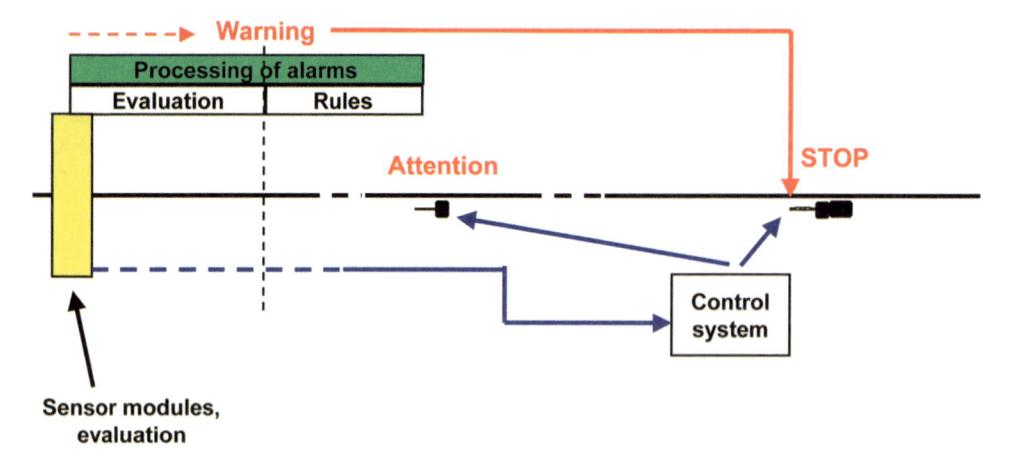

Figure 14.9: Calculation of position according to a predefined stopping point

Because of the different times for response, the nearest locations of different components will vary. But if several sensor systems are required for the same risky element, due to other practical reasons (power supply, network availability) the sensors will be concentrated at the same place. For determination of the installation location, sensor components with the largest processing time will be relevant.

When finalising the introduction of management operation systems and using sensor components for the supervision of rolling-stock, a small number of station inspectors will be responsible for a relatively large number of train inspections. The idea of visualising alarms via the control system can be rejected by the infrastructure operator. During peak times, it might be possible that the necessary measures cannot be initiated on time. Therefore, the integration of inspection systems (e.g. by usage of Checkpoints) into the control system and/or electronic interlocking is a very important goal.

Another operational issue deals with the aspect of errors or malfunction of a sensor component. On a double-track section it is possible to use the opposite track if it is also equipped. This will lead to additional delays of the passing trains but will fulfil safety aspects. Also, results from some formerly passed sensor can be extrapolated if the rail car property has almost static character. Moreover staff could do the job of train observation for this period again. In any case, these procedures must be agreed by the national authority and should guarantee safe operation.

In Russia, the hazard alert systems used are manually integrated into operations control at every point of its application. The basic system is the hot box detection system. Other systems (e.g. rubbing wheel detection, derailment detection, defective wheel detection, silhouette measuring, overload measuring etc.) can be added to the basic component.

14.3.2 Classification of Risky Elements

The choice of location for wayside train monitoring is depending on the elements of infrastructure which are to be protected from hazard situations. Therefore it is necessary to define risky elements in a railway network. For the classification process following elements have to be specified by individual parameters:
- Bridge
- Sequence of curves
- Points (slip crossings)
- Gradient
- Tunnel
- Highly curved section
- Changes in superstructure

Bridges have two interesting aspects for this definition: One, the height and the length which can be combined into a product that is one parameter for risk. This specification contains on the one hand the short bridge over a deep valley and also on the other hand a long section on arches. The second criterion is the construction type: if the running track is the same as before and after the bridge, there will not be any dynamic influences on the train, otherwise the dynamic forces can lead to a derailment. In general, for prevention the clearance profile should be checked and for damage reducing one derailment detection should be installed.

A **sequence of curves** with different orientation can lead to derailments. The rail car properties which are indicators for that kind of risk should be checked before going into operation. So it is recommended to do this in shunting yards or while entering a network.

Points are necessary for building railway networks. For the derivation of measures, it is necessary to distinguish between the locations of one element. Points in shunting yards will not be protected by derailment sensors, but points on high-speed lines are a good example for a risky element of infrastructure which should not be passed with a derailed axle.

A **gradient** section is an additional stress for brakes. Depending on its severity, the stress for brakes will vary. Because of the increased braking, the probability of blocked brakes will be higher in such sections than in plain sections. For avoiding this fault state graduated braking of heavy freight trains is usual. Moreover the temperature of brakes can be monitored by wayside hot brake detection.

If a line is adapted for higher speed in many regions of Europe the percentage of tunnel sections will increase. In comparison to other parts of a railway network the probability for an accident is lower, because the operation process is much more simple (e.g. no shunting) although the severity is rising in case of an accident. There is a tradition in Europe for tunnel safety concepts which also include hazard alert systems. For the calculation of tunnel risk the density of trains, the mix of trains (passenger, freight) and many other parameters are taken into account. Without an expert's report on tunnel safety it is not possible to start operation any more. So for each tunnel a couple of specific arrangements have to be evaluated under the boundary conditions of cost and benefit.

Highly Curved sections can be defined as sections with over 10 % of curves with a radius below 300 metres. The rail car properties which can cause trouble on such sections will undergo preventative checks at the shunting yard or when entering the network.

Changes in superstructure are discontinuities of the stiffness of superstructure, which may lead to dynamic movements of the train in longitudinal direction. Rail cars with instabilities should also be checked preventive next to a shunting yard or to a border of an infrastructure manager. So if one train is prepared at a shunting yard, the first possibility for checking this train is in departure sidings of the shunting yard. On the other side entering trains should also be checked, so it would be possible to stop damaged cars. For economic reasons the number of locations should be a minimum and must therefore have regard to the local situation of a shunting yard. The same ideas must be taken into account for network borders.

14.3.3 Strategies for Infrastructure Manager

If hazards in railway operation are analysed, measures against them have to be worked out by responsible staff. The following measures are common (Schäbe 2001):
- Elimination of hazard potential
- Avoiding circumstances which might activate any underlying hazards
- Reducing occurrence probability of consequences

This approach is event driven. First step is to try to eliminate a hazard. In the second step the trigger or its probability should be prevented. The remaining risk can also be reduced by measures against the consequences of the initial event. Therefore the distribution of risk on the elements leading to one event must be known to be able to reduce the risk at the right point. It is not possible to eliminate all hazards, because some are in responsibility of another company. So an infrastructure manager is only able to avoid hazards created by the infrastructure or to reduce consequences for the infrastructure. Therefore a combination of event-avoiding and damage-reducing wayside train monitoring components must be developed which fulfils the requirements of cost and benefit.

Another way to determine safety measures is given by the scope:
- Measures which are only affecting a special section of a line or a few rail cars
- Measures which are affecting all lines of an infrastructure company
- Measures which are affecting the whole railway network

Generally measures are more efficient if the scope is bigger. Measures of the first group have often a bad cost-benefit-ratio (typically construction measures). Localised measures always have a disadvantage in comparison with other measures because they are located at one special place and are only working there. Furthermore, there are differences if the measures are divided into construction, technological and organisational measures. In this context the durability is different in these three groups of measures. Construction measures are always representing the state-of-art for a long period. Technological appliances can be changed more easily by newer ones. Organisational rules can be adjusted in short time. So hazard alert systems for railway operation are located in the group of technological measures.

Finally it is a question of cost-benefit for all kind of hazard alert systems to get the best combination of reasonable measures. Moreover, the responsibility for measures has to be checked according to national legislation and organisation. Only in this way it is possible to design a promising strategy for the application of any hazard alert system. In practical usage lower maintenance costs for hazard alert systems are achieved by means of self-diagnosis and remote diagnosis.

References

(Akita 1985) Akita, K.: Practical Use of Computerized Interlocking System 'SMILE' in JNR. In: Quarterly Reports of RTRI, 1985

(Akita/Nakamura/Wanatabe 1987) Akita, K.; Nakamura, H.; Wanatabe, T.: Solid-State Interlocking in Railway Signalling, SMILE and µ-SMILE. In: Quarterly Reports of RTRI, 1987

(Akuzawa 1982) Akuzawa, M.: Application of Train Sorting System to the Signalling Equipment of Commuter Railway. In: Japanese Railway Engineering 4/1982

(Althaus 1994) Althaus, H.: Linienförmiges Zugbeeinflussungssystem ZSL 90. In: Signal+Draht 5/1994

(Altmann et al. 2007) Altmann, K.; Lenk, M.; Hetzel, R.: Das Betriebs- und Leitsystem für Umschlagbahnhöfe im Kombinierten Verkehr. In: ETR, issue 6/2007

(Ansaldo 2002) Ansaldo Signal/Union Switch & Signal: UM71 – Failsafe Train Running. Australia, 2002

(Ansaldo 2008) AnsaldoSTS: ACC Technical Specification. 2008

(Aquasys 2007) www.aquasys.at

(Arnold 1987) Arnold, H.-J. et al.: Eisenbahnsicherungstechnik, 4th edition. Transpress, Berlin 1987

(ASE 2007) http://www.ase-industry.com

(Bailey et al. 1995) Bailey, C. (ed.); Institution of Railway Signal Engineers (IRSE): European Railway Signalling. Adam & Clark Black, London 1995

(Barwell 1983) Barwell, F. T.: Automation and Control in Transport. 2nd ed. Pergamon Press, Oxford/New York 1983

(Belov/Geršenzon/Kotlecov 2003) Belov, V.; Geršenzon, M.; Kotlecov, D.: Vnedrenie sistemy avtomatičeskoj identifikacii podvižnogo sostava na Rossijskich železnych dorogach. In: Železnye dorogi mira. Moscow, Nr. 7-2003

(Bianchi 1985) Bianchi, C.: Die Führerraumsignalisierung auf der Direttissima Rom-Florenz. In: Signal+Draht, 1+2/1985

(Bombardier 2003) Bombardier Transportation. Audio Frequency Track Circuit Style TI21-4, Technical Manual. 2003

(Brux 2002) Brux, G.: Brand- und Katastrophenschutz in Tunneln der Neubaustrecke Köln-Rhein/Main. In: Der Eisenbahningenieur 4/2002

(Bryleev/Šišljakov/Kravcov 1966) Bryleev, A.; Šišljakov, A.; Kravcov, Ju.: Ustrojstvo i rabota rel'sovych cepej. Moscow 1966

(CENELEC 1999) CENELEC: Railway applications – Specification and demonstration of reliability, availability, maintainability and safety (RAMS). EN 50126:1999

(CENELEC 2001) CENELEC: Railway applications – Communications, signalling and processing systems – Software for railway control and protection systems. EN 50128:2001

(CENELEC 2003) CENELEC: Railway applications – Communications, signalling and processing systems – Safety related electronic systems for signalling. EN 50129:2003

(Česke Drahy 1998) Česke Drahy: Vorschrift für die Verwendung der Signale bei der Organisation und Durchführung des Betrieb, German translation. 1998

(Christov 1991) Christov, Ch.: Osnovy na osiguritelnata technika. Technika, Sofia 1991

References

(DB Netz 2001) DB Netz: Punktförmiges Zugbeeinflussungssystem PZB 90: Systembeschreibung. Frankfurt(M) 2001

(DB Netz 2006) DB Netz: 301 DS/DV – Signalbuch. Frankfurt (M) 2006

(Dmitriev/Minin 1992) Dmitriev, V.; Minin, V.: Sistemy avtoblokirovki s rel'sovymi cepjami tonal'noj častoty. Transport, Moscow 1992

(Dmitriev/Serganov 1988) Dmitriev, I.; Serganov, I.: Osnovy železnodorožnoj avtomatiki i telemechaniki. Moscow 1988

(Doswell 1957) Doswell, P. C.: Single Line Control (British Practice). IRSE Green Booklet No. 4, 2nd edition, London 1957

(Eichenberger 2007) Eichenberger, P.: Kapazitätssteigerung durch ETCS. In: Signal+Draht 3/2007

(Eisenbrand 2001) Eisenbrand, E.: PHOENIX MB – Die neue Dimension in der Heißläuferortung. In: Signal+Draht 2001

(European Commission 2004) European Commission: Directive 2001/16 – Interoperability of the trains-European conventional rail-system: Draft Technical Specification for Interoperability, Control-Command and Signalling Sub-System (http://europe.eu.int/comm/transport/rail/interoperability/doc/ccs-tsi-en-annex.pdf) 2004

(Fenner/Naumann 1998) Fenner, W.; Naumann, P.: Verkehrssicherungstechnik. Siemens – Publicis Corporate Publishing, Erlangen 1998

(Fenner/Naumann/Trinckauf 2004) Fenner, W.; Naumann, P.; Trinckauf, J.: Bahnsicherungstechnik. Siemens – Publicis Corporate Publishing, Erlangen 2003

(Fricke/Pierick 1990) Fricke, H.; Pierick, K.: Verkehrssicherung. Stuttgart, Teubner 1990

(Friesen/Uebel 1999) Friesen, W.; Uebel, H.: Automatisierter Betrieb im Nahverkehr – Erfahrungen von Alcatel. In: Signal+Draht, 10/1999

(Funkwerk 2008) http://www.funkwerk-it.com

(Gertler/Stolzenberg 1994) Gertler, F.; Stolzenberg, H.-J.: Das Zugbeeinflussungssystem ZUB 123. In: Signal+Draht, 11/1994

(Goldsbrough 1961) Goldsbrough, J. V.: IRSE Green Booklet No 22: Route Control Systems; The S.G.E. Route Relay Interlocking System (British Practice). 1958

(Guilloux 1990) Guilloux, J.-P.: Das Signalsystem der Hochgeschwindigkeitsstrecken in Frankreich. In: Signal+Draht, 1+2/1990

(Hagelin/Stridh 1997) Hagelin, G.; Stridh, A.: Signalisierung für höhere Geschwindigkeiten – eine Fallstudie. In: Signal+Draht 4/1997

(Hahn 2006) Hahn, M.: Analyse der Sicherung europäischer Bahnübergänge. TU Dresden 2006

(Hájek 2006) Hájek, Z.: Zařízení pro komplexní automatizaci spádovišť – seznámení s filozofií, 2. seminár železničnej zabezpečovacej techniky. Vyhne 2006

(Hall 2000) Hall, S.: Broad survey – The history and development of railway signalling in the British Isles. Friends of the National Railway Museum, York 2000

(Hall 2001) Hall, S.: Modern Signalling Handbook, 3rd edition. Ian Allan Publishing 2001

(Hansen/Pachl 2008) Hansen, I.; Pachl, J. (editors): Railway Timetable & Traffic. Eurailpress, Hamburg 2008

(Hawkes 1969) Hawkes, J.: Circuits for Colour Light Signalling. IRSE Green Booklet No. 15, 1969

(Hecht/Schirmer 2001) Hecht, M.; Schirmer, A: Versuche zur Diagnose von Entgleisungen. ZEV Glasers Annalen 2001

(IEC 2001) CENELEC: Functional safety of electrical/electronic/programmable electronic safety-related systems – Part 4: Definitions and abbreviations. EN 61508-4:2001

(Igarashi/Siomi 2006) Igarashi Y.; Siomi S.: Development of Monitoring System for Electric Switch Machine. In: Quarterly Reports of RTRI, 2/2006

(Inteletrack 2007) http://www.inteletrack.co.za/d_e_d.htm

(IRSE 1980) Institution of Railway Signal Engineers (IRSE): Railway Signalling. A&C Black, London 1980

(IRSE 1991) Institution of Railway Signal Engineers (IRSE): Railway Control Systems. A&C Black, London 1991

(IRSE 1999) Institution of Railway Signal Engineers (IRSE): Introduction to Signalling. London 1999

(IRSE 2008) Institution of Railway Signal Engineers (IRSE): Introduction to North American Railway Signaling. Simmons-Boardman Books, Omaha 2008

(Ivančenko et al. 2002) Ivančenko, V.; Kovalëv, S.; Šabel'nikov, A.: New Information Technologies : Control System for Automation of Braking-up – Forming Trains Process, Rostov-on-Don 2002

(Jones 2008) Jones, A.: Modernisation of Cambrian Lines. http://www.signalbox.org/branches/aj/index.htm; 2008

(Judge 2007) Judge, T.: Yard management gets smarter. In: Railway Age, November 2007

(Koetting 2007) Koetting, H.; Metschulat, H.; Weber, S.: www.stellwerke.de, 2007 (last update)

(Konarev 1994) Konarev, N.: Železnodorožnyj transport: Ènciklopedija. Bol'shaya Rossijskaja Ènciklopedija, Moscow 1994

(Kondo 1980) Kondo, R.: Introduction of Advanced Type Automatic Train Stop System. In: Japanese Railway Engineering 4/1980

(Kononov/Lykov/Nikitin 2003) Kononov, V.; Lykov, A.; Nikitin, A.: Osnovy proektirovanija èlektričeskoj centralizacii promežutočnych stancij. Maršrut, Moscow 2003

(Kravcov/Nesterov/Lekuta 1996) Kravcov, Ju.; Nesterov, V.; Lekuta, G.: Sistemy železnodorožnoj avtomatiki i telemechaniki. Transport, Moscow 1996

(Kusche 1984) Kusche, W.: Stellwerks- und Blockanlagen – Gleisbildstellwerke, 2nd edition. Transpress, Berlin 1984

(Lazarevic et al. 2007) Lazarevic, N.; Khoudour, L.; El Koursi, El M.; Machy, C.: An Intelligent Level Crossing : Technical Solutions for Improved Safety and Security. Presentation and paper on Transport 2007 in Sofia

(MacFarlane 2004) MacFarlane, I.: Railway Safety – Interlocking and Train Protection. Engineers Media, Crows Nest (Australia) 2004

(Makowski 1992) Makowski, S.: Stand und Entwicklungsperspektiven der Zugbeeinflussungsanlagen bei den PKP. In: Signal+Draht, 12/1992

(Maly et al. 2001) Maly, H.; Säglitz, M.; Klose, C.; Ullrich, D.; Kolbasseff, A.: Neue Onboard- und stationäre Diagnosesysteme für Schienenfahrzeuge des Hochgeschwindigkeitsverkehrs. In: ZEV Glasers Annalen 125, Georg Siemens Verlagsbuchhandlung 2001

(Mandola 1992) Mandola, I.: Das Zugbeeinflussungssystem der MAV. In: Signal+Draht, 12/1992

References

(Marioff 2007) http://www.marioff.de

(Maschek 1996) Maschek, U.: Analyse zur Gestaltung elektronischer Stellwerke. Diploma Thesis at TU Dresden. Dresden 1996

(Maschek/Lehne 2005) Maschek, U.; Lehne, U.: Das Mechanische Stellwerk der Bauform "Einheit". TU Dresden, Dresden 2005

(Mironov 2006) Mironov, A.: New potentialities of KTSM and ASK PS. In: Automation, Communications, Information Science 12/2005

(Mironov et al. 2002) Mironov, A.; Tagirov, A.: Use of complexes KTSM in modern conditions. In: Automation, Communications, Information Science 9/2002

(Mittermayer et al. 2005) Mittermayr, P.; Presle, G.; Stephanides, J.; Weilinger, W.: Die neuen Infrastrukturmessstellen der ÖBB – Kontinuierliche Messungen und Auswertungen. In: ZEV Glasers Annalen, Georg Siemens Verlagsbuchhandlung 2005

(Modern Railways 2004) The search for low-cost crossings. In: Modern Railways, June 2004

(Mraz 1992) Mraz, J.: Die Zukunft von Zugbeeinflussungsanlagen bei der CSD. In: Signal+Draht, 12/1992

(Naumann/Pachl 2002) Naumann, P.; Pachl, J.: Leit- und Sicherungstechnik im Bahnbetrieb. Fachlexikon. Tetzlaff, Hamburg 2002

(Nester et al. 2007) Nester, T.; Schöbel, A.; Drabek, U.; Kirnbauer, R.; Rachoy, C.: Hochwasserwarnystem für Eisenbahnstrecken. Conference 21. Verkehrswissenschaftliche Tage, Dresden 2007

(Nishibori/Sasaki/Hiraguri 2002) Nishibori, N.; Sasaki, T.; Hiraguri, S.: Development of Train System by Microwave Balises. In: Quarterly Repoerts of RTRI, 4/2002

(Nock 1982) Nock, O. S.: Railway Signalling – A Treastise on the Recent Practice of British Railways. A&C Black, London 1982

(NORAC 2003) NORAC: Rules of the Operating Department. 8th edition, 2003

(Nuctech 2007) www.nuctech.com

(OAO RZD 2007) OAO RZD, VNIIAS. Pilotnyj proekt po primeneniju sputnikovych technologij na opytnom učastke Moscow-Klin. 2007

(Oehler 1981) Oehler, K.: Eisenbahnsicherungstechnik in der Schweiz – die Entwicklung der elektrischen Einrichtungen. Birkhäuser, Basel 1981

(Ohta 2005) Ohta, M.: Level Crossing Obstacle Detection System Using Stereo Cameras. In: Quarterly Reports of RTRI, 2/2005

(OST 32.17-92) Otraslevoj standart OST 32.17-92. Bezopasnost' železnodorožnoj avtomatiki i telemechaniki. Osnovnye ponjatija. Terminy i opredelenija.

(Pachl 2000-I) Pachl, J.: Betriebliche Rückfallebenen auf Strecken mit selbsttätigem Streckenblock. In: Signal+Draht 7+8/2000

(Pachl 2000-II) Pachl, J.: Zugbeeinflussungssysteme Europäischer Bahnen. In: Eisenbahntechnische Rundschau 11/2000

(Pachl 2002) Pachl, J.: Railway Operation and Control. VTD Rail Publishing, Mountlake Terrace 2002

(Piastowski 1960) Piastowski, J.: Das neue Lichtsignal-Einheitssystem der OSShD-Mitgliedsbahnen. Zeitschrift der OSShD 5/1960

(Railtrack 1994) Railtrack: General Information on Track Circuits. 1994

(Railways Museum St. Petersburg) Railways Museum in St. Petersburg, Exposition of Block Systems

(Retiveau 1987) Retiveau, R.: La signalisation ferroviaire. Presses de l'école nationale des Ponts et chaussées, Paris 1987

(Reznikov 1985) Reznikov, Ju.: Êlektroprivody železnodorožnoj avtomatiki i telemechaniki. Transport, Moscow 1985

(RFNIIAS 2008) RFNIIAS: http://www.rfniias.ru, 2008

(Rottensteiner 2003) Rottensteiner, U.: VAE-HOA 400 DS – Heißläuferortungsanlagen für finnische Hochgeschwindigkeitsstrecken. In: Signal+Draht 2003

(RSSB 2004) Rail Safety and Standards Board (RSSB): Rule Book, Module S (Signals), 2004

(Russian Ministry of Transport 2004) Russian Ministry of Transport: Instrukcija po signalizacii na železnych dorogach Rossijskoj Federacii, Moscow 1999

(Sapožnikov et al. 1995) Sapožnikov, V.; Sapožnikov, Vl.; Christov, Ch.; Gavzov, D.: Metody postroenija bezopasnych mikroêlektronnych sistem železnodorožnoj avtomatiki. Transport, Moscow 1995

(Sapožnikov et al. 1997-I) Sapožnikov, V.; Sapožnikov, Vl.; Talalaev; V. et al.: Sertifikacija i dokazatel'stvo bezopasnosti sistem železnodorožnoj avtomatiki. Transport, Moscow 1997

(Sapožnikov et al. 1997-II) Sapožnikov, Vl.; Ëlkin, B.; Kokurin, I.; Kondratenko, L.; Kononov V.: Stancionnye sistemy avtomatiki i telemechaniki. Transport, Moscow 1997

(Sapožnikov et al. 2003) Sapožnikov, V.; Sapožnikov, Vl.; Samanov, V.: Nadežnost' sistem železnodorožnoj avtomatiki, telemechaniki i svjazi. Maršrut, Moscow 2003

(Sapožnikov et al. 2006) Sapožnikov, Vl.; Kokurin, I.; Kononov, V.; Lykov, A.; Nikitin, A.: Êkspluatacionnye osnovy avtomatiki i telemechaniki. Maršrut, Moscow 2006

(Sapožnikov et al. 2008) Sapožnikov, Vl.; Kononov, V.; Kurenkov, S.; Lykov, A.; Nasedkin, O.; Nikitin, A.; Prokof'ev, A.; Trjasov, M.: Mikroprocessornye sistemy centralizacii. GOU „Učebno-metodičeskij centr po obrazovaniju na železnodorožnom transporte", Moscow 2008

(Sapožnikov/Gavzov/Nikitin 2002) Sapožnikov, V.; Gavzov, O.; Nikitin, A.: Koncentracija i centralizacija operativnogo upravlenija dviženiem poeždov. Transport, Moscow 2002

(Sasaki 1986) Sasaki, T.: Development of Electronic Blocking System. In: Japanese Railway Engineering 1986

(Sasaki/Wakabayashi 1989) Sasaki, T.; Wakabayashi, T.: Development of a New Electronic Blocking System. In: Quarterly Reports of RTRI 1989

(Sasse 1941) Sasse, W.: Das Belgische Signalwesen. In: Zeitschrift für gesamte Eisenbahnsicherungs- und Fernmeldewesen (Das Stellwerk) 1941

(Schäbe 2001) Schäbe, H.: Neue Ansätze zur Systemsicherheit in der Bahntechnik. In: Eisenbahntechnische Rundschau 2001

(Schmitz 1962) Schmitz, W.: Das Spurplanstellwerk Sp Dr S60. In: Signal+Draht 1962

(Schöbel 2005) Schöbel, A.: Zur Frage der Standortwahl von Zuglaufüberwachungseinrichtungen. Dissertation, Technische Universität Wien. Begutachter: N. Ostermann. E. Kopp 2005

(Schöbel et al. 2006) Schöbel, A.; Pisek, M.; Karner, J.: Hot box detection systems as a part of automated train observation in Austria. Conference EURNEX – ZEL 2006, Zilina 2006

(Schubath/Grotheer 2002) Schubath, S.; Grotheer, U.: Neues Simis-W-Stellwerk in Polen. In: Signal+Draht 2002

References

(Shannon/Short 2004-2005) Shannon, I.; Short, R.: Interlocking Developments, Proceedings of the IRSE 2004-2005

(Šeluchin 2005) Šeluchin, V.: Avtomatizacija i mechanizacija sortirovočnych gorok. Moscow 2005

(Siemens 1978) Siemens AG: Gleisbildstellwerk SpDrS (System 60). Braunschweig 1978

(Siemens 2001) Siemens Transportation Systems: Systembeschreibung Mikrocomputer-Achszählsystem Az S 350 U. 2001

(Siemens 2003) Siemens AG: Siemens und Eisenbahnsignaltechnik. 2003

(Siemens 2006) Siemens Transportation Systems: LED-Signalgeber für Kombinationssignale der DB AG : Wirtschaftliche Betriebsführung durch Einsatz der LED-Technologie. 2006

(Siemens 2007) http://www.siemens.com

(Siemens 2008) Siemens Transportation Systems: The MSR 32 Microcomputer System. Greater Efficiency and Safety for Freight Transport. 2008

(SST 2007) http://www.sst-online.de

(Stadlbauer et al. 2007) Stadlbauer, R.; Schöbel, A.; Karner, J.: Wayside Derailment Detection and its Integration in the Operation Management. Conference EURNEX – ZEL 2007, Zilina 2007

(Studer/Marti 2008) Studer, C.; Marti, M.: Paradigmenwechsel bei Eisenbahnsignalen in der Schweiz. In: Signal+Draht 6/2008

(Such 1956) Such, W. H.: Mechanical and Electrical Interlocking. IRSE Green Booklet No. 3, 2nd edition, 1956

(Sünder et al. 2006) Sünder, M.; Knoll, B.; Maly, T.; Schöbel, A.: Checkpoint systems and their integration into solid state interlockings for automatic train supervision. IRSE – ASPECT 2006, London 2006

(Suwe 1988) Suwe, K.-H.: Signaltechnik in Japan. In: Signal+Draht 6/1988

(Švalov et al. 2005) Švalov, D.; Šapovalov, V.: Systems for diagnosis of rolling-stock: textbook for technical schools and colleges of railway transport. Maršrut, Moscow 2005

(TDJ 2008) TDJ: http://www.chinamet.com, 2008

(Tenzler 2005) Tenzler, S.: Vergleichende Betrachtungen zu Sicherungstechnik und sicherungstechnischen Planungsabläufen in Russland und Deutschland. Diploma theses at TU Dresden, 2005

(Theeg/Maschek 2005) Theeg, G.; Maschek, U.: Analyse Europäischer Signalsysteme. In: Signal+Draht 6/2005

(Theeg/Vincze 2007) Theeg, G.; Vincze, B.: Comparison of European Train Protection Systems. In: Signal+Draht 8/2007

(Ugajin et al. 1999) Ugajin, H.; Shiroto, H.; Fujinami, K.; Omino, K: Two Greens – A New Signal Aspect for High-Speed Train on Wayside Signalling. In: Quarterly Report of RTRI, 1/1999

(UIC 2003) UIC: Implementing the European Train Control System ETCS – Opportunities for European Rail Corridors. 2003. gsm-r.uic.asso.fr/meetings/ertms/ertms_2003/docs/etcs_report_1.pdf

(UIC code 732) UIC code 732: Principles for signalling trains routes using wayside signals. 3rd edition, Paris 2002

(UIC code 734) UIC code 734: Adaptation of safety installations to high-speed requirements. 2nd edition, Paris 2004

References

(UN/ECE 2001) UN/ECE: Terminology on Combined Transport. Prepared by the UN/ECE, the European Conference of Ministers of Transport (ECMT) and the European Commission (EC). United Nations, New York/Geneva 2001

(Unisig 2006) Unisig: ERTMS/ETCS – Class 1 Subset-026: System Requirement Specifications. Version 2.3.0, 2006

(Unisig 2008) Unisig: ERTMS/ETCS – Class 1 Subset-026: System Requirement Specifications. Draft version 3.0.0, December 2008

(Uzdin et al. 2002) Uzdin, M.; Efimenko, Ju.; Kovalev, V.: Železnye dorogi – obščij kurs. Informacionnyj centr „Vybor", St. Peterburg 2002

(Vincze/Tarnai 2006) Vincze, B.; Tarnai, G.: Evolution of train control systems. Budapest University Technology and Economics, 2006

(Vlasenko 2006) Vlasenko, S.: Analiz effektivnosti peregonnych sistem avtomatiki i telemechaniki na železnych dorogach mira. Sputnik +, Moscow 2006

(Watanabe et al. 1999) Watanabe, I.; Ushijima, Y.; Fukuda, M.; Takashige, T.: Development of Digital ATC System. In: Quarterly Report of RTRI 1/1999

(Watanabe/Takashige 1989) Watanabe, I.; Takashige, T.: Moving Block System with Continuous Train Detection Utilizing Train Shunting Impedance of Track Circuit. In: Quarterly Reports of RTRI, 4/1989

(Wennrich 1997) Wennrich, R.: Der Elektronische Token für die Ägyptischen Eisenbahnen. In: Signal+Draht 11/1997

(Wenzel 2006) Wenzel, B.: Vergleich des Aufbaus, der Funktionen und der Datenverarbeitungsvorgänge zwischen LZB mit CIR-ELKE und ETCS Level 2. Seminary Wort at TU Dresden, 2006

(White 2003) White, T.: Elements of Train Dispatching – Vol. 1. VTD Rail Publishing, Mountlake Terrace (USA) 2003

(wikipedia) Wikipedia (www.wikipedia.de)

(Winter et al. 2009) Winter, P. et al.: Compendium on ERTMS. DVV Media/Eurailpress, Hamburg 2009

(Witt 2007) www.ewitt.de/downloads/arzd.pdf

(Xu Zhengli 2003) Xu Zhengli: The Application of the Controllable Retarders and the Boosters for the Speed Control in Marshalling Yard. Conference ZEL 2003, Zilina 2003

(Yamanouchi 1979) Yamanouchi, S.: Safety and ATC of Shinkansen. In: Japanese Railway Engineering 2/1979

(Zářecký 2008) Zářecký, S.: The Newest Trends in Marshalling yards automation. In: Transport Problems – Problemy Transportu, Tom 3 Volume 4 Part 1, 2008

(Zoeller 2002) Zoeller, H.-J.: Handbuch der ESTW-Funtionen. Eurailpress, Hamburg 2002.

Glossary

Railway terms often vary between countries. For this book, the authors have therefore tried to find terms and definitions which are internationally understandable and appropriate. Here, the terms used in this book are explained. They might deviate from those with which the reader is familiar.

Words in *italics* refer to a separate entry in the Glossary.

A

Absolute Block System	Form of *fixed block system* where, under normal operation, only one train is permitted in any one block section at any one time.
Absolute Stop	The stop aspect of a signal, which must not be passed by the train without special permission from the *signaller*. It is used where *movable track elements* have to be protected and in *absolute block system*s.
Acceptance	In the context of a *line block* system with *neutral direction*, this is the permission given from one *train sequence station* to another to send a *train*.
Active Level Crossing	A *level crossing* where stationary technical equipment or persons actively signal the road user if a *train/shunting movement* is approaching.
Alternative Routes	If for an *entrance-exit (NX) operation* two or more routes are possible between the same entrance and exit points, these are referred to as alternative routes. One of these is usually a *priority (preferred) route*. Depending on the technology, route selection may be manual or designed to default automatically to the priority route.
Approach Aspect	North American term for *Caution Aspect*.
Approach Control	This is a British dynamic form of *speed restriction warning*. Signals show restrictive aspects (e.g. to *Stop* or *Caution*) to make the driver brake, and are then cleared when the *train* has reached a defined approach point.
Approach Locking	Performing *irreversible route locking* automatically upon detection of the approaching train.
Approach Sight Triangle	At a *passive level crossing*, the area which must be clear of obstacles to enable road users to see an approaching *train/shunting movement* in time, taking into account the defined speed of each.
Approach Time	The running time between a signal that provides an *approach aspect* and the following signal.
Approach to	A position along the track which is reached by the train before a defined item of equipment, such as a signal. An alternative term is '*in rear of*'.

Glossary

Area of Conflict	The areas where the *clearance profiles* of two tracks, or at a *level crossing* of the railway and the road, overlap, and which can only be assigned exclusively to one of them at any one time.
Aspect	See *signal aspect*.
ATC	Abbreviation for Automatic *Train Control*.
ATO	Abbreviation for *Automatic Train Operation*.
ATP	Abbreviation for Automatic *Train Protection*.
Attentiveness Check	A function of *train protection systems* where, in certain situations (e.g. when passing a *distant signal* at *caution*), the driver is required to confirm his attentiveness and that he is ready to undertake the required actions (e.g. braking).
Automatic Equipment Identification (AEI)	AEI provides a direct data connection between vehicles (wagons, locomotives) or loading units (containers, swap bodies) and the information systems. With that, the objects can be identified in the network and detailed information forwarded.
Automatic Point Setting	A feature which automatically commands all required *movable track elements* to take up their appropriate positions when a *route* has been selected.
Automatic Route Calling	An automation function which allows the consecutive passing of several *trains* through the same *route* by automatically repeated calling of the same route for each train.
Automatic Route Setting	An automation functions where *trains* are routed into different tracks according to stored information such as timetables, *priority* and *alternative routes*.
Automatic Signal	A signal, usually a *block signal,* which is controlled automatically by the passage of trains and does not need any action by the *signaller* to set or restore it.
Automatic System	A system (e.g. for *line block*, *route interlocking* or train driving) where defined processes can be operated without the active involvement of a person.
Automatic Train Operation	An advanced form of *train control* where in normal operation the train can be driven automatically without the active involvement of a driver.
Automatic Train Routeing	Another term for *automatic route setting*.
Autonomous Level Crossing	An *active level crossing* which, in contrast to the *route-dependent level crossing*, is controlled and supervised independently from *routes*.
Auxiliary Signal	Signal to issue a *movement authority* in *degraded mode operation* which is used to substitute a written instruction in case the normal *signal* cannot be cleared. There are two main types: those which require *movement on sight* and those which do not.

Glossary

Availability	The ability of a product to be in a state to perform a required function under given conditions at a given instant of time, or over a given time interval, assuming that the required external resources are provided.
Availability Redundancy	A principle where the same components are provided in duplicate (or more) to enable the continuation of that function in the case of *failure*. Availability redundancy is often combined with *safety redundancy*.
Axle Counter	A device for *track clear detection*, based on the counting of axles entering and leaving a defined portion of track and comparing these values.

B

Balise	A trackside spot element used in *train protection* and *cab signalling* systems that transmits information to passing trains, using the transponder principle.
Beyond	A position along the track which is reached by the train after a defined point such as a trackside signal. An alternative term is '*in advance*'.
Bi-Stable Relay	A relay with two stable positions which remains in the last position in the case of power failure.
Block Information	Logical principle in *line block* systems which is based on the exchange of arrival and departure messages. Also a term for the messages exchanged between *block stations*.
Block Section	A section of track in a *fixed block system* which a train may enter only when it is not occupied by other trains or vehicles.
Block Signal	A signal that governs train movements into a *block section*.
Block Station	An evaluation unit including the related signals which limits *block sections*.
Block System	A short term for *line block* system.
Blocking	In a *tokenless block system,* the reservation of a section of line for a certain train.
Blocking Time	The time interval in which a section of track is allocated exclusively to one train and is therefore blocked to other trains.

C

Cab Signalling	A signalling system that displays *signal indications* on the driver's console.
Call(ing) on Signal	A signal in Britain and some other countries which is used for two functions: 1. as *auxiliary signal* 2. to authorise a *movement on sight* by a *train* in regular operation

Glossary

Cascade Interlocking	Logical principle and the related form of *interlocking machine* where a route is built up on the principle that each element of the route locks the element beyond.
Catch Points	A *movable track element* that is designed as an incomplete *set of points*. The purpose is to protect *route(s)* in advance by derailing any unauthorised movement that would otherwise conflict with those route(s). For the protected route(s), this would be mainly classed as *flank protection*.
Caution Aspect	A *signal aspect* for trains, in most countries indicated by a yellow or double yellow aspect, that informs the driver that he is at or within service braking distance of the *stop aspect* ahead and hence requires him to start braking if he is not already doing so.
Centralised Block System	A line block system in which the block information for several *block stations* is processed in a central place.
Centralised Traffic Control (CTC)	An operation control system in which the local *interlockings* are remotely controlled from a central place.
Clamp Lock	The most frequently used mainly *trailable point* locking mechanism, which clamps the onlaying blade to the stock rail.
Clear a Signal	Alternative term for *'open a signal'*
Clear Aspect	A signal aspect which permits the driver to pass that signal and indicates that the signal beyond does not require the driver to stop, either. However, a *speed restriction* or *speed restriction warning* can be connected to it. In most countries a clear aspect is displayed by one or more green lights.
Clearance Profile	The space around a track which must be free from obstruction to enable the safe movement of a rail vehicle on that track. Different clearance profiles may be required for different types of rolling stock, and such profiles differ between countries.
Clearing Point	A point beyond a signal which a train must have cleared completely before a signal may be *closed* (restored to danger) and the signal in rear be *opened* (cleared).
Close a Signal	Switching process to restore a signal to its most restrictive *aspect*.
Colour Light Signal	A *light signal* which displays the aspects by different colours.
Combined Signal	Combination of *main* and *distant signal* functions in one trackside signal.
Composite Fail-Safe	Strategy to achieve *fail-safe* by processing the safety-related action in at least two redundant channels and comparing the results.
Conflicting Area	See *Area of Conflict*.
Consecutive Routes	Routes in the same direction which can be used by the same *train/shunting movement* in sequence without stopping, with the exit from the one route being the entrance to the other. In some cases this has to be prohibited by *special route interlocking*.

Glossary

Continuous Speed Control Method
: A *retarding concept* with permanent or frequently repeated control and influence on the speed of free running wagons (or groups of wagons) during the *gravity shunting* process.

Control Length of a Signal
: The length of track beyond a *trackside signal* that must be clear before the signal can be *cleared* for a train movement.

Controlled Signal
: A signal that is locally or remotely controlled by an operator or by an external automation system (e.g. *Automatic Route Setting, Automatic Train Routeing*). It is the opposite of an *Automatic Signal*. A controlled signal which can be set by the signaller to automatic mode is known as a semi-automatic signal.

D

Degraded mode operation
: Alternative technical or procedural methods of safety in case the normal technical equipment or procedures cannot be used due to a technical defect or non-applicability to the particular case. Degraded mode operation often implies a lower level of safety and/or operational disadvantages.

Departure Signal
: Alternative term for *station exit signal*.

Derail
: North American term for a *derailer* (called 'block derail') or a set of *catch points* (called 'split point derail').

Derailer
: A *movable track element* which is designed as a block to be set on one or both rails and whose purpose is to derail an unintendedly moving rail vehicle. The main use is *flank protection* for another authorised movement.

Destination Signal
: An *interlocking* signal at the exit of an interlocked *route*. It is also called the exit or target signal of that *route*.

Diagnostic System
: A system or sub-system that supervises other technical systems like infrastructure elements (points, signals), *interlocking* systems, shunting technical facilities (retarders, handling systems, brake test facilities), information and communication systems. It detects and records faults and gives appropriate messages to support maintenance.

Diamond Crossing
: A piece of infrastructure where two tracks cross without any ability for rail vehicles to transfer from one to the other track. Some diamond crossings have movable parts, in which case they belong to *movable track elements*.

Dispatcher
: An employee who supervises train movements on a line or within a certain area. In centralised operation, the dispatcher can be the same person as the *operator* and/or *signaller*.

Distant Signal
: A signal that provides a *caution aspect* to a *main signal* ahead. A distant signal does not define a *block section*.

Disturbance
: Deviation from the specified performance of a system due to external influences.

Glossary

Dual Protective Points	A *set of points* which momentarily is requested for *flank protection* from both trailing ends at the same time, but can give *flank protection* only to one side.
Dynamic Speed Profile	Based on the *static speed profile*, a continuous definition of the possible speed. This considers also the necessary braking processes for a required stop or a reduced speed in advance.

E

Electric Interlocking	1) Alternative (mainly Eastern European) term for *relay interlocking*. 2) Alternative (North American) term for a certain kind of *electro-mechanical interlocking*.
Electro-Mechanical Interlocking	An *interlocking machine* where the functions of *interlocking* and *field element* control are performed partly by mechanical and partly by electrical functions.
Electro-Pneumatic Interlocking	An interlocking machine that controls *field elements* pneumatically.
Electronic Interlocking	*Interlocking machine* where the *interlocking* function is performed by electronic logic circuits respectively by the software processed by these.
Element Control	All functions of control and supervision of *field elements* (e.g. *movable track elements*, *track clear detection*, *signals*) below the *interlocking* level.
End of Authority	The position where a *movement authority* ends. In *trackside signal*ling, it is marked by a *main signal* displaying a stop aspect, or comparable devices.
End of Train Detection System	A system, either train-based or track-based, which automatically detects the presence of the end of the train to prove train completeness.
Entrance Signal	A signal at the entrance of an interlocked *route* or a *block section*.
Entrance-Exit (NX) Operation	Frequently used form of *operation control* in *relay* and *electronic interlocking*. A route is selected and automatically set when the signaller operates two elements, one representing the *entrance* (start), the other the *exit* (target) of the route.
Entry Signal	Alternative term for the *home signal* of a *station area*.
Error	Deviation from the intended design which could result in unintended system behaviour or failure.
European Train Control System (ETCS)	A standardised European system for *train protection* and *cab signalling*.
Exit Signal	1) An *interlocking signal* that governs *train* movements to leave a *station track* and enter the *open line*. It is also called a station exit signal.

Glossary

	2) North American term for the opposing *interlocking signal* a train passes when leaving interlocking limits.
	3) An *interlocking signal* at the exit from an interlocked route. It is also called a destination or target signal.
External Object	All objects except *rail vehicles* which might (legitimately or otherwise) be inside the *clearance profile* of the railway.

F

Facing Move(ment)	Movement over a *set of points* in the direction in which the tracks diverge.
Fail Safe Principle, Fail Safety	A basic principle of *safe systems*, in which any kind of assumable failure must lead to a safe situation. This may result in severe traffic disruption or other inconveniences.
Failure	Deviation from specified performance. A failure is the consequence of a *fault* or *error* in a system.
Fallback Level	A particular level of *degraded mode operation*.
Fault	Abnormal condition that could lead to an *error* or a *failure* in a system. A fault can be random or systematic.
Field Elements	Summary term for all signalling devices which are connected with the *interlocking machine* and are situated outside it, e.g. *points*, *signals*, *track circuits* and *axle counters*.
First Class Relay	Ancient term for *type-N-relay*.
Fixed Block Systems	Form of *movement in space intervals* in which the track behind a train is cleared sequentially in the form of fixed *block sections*.
Flank Area	The portion of track which is situated between an element which gives *flank protection* to the *route* and the *route* itself.
Flank Protection	Protection of a train against a sideways collision movement.
Fleeting	An automation function which allows the passing of several consecutive trains through the same *route* without intervention from the *signaller*. The signals are used like automatic *block signals*. It is also known as Auto-Working Facility.
Following Protection	Protection of a train against collision from another movement in the same direction.
Fouling Point	That location at a set of *points*, *diamond crossing, slip crossing* or other track element with two overlapping tracks, up to which a rail vehicle can stand on the one track without being foul of a movement on the other track.
Free Shunting	A method of *shunting* without interlocked shunting *routes*, where the staff have full responsibility for safety.
Front Protection	A feature of *route interlocking* that is provided by some railways to prevent any unauthorised opposing movement affecting that route. It is analogous to *flank protection*.

G

Gradient	The rate at which the track is inclined up (+) or down (-).
Gravity Shunting	A principle of shunting where freight trains are split up and new trains made up based on wagons rolling freely downhill.
Ground Staff	Summary term for all railway employees whose workplace is on the trackside, e.g. *signallers*, *dispatchers*, gatekeepers and train observers.

H

Hazard Alert System	Track-based and/or vehicle-based equipment for the monitoring of *faults* on the train or trackside to avoid dangerous events or to reduce the damage caused by such an event.
Headway	The time interval between two successive trains.
Home Signal	1) A signal governing the entrance to an *interlocking area*. 2) A signal governing the entrance to a *station area*.
Hump Process Control System	Control system for a hump yard, which is actually an *electronic interlocking* with extensive automation functions and serves for the optimum control of the hump process.

I

In Rear of	Alternative term for *approach to*.
Inherent Fail Safety	Strategy to achieve *fail safety* by the characteristics of materials. The probability of dangerous failure of the components is extremely low and does therefore not need to be assumed.
Insulated Rail Joint (IRJ)	A joint of two sections of rails with insulation between. It is used as a boundary for *track circuits*.
Interlocking	An arrangement of *points* and *signals* interconnected in a way that the switching of these elements can only be performed in a proper and safe sequence.
Interlocking Area	The area of control of one *interlocking machine*.
Interlocking Machine	A machine which performs *interlocking* in a specific installation.
Interlocking Plant	North American term for *Interlocking Area*.
Interlocking Signal	A signal that governs a train route inside an interlocking area.
Intermediate Interlocking/ Station Signal	An *interlocking signal* that is neither a home signal nor one which governs a route to leave the home signal limits.
Intervention	A function of *train protection systems* where safe behaviour is enforced by the technical system, e.g. braking the train in the case of excessive speed. Intervention requires *supervision* as a precondition.
IRJ/IJ	Abbreviation for *Insulated Rail Joint*
Irreversible Route Locking	Step of *route locking* which, for the prevention of *route release* under a moving train, can only be cancelled by the detection of the train's position, unless methods of *degraded mode operation* are used.

Glossary

L

Leaving Signal — North American term for a signal that governs train movements to leave a siding, yard, or branchline onto a main line.

Level Crossing — A crossing of the railway and a road at grade.

Light Signal — A form of optical *trackside signal* where *signal aspects* are indicated by lights in different colours *(Colour light signals)* and/or geometrical arrangements (Position light signals). Optionally, this can be combined with the selective blinking of one or more lights.

Limit of Shunt Board/Signal — A fixed sign or a signal that marks the *shunting limit*.

Line Block — Summary term for all *technical safety systems* on *open lines* to guarantee safety against *following* and *opposing movements*.

Local Operation Area — A *shunting* area which can be selectively controlled either centrally from the interlocking or locally, e.g. by the driver or the shunting staff.

Locally Operated Point Switches (LOPS) — An arrangement of electric *points* control in marshalling yards. This includes simplified *interlocking* functions where the path is selected by the shunting vehicles or by shunting staff. The term is used in North America, but similar arrangements are also used in Europe.

Locking of a Route — See *route locking*.

Loop — A *main track* for passing and overtaking trains.

LOPS — Abbreviation for *locally operated point switches*.

M

Main Signal — A signal that authorises the movement of a train to enter a given section of track.

Main Track — A track used for regular *train movements*.

Maintainability — The probability that a given active maintenance action, for an item under given conditions of use, can be carried out within a stated time interval. The maintenance needs to be performed under stated conditions and using stated procedures and resources.

Management Information System (MIS) — A system or process that provides the information necessary to manage an organisation effectively (in this context a railway company or a business unit like a marshalling yard). This system and the information it generates, together with financial information, are generally considered essential components of business decisions.

Mechanical Interlocking — *Interlocking machine* based on the use of mechanical locks and mechanical connections to *field elements*.

Mono-Stable Relay — A relay with one stable position which is taken by the relay in the case of power failure.

Glossary

Movable Point Crossing	North American term for a *diamond crossing* with movable frogs.
Movable Track Element	A piece of infrastructure with moving parts which interrupts the continuous rail on which trains run, or which protrudes into the *clearance profiles* of the track. Therefore, it needs particular safeguards.
Movement Authority	The permission given to a *train* to move up to a defined position. This position is called *End of Authority*.
Movement in Space Intervals	A principle of operation in which *trains* are separated by dividing the track into portions, such that one portion can only be occupied by one *train* at the same time.
Movement in Time Intervals	A principle of operation in which *trains* are separated by a defined time interval between the departure of two consecutive *trains*.
Movement on Sight	A principle of operation in which the driver must be able to stop at any time within the track length he/she can observe, and therefore must not exceed an appropriate speed. Versions are *movement to half the sighting distance* and *movement to the full sighting distance*.
Movement to Half the Sighting Distance	A form of *movement on sight* where the train driver must assume an *opposing movement* approaching at the same speed, and must therefore be able to stop at any time within the half length of the track he can observe.
Movement to the Full Sighting Distance	A form of *movement on sight* where the train driver does not need to assume an *opposing movement*, but must be able to stop at any standing obstacle appearing in his sight.
Moving Block System	A form of *movement in space intervals* where the required space in front of a train is not defined by fixed track locations but is moving with the train. The faster the train is moving, the greater the distance (and hence the time) it needs to stop. Thus, the length of the space interval depends on the speed of the train.
Moving Unit	A unit of one or more *rail vehicles* which moves along the railway. Moving units can be distinguished into *trains* and *shunting movements*.

N

Neutral Direction	Logical principle of *opposing protection* in *block systems* where in *normal position* no train can enter the *train sequence section*, and each movement has to be agreed between the neighbouring *train sequence stations*.
Non Signal-Controlled Operation	An operating procedure in which the traffic is controlled by verbal or written authority (a *non-technical safety system*). Also known as an unsignalled or hand signalled move. A non signal-controlled operation may be combined with a signalling system as a *safety overlay*.

Glossary

Non-Technical Safety System	A system where safety is achieved non-technically, e.g. by persons obeying rules.
Normal Position	Originally the position of a lever e.g. in mechanical interlocking where the lever rests in the machine and the function attached to it is currently not activated. In the wider sense (in which it is used in the book) that position of points to which the points are usually returned if unused. Also, in more modern technologies, the initial position of a function when no train is in proximity.
NX-Operation	Abbreviation for *Entrance Exit Operation*.

O

Open a Signal	Switching process to switch a signal from its most restrictive to a less restrictive *aspect*; also called 'to clear a signal'.
Open line	Main tracks outside *station areas* (not used in North American and British signalling).
Operation Control	The functions above the *interlocking* level for the selection of *routes* or other command input by the *signaller/operator*, or by automatic systems.
Operator	A staff member who controls train movements. Summarized term for *signaller* and *dispatcher*.
Opposing Protection	The protection of a *train/shunting movement* against collision with a movement from the opposite direction.
Opposing Routes	Routes from opposite directions which lead into the same track, and which require *special route interlocking* to achieve *opposing protection*.
Overlap	A certain length of track *beyond* a *signal* that must be clear for a *train movement* to be given a *movement authority* up to that *signal*, and which must remain locked whenever such a *movement authority* has been given.

P

Passive Level Crossing	A *level crossing* which always appears to the road user in the same way, without any indication of whether a rail vehicle is approaching or not. The road user must therefore take full responsibility for the safe use of the crossing.
Path	A way the *train* or *shunting movement* unit can take according to the positions of *movable track elements*.
Permissive Block System	A method of working of a *line block system* which allows a *train movement* into an occupied *block section* if it *moves on sight*.
Permissive Stop	A stop aspect of a signal which may be passed by the driver on sight after stopping. It is used in *permissive block systems*.

Glossary

Placed Direction	The logical principle of *opposing protection* in *block systems*, where at any one time either of the two neighbouring *train sequence stations* has the permission to send trains into the *train sequence section* and thus the other has not. That direction can be exchanged.
Point Machine	A machine in the field equipment for the physical operation, locking and detection of *points* or other *movable track elements*.
Point Zone	The zone behind an *interlocking signal* where, due to *movable track elements* located there, a speed restriction displayed at the signal is valid.
Points	An assembly of rails, movable point blades and a frog, which effect the tangential branching of tracks and allows trains or vehicles to run over one track or another. Points are the probably most frequently found *movable track element*. In North America, the term 'switch' is used for the whole, whereas 'points' mean the blades only.
Preliminary Caution Aspect	A *signal aspect* which warns of a *caution aspect* at the next signal and therefore requires the *train* to start braking under certain circumstances.
Priority Route	If after an *entrance-exit operation* two or more *routes* are possible between the same entrance and exit, the priority route is that given preference in selection.
Proceed Aspect	Any *aspect* which allows to a driver to pass the *signal*. This includes any *caution aspect* as well as the *clear aspect*.

R

Rail vehicle	Any rolling stock which is designed to move on a railway and has permission or is licensed so to do.
Range of Vision	1) In context with optical *trackside signals*, the distance over which, based on optical parameters, the signal is visible to the driver from a defined direction. 2) The distance in front of a train or shunting movement which is visible to the driver. Either of these may need to take into account the effects of darkness, fog and falling snow.
Reactive Fail Safety	Strategy to achieve *fail-safety* by supervision of components and active switching to a safe state in case of failure.
Registration Method	A method where safety critical occurrences, in many cases also *special commands* in *degraded mode operation*, are registered to ensure the appropriate behaviour of those responsible.
Relay Interlocking	*Interlocking machine* where the *interlocking* function is performed by relay logic circuits.
Relay Set	In some types of *relay interlocking* (mainly with *topological* logic): A standardised and pre-manufactured group of relays and other electrical elements which represent a defined kind of element in the track layout.

Glossary

Reliability	The probability that an item can perform a required function under given conditions for a given time interval.
Replace a Signal	Alternative term for *close a signal*.
Restore a Signal	North American term for *close a signal*.
Restrictive Permissive Stop	An intermediate solution between *absolute* and *permissive stop* where the driver is permitted to pass the signal in own responsibility only in certain defined cases.
Retarding Concept	Arrangements of trackside braking to make gravity shunting processes at marshalling yards safe and efficient. Parts of this concept are the optimisation of choice and arrangement of retarders. The two basic retarding concepts are the *continuous speed control method* and the *target shooting method*.
Reverse Position	For *points*, that end position which is not the *normal position*.
Reversible Route Locking	The step of *route locking* which simply provides a dependence between *movable track elements* and *signals* and can be cancelled by the *signaller* without using methods of *degraded mode operation* if the signal is at stop.
Risk	Product of the frequency, or probability, and the consequence of a specified hazardous event.
Route	A technically protected *path* for safe *train* and *shunting movements*.
Route Conflict	A conflict between two called routes (e.g. points required in different positions) which prevents both from being set at the same time.
Route Dependent Level Crossing	An *active level crossing* which is included in a *route* that way that a *main signal* at Proceed guarantees to the train driver that the level crossing is safely closed or is ready to close safely for road traffic.
Route Locking	A *route* can become locked if all elements which belong to the route (including *overlap*, *flank protection* and *start section*, if required) are locked in the proper positions. It is a precondition to issue a *movement authority*. Route locking is typically performed in two sequential steps: *reversible route locking* and *irreversible route locking*.
Route Queuing	An automation function which, if a requested *route* is currently not available due to a *route conflict*, automatically repeats the route request until the route becomes available.
Route Release	Unlocking of a route, in normal operation after the *train/shunting movement* has passed the route or the respective *route section*. This is the counterpart to *route locking*. In *degraded mode operation*, route release can also be performed manually by the *signaller*, also if the train has not (yet) used the route.
Route Section	See *sectional route release*.

Route Signalling	A signal system, the signal indications of which display only information about the *path* to be taken through the interlocking, with no direct information about the maximum speed permitted.
Running Path	That part of a *route* which is legitimately traversed by the *train* or *shunting movement*.

S

Safe System/Component	See *vital system/component*.
Safety	Functional safety within the system that protects against hazardous consequences caused by technical failure and human *error*.
Safety Functions	Summary of all functions which are responsible for safety and have therefore to fulfil strict requirements regarding their correct functioning (e.g. *fail-safety*).
Safety Overlay	An additional technical support in a *non-technical safety system* which reduces the probability or the consequences of dangerous human *error*. In contrast to *technical safety systems*, the safety responsibility basically remains that of the human.
Safety Redundancy	A principle to achieve *composite fail-safety* in which each process is calculated at least twice and the results compared. It is found particularly in electronic systems. The respective action is undertaken only in case of agreement, otherwise the system goes to a safe (often obstructing) status. Safety redundancy can be combined with *availability redundancy*.
Second Class Relay	Ancient term for *type-C-relay*.
Secondary Tracks	Tracks which are not used for regular *train* movements, only for *shunting*.
Sectional Route Release	The splitting of a *route* into two or more sections, with each section being individually released behind the *train/shunting movement*.
Security	Protection against hazardous consequences caused by wilful unreasonable human actions, e.g. crime, terrorism or vandalism, but also wilful careless actions.
Set of Points	Singular form of *points*.
Shunt(ing) Signals	A fixed signal that authorises *shunting movements*.
Shunting Limits	A position along the track, usually demarcated by some form of signing, which limits an area where *shunting* is permitted and which therefore must not be passed by *shunting movements*.
Shunting Movement	All movements with *rail vehicles* other than *train* movements. Shunting movements are undertaken at low speed and over short distances.
Siding	1) Another term for a *secondary track*. 2) North American term for a *loop*.

Glossary

Sighting Distance	The distance, measured along the *path* of the train, from the *sighting point* to the *trackside signal*. In contrast to the *range of vision*, this takes into account obstructions such as bridges and track curvature.
Sighting Point	The furthermost point on the *approach to* a *trackside signal* from where the driver can reliably read the *signal aspect* including its associated subsidiary aspects. This needs to consider also obstructions to vision such as bridges and track curvature.
Signal Aspect	Any valid display of a *trackside signal*.
Signal Box	Alternative term for *interlocking machine*, mainly used in *mechanical interlocking*.
Signal Clearing/Opening	See *clear a signal*.
Signal Dependence	Basic logical principle of *interlocking* whereby a *main signal* must not be cleared unless the *movable track elements* beyond the signal are locked in safe positions.
Signal Indication	The information content that is connected with a *signal aspect* by operational rules.
Signal Release	Returning a signal to stop after the front end of the *train/shunting movement* has passed it and the driver can no longer see it; to protect the train from a following movement.
Signal-Controlled Operation	A form of operation in which train movements are governed by signal indication.
Signaller	An employee who operates signalling apparatus. The signaller can also be *operator* and/or *dispatcher*.
Simple Route Interlocking	The locking of two *routes* against each other as a direct result of a *movable track element* being required in different positions by these two routes.
Slip Crossing	A *movable track element* where two tracks cross and where, in contrast to *diamond crossings*, *rail vehicles* can transfer from one track to the other. A further distinction is whether that transfer can take place on both sides (double slip) or one side only (single slip).
Special Command	All command inputs by staff in the operation of an *interlocking* system, which can be potentially dangerous. Special commands are often used in *degraded mode operation*, but also in other situations where safety critical actions of staff cannot be completely eliminated. Special safety precautions apply for the input of these commands, such as *registration method* and/or *time delay method*.
Special Route Interlocking	The locking of two *routes* against each other which do not need any *movable track element* in a different position. Examples are *opposing* and, in some cases, *consecutive routes*.

Glossary

Speed Restriction Aspect	A *signal aspect* which defines a speed restriction which must not be exceeded when passing the *trackside signal*.
Speed Restriction Warning Aspect	A *signal aspect* which defines a *speed restriction aspect* beyond and requires the driver to slow down appropriately.
Speed Signalling	A signal system, the indications of which convey information about the maximum speed through the interlocking and not necessarily the route to be taken.
Splitting of Trains	An operational method where one train is split into two or more trains.
Spring Points	Special form of *trailable points* which have a preferred end position, are regularly opened by the wheels of trains and return automatically to the preferred position in the absence of a rail vehicle. The return can be performed by a spring, or electrically in combination with *track clear detection*.
Start Section/Zone	That portion of a *route* which lies in rear of the *route entrance signal*, but is occupied by a train waiting at that signal, or will become occupied by moving from the waiting place to the signal. In some cases, elements in the start zone have to be considered for route *interlocking* and speed control.
Static Speed Profile	A profile which continuously defines the permitted speed of a train over the whole length of its *path*. The permitted speed at a defined location is the minimum of all possible speed restrictions at that location.
Station	1) A place designated in the timetable, by name. 2) A place with a platform stop for passenger trains. 3) A short term for a *station area* (not used in North American and British signalling).
Station Area	An arrangement of station tracks limited by opposing *home signals* (not used in North American and British signalling).
Station Exit Signal	An *interlocking signal* that governs train movements to leave a *station track* onto the *open line*.
Station Throat	Area at one end of a typical *station* where the *open line* tracks branch out into the *station track*s.
Station Track	A *main track* limited by successive interlocking signals within the same *station area* (not used in North American and British signalling).
Stop Aspect	Any *signal aspect* which requires the driver to stop at the signal. In most countries it is displayed by one or two red lights.

Glossary

Supervision	1) In the context of *train protection systems*, this is a function in which certain parameters of the moving *train/shunting* (particularly the speed) are supervised and compared with a permitted value. 2) In the context of *movable track elements*, supervision is the detection of the current position of the element and evaluation of this information for *interlocking* purposes.
Switch	North American term for a *set of points*.

T

Tabular Interlocking	Logical principle and the related form of *interlocking machine*, where all *routes* are pre-defined in a route locking matrix, including the information about which elements have to be locked for the route.
Target Section	Last portion of tracks of the *running path* of a *route* which remains occupied if the train/shunting has to stop at the route *exit signal*, even if the route has already been released.
Target Shooting Method	A *retarding concept* of *gravity shunting* where the wagon, after leaving the last retarder, has exactly the proper speed to achieve the desired position in classification track. That means different speeds depending on the changing distances between last retarder and target position.
Target Signal	Another term for *exit* or *destination signal*.
Technical Safety System	A system where safety is achieved by technology.
Time Delay Method	A frequently used method, particular in *degraded mode operation* and for other *special commands*, where a time delay is applied between commanding a function and its actual performance. This assumes that a present or approaching train has left the dangerous area, has come to a stop, or has shown its presence by occupying more track sections during this time.
Token Block System	A safety system for *open lines* using *block information*, based on the principle that the exclusive right to enter a defined section is given by the presence of a dedicated physical object on the train.
Tokenless Block System	A safety system for *open lines* using *block information*, where the right to enter a defined block section is managed without physical objects, but by exchange of *block information* between *block stations*.
Topological Interlocking	Logical principle and the related form of *interlocking machine* where *routes* are defined based on neighbourhood connections in the track layout, and each route is searched anew in the track layout upon request.
Track Circuit	A frequently used device for *track clear detection*, based on the detection of a current flowing between the rails via the axles.

Glossary

Track Clear Detection	Summary term for all technical equipment and procedures which serve to prove the tracks clear.
Track Occupancy Detection	Another term for *track clear detection*.
Trackside Signal	A signal that stands at the trackside and displays information to drivers or other persons.
Trailable Points	*Points* which can be forced to open by a *trailing movement* without damage to the point structure, *point machine* or vehicle.
Trailing Move(ment)	Movement over a *set of points* in the direction in which the tracks converge.
Train Control System	A *train protection system* with continuous guidance of the driver by cab signal.
Train Movement	A locomotive or self-propelled vehicle, alone or coupled to one or more vehicles, with authority to operate on a *main track* in accordance to rules specified for train movements.
Train Protection System	A *technical safety system* or *safety overlay system* which supervises the driver and enforces compliance with the rules, particularly stopping at an *end of authority* and obeying speed restrictions.
Train Sequence Section	The section between two neighbouring *train sequence stations*, within which the sequence of trains cannot be changed, and only trains proceeding in the same direction are permitted at the same time.
Train Sequence Station	A location (e.g. *station*, junction) where, in contrast to simple *block stations*, the sequence of trains can be changed. *Opposing protection* in a *block system* is performed between the train sequence stations.
Train Staff	1) Summary term for all railway employees whose normal workplace is on the train e.g. drivers, co-drivers, conductors etc. 2) In a *token block* system a device (a staff) that authorises the entrance into an assigned *block section*.
Train Stop	Trackside device in *train protection systems* used to implement the *train trip* function.
Train Trip	An *intervention function* which is used to apply brakes to the train with the maximum available braking force until a complete stop is achieved. The term is used particularly in context with *ETCS*.
Transferred Flank Protection	If an element is requested for *flank protection*, but cannot give it at present, it forwards the request to another element further beyond.
Trap Points	A *set of points* with short diverging track which ends at a buffer stop or in a sand drag. The main use is *flank protection*; in some cases it is also used for *opposing protection*.

Type C (Controlled) Relay	A signalling relay which can only considered to be safe if its dropping down is supervised.
Type N (Non-Controlled) Relay	A signalling relay which, due to particular technical characteristics, can be assumed to drop down safely, the dropping down of which therefore does not need to be supervised.

U

UMLER	North America only. The Universal Machine Language Equipment Register (UMLER) is a database which contains a distinct combination of the reporting mark and number of each locomotive and freight wagon (including containers and highway truck trailers). This system cooperates with *Automatic Equipment Identification*.
Unblocking	In a *tokenless block system*, this is the technical clearing of the block after the train has cleared the related section.

V

Vehicle Signal	A signal that is attached to a *rail vehicle* and gives information to drivers of other rail vehicles, *ground staff* and other persons.
Vital System/Component	A system or component whose purpose is to fulfil safety functions in the railway system and which therefore has to be designed to fulfil particular requirements, e.g. *fail-safety*.

W

Warning	In the context of *train protection systems*, a function where a signal is given to the driver in a selected, possibly dangerous situation (e.g. excessive speed) to direct his attention.
Warning Time	At a *level crossing*, the advance warning time for the road user before a train reaches the crossing.
Wayside Signal	North American term for *trackside signal*.

Y

Yard	An arrangement of *secondary tracks* used for making up trains, storing cars and trains and other purposes.
Yard Limits	North American close equivalent for *Shunting Limits*.
Yard Management System	A system which controls a marshalling yard at high levels of automation. It consists of *hump process control* systems, *diagnostic systems* and *management information systems*.

Explanation of Symbols in Track Layout Schemes

The symbols which are used in track layout plans differ much between the countries. The authors of the book have agreed on common symbols which are valid only for this book. In the following they are listed.

Symbol	Description
	track (general)
	track with a route set on it
	signal for trains (general)
	signal at 'Clear', green light
	signal at 'Caution', yellow light
	signal at 'Stop', red light
	signal, white light
	signal, flashing light
	signal, speed restriction aspect
	signal, speed restriction warning aspect
	shunting signal
	points (general)
	points in straight position
	points in branching position
	single slip crossing
	double slip crossing
	derailer (general)
	derailer in passing position
	derailer in derailing position
	insulated rail joint (track circuit on one side)

Explanation of Symbols in Track Layout Schemes

Symbol	Description
—⊥—	insulated rail joint (track circuits on both sides)
—••—	axle counter
—•—	spot wheel detector (no axle counter)
—▬—	balise
—▭—	other spot transmitter of train protection system
—⊓—	platform
—⊐	buffer stop
(level crossing symbol)	level crossing
—▬▶—	train
—▭—	parked rail vehicle

The Editors

Dipl.-Ing. Gregor Theeg
Studied Transport Engineering at the Technische Universität Dresden (Germany) with several practical training periods in different countries (England, Spain, Japan). Since 2004, he has been a scientific staff member at Technische Universität Dresden on the Chair of Railway Signalling and Transport Safety Technology, where he has worked on projects such as studies on signalling rules, interlocking systems and technical analysis, as well as giving lectures and training courses in Germany and Saudi Arabia. From 2009, he has been working for Ansaldo STS on ETCS projects.

Dr. Sergej Vlasenko
Studied Railway Infrastructure at the State Transport University in Omsk (Russia). In 1992-1995 he was a manager in the infrastructure servicing department of the Astana region of the railways of Kazakhstan. From 1995 to 1997, he was a researcher at the State Transport University in St. Petersburg. After graduation as Doctor of Engineering, he has occupied the chair of Railway Automation and Telemechanics at the Omsk State Transport University. From 2007 he has been working at the Rail Automation division of Siemens in Brunswick (Germany).

The Other Authors

Dr.-Ing. Enrico Anders
Studied Transport Engineering at the Technische Universität Dresden (Germany). In 2001-2006 he was a scientific staff member at the Technische Universität Dresden, focussing on safety analysis research. In 2006, he began work for Thales Rail Signalling Solutions GmbH (formerly part of Alcatel), specialising on specifying the requirements for electronic interlocking. Since 2008, he has been a member of the safety assessment centre of Thales Rail Signalling Solutions GmbH.

Prof. Dr.-Ing. Thomas Berndt
Studied Transportation Systems Engineering at 'Friedrich List' University of Transport and Communications in Dresden. From 1981 to 1985 he was a researcher and in 1985 graduated as Doctor of Engineering at 'Friedrich List' University of Transport and Communications. From 1986 to 1991 he worked as an engineer in the electronics industry. From 1991 to 1996, he was Senior Consultant at Computer Sciences Corp., specialising in consulting and software development for railway business information systems. Since 1996, he has been Professor of Railway Engineering at Erfurt University of Applied Sciences.

The Authors

Prof. Dr. Igor Dolgij
In 1972 graduated from the State Transport University in Rostov-on-Don (Russia). In 1980 he completed his doctoral thesis in the field of automation of traffic control processes at railway stations. From 1974 to 1981 he was assistant lecturer, then associate professor, and professor since 2001. From 2005 he has been Head of the Railway Automation and Telemechanics Department at the State Transport University in Rostov-on-Don. Since 1983 he has been Chief of research laboratory Railway Traffic Control Systems.

Prof. Dr. Vladimir Ivančenko
In 1965 graduated from the State Transport University in Rostov-on-Don (Russia). From 1955-1971 he worked in railway transport and in 1973 defined his doctoral thesis. In 1988 he completed his Doctor of Science thesis in the field of automation of shunting processes. From 1973 to 1988 he was associate professor, Head of the Microprocessor Information-Management Systems department. In 1988-1999 he was Chief of the Rostov Branch of the Research Institute Željdoravtomatizacija and since 1999 he has been professor of the Railway Automation and Telemechanics Department at the State Transport University in Rostov-on-Don.

Dr. Andrej Lykov
Graduated from Petersburg State Transport University in 1993 with a diploma in Railway Signal Engineering and Communications. In 1997 he began teaching at the Railway Automation and Telemechanics Department of Petersburg State Transport University. Since 2002 he has been deputy chief of the department. He earned his Doctorate in 2006. He is an associate professor, specialising in interlocking systems and technical diagnostics.

Ing. Peter Márton, PhD.
Graduated in Operation and Economy of Railway Transport at the University of Žilina (UNIZA), former Vysoká škola dopravy a spojov Žilina. He completed his dissertation at UNIZA in 2004. In 2001 and in 2007 he had short term research stays at the Technische Universität Dresden. Since 2002 he has been lecturer at the Faculty of Management Science and Informatics, UNIZA. He specialises in the operation of freight railways and is co-author of a simulation tool of transportation terminals operation, and several simulation studies in Slovakia, Germany, Switzerland and China. He is a Member the Transport Section of Slovak Scientific and Technological Society.

Dr.-Ing. Ulrich Maschek

Apprenticeship as Signal Maintainer. Studied Electrical Engineering at the Technische Universität Dresden. From 1996 he worked as layout engineer for signalling systems. From 1998 worked as a researcher at the Technische Universität Braunschweig. Graduated as PhD in 2002 on Data Modelling for the Planning of Interlockings. Since 2002, Senior Academic Assistant at the Chair of Railway Signalling and Transport Safety Technology at the Technische Universität Dresden. He is also Director of Studies in Railway Signalling at the Wilhelm Büchner Hochschule and trainer of the Siemens Rail Automation Academy and Dresden International University.

Dott. Giorgio Mongardi

Gained a degree in electronics at the Polytechnic of Turin in 1977. With Ansaldo since 1979, he was involved in the development of Computer Based Interlocking (ACC). From 1990 to 1997, head of the Safety Critical Software team, and in 1995-1996 was responsible for 13 ACC installations on the Chinese Railways. From 1997 he was head of a Signalling Engineering Unit; and since 2007 engineering manager for businesses on foreign markets. He is the author of papers at scientific and railway conferences on Interlocking Systems, Methodologies for Design and Verification and Validation for Safety Critical Applications.

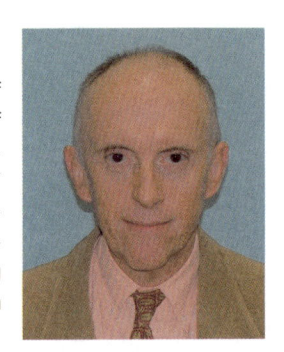

Dr. Oleg Nasedkin

In 1985 graduated in the electrical engineering faculty of the State Transport University in Leningrad (now St. Petersburg) where he specialised in automation, telemechanics and communication. In 1991 he completed his postgraduate dissertation study. In 1995 he was appointed as the head of the test laboratory for devices and systems of railway automation. In 1990-2000 he took part in the working out of standard documents on safety of railway automation. Since 2002 he has been the head of the test centre of railway automation.

Prof. Dr. Aleksandr Nikitin

In 1984 he graduated in the electrical engineering faculty of the State Transport University in Leningrad (now St. Petersburg). From 1984 to 1986 he was signalling engineer in the Zhitomir Signalling Department of the South-West Railway in USSR. From 1986 to 1989 he was a postgraduate, and completed his first dissertation. From 1991 to 2004 he was head of the Research Laboratory at the Petersburg State Transport University. In 2003 he defended his PhD thesis. Since 2005, he has been head of the Computer-based Railway Technologies Centre and Professor of Railway Automatiion and Telemechanics Department of this university.

The Authors

Prof. Dr.-Ing. Jörn Pachl, FIRSE

Studied Transportation Systems Engineering at the Dresden 'Friedrich List' University of Transport and Communications. From 1989 to 1991 he was researcher at that university followed by a management position in the infrastructure planning department of the Hamburg region of German Railways from 1991 to 1996. In 1993 he received a doctor's degree in engineering from Technische Universität Braunschweig. From 1996 he has been Professor at that university and head of the Institute of Railway Systems Engineering and Traffic Safety. Prof. Pachl is author and co-author of several textbooks on railway operation and signalling.

Prof. Dr. Valerij Sapožnikov

Received his MSc degree 1963, the PhD degree 1968 and the doctorate 1980, all in electrical engineering from the State Transport University in Leningrad (now St. Petersburg, Russia). In 1982 he became a professor and in 1989 a vice president of this university. He is an author of numerous scientific papers and books in the field of synthesis, diagnosis, self-checking circuits and control systems for railway transportation. He is a member of the Russian Transportation Academy.

Prof. Dr. Vladimir Sapožnikov

Received his MSc degree 1963, PhD degree 1969 and his doctorate 1984, all in electrical engineering from the State Transport University in Leningrad (now St. Petersburg, Russia). In 1987 he became a professor and in 1991 the head of the Railway Automation and Telemechanics Department of this university. He is an author of numerous scientific papers and books in the field of synthesis, diagnosis, self-checking circuits and control systems for railway transportation. He is a member of the Russian Transportation Academy.

Dr. Andreas Schöbel

Joined the Institute for Railway Engineering, Traffic Economics and Ropeways at Technische Universität Wien in May 2002 as a graduate civil engineer. Since 2003 he has worked on the implementation of wayside train monitoring systems and their integration into interlocking systems. He presented his PhD thesis in 2005 on the positioning of wayside train monitoring systems in a railway network. He is a lecturer on the simulation of railway operation.

The Authors

Dipl.-Ing. Eric Schöne
Studied Transport Engineering at Technische Universität Dresden, with professional practical training at Deutsche Regionaleisenbahn GmbH, Berlin. From 2005 he worked as assistant to the management at infrastructure division of Deutsche Regionaleisenbahn GmbH. From 2006 he worked as a member of academic staff at Chair of Railway Signalling and Transport Safety Technology at Technische Universität Dresden, focus on research into level crossing safety.

Dr. Dmitrij Švalov
In 1981 he graduated from the State Transport University in Rostov-on-Don (Russia). In 2001 he completed his doctoral thesis in the field of technical diagnostics. From 1989 to 1991 he was an engineer at the research laboratory Railway Traffic Control Systems. From 1992 to 2000 he was assistant lecturer and since 2001 associate professor, vice-Head of the Railway Automation and Telemechanics Department at the State Transport University in Rostov-on-Don.

David Stratton MA CEng MIRSE
Studied Electronic Engineering at the University of Cambridge with professional training at AEI (Manchester) Ltd. From 1967, he worked on safety-critical remote control systems and jointless track circuits for electrified railways. From 1981 he worked for GEC on the development of Solid-State Interlocking (SSI) and its introduction into Belgium and France. From 1997 he worked for ALSTOM on specifying requirements for Smartlock 400, a new generation computer-based interlocking to supersede SSI, particularly on specific application design language and tools.

Dipl.-Inform. Heinz Tillmanns
Studied Computer Science at the Technische Universität Berlin. From 1983 he worked as software engineer, from 1990 for Alcatel in the area of human machine interfaces for telecommunication management systems. From 1996 he focussed on requirements engineering and change management. In 2004 he started work for Thales Rail Signalling Solutions GmbH (a former part of Alcatel) on defining system architecture and specifying requirements for a new generation of computer-based interlockings called LockTrac 6151, mainly used in export markets.

The Authors

Prof. Dr.-Ing. Jochen Trinckauf, FIRSE
Studied Transport Systems Engineering at 'Friedrich List' University of Transport and Communications in Dresden. After a period of scientific research at this university from 1979 to 1983 he graduated as Doctor of Engineering. From 1984 to 1990 he worked with Coal Mining Railway Network in Senftenberg, Germany, and participated in the development of a centralised operation control system. He was employed by different industrial and engineering companies of railway branch since 1990, as managing director, too. From 1998 he joined the Technische Universität Dresden as Full Professor.

Dipl.-Ing. Carsten Weber
Studied Transport Engineering at Technische Universität Dresden (Germany) became scientific staff member at Technische Universität Dresden in 2003. He has worked for different project such as studies on operator workload in electronic interlocking, creation of simplified interlocking systems for secondary lines, traceable risk analysis for secondary lines, and technical analysis, as well as training courses, for several users of railway systems. He is now a self-employed consultant for railway operation and interlocking systems.

Thomas White
Began railway career on the Baltimore and Ohio Chicago Terminal Railroad in 1967 as interlocking operator, then served as train dispatcher and assistant chief train dispatcher, subsequently holding these positions on other US railroads. In 1991, he was assigned by Burlington Northern Railroad to network planning, schedule development, capacity management, new passenger service planning, control and network management systems development, and information system implementation. He has been a railway operations consultant for Transit Safety Management, Inc. since 1997.

Index

Symbols

2*(2oo2) system 282
2 out of 2 (2oo2) system 36, 281
2 out of 3 (2oo3) system 36, 282

A

Absolute braking distance 57
ACC (Apparato Centrale con Calcolatore) 300ff
Acceptance 101
ALS-EN 232
ALSN 228ff
Amplitude attribute 337
Approach
 ~ distance 56, 190ff, 373, 387ff
 ~ locking 41, 88, 93
 ~ sight triangle 371ff
 ~ signal clearing 93
 ~ time 54ff, 372
Area overview picture 284
ASFA 227
ATB-EG 231
ATB-NG 236
ATC 126, 208, 233f, 318, 325
ATO 212f
ATP 55, 208
ATS 208
ATS-P 225f
Attentiveness check 210
Automatic
 ~ block 307, 310, 313, 316
 ~ point setting 75, 93
 ~ route calling 94
 ~ route setting 94
 ~ signal 43
 ~ Train Control (ATC) 126, 208, 233f, 318, 325
 ~ Train Operation (ATO) 212f
 ~ Train Protection (ATP) 55, 208
 ~ train routeing 94
 ~ Train Stop (ATS) 208
 ~ Warning System (AWS) 221
AWS 221
Axle counter 119, 143ff
AzL 119

B

BACC 232
Back locking 88
Balise 214, 216, 235ff, 242ff
Balise group 244
Barriers 376f, 378f
 full ~ 376
 half ~ 376
Bidirectional signalling 47

Block 94ff, 306ff
 absolute ~ 98
 automatic ~ 307, 310, 313, 316
 centralised ~ 306, 321ff
 decentralised 306, 307ff
 electric token ~ 99, 309f
 fixed ~ 41, 43, 52f, 61, 97
 ~ information 62f, 94
 ~ instrument 40, 310, 311
 manual ~ 310ff
 ~ marker board 56
 moving ~ 56f, 326f
 permissive ~ 98, 106f
 Radio Electronic Token ~ (RETB) 99, 323f
 relay ~ 313ff
 ~ section 52ff, 94f
 semi-automatic ~ 307f, 310, 313ff
 ~ Siemens & Halske 311ff
 ~ station 95
 telephone ~ 95, 306, 310
 token ~ 97f, 99, 308
 tokenless ~ 98, 100ff
Blocked brake detection 399
Blocking 100ff
Blocking time 53ff
Blocking time stairway 55
Braking supervision 210

C

Cab signalling 55, 56ff, 138, 180, 208ff, 228, 244
Centralised Traffic Control (CTC) 42, 60, 330, 343ff
Checkpoints 398f
Clamp lock 163f
Clearance profile 19, 115
Clearing point 53, 96
Clearing time 54, 372
COMBAT 125, 316
Conditional locking 63, 66, 257
Control length 52f, 96
Control loop 18ff
Controller → Operator
Countdown markers 371
Crocodile 221
Crossover 46, 153f
CSS 231
CTC 42, 60, 330, 343ff

D

Dark territory 58
Dead-man's handle 210
Degraded mode operation 25, 28f, 67, 107ff, 197, 250, 284, 389
Dependence between points and signals 61, 70, 85

443

Index

Derailer 155
Derailment detection 132, 400
Detector
   galvanic ~ 120
   hydraulic ~ 117
   inductive ~ 119
   infrared ~ 121f
   laser ~ 121f
   mechanical ~ 117, 120
   pneumatic ~ 117
   radar ~ 121f
Diagnostic system 283, 365
Diamond crossing 152
Differential GPS 126f
Digital ATC 126, 238ff
Direct Traffic Control (DTC) 58, 95, 306
Direction permission 101f, 312, 315f
   neutral ~ 98, 101, 313, 315, 316
   placed ~ 98, 101f, 311f
Dispatcher 23, 41, 59f, 306f, 328ff
Diversity 282
Dnepr 232
Driver-Machine-Interface (DMI) 244
Dynamic weighing systems 401

E

Ebicab 236
Ebilock 294ff
Electric motive power unit
   entry of ~ into inappropriate area 73
Element dependence 63
   conditional bidirectional locking 63, 65ff
   coupled elements 63
   simple bidirectional locking 63, 65
   unidirectional locking 64
End of authority 246
End of train detection 127f
Entrance-exit operation 75, 269
Equipment identification
   automatic ~ (AEI) 363
   local ~ 362
ETCS 236, 239, 240ff, 325f
   levels 242f
   operation modes 249ff
Eurobalise 242, 244
Euroloop 242, 245
European Train Control System → ETCS
European Vital Computer (EVC) 243
Euroradio 242, 245
EVC 243
EVM 231
Extendable overlaps 82
External object 113, 115, 121, 387

F

Facing move 151
FFB 323
Fire fighting system 403
Fire recognition 403
Fixed brake detection 399
Flank area 68, 77
Flank movement 76f
Flank protection 62, 76ff, 157, 273f
   branched ~ 78
   remote ~ 78f
   transferred ~ 77f
Fleeting 94
Flood warning 404f
Following protection 62, 85, 100ff
Fouling point 149f
   ~ indicator 149f
Four-aspect-signalling 192
Frequency attribute 338
Front protection 68, 79f
Funkfahrbetrieb (FFB) 323

G

GLONASS 238
GPS 126, 238
GS II DR 171ff
Guidance speed 209

H

Handling system 360f
Headway 55
Holding the route 86
Home signal limits 46ff
Hot box detection 399

I

Individual point operation 75
Indusi 223ff
Infill function 236, 242, 245
Insulated rail joint 132ff
   oversize ~ 156ff
Integra Signum 222
Interlocking 61ff, 252ff
   ~ area 44ff, 341
   cascade ~ 89
   ~ districts 341
   electro-mechanical ~ 303f
   electronic ~ 253, 280ff
   electro-pneumatic ~ 303f
   free-wired ~ 267, 270ff
   geographical ~ 90, 267f, 271ff, 276ff, 284
   ~ limits 44ff
   mechanical ~ 253ff

pushbutton ~ 268
relay ~ 253, 263ff
relay-electronic ~ 305
~ station 44ff, 58
tabular ~ 89, 267, 270f
tappet ~ 256f
topological ~ 90, 267f, 271ff, 276ff, 284
~ tower 44ff, 58
Intervention 211f

J

Junction 46

K

Key lock 167
KHP 227
KLUB-U 238
KVB 236

L

L90 291ff
Left-track operation 47
LEU 242, 244
Level crossing 369ff
    active ~ 370, 375ff
    autonomous ~ 384
    ~ in proximity to road junction 390f
    passive ~ 370, 371ff
    route dependent ~ 384
Lime Street control 73
Limit of shunt board 49f
Lineside Electronic Unit 242, 244
Local operation area 92
Local-electrical Operated Point Switches (LOPS) 302f
Long routes 94
Loop (track) 42
Loop (cable ~) 217, 245
LS 231
LZB 238ff, 325

M

Maintainability 25f
Management information system 365
Marshalling yard 351ff
    automation 354f
    control 355ff
    function 351f
    layout variants 352ff
    retarding concept 359f
    use of sensors 362f
Master button 177f, 268f
Movable frog 152

Movable track element 149ff
    coupled ~s 63f
Movement
    ~ authority 51ff, 105, 188ff, 246f
    ~ on sight 72, 98, 106f, 110f, 180, 188, 250
    ~ to half the sighting distance 72
    ~ to the full sighting distance 72
Moving units on crossings 113, 387

N

Neutral direction 98, 101, 313, 315, 316
Non signal-controlled operation 58, 67, 95, 306f
NX operation 75, 269

O

Obstacle 113, 121ff
Odometry 126, 211, 244
One-train-staff system 99, 308
Open line 46
Operation areas 341
Operator 58ff, 328, 339ff
Opposing protection 62, 95, 98, 101
Opposing shunting movements 92
Overlap 52f, 68, 79f, 96ff
    extendable ~s 82
    shared ~s 80f
    swinging ~s 82
Override 225, 233, 250

P

Path 66
Permissive driving 110f, 188
Phantom light 182f
Pilotman 98, 308
Placed direction 98, 101, 311ff
Point circuitry 167
    five-wire ~ 168
    four-wire ~ 171
Point locking mechanism 163ff
Point machine 157ff
    electric ~ 159ff
Points 149ff
    catch ~ 155
    dual control ~ 92
    dual protective ~ 78f
    end position of ~ 150
    high speed ~ 152
    non-trailable ~ 160, 162, 167
    simple ~ 149ff
    spring ~ 151
    trailable ~ 160, 163
    trap ~ 155
    wye ~ 152

Index

Polar attribute 338
Positive train control 126
Poste à Logique Paramétrisée (PLP) 285
Poste d'Aiguillage Informatisée (PAI) 285
Pulse-width attribute 337f
PZB 90 223ff

R

Radio
  ~ Block Centre (RBC) 242f, 245
  ~ communication 218f, 242ff
  ~ Electronic Token Block (RETB) 99, 323f
  ~ infill unit 242, 245
Rail circuit → Track circuit
Range of vision 184
RBC 242f, 245
RB II 60 313f
Recording 109
Redundancy 28, 36, 281f
Relative braking distance 57f
Relay 263ff
  AC immune ~ 265
  biased ~ 265
  bi-stable ~ 265, 266
  double wound ~ 265
  first-class-~ 264
  magnetic latching ~ 266
  mechanically latching ~ 266
  mono-stable ~ 265
  polyphase ~ 135f
  second-class-~ 264
  signal ~ 263ff
  slow pick up ~ 265
  slow release ~ 265
  third-class-~ 264
  toggle ~ 266
  type-C-~ 264
  type-N-~ 264
  vane ~ 135f, 265
Relay contact
  back contact 265
  closer 265
  front contact 265
  opener 265
  two way contact 265
Release time 54
Reliability 25ff
Right-track operation 47
Risk 26
Route 62, 66ff
  alternative ~ 75
  automatic ~ calling 94
  ~ calling 84
  ~ checking 84
  conflicting ~s 71
  consecutive ~s 71
  emergency ~ release 108f
  entrance of a ~ 67, 75, 94
  exit of a ~ 67, 75, 95
  irreversible ~ locking 86ff
  ~ locking 70ff, 84
  long ~s 94
  opposing ~s 71
  prevention of premature release 72
  priority ~ 75, 79
  ~ queuing 94
  ~ release 74, 85
  reversible ~ locking 85ff
  sectional ~ realease 85
  shunting ~ 67, 91ff
  ~ signalling 193, 200f
  ~ supervision 84f
  time delayed ~ release 109
  train ~ 67ff
  train operated ~ release (TORR) 88
RPB GTSS 315f
Running line 42
Running movement 48
Running path 68

S

Safety 24ff, 30ff, 393
  ~ comparison 36
  ~ overlay 306f
SAIPS 124f
Satellite positioning 126f, 238, 362
Saxby & Farmer 255ff
Scissors crossing 154
Security 24, 395
SELTRAC 326
SGE 1958 270f
Short track
  entry into ~ 73
Shunting limits 49ff, 250
Shunting movement 49ff
Sidings 42
Sighting distance 184, 372
Signal 179ff
  approach ~ 43, 189, 195
  automatic ~ 43
  auxiliary ~ 109ff, 197
  block ~ 44, 46, 94ff, 188
  bulb lamp ~ 182ff, 185
  cab ~ 180
  call-on ~ 197
  colour light ~ 181, 187f, 197
  colour position light ~ 181, 197
  combined ~ 189, 195f
  controlled ~ 43
  distant ~ 43, 189, 195
  exit ~ 46
  hand ~ 180, 197, 389
  home ~ 45ff

interlocking ~ 44ff
intermediate block ~ 44, 47
intermediate interlocking ~ 46f
junction indicator 200
LED ~ 183, 186f
light ~ 180ff
limit of shunt ~ 49
main ~ 43, 188ff, 195f
mechanical ~ 180, 198f
multi-unit ~ 182f
negative ~ 180
position light ~ 181, 197
positive ~ 180
route indicator 73, 200
searchlight ~ 182f
section ~ 45
semi-automatic ~ 43
shunting ~ 43, 196f
~ board 181, 184f
~ box 44f, 254ff
~ selection 84
supervision ~ 385ff
trackside ~ 180ff
vehicle ~ 180
Signal aspect 68
  absolute stop 188
  advanced approach 192
  approach 43, 189
  caution 43, 189
  clear 189
  permissive stop 188
  preliminary caution 192
  speed cancellation 194
  speed restriction 194
  speed restriction warning 194
  stop 188ff
  warning 43, 189
Signal system
  H/V ~ 198f
  Ks ~ 204
  NORAC ~ 205ff
  OSŽD ~ 201f
Signalisierter Zugleitbetrieb (SZB) 323
Signaller 60, 328ff
Signum 222
Silhouette measuring system 402f
Simis 291ff
Slip crossing 152
Slotted control 258
SMILE 289ff
Solid State Interlocking (SSI) 285ff
SpDrS60 271ff
Specific Transmission Module (STM) 243, 244
Speed profile
  dynamic ~ 209
  Most Restrictive ~ (MRSP) 247
  static ~ 209, 247

Speed restriction 68f, 192ff
  local validity of ~ 68f
Speed signalling 193f
Speed supervision 211
Spring points → Points
SSI → Solid State Interlocking
St. Andrew's cross 370f
Start section 68f
Station 44ff
  ~ limits 45f
  ~ throat 152
STM → Specific Transmission Module
Supervised location 246
Swinging overlaps 82
Switch, Switchpoints → Points
SZB 323

T

Target speed 209
TBL 236
Three-aspect-signalling 189
Three-channel-structure 36, 281f
Time spacing 41, 51
Timetable 48
Timetable & train order 41
Token block 97, 99, 308f
TPWS 227
Track
  ~ clear detection 72, 77, 115f, 128ff, 143ff
  main ~ 42f
  secondary ~ 42
  Station ~ 44
Track circuit 120f, 128ff
  alternating current (AC) ~ 130, 139
  audio frequency 140ff
  berth ~ 103
  direct current (DC) ~ 130, 139
  impulse ~ 130, 139f
  ~ interrupter 132
  jointless 140ff
  normally closed ~ 128ff
  normally open ~ 128ff
  radio frequency ~ 130
Track Warrant Control (TWC) 58, 95, 306
Traction return currents 132ff
Trailing move 151
Train
  ~ describer system 94
  ~ movement 48f
  ~ protection 208ff
  ~ separation 51ff
  ~ sequence section 95
  ~ sequence station 95

Index

~ staff and ticket system 99, 308
~ trip 248, 250
Trainstop 210, 220f
Transmission
 linear ~ 216ff
 spot ~ 214ff
Transponder balise → Balise
Traverser 153
Turntable 153
TVM 300 234f, 325
TVM 430 238f, 325
TWC 58, 95, 306
Two-aspect-signalling 189
Two-channel-structure 36, 281f

U

UBRI 276ff
UMLER 363
Unblocking 98, 100ff
 continuous ~ 98, 316
 singular ~ 98, 309
Unidirectional locking 64

Unidirectional signalling 100
Unit-Block Relay Interlocking (UBRI) 276ff

W

Warning distance → Approach distance
Warning time 376, 387
Weighing systems 401ff
Wind measuring system 404

Y

Yard 42
Yard limits 50f
Yard management systems 363ff

Z

Zoom picture 284
ZP 43 119, 144
ZSL 90 237
ZUB 235ff

Advertisements

Frauscher GmbH, St. Marienkirchen 16

ISV Ingenieurges. für Schienenverkehrstechnik mbH, Berlin 249

Keymile GmbH, Hanover 33

Rahmann GmbH, Wuppertal 23

Siemens AG, Erlangen 13

TÜV Süd Rail GmbH, Munich 59